Immunocytochemistry

Immunocytochemistry

Practical Applications in Pathology and Biology

Edited by
Julia M. Polak and
Susan Van Noorden

with a Foreword by
A. G. E. Pearse

and an Introduction by
L. A. Sternberger

WRIGHT · PSG

Bristol London Boston
1983

Published by:
John Wright & Sons Ltd, 823–825 Bath Road, Bristol BS4 5NU, England.
John Wright PSG Inc., 545 Great Road, Littleton, Massachusetts 01460, U.S.A.

First published, 1983
Reprinted, 1984

British Library Cataloguing in Publication Data
Immunocytochemistry.
1. Immunocytochemistry 2. Cytochemistry
I. Polak, Julia M. II. Van Noorden, Susan
574.2'9 QR183.6

ISBN 0 7236 0669 2

Library of Congress Catalog Card Number: 82-50782

Typeset and printed in Great Britain by
John Wright & Sons (Printing) Ltd. at the Stonebridge Press, Bristol BS4 5NU

Preface

In 1941 the genius of Albert H. Coons[1] (*Fig.* 1) planted the seed of a new way of localizing tissue constituents. His revolutionary concept, known variously as 'immunohistochemistry' or 'immunocytochemistry', took root so firmly that its offshoots now reach to the distant corners of all the fields of biology and diagnostic pathology from cell membranes to tumours.

The subject's literature has burgeoned likewise, and is scattered as widely as its many applications. Thus scientists seeking the best method of solving a particular problem and wondering whether immunocytochemistry might be the answer have had a hard struggle to find reliable information on the latest methods, all propounded eagerly by their inventors but not necessarily right for the job in hand.

It is therefore opportune to present a compact and comprehensive book giving in one volume an account of how immunocytochemical methods may be applied to many problems of pathology and biology, showing the variety of methods in use today and providing enough practical detail to inspire the nervous or uninitiated reader with confidence to try them.

We do not attempt to discuss the principles of antigen–antibody interactions as applied to immunocytochemistry because this has recently been admirably done[2] by Ludwig Sternberger (*Fig.* 2), who has himself graciously agreed to write the Introduction to this book. Following the introduction of enzymes as antibody markers by Paul Nakane[3] (*Fig.* 3), Sternberger created the unlabelled antibody–enzyme bridge procedures and the peroxidase–anti-peroxidase complex[4] which enormously increased the sensitivity of the method. Under his influence, immunocytochemistry has grown from an interesting but not always reliable technique into an essential tool to help us not only in diagnosis but also in understanding more about basic inter- and intracellular processes. In fact, the entire discipline of histochemistry, which used to consist largely of the localization of enzymes by their reactions in situ, has been altered by the expansion of immunocytochemical methods, as a glance at the journals dealing with histochemistry will show. Even enzymes are now localized as often as not by their immunocytochemical reactivity, rather than by their biochemical actions.

Sternberger's own recent work on the relationship between the different cell types of the nervous system uses the newest immunocytochemical techniques made possible by Cesar Milstein's (*Fig.* 4) concept of monoclonal antibodies of absolutely uniform specificity and is fully described in his introductory chapter.

Fig. 1. Albert H. Coons (1912–1978).

Fig. 2. Ludwig A. Sternberger.

Fig. 3. Paul K. Nakane.

Fig. 4. Cesar Milstein.

Some pathologists argue that immunocytochemistry, though of great value as a research tool, currently has rather few day-to-day practical applications in diagnosis, and that retrospective confirmation by immunocytochemistry of a diagnosis made on clinical, morphological and biochemical evidence is not vital to the patient. This viewpoint is put forward in Chapter 18, together with the author's acknowledgement of the very real use for immunocytochemical techniques in diagnosis of renal disease, lymphomas, skin diseases and the identification of organisms. Other chapters (11, 14, 15, 17, 19, 20) make abundantly clear the view that immunocytochemistry has a wide application in these and other branches of diagnostic histopathology. Indeed, the advances in technique that have come about during the past decade will surely soon be outdated by further improvements in the sensitivity and specificity of methods of localizing substances in tissues, defining diseases and contributing to their cure.

The authors of this book have been chosen particularly because they are all active exponents of the technique. They daily use up-to-date immunocytochemical methods in their routine diagnostic and research laboratories; they innovate, adapt and experiment with modifications and have themselves made large contributions to the available methodology and applications. A certain degree of overlap between the chapters was inevitable but we hope there is not too much. Each chapter stands by itself but references to other chapters are included. Some of the chapters deal mainly with techniques, and a technical appendix follows many of the others. Although several authors may use the same technique (e.g. the PAP method), there are so many variations in practice that we thought it useful for authors to describe their preferred variants, partly to show that there are more ways than one of going about this very flexible method and partly to suggest the sort of modification that might be tried. We hope that people will read the whole book, enjoy the reviews and get an idea of what is worth doing, how to do it and where to look for further information.

The whole point of immunocytochemistry lies in its visual impact and to this end we have aimed at high quality photographic reproduction. We have been able to produce colour prints as well, where they are essential for the appreciation of results, for example in double staining methods. We are most grateful to the many commercial firms whose generous financial help has made the colour reproduction possible (*see* p. 372).

The idea for this book arose as a result of our annual practical course in immunocytochemistry at the Royal Postgraduate Medical School. The number of applicants far exceeds the available places and it seemed to us that there was an urgent demand for information on the usage of immunocytochemistry from practitioners of every branch of biological science. Needless to say, the idea could not have grown into the present reality without the dedicated cooperation of all our authors, many of whom have also been kind enough to participate in the courses. We have indeed been fortunate in their enthusiastic response to our requests that they write chapters for this book. We owe them our heartfelt thanks for their toleration of our exigent instructions that they keep to the deadline, and above all for their hard work and splendid achievements.

Julia M. Polak
Susan Van Noorden

REFERENCES

1. Coons A. H., Creech H. J. and Jones R. N. Immunological properties of an antibody containing a fluorescent group. *Proc. Soc. Exp. Biol. Med.* 1941, **47**, 200–202.
2. Sternberger L. A. *Immunocytochemistry*, 2nd ed. 1979. New York, John Wiley & Sons.
3. Nakane P. K. and Pierce G. B. Jr. Enzyme-labelled antibodies: preparation and application for the localization of antigen. *J. Histochem. Cytochem.* 1966, **14**, 929–931.
4. Sternberger L. A., Hardy P. H. Jr, Cuculis J. J. and Meyer H. G. The unlabelled antibody–enzyme method of immunohistochemistry. Preparation of soluble antigen–antibody complex (horseradish peroxidase-antihorseradish peroxidase) and its use in the identification of spirochetes. *J. Histochem. Cytochem.* 1970, **18**, 315–333.

List of Contributors

Z. Abdulaziz
Department of Haematology, John Radcliffe Hospital, Oxford OX3 9DU, UK. *Chapter 7*

B. Bhogal
St John's Hospital for Diseases of the Skin, 5 Lisle Street, Leicester Square, London WC2H 7BJ, UK. *Chapter 17*

M. M. Black
St John's Hospital for Diseases of the Skin, 5 Lisle Street, Leicester Square, London WC2H 7BJ, UK. *Chapter 17*

S. R. Bloom
Department of Medicine, Royal Postgraduate Medical School, Hammersmith Hospital, Du Cane Road, London W12 0HS, UK. *Chapter 11*

G. F. Bottazzo
Department of Immunology, The Middlesex Hospital Medical School, London W1P 9PG, UK. *Chapter 19*

M. Chilosi
Istituto di Anatomia e Istologia Patologica, Università di Padova, Verona, Italy. *Chapter 13*

A. C. Chu
Department of Dermatology, Hammersmith Hospital, Du Cane Road, London W12 0HS, UK. *Chapter 17*

B. M. Dean
Department of Immunology, The Middlesex Hospital Medical School, London W1P 9PG, UK. *Chapter 19*

D. J. Evans
Department of Histopathology, Royal Postgraduate Medical School, Hammersmith Hospital, Du Cane Road, London W12 0HS, UK. *Chapters 16 and 18*

B. Falini — Department of Haematology, John Radcliffe Hospital, Oxford OX3 9DU, UK. *Chapter 7*

M. Goldstein — Department of Psychiatry, New York University Medical Center, New York, NY 10016, USA. *Chapter 9*

P. U. Heitz — Department of Pathology, University of Basel, Schönbeinstrasse 40, CH-4003 Basel, Switzerland. *Chapter 20*

E. Heyderman — Department of Histopathology, St Thomas's Hospital Medical School, London SE1 7EH, UK. *Chapter 15*

S. Hobbs — Department of Immunology, Royal Free Hospital School of Medicine, Pond Street, London NW3 2QG, UK. *Chapter 13*

P. Isaacson — Department of Pathology, University College Hospital, Gower Street, London WC1E 6AU, UK. *Chapter 14*

G. Janossy — Department of Immunology, Royal Free Hospital School of Medicine, Pond Street, London NW3 2QG, UK. *Chapter 13*

K. R. Jessen — Department of Zoology, University College, London University, Gower Street, London WC1E 6BT, UK. *Chapter 10*

H. W. J. Joosten — Department of Anatomy and Embryology, University of Nijmegen, PO Box 9101, 6500 HB Nijmegen, The Netherlands. *Chapter 9*

D. Y. Mason — Department of Haematology, John Radcliffe Hospital, Oxford OX3 9DU, UK. *Chapter 7*

J. De Mey — Laboratory of Oncology, Janssen Pharmaceutica Research Laboratories, B-2340 Beerse, Belgium. *Chapters 3 and 6*

R. Mirakian — Department of Immunology, The Middlesex Hospital Medical School, London W1P 9PG, UK. *Chapter 19*

P. Ordronneau — Department of Anatomy, University of North Carolina, Chapel Hill, North Carolina 27514, USA. *Chapter 12*

B. Penke — Department of Medical Chemistry, University of Szeged, Dom Tér 8, 6720 Szeged, Hungary. *Chapter 9*

P. Petrusz — Department of Anatomy, University of North Carolina, Chapel Hill, North Carolina 27514, USA. *Chapter 12*

J. M. Polak — Department of Histochemistry, Royal Postgraduate Medical School, Hammersmith Hospital, Du Cane Road, London W12 0HS, UK. Editor, *Chapters 2 and 11*

B. A. J. Ponder — Institute of Cancer Research, Royal Cancer Hospital, The Haddow Laboratories, Clifton Avenue, Sutton, Surrey SM2 5PX, UK. *Chapter 8*

L. W. Poulter — Department of Immunology, Royal Free Hospital School of Medicine, Pond Street, London NW3 2QG, UK. *Chapter 13*

R. Pujol-Borrell — Department of Immunology, The Middlesex Hospital Medical School, London W1P 9PG, UK. *Chapter 19*

R. C. Richards — Department of Histology and Cell Biology (Medical), University of Liverpool, PO Box 147, Liverpool L69 3BX, UK. *Chapter 5*

W. A. Scherbaum — Department of Immunology, The Middlesex Hospital Medical School, London W1P 9PG, UK. *Chapter 19*

G. J. Seymour — Department of Immunology, Royal Free Hospital School of Medicine, Pond Street, London NW3 2QG, UK. *Chapter 13*

H. Stein — Institute of Pathology, University of Kiel, Kiel, GFR. *Chapter 7*

H. W. M. Steinbusch — Department of Anatomy and Embryology, University of Nijmegen, PO Box 9101, 6500 HB Nijmegen, The Netherlands. *Chapter 9*

L. A. Sternberger	Center for Brain Research, University of Rochester School of Medicine, 601 Elmwood Avenue, Rochester, New York 14642, USA. *Chapter 1*
M. Szelke	Department of Chemical Pathology, Royal Postgraduate Medical School, Hammersmith Hospital, Du Cane Road, London W12 0HS, UK. *Chapter 4*
S. Van Noorden	Department of Histopathology, Royal Postgraduate Medical School, Hammersmith Hospital, Du Cane Road, London W12 0HS, UK. Editor, *Chapter 2*
J. Varga	Department of Medical Chemistry, University of Szeged, Dom Tér 8, 6720 Szeged, Hungary. *Chapter 9*
A. A. J. Verhofstad	Department of Anatomy and Embryology, University of Nijmegen, PO Box 9101, 6500 HB Nijmegen, The Netherlands. *Chapter 9*
D. H. Wright	Department of Pathology, Southampton General Hospital, Tremona Road, Southampton SO9 4XY. *Chapter 14*

ERRATA

On page 34 9. line 4.	(0·015–0·3 per cent . . .) should read (0·015–0·03 per cent . . .)
Opposite page 207.	*Plates 29a* and *29b* are reversed. The caption for *Plate 29a* is correct but applies to the plate printed as *29b* and vice versa
On page 373.	The additional firm given below should be included in the acknowledgements: RIA (UK), 7 Whitworth Road, Armstrong Estate, Washington, Tyne and Weare NE37 1PP

Contents

List of Plates

Foreword

by Professor A. G. E. Pearse
Department of Histopathology,
Royal Postgraduate Medical School,
Hammersmith Hospital, London.

This book is written to good purpose and its publication is timely. Suddenly the pace of advance in the twin fields of immuno- and affinity cytochemistry has accelerated to such effect that the great majority of newcomers have been left floundering. So here gathered together are some twenty chapters, by something more than twenty authors, all of whom are currently engaged in the effective use of one or other of the main technologies encompassed in the title.

Sternberger's 'Introduction' sets the tone and purpose and the chapters which follow can be divided, loosely, into three main groups. A single overview chapter covers the field of immunocytochemistry, the succeeding five chapters are devoted to more or less pure technology. These are backed up by six which are more purely cytochemical, albeit applied to scientific problems as opposed to disease processes. Finally there are seven chapters dealing with different pathological topics, most of them relatively restricted.

It is clear that multiple author works of this kind depend to no mean extent on the judgement of the editors and on their skill in persuading their chosen authors to write both swiftly and cogently on their selected topics. In this book is reflected that skill and I foresee that it will be a trailblazer and the forerunner of many similar productions. But the field is wide enough and competition is the essence of advance. I wish the editors, and their galaxy of authors, every success.

1

Introduction
with emphasis on brain immunocytochemistry

L. A. Sternberger

THE DEVELOPMENT AND SCOPE OF IMMUNOCYTOCHEMISTRY

When immunofluorescence was introduced by Coons,[1] it was considered a difficult technique. As with many great discoveries, the concept predated its applicability. It was only through the painstaking work of Coons and his associates, and of others, that immunofluorescence became applicable as a routine technique in the 1960s.

Immunocytochemistry had many difficulties even as late as ten years ago. In general, the technique was restricted to visualization of those antigens in tissues that were known to exist in their locations. Immunocytochemistry was not yet trustworthy enough to localize components without resorting to other, independent evidence. It is only during the last decade or so that immunocytochemical discovery began to predict general biochemical evidence. An example of this trend was the discovery of myelin-associated glycoprotein in the central as well as the peripheral nervous system by Nancy Sternberger et al.,[2] even though previously it had been thought that on the basis of biochemical separation techniques, this antigen only existed in the central nervous system. However, subsequent to the immunocytochemical finding, a re-examination of the biochemical evidence by Figlewicz et al.[3] with double labelling technique disclosed the antigen as a shoulder of another peak in the peripheral nervous system as well, and its isolation from this fraction was thus guided. Today, immunocytochemistry may go a step further. Today we may perhaps venture to ask the question whether by the use of immunocytochemistry we might discover antigens that are so scarce in concentration and distribution that other biochemical techniques cannot disclose them.

In the meantime, attempts have been abundant to improve methodology. A number of markers for electron or light microscopy were introduced to supplement immunofluorescence. Peroxidase with suitable substrates became popular for light and electron microscopy, and ferritin, colloidal gold and haemocyanin were used in electron microscopy. If one averages the results obtained in many investigations in which sensitivities of these markers were compared, one will conclude that such markers as fluorescence or peroxidase are equally sensitive. When special precautions are taken and carefully controlled procedures are used, truly magnificent micrographs of nervous tissue can be obtained by immunofluorescence as illustrated in the work of Hartman,[4] Hökfelt et al.,[5] Polak and Bloom,[6] and others.

Although sensitivity is not primarily dependent on the nature of the marker, an increase of sensitivity can be obtained by the manner in which the marker is employed. The antiserum containing the second antibody (the anti-immunoglobulin) used in immunocytochemistry, does not only react specifically with the antigen-bound first antibody, but non-specifically with tissue even in the absence of first antibody. When the antibodies of the second antiserum are covalently labelled, both the specific and the non-specific antibodies are labelled and, therefore, reaction of the non-specific antibodies will produce background contaminant. The unlabelled antibody method avoids this complication, irrespective of whether the marker is peroxidase–anti-peroxidase (PAP), ferritin–anti-ferritin or, conceivably, fluorescent albumin–anti-albumin complex. The reason for this increase of sensitivity is the sequential reaction with each of the combining sites of the second antibody as discussed in more detail previously.[7] Although antibodies in antisera are heterogeneous, consisting of mixtures of more and less specific antibodies, each single antibody molecule exhibits identical specificity in both combining sites. The use of PAP which is an affinity-purified reagent and which, in addition, can react with peroxidase only to the extent that it is truly an anti-peroxidase-containing reagent, provides a dual amplification of sensitivity. This amplification is not due to easier detectability of the peroxidase marker, but, rather, to the low background obtained and the resulting high signal-to-noise ratio. Because of the increase of sensitivity, the first antibody can be diluted extensively. Since the affinity of antibodies in their reaction with specific antigen is higher than the affinity for cross-reaction with non-specific antigens, high dilution of primary antibody is essential to minimize non-specific reactions. It is safe to say that immunocytochemistry with antibodies diluted 1 : 100 or less is almost certainly beset with much non-specific reaction.

The most important problem in current immunocytochemistry appears to be specificity of the primary antibody. Since immunocytochemistry is often more sensitive than other immunological methods, except perhaps radioimmunoassay, other immunological methods cannot be used to test the specificity of the primary antibody. Radioimmunoassay is not suitable either, since it measures the specificity of a defined antigen and not that of antibodies. Radioimmunoassay ignores those antibodies with which the defined antigen is unreactive. Since the antigen in immunocytochemistry is in tissue, generally undefined and often unknown, absolute specificity cannot be ascertained.

There are, however, three methods which make specificity plausible or alternatively, prove non-specificity:

1. Use of the primary antibody in high dilution.
2. Consistency of staining patterns.
3. Comparison of staining intensities with different antisera and antibodies in different regions of the same section. If the antigen is the same, and is held by

any carrier or receptor in the same conformation in two different regions of a section, the ratio of staining intensities with both antisera or both antibodies should be identical.

In the case of haptens (such as many neuropeptides), successful blocking of antisera by admixture with the hapten has been shown to be a test for cross-reactivity and not for specificity.

IMMUNOCYTOCHEMISTRY IN BRAIN RESEARCH: THE USE OF MONOCLONAL ANTIBODIES

Immunocytochemistry has many uses in brain research. Looking at the brain as an organ is an oversimplification. Each region of the brain, each group of cells, and perhaps each neurone seem to be capable of differentiation and specialization throughout life. If we assume that neuronal specialization finds a biochemical expression, we should be able to differentiate varying brain regions and perhaps individual neurones or groups of neurones by identification of characteristic substances for each, or permutations of characteristic substances for each. Heterogeneous nuclear RNA (hnRNA) complexity of the brain exceeds that of other organs, as studied by hybridization with non-repetitive DNA.[8,9] These studies give us an indication of the diversity of the brain, but at present no practical means are available to isolate quantities of a single hnRNA or combination of hnRNAs characteristic of individual neurones or neuronal groups, either for the purpose of protein synthesis or synthesis of cDNAs (complementary DNAs) specific to an individual cell. Nor are there available biochemical methods for isolating a protein found in only a few cells in the brain, and, within these cells, only as a minor component.

An alternative approach would be the formation of antibodies to whole brain or semipurified brain with the task of absorbing all antibodies reactive with constituents of brain common to most neurones and leaving unabsorbed those antibodies specific for individual neurones. This approach is not feasible not only because it would require absorption of the bulk of total antibodies, leaving behind only a minimum fraction of antibody, but also because it would require knowledge and availability of the very antigens we desired to eliminate from reactivity with our final antibody preparation.

The advent of monoclonal antibodies[10] circumvents these difficulties. No longer is it necessary to absorb antiserum that contains multiple species of antibodies; instead, the monoclonal antibody technique selects individual antibody-forming cells and, since a single cell only produces one species of antibody, the clone produces an antibody of homogeneous specificity and constant affinity. However, the standard selection procedure in the production of monoclonal antibodies requires purified antigens to select appropriate antibody-forming clones. On the other hand, if one were to use immunocytochemistry as a primary method for assaying supernatants from spleen cell–myeloma fusions, one would avoid the need for purified antigen. Using either more or less defined antigens for immunization, or whole brain homogenate or fractions thereof, many workers have pioneered this approach.[11–19] Our work tends to confirm many of their conclusions. The use of mouse PAP permitted optimal fixation which is necessary for obtaining the maximum discrimination in immunocytochemical assay. This has enabled us to observe small differences in reactivities with antibodies that otherwise exhibit a similar pattern and thus seem to react with given groups of antigens. Analysis of a series of such antibodies has suggested the possibility of microheterogeneity in brain-specific antigens.[20] Our continuing studies in this area are aimed at exploring further the possibility of such heterogeneity or, alternatively, lead to different explanations of the observed phenomena. If heterogeneity can be confirmed, it may suggest further explorations towards its relationship to complexity of the brain and specificity of function among differentiated or experienced neurones or neuronal groups.

In using immunocytochemistry of sagittal paraffin sections of rat brain with attached meninges, trigeminal nerve and pituitary as primary assay of supernates from fusion and cloning microculture wells, we discarded all antibodies that reacted with surrounding tissues as well as brain and kept for further studies those antibodies that reacted only with brain and were, therefore, presumably more or less brain-specific. Among 135 hybridomas produced in our initial experiments, 81 were apparently brain-specific. Forty-four of these reacted with non-neuronal elements and 37 with neuronal elements.

The non-neuronal antibodies localized only limited structures, such as oligodendrocytes, basement lamina of brain endothelial cells or certain cells in the adenohypophysis. However, other structures, although abundantly represented in the antigen used for immunization of the BALB/c mice were not represented among the clones. Each one of the structures visualized was represented by a number of antibodies, thus defining several groups of antibodies reactive with non-neuronal brain elements. We may expect that each, or at least most, antibodies reactive with an antigen of a given group, such as basement lamina, were different, and were reactive either with a different determinant in the basal lamina antigen or with different regions of given determinants and also with different affinities for given determinants. Nevertheless, each antibody in its group gave identical immunocytochemical localization. This, of course, is the expected result, despite diversity of antibodies, if the antigen visualized is the same in every brain region examined. For instance, it appears that the antibody to basal lamina is specific to a common blood–brain barrier antigen, since it is localized throughout the brain but not in peripheral capillaries.

In contrast to the identical localization exhibited among antibodies of the non-neuronal groups, the anti-neuronal antibodies exhibited diversity of localization. Nevertheless, among the anti-neuronal antibodies again certain similarities were found which permitted their assignment to given groups. Among these groups three major patterns could be distinguished on morphological grounds. We have termed the groups so defined neurofibrillar, perikaryal neurofibrillar (previously called 'neuronal') and synapse-associated (previously called 'cerebellar'). The common characteristics of the antibodies of the neurofibrillar group was their staining of fibres known to be rich in neurofilaments. Each antibody in this group stained basket cell fibres as they surrounded Purkinje cell bodies. However, this was the only common characteristic to all of the anti-neurofibrillar antibodies. Each antibody stained other structures besides basket cell fibres throughout many regions of the brain, but the kinds of other structures stained were different with each antibody. Some had a wide and others a more restricted distribution, though in no case was the distribution exceedingly sparse. The broadest distribution was exhibited by antibody 02-40, which appeared to stain intensively nearly every major fibre tract and also many large perikarya including Purkinje cells, motor neurones in the spinal cord, cells in the reticular formation, the mesencephalic nucleus of the trigeminal nerve and pyramidal cells in the cerebral cortex. Two other antibodies of the neurofibrillar group also stained cell bodies but their number was less than those revealed by 02-40. These antibodies, for instance, failed to stain Purkinje cell bodies, although they still stained pyramidal cells in the cortex and motor neurones in the spinal cord. Six additional neurofibrillar antibodies stained neurofibrils without staining any cell body. Even the neurofibrillar staining was diverse, often revealing a slightly different distribution of neurofibrils, usually more in the mesencephalon and rhombencephalon than in the diencephalon and cerebral hemispheres (*Plate* 1).

The perikaryal-neurofibrillar antibodies were characterized by staining individual neuronal cell bodies and both their axonal and dendritic projections. Projection staining occurred over long distances. However, only selected neurones were stained, and thus these antibodies never revealed the staining of a whole fibre tract such as was observed in the neurofibrillar pattern. An exception was the staining of tangential fibres in layer I of the cerebral cortex, that remained unstained with an antibody from the neurofibrillar group. Nearly all pyramidal cell bodies were stained with this antibody. Most perikaryal-neurofibrillar antibodies stained a different group of cells, although all of them included in their staining patterns cell bodies of Purkinje cells and of the red nucleus (*Plate* 1). A few of the antibodies stained stellate cells in the cerebral cortex and one of them (02-135) stained pyramidal cells of area CA2 of the hippocampus, but not of area CA1c.

Most of the antibodies reactive in a synaptosomal pattern, and therefore classified as belonging to a synapse-associated group, exhibited a broad distribution. The broad distribution included staining of most gray matter throughout the brain and intensive staining of glomeruli in the granule layer of the cerebellum as well as staining of the cerebellar cortex, occasionally with inclusion of weakly staining Purkinje cells. Large neurones in the brain stem as well as perikarya in the trapezoid body were decorated on their surfaces with small thickenings, but the cytoplasm of the cells proper was not stained. Some of these antibodies resembled the cerebellar distribution described by Matthew et al.[18] with a monoclonal antibody selected after immunization with a synaptic vesicle preparation. Other antibodies of the synapse-associated groups revealed a more restricted pattern. One of them failed to stain cerebellar glomeruli but seemed to reveal parallel fibres as fine projections from granule cells expanding in the molecular layer. Another antibody (02-172) was restricted to the molecular layer of the cerebellum without staining any other region of the brain.

Thus, there was a basic difference between localization of the anti-neuronal and anti-non-neuronal antibodies. While, in each group of anti-non-neuronal antibodies localization was identical, individual antibodies from each group with neuronal reactivities revealed different staining patterns. We interpreted this staining pattern diversity as a result of heterogeneity of the antigens visualized. Thus, if we look at neurofilament protein or proteins closely associated with neurofilament, not only as a structural protein, but also as a protein that may undergo diversification during differentiation or neuronal experience, it could exhibit different antigenic determinants in different cells and in different fibres, and these determinants would vary during development. Conceivably, part of neurofilament protein may be common and its basic structure may be genetically and invariantly determined. Such constant regions may perhaps include those necessary for its assembly as a triplet. Constancy may also express itself as its total molecular weight. However, within generally constant regions, substitution of individual amino acids or groups of amino acids could occur during development and throughout life, thus giving neurofilaments in various areas of brain increasing diversity. We assume that the basic structure of neurofilament-associated proteins, that is the 200, the 150 and the 68 kilodalton subunits, as well as perhaps some other subunits of the neurofilament complex, are genetically determined by a number of exons. A given exon may occur in several subunits of the triplet, thus contributing to structural and possibly antigenic homology among the subunits, and perhaps aiding in their post-translational assembly as a triplet.

We also assume that the changes that occur in brain antigens, including neurofilament antigens, during differentiation and perhaps during experience throughout life, are small, and may be either genetic or post-translational. A post-translational change has become more plausible since the discovery by Lindquist et al.[21] of 4S RNA in axonal projections that was found to be identical with tRNA from the cell bodies. Since axons do not contain ribosomes, the function of this axonal tRNA is not likely to be pre-translational, and the possibility that it mediates substitution of amino acids after protein assembly has become tempting. In any event, such substitutions would afford only minor changes to the total structure of neurofilaments, yet they may have marked effects on their antigenic reactivities. As long as these changes are minor, any structural function of neurofilaments would not be interfered with. However, other possible functions of neurofilaments could be strongly affected. Finally, just as other large proteins, such as pro-opiocortin, are broken down to a variety of active principles, one cannot exclude the possibility that neurofilament antigens, once degraded, may also yield peptides that do not go to waste, but may fulfill specific transmitter or modulator functions. We have termed the antigens that are visualized in a given neuroanatomical pattern 'group antigens', and each variation within these antigens, as detected by individual antibodies, 'neurotypes'.[22] Thus, proteins of the neurofilament triplet would all belong to the group of antigens visualized by its given set of antibodies. However, each heterogeneous modified neurofilament antigen visualized by only one antibody from this group would be called a 'neurotype' of the neurofibrillar group.

We have also assumed that the small changes in antigens that increase their heterogeneity during the development of neurotypes do not necessarily increase the complexity of the molecule. It is conceivable that the original primitive molecule is rich in amino acids of functional importance and high antigenicity, while differentiation may be more a process of elimination of some of the high activity amino acids and substitution by less relevant amino acids, rather than a process of insertion of additional highly active amino acids. We made this assumption because of the rich endowment of specific substances in malignant tissue and undifferentiated cells. Thus neuroblastoma cells[23] contain enzymes for the synthesis of acetylcholine as well as noradrenaline. Embryonic superior cervical ganglion cells acquire *in vitro* cholinergic properties while maintaining characteristics of adrenal neurones, though in the intact ganglion this coexistence is, at best, rare.[24] Again, endocrine tumours of the lung are liable to produce profuse amounts of ACTH and other hormones even though the normal lung does not produce ACTH.[25] Thus, differentiation may be looked upon as a process of specific elimination, and de-differentiation as a return to non-specific diversity. The assumption would appear to make evolutionary sense since the unicellular organism requires all the factors necessary for survival, while the multicellular organism may divide the task among different cell systems.

We have pursued the following approaches to examine these concepts.

1. Nancy Sternberger examined the reactivities of five anti-neuronal monoclonals with normal and pathological nervous tissue, and found even greater heterogeneity than in rat tissue.

2. If the undifferentiated antigen is more general in its expression of antigenic determinants and if differentiation is the expression of only restricted determinants in given regions, and their elimination in other regions, then the distribution of neurotypes should be more general in young animals than in mature animals.[22]

In each neuronal group there were some antibodies which had a wider distribution in adult animals and others which had a narrower, more specific, distribution. Antibody 02-40 from the anti-neurofibrillar group had the widest distribution in its group and visualized all major neurofilament areas; it developed in parallel with the anatomical development of the brain. In young animals, it localized few cell bodies and nearly no fibres, and as the brain developed, its cell body and fibre localization became more and more pronounced. Although this antibody did not, as revealed in the developmental studies, react with every neurofibril, it did visualize a large population of them. Apparently, it reacted with an antigenic determinant that encountered little modification during differentiation, and thus was available in most areas of the adult brain, revealing itself in a distribution that grew as the brain grew.

Most of our antibodies had much more restricted distribution than 02-40. The restricted adult distribution was reflected in development, for antibodies that had a restricted adult distribution had a widespread distribution in the postnatal animal. However, as the animal matured, some areas localized in the developing brain were eliminated from visualization in the adult brain. Thus, the anti-neurofibrillar group antibody 03-44 stained intensively basket cell fibres as well as white matter fibres of the cerebellum in the developing animals, but, in the young adult, the white matter staining nearly all disappeared, while the basket cell fibre staining increased in intensity.

3. To reveal the nature of the antigens visualized by the anti-neurofibrillar and other groups of antibodies, Eng et al.[26] examined their reaction on immunoblot-transferred purified bovine neurofilament proteins. In this technique, the antigen is electrophoresed in one or two dimensions in acrylamide gel and is then electrophoretically transferred onto cellulose nitrate. In the single dimension mode, the cellulose nitrate paper is cut into strips, and each strip is stained by the PAP technique with a different monoclonal antibody (*see* Chapter 3). Eng found that the widely distributed neurofilament pattern 02-40 antibody reacted with the isolated 200 as well as the 150 kilodalton subunit of the neurofilament triplet. The antibodies that exhibited immunocytochemically restricted neurofibrillar staining patterns, as well as the antibodies from the other two groups, failed to react with the isolated neurofilament proteins.

On the other hand, when we purified a neurofilament preparation from rat cerebellum and electroblotted it after electrophoresis in sodium dodecylsulphate (SDS) gel, we found that a number of antibodies from the neurofibrillar group reacted with the 200 kilodalton band.[29] In addition, they reacted with up to three faster bands, including a 150 kilodalton band. Each antibody tested reacted, however, with a different combination of these faster bands. We interpret these findings as suggestive of heterogeneity in neurofilament protein. This appears to be a microheterogeneity, possibly the result of substitution of a few amino acids. The overall structure and size of neurofilament components is not affected, thereby, but the charge might be. Interestingly, neurotypes co-migrating with neurofilament subunits were not only revealed by the anti-neurofibrillar pattern antibodies, but also by those staining in perikaryal patterns and therefore we renamed these patterns 'perikaryal-neurofibrillar'. There was no overlap in staining between anti-neurofibrillar and anti-perikaryal-neurofibrillar antibodies (*Plate* 1). Their localization was mutually exclusive.

When brain extract is obtained without precautions against protein breakdown, additional bands corresponding to smaller fractions are visualized by monoclonal antibodies on electroblot transfer. Each antibody had its own individual pattern of localizing smaller fractions, suggesting a high degree of heterogeneity of these fractions. This finding would tend to support the heterogeneity (neurotypy) of regions within neurofilament protein.

The diversity of the breakdown products recalls the universality of the Scharrer concept of neurosecretion.[27,28] These authors have postulated that neurosecretion is a fundamental system (the peptidergic system), which is required for intercellular communication, whether this be via structured nerves or via endocrine interaction. It has been demonstrated to be essential and, in principle, similar in vertebrates and in invertebrates. In confirmation of the importance of this principle, a great diversity of peptides has recently been discovered. Perhaps the diversity of small fragment peptides as well as their large molecular precursors as identified by monoclonal antibodies may illustrate the applicability of the Scharrer concept, not only to cellular communication in general, but also to the functional specialization of the differentiated and educated brain.

Immunocytochemical techniques appear to be essential to broaden such concepts and generate new ones. The techniques and results discussed in the following pages illustrate the maturity which immunocytochemistry has recently achieved and the opportunity for carefully controlled immunocytochemistry to gather data independently of those obtainable by other approaches.

REFERENCES

1. Coons A. H. and Kaplan M. H. Localization of antigens in tissue culture cells. II. Improvements in a method for the detection of antigen by means of fluorescent antibody. *J. Exp. Med.* 1950, **91**, 1–13.
2. Sternberger N. H., Quarles R. H., Itoyama Y. and Webster H. DeF. Myelin-associated glycoprotein demonstrated immunocytochemically in myelin and myelin-forming cells of developing rat. *Proc. Natl Acad. Sci. USA* 1979, **76**, 1510–1514.
3. Figlewicz D. A., Quarles R. H., Johnson D., Barbarash G. R. and Sternberger N. H. Biochemical demonstration of the myelin-associated glycoprotein in the peripheral nervous system. *J. Neurochem.* 1981, **37**, 749–758.
4. Hartman B. K. Immunofluorescence of dopamine-β-hydroxylase. Application of improved methodology to the localization of the peripheral and central noradrenergic nervous system. *J. Histochem. Cytochem.* 1973, **21**, 312–332.
5. Hökfelt T., Johannson O., Ljungdahl Å., Lundberg J. M. and Schultzberg M. Peptidergic neurons. *Nature* 1980, **284**, 515–521.
6. Polak J. M. and Bloom S. R. Immunocytochemistry of regulatory peptides. Chapter 11 of this volume.
7. Sternberger L. A. *Immunocytochemistry*, 2nd ed. New York, John Wiley and Sons, 1979.
8. Grouse L. D., Schrier B. K., Bennett E. L., Rosenzweig M. R. and Nelson P. G. Sequence diversity studies of rat brain RNA: effects of environmental complexity on rat brain RNA diversity. *J. Neurochem.* 1978 **30**, 191–203.
9. Kaplan B. B., Schachter B. S., Osterburg H. H., de Vellis J. S. and Finch C. E. Sequence complexity of polyadenylated RNA obtained from rat brain regions and cultured rat cells of neural origin. *Biochemistry* 1978 **17**, 5516–5524.
10. Köhler G. and Milstein C. Derivation of specific antibody-producing tissue culture and tumor lines by cell fusion. *Eur. J. Immunol.* 1976 **6**, 511–519.
11. Barnstable C. J. Monoclonal antibodies which recognise different cell types in the rat retina. *Nature* 1980 **286**, 231–235.
12. Trisler G. D., Schneider M. D. and Nirenberg M. A topographic gradient of molecules in retina can be used to identify neuron position. *Proc. Natl Acad. Sci. USA* 1981 **78**, 2145–2149.
13. Vulliamy Y., Rattray S. and Mirsky R. Cell surface antigen distinguishes sensory and autonomic peripheral neurons from central neurons. *Nature* 1981, **291**, 418–420.
14. Cohen J. and Selvendram S. Y. A neuronal cell surface antigen is found in the CNS but not in peripheral neurons. *Nature* 1981, **291**, 421–423.
15. Zipser B. and McKay R. Monoclonal antibodies distinguish identifiable sets of neurons in the leech. *Nature* 1981, **289**, 549–554.
16. Pruss R. M., Mirsky R., Raff M. C., Thorpe R. and Anderton B. H. A monoclonal antibody that recognizes a determinant present on common as well as class-specific intermediate filament units. *Cold Spring Harbor Reps. Neurosci.* 1981, **2**, 125–132.
17. Stallcup W., Levine J. and Raschka W. Monoclonal antibodies to the NG2 marker. *Cold Spring Harbor Reps. Neurosci.* 1981, **2**, 39–49.
18. Matthew W. D., Reichard L. F. and Tsavaler L. Monoclonal antibody to synaptic membranes and vesicles. *Cold Spring Harbor Rep. Neurosci.* 1981, **2**, 163–180.
19. Lampson L. A. Expression of the major histocompatibility antigens in the nervous system. *Cold Spring Harbor Reps. Neurosci.* 1981, **2**, 69–72.
20. Sternberger L. A., Harwell L. W. and Sternberger N. H. Neurotypy: regional individuality in rat brain detected by immunocytochemistry with monoclonal antibodies. *Proc. Natl Acad. Sci. USA* 1982, **79**, 1326–1330.
21. Lindquist T. D., Ingoglia N. A. and Gould R. M. 4S RNA is transported axonally in normal and regenerating axons of the sciatic nerve of rats. *Brain Res.* 1981, **230**, 181–194.
22. Goldstein M. E., Sternberger N. H. and Sternberger L. A. Developmental expression of neurotypy revealed by immunocytochemistry with monoclonal antibodies. *J. Neuroimmunol.* 1982, in press.
23. Ross R. A., Biedler J. L., Spengler B. A. and Reis D. J. Neurotransmitter synthesizing enzymes in 14 human neuroblastoma cell lines. *Cell. Mol. Neurobiol.* 1981, **1**, 301–312.
24. Higgins D., Iacoritti L., Joh T. H. and Burton H. The immunohistochemical localization of tyrosine hydroxylase within rat sympathetic neurons that release acetylcholine in culture. *J. Neurosci.* 1981, **1**, 126–131.

25. McDowell E. M., Wilson T. S. and Trump B. F. Atypical endocrine tumors of the lung. *Arch. Pathol. Lab. Med.* 1980, **105**, 20–28.
26. Eng L. F., Sternberger N. H. and Sternberger L. A. A monoclonal antibody that reacts with two neurofilament proteins. *Trans. Am. Soc. Neurochem.* 1982, **13**, 108.
27. Scharrer E. and Scharrer B. Secretory cells within the hypothalamus. *Res. Publ. Assoc. Res. Nerv. Ment. Dis.* 1939, **20**, 170–193.
28. Scharrer B. Peptidergic neurons: facts and trends. *Gen. Comp. Endocrinol.* 1978, **34**, 50–62.
29. Sternberger N. H. et al. in preparation.

2

Immunocytochemistry Today

techniques and practice

S. Van Noorden
and J. M. Polak

Our aim in this chapter is to provide an introduction to immunocytochemical methods in current use for light and electron microscopy, with enough practical information to set potential immunochemists on the right path to perfect preparations. We do not propose to discuss in detail the theoretical basis of immunocytochemistry and its techniques because this is fully and admirably dealt with in several texts, notably *Immunocytochemistry* by L. A. Sternberger.[1] Very useful discussions will be found in Nairn's *Fluorescent Protein Tracing*[2] and articles by Vandesande,[3] Petrusz et al.,[4] Larsson[5] and Ordronneau et al.[6]

We make no apology for our bias towards peptide immunocytochemistry, which has been our own main activity. Immunocytochemistry of peptides, which is now very widely practised, has led to greatly increased understanding of hormones and neurohormonal systems (*see* Chapter 11). In addition, the difficulties of ensuring specificity of reaction in this area have opened the eyes of immunocytochemists to problems of cross-reactivity and stimulated the invention of methods of overcoming them by using fragments of the molecule (*see* Chapter 4) for controlling the results. In this and many other ways experience with peptide immunocytochemistry has contributed enormously to the range and specificity of methods applicable to all fields.

Immunocytochemical innovations appear with encouraging frequency, and it is expected that the methods we describe will soon be superseded by even better ways of localizing substances in tissues.

1. DEFINITION

Immunocytochemistry is the identification of a tissue constituent *in situ* by means of a specific antigen–antibody reaction tagged by a visible label.

2. HISTORY

The first attempts to label antibodies were made with ordinary dyes,[7] but these were unsatisfactory because the label was not sufficiently visible under the microscope. Immunocytochemistry did not come into its own until A. H. Coons and his associates devised the idea of localizing substances in tissues by means of specific antibodies labelled with a fluorescent dye, fluorescein isocyanate. At first[8] the specific antibody itself was labelled (direct method), and later[9] the more sensitive and versatile indirect method was introduced. Fluorescein isocyanate was replaced by the isothiocyanate[10] which is simpler to conjugate to antibodies and has remained the fluorescent label of choice. Fluorescein compounds, even in the low concentrations demanded by the ratio of optimal antibody reactivity to label, yield a very bright apple-green fluorescence on irradiation in the near ultraviolet (maximum absorbance at excitation wavelength 490 nm, emission at 520 nm). A good alternative or additional label for use in double immunofluorescence staining is rhodamine isothiocyanate which fluoresces red. The immunoreactivity of the antibody is somewhat impaired by the fluorescent label and indeed by conjugation with other labels. An optimal ratio of antibody to label must be found.

The disadvantages of the fluorescence methods are that they require a fluorescence microscope, background details are difficult to appreciate and the preparations are not permanent because the fluorescent labels will not withstand dehydration. Nevertheless, the speed and simplicity of the fluorescent antibody methods have ensured that they remain popular, particularly for diagnosis of immunoglobulin deposits in renal glomeruli (*see* Chapter 18), autoimmune diseases (*see* Chapter 14) and staining of unfixed cryostat sections where the structure of the tissue is insufficiently preserved for a good conventional appearance.

The search for other light microscopical labels resulted in the widespread use of enzymes, particularly peroxidase, introduced by Nakane and Pierce[11] and latterly glucose oxidase[12] (*see also* Chapter 7) and alkaline phosphatase, used for enzyme-labelled immunosorbent assay (ELISA)[13] and also used in unlabelled antibody methods in combination with an antibody raised to the enzyme[14] (*see also* Chapter 7). Enzyme labels are developed histochemically at the end of the antigen–antibody reaction and yield intensely coloured end-products which can be viewed in an ordinary light microscope. A further advantage is that the reactions can be adapted to the electron microscope by using suitably prepared material and making the end-product electron dense, usually by osmication.

Other labels that have particular uses are, for electron microscopical immunocytochemistry, ferritin,[15] and for both light and electron microscopical immunocytochemistry, various radioactive labels, either on the antigen, labelling the antibody before reaction,[16] or incorporated into the antibody,[17] and the newly developed colloidal gold particles (*see* Chapter 6). Antibodies may also be labelled with unrelated antigens, visualized by a second antigen–antibody reaction (hapten sandwich technique).[18] The reasons for the evolution of new labels are the continuing search for greater specificity and sensitivity of reaction, together with the possibility of identifying two or more differently labelled antigens in the same preparation.

3. USES OF IMMUNOCYTOCHEMISTRY

The uses of the method in diagnostic histopathology are legion and likely to expand further, as is clearly brought out in other chapters of this book. A short list would include diagnosis of tumours by specific antigen markers and identification of lymphoma types, foreign organisms, autoimmune diseases, abnormal hormone production and cellular constituents. In experimental work the scope is even wider and would include hormone receptor studies, phylogenetic comparison studies and quantification of the dynamics of cells, for example, in hormone production and secretion.

4. METHODS

A brief outline of the methods commonly used will be given here. A detailed description of some of them may be found in the Appendix.

The methods are described as for light microscopy but many are suitable for electron microscopical immunocytochemistry (*see* Section 10). The descriptions refer to tissue sections on slides but this term should be understood to include cell cultures, suspensions or smears.

4.1. Direct Methods

This type of reaction was the first to be developed. The labelled primary antibody is applied to the section and identifies the antigenic sites.

4.2. Indirect Method

This is more sensitive than the direct method, i.e. the same number of primary antibody molecules bound to the tissue antigen will be more intensely labelled in comparison with the direct method or the same amount of antigen can be visualized by a lower concentration of primary antibody. The primary, unlabelled, antibody is applied to the section, and the excess is washed off with buffer. A second, labelled, antibody from another species, raised to the IgG of the animal donating the first antibody, is then applied, the first antibody now acting as an immunoglobulin antigen. The primary antigenic site is thus revealed. It is unresolved whether the method's increased sensitivity is due to each primary antibody molecule binding several molecules of the second, labelled, antibody, or simply to the hyperimmunity and therefore higher titre of the second antibody. Whatever the reason, enhancement of the label intensity is achieved in comparison with the direct method (*see* the Appendix, p. 34).

Another major advantage of the indirect method is that, provided that the primary antibodies have been raised in the same animal species, one labelled antibody to IgG of that species may be used as the second layer to locate any number of primary antigens without the necessity of labelling each primary antibody. The label used may be fluorescent (fluorescein, rhodamine), enzymic (usually peroxidase), colloidal gold or any other.

4.3. Unlabelled Antibody–Enzyme Methods: bridge techniques

As the name implies, a bridging antibody (unconjugated) is used between the primary antibody and the final antibody, which is also unconjugated and is raised to horseradish peroxidase in the same species as the primary antibody. This layer is followed by horseradish peroxidase which is then bound by an antigen–antibody reaction and can then be developed. The avoidance of chemical conjugation means that the immunological reactivity of all the antibodies is kept to the maximum. For discussion *see* Ordronneau et al.[6]

4.3.1. *Single Bridge*

The first layer might be rabbit anti-primary antigen, the second layer, unlabelled (goat) anti-rabbit IgG, the third layer, rabbit anti-horseradish peroxidase and the fourth layer, horseradish peroxidase, which is then developed by a histochemical method such as the diaminobenzidine and hydrogen peroxide method of Graham and Karnovsky.[23]

4.3.2. *Double Bridge*

This involves repeating the second layer of (goat) anti-rabbit IgG after the rabbit anti-peroxidase, and then repeating the rabbit anti-peroxidase before the application of the peroxidase and its development.[24] A build-up of peroxidase binding sites is thus achieved.

4.3.3. *Peroxidase–Anti-peroxidase (PAP)*

The development of this method by Sternberger[1,25] has had a great impact on immunocytochemistry.

The modification of his bridge technique made by Sternberger was to combine the anti-peroxidase with peroxidase before applying it to the tissue. The combination produces a very stable cyclic complex with three peroxidase molecules to two antibody molecules. The complex is separated from unreacted antibody and peroxidase before use. It acts as a third layer antigen, the first layer being rabbit anti-tissue antigen, the second layer being unconjugated (goat) anti-rabbit IgG, in excess so that one of the two identical binding sites (Fab portion) is free to combine with the third layer, the rabbit PAP complex. There is thus a high ratio of peroxidase label to primary antigen (*Fig.* 2.1). The rabbit PAP molecules will react only with the anti-rabbit IgG of the second layer, so that provided there is no unwanted binding to the tissue of the second layer this is a very specific and background-free technique. The increased amount of label allows the primary antibody to be highly diluted which has the advantage of reducing unwanted

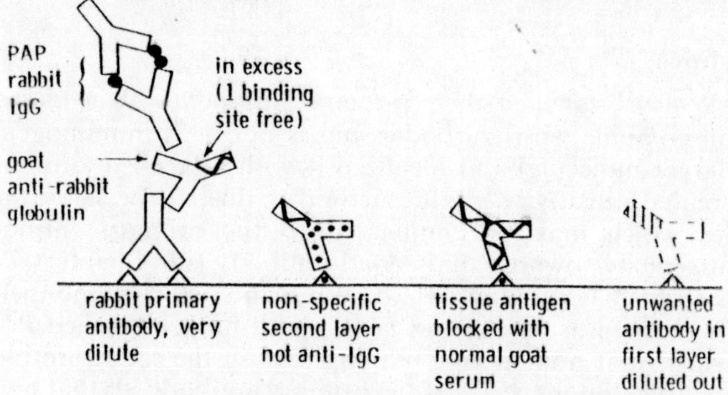

Fig. 2.1. Peroxidase-anti-peroxidase (PAP) method.

background staining (*see* Section 5.5.6), of allowing preferential attachment of high affinity antibodies (*see* Section 5.1.3) and of decreasing the possibility of dissociation of the antibody from the tissue antigen because both combining sites of each antibody molecule are bound. For even greater intensity, the second and third layers of this method may be repeated.[26]

Although in fact the end-results of the PAP method are not always better than those of the unlabelled antibody enzyme bridge techniques on which it is based,[6] the convenience of cutting out one staining step by use of a peroxidase-anti-peroxidase complex and the high degree of purification and specific binding of this final layer provide a very clean and sensitive method which is still the most popular one in use (*see* the Appendix, p. 35).

4.4. Labelled Antigen

A method of high specificity that involves the labelling of an antigen which is then bound to its specific antibody before application to the tissue; the labelled antigen and the antibody are mixed in such proportion that one combining site of the antibody is left free to bind to the tissue antigen (*Fig.* 2.2). Mason and Sammons[27] used an enzyme label, Larsson and Schwartz[16] a radioactive one. The advantage of this method is that, automatically, only specifically attached antibody is labelled, provided that potential non-specific binding sites have been blocked (*see* Section 5.5). The disadvantages are that individual antigens must be labelled which can be expensive and laborious and that it may involve the use of autoradiographic techniques. Antigens have also been labelled with gold particles,[28] which partly overcomes these objections, at least for electron microscopical immunocytochemistry.

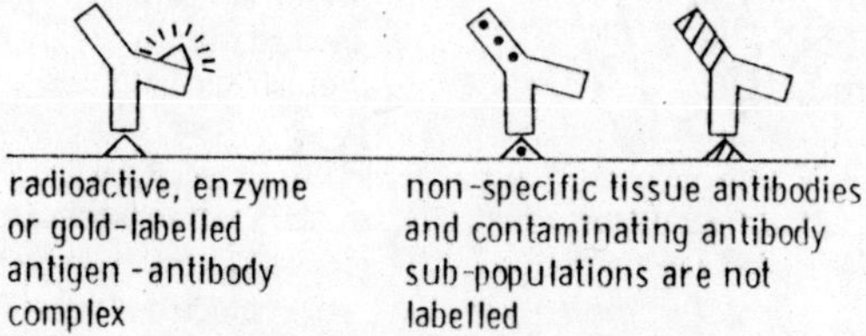

Fig. 2.2. Labelled antigen method.

4.5. Hapten Sandwich

Cammisuli and Wofsy[18] labelled their primary antibody with a hapten (small molecule that can combine with antibodies but is not itself immunogenic unless coupled with a larger molecule) and localized it with a labelled antibody to the hapten. The increased sensitivity of the method is due to the large number of hapten molecules which may be conjugated to the primary antibody, thus increasing its antigenicity towards the second antibody (*see* Chapters 7 and 13). This method has recently been adapted[29] by using an unlabelled monoclonal (*see* Section 5.3.3) anti-hapten as the second layer and a hapten-labelled PAP as the third layer. Provided that non-specific binding sites in the tissue are blocked by normal serum from the donor species of the primary antibody, so that non-specific hapten-labelled antibodies cannot react, this method should provide an intensely labelled preparation, free from background staining.

4.6. Avidin–Biotin Methods

Two new reagents used in immunocytochemistry[30] are a large glycoprotein, avidin, from egg white and a small protein, biotin, which is a vitamin found in egg yolk, among other sites. Avidin has a very high affinity for biotin; one molecule can bind four biotin molecules. Biotin will also bind to the Fc portion of immunoglobulins, each biotin molecule having one binding site but several biotin molecules being bound to one immunoglobulin molecule. Both avidin and biotin may be labelled with fluorescent, enzyme, ferritin or gold labels. Numerous combinations of avidin, biotin and antibody may be built up. A technique using a preformed avidin–biotin complex has been claimed to equal or even to exceed the PAP method in sensitivity.[31] In this method the first layer is rabbit primary antibody, the second layer is biotinylated goat anti-rabbit IgG, and the third layer is an avidin–biotin complex, the avidin having been reacted with biotinylated peroxidase in such proportion that three of the biotin-binding sites are taken up by biotinylated peroxidase, leaving one site per molecule free to react with the biotin on the second layer antibody. A large amount of label is thus localized over the original antigenic site (*Fig.* 2.3). With judicious combination of avidin, biotin and antibody it should be possible to create a widely branching complex and build up

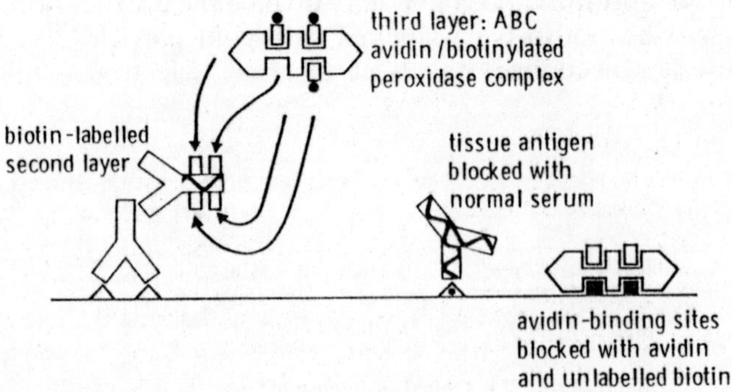

Fig. 2.3. Avidin–biotin complex (ABC) method.

very high amounts of label to increase the sensitivity even further. A blocking method (*see* Section 5.5) to prevent unwanted avidin from binding to biotin in the tissue was devised by Wood and Warnke.[32] Unconjugated avidin is applied to the tissue to occupy the biotin sites, and is then saturated with unlabelled biotin, preventing any attachment of the avidin-labelled biotin complex.

4.7. Protein A

Protein A is derived from the cell wall of a bacterium, *Staphylococcus aureus*, and has the property of attaching to the Fc portion of immunoglobulin molecules. It, too, may be labelled with any suitable marker such as fluorescein isothiocyanate or colloidal gold particles and is used as a second layer in both light and electron microscopical immunocytochemistry to localize the primary antibody (*see* Roth et al.[33] and Chapter 6).

5. ESSENTIAL CONDITIONS FOR IMMUNOCYTOCHEMISTRY

The aim of the method is to achieve consistent and unequivocal staining of the substance in question with little or no background staining and all modifications of the original technique have been designed to increase the specificity and sensitivity of the reaction.

5.1. Primary Antibody

The primary antibody must be specific and of high affinity, avidity and titre.

5.1.1. *Specificity*

Specificity is achieved first of all by immunizing with as pure an antigen as possible, preferably synthetic, to avoid its contamination with unknown antigens or unwanted tissue constituents. Nevertheless, any serum taken from an immunized animal will contain numerous natural antibodies and antibodies to any carrier protein and conjugating molecules used in immunization. These will result in unwanted antibodies binding to the tissue and purification of the antibody by affinity absorption may be necessary (*see* Section 5.3 and Chapters 3 and 15). Numerous controls are also necessary in staining (*see* Section 6). A further problem connected with specificity relates to the possibility of cross-reactivity. Unless an antibody reacts with an antigenic sequence that is unique to one substance, there is always the chance of a reaction to a similar sequence in another substance. This problem has been of particular hazard in peptide immunocytochemistry because of the many 'families' of different regulatory peptides related by their molecular structure (*see* Section 6 and Chapters 3 and 4). If the titre, or concentration, of the antibody in the serum is high enough, the primary antibody can sometimes be diluted to such an extent that most of the unwanted reactions become negligible and purification is not necessary. An additional advantage of high dilutions is economy in the use of precious antisera.

5.1.2. *Affinity*

Affinity is the three-dimensional fit of the antibody molecule to its specific antigen, and depends on the size and shape of the antibody sequence with which the antibody is reactive. Antigenic sites usually consist of a sequence of 4 to 7 amino acids. The shorter the sequence, the lower will be the affinity of the antibody and the higher its chances of finding similar sequences in other substances (cross-reactivity, *see* 5.1.1. *above*). The longer the sequence, the higher will be the affinity and the lower the chances of dislodging the antibody from its antigen during the rigorous washing processes of the immunocytochemical technique.

5.1.3. *Avidity*

Avidity is a related property referring to the heterogeneity of the antiserum which will contain various antibodies reacting with different areas of the antigen molecule. A specific but multivalent antiserum is less likely to be removed during the washing process than a monovalent antibody because although each antigen–antibody reaction is in a permanently dynamic state it is unlikely that all the different bonds will be broken at once, and enough antibodies will remain attached to be visible (if labelled) or to react with the next layer. The larger the antigen, the more immunogenic sites it will possess and the greater will be the valency of the antisera raised to it. Unfortunately the conditions leading to production of avid antibodies of high affinity and titre are not known.

5.2. Secondary Antibody

All the above considerations, and those in the next section, apply equally to second layer antibodies.

5.3. Methods of Purifying Antibodies

Several methods have been devised for ridding antisera of unwanted and non-specific reactivity.

5.3.1. *Affinity Purification*

Purification of a mixed antiserum may be achieved by affinity absorption. The pure antigen is absorbed onto a solid phase such as cyanogen bromide-activated Sepharose beads; the coated beads are then mixed with the antibody, which reacts with the antigen. Non-attached proteins are washed off and the specific antibody is eluted from its antigen by washing with a low or high pH buffer. For the details of the method *see* Chapter 3 and Reference 34. The disadvantage of this method is that antibodies of high affinity and avidity that are most useful for immunocyto-chemistry are, by their very nature, difficult to dissociate when bound to specific antigen on the beads. The antibody that is eluted is therefore likely to be of low affinity, although this may be compensated for by its purity.

Absorption of an antibody onto antigen-coated beads may also be used as a negative control for staining (*see* Section 6.2.2).

Another way in which affinity absorption may be useful is in removing a known contaminant from an antibody. For example, polyvalent antiserum to follicle-stimulating hormone (FSH) will contain antibodies both to the β subunit, specific to the FSH molecule, and to the α subunit, shared by FSH, luteinizing hormone (LH), thyroid-stimulating hormone (TSH) and human chorionic gonadotrophin (HCG). Absorption of this antibody on, for example, HCG-coated beads will remove the confusing antibodies to the α subunit, leaving only the antibodies specific to the β subunit of the molecule that is not shared with the other three peptides. The unwanted tissue antibodies will of course remain after this treatment.

5.3.2. *Absorption on Tissue Powders*

Absorption on tissue powders or homogenates from tissue not containing the antigen in question, has been used to get rid of unwanted tissue reactions. In some cases this may be successful, but random non-specific absorption of protein does reduce the amount of antibody available, which is a serious consideration.

5.3.3. *Monoclonal Antibodies*

Milstein and his colleagues[35] have devised an ingenious method of producing pure and reproducible antibody. Mice of a certain strain, e.g. BALB/c, are immunized, and when antibody is being produced the mice are killed and the dissociated cells of their spleens, source of the antibody-producing plasma cells, are mixed with cultured myeloma (plasmacytoma) cells from the same strain of mice, that can be kept indefinitely in culture. The spleen cells fuse with the myeloma cells, producing hybrid cells that retain the properties of both parent cells. Thus they will continue to grow and divide in culture while producing antibodies. Naturally, the hybrid cell population will produce as many assorted antibodies as the original immunized mouse. To obtain pure antibodies, the culture must be divided and the culture medium from each subculture must be tested for antibody production. Only cultures producing the wanted antibody are kept, and these are further divided and tested until clones derived from a single cell and producing specific antibody are achieved. The culture medium from these clones may then be used as a monospecific antibody to the original antigen. The cloned cells may be frozen and stored until supplies of antibody are needed. The universal and continual availability of one particular antibody, free from unwanted interference, is an enormous advantage and justifies the time and effort required in its production. Monoclonal antibodies are increasingly widely used in all fields of immunocytochemistry and many are now obtainable commercially, as are the necessary second layer anti-mouse immunoglobulins. Monoclonal rat antibodies are also available. Monoclonal antibodies are of course monovalent: if cross-reactivity (*see* Section 6.3) is still a problem the use of several differently monovalent monoclonal antibodies may help to define the antigen being stained.

5.4. Tissue Antigen

5.4.1. *Fixation*

The antigen to be localized must be insoluble or must be made insoluble by being fixed in the tissue, but it must also be available to the antibody in an immunoreactive form. Tissue preparation methods will depend largely on the nature of the antibody to be stained. The best morphological preservation is usually seen after fixation by routinely used fixatives such as formaldehyde or glutaraldehyde which produce inter- and intramolecular cross-linkages among proteins in the tissue. As long as the fixation process does not destroy the antigenicity of the substance to be localized, this is the method of choice. However, some antigens, such as surface antigens on lymphocytes, are made partially or completely non-immunoreactive by fixation, or by the paraffin-embedding process, and in such cases unfixed cryostat sections of fresh-frozen tissue are necessary (*see* Chapter 14) or, of course, suspensions of cells. Whole organisms such as viruses are insoluble and need not be fixed, but if they are to be identified in context some tissue fixation is desirable.

It used to be thought that antigenic immunoreactivity would survive only very gentle fixation, for example by a low concentration of formaldehyde or specially developed weakly cross-linking reagents such as diethyl pyrocarbonate and *p*-benzoquinone[36] (*see* the Appendix, p. 35).

However, although these fixation methods are still extremely useful, the improved sensitivity of modern immunocytochemical methods such as the PAP method, now frequently allows the use of formaldehyde in its routinely used concentration. This has proved very beneficial in retrospective studies of paraffin-embedded histological material. Even the more powerful cross-linking reagent, glutaraldehyde, followed by osmium tetroxide, can sometimes be used for electron microscopical immunocytochemistry.

For every new antigen to be immunostained, a variety of fixation methods should be tried. A table of fixation guidelines is given in the Appendix, p. 35).

5.4.2. *Protease Digestion*

A further useful advance has been provided by the technique of 'digesting' formalin-fixed sections with a protease such as trypsin or pronase before applying the first antibody.[37] The mechanism is not understood, but it seems that some of the protein cross-linking caused by the fixative is released by the protease, making more antigenic sites available to the antibody (*see* Appendix, p. 35 and Chapter 14).

5.5. High Specific Reaction: Low Background Staining

Background staining due to many causes has bedevilled immunocytochemistry since its beginnings and although the experimenter usually thinks he knows what is specific staining and what is non-specific background staining, he may have difficulty in convincing an unbiased observer. The most satisfactory way of eliminating unwanted staining is to use affinity-purified or monoclonal antibodies throughout, although even these may still suffer from cross-reactivity (*see* Section 6.3). Some other ways of minimizing the problem are described below.

5.5.1. *Removal of Background Fluorescence*

High background staining was originally found to be due to unreacted fluorescein isothiocyanate in the conjugated antibody solution. This could be removed by running the mixture on a Sephadex G-25 column.[38] Fluorescein-conjugated antibodies bought from commercial suppliers should have been purified already by this means.

5.5.2. *Adsorption on Tissue Powders or Homogenates*

Some of the reaction of unwanted antibodies to tissue constituents and non-specific adsorption to tissue may be removed by absorption against common tissue components such as albumin, which may indeed have stimulated antibody formation if it was used as a carrier protein during immunization. Adsorption with homogenates or acetone-dried powder of 'inert' tissue such as liver (i.e. not containing the antigen under investigation, but containing normal components of tissue) may also be helpful (*see* Section 5.3.2). For details of the technique *see* Nairn.[2]

5.5.3. *Blocking of Reactive Sites on Tissue Sections by Normal Serum Prior to Staining*

This is an essential step in all the methods unless affinity-purified or monoclonal antibodies are exclusively used or protein A is employed. Protein A reacts with the Fc portions of immunoglobulins which are even more likely to be present in the normal serum used for blocking than in the tissue itself.

The principle is that non-immune serum from the same species as that donating the second layer antibody in the indirect or PAP method is applied for 10 to 30 minutes at the beginning of the procedure and sticks to protein-binding sites either by non-specific adsorption or by binding of specific but unwanted serum antibodies to antigens in the tissue. Because the non-specific adsorption reactions are likely to be of low affinity the serum is not washed off, merely drained, before the primary antibody is applied and its presence prevents further non-specific or unwanted specific attachment by the primary antiserum. Logically, it should be re-applied after washing off the first layer and before applying the second layer to prevent non-specific attachment by the second layer, but in practice this is not always necessary. If a direct technique is being used the blocking serum should be from the species providing the primary antibody.

5.5.4. *Addition of Protein to Antibody Solution*

Inclusion in the primary antibody solution of 1 per cent of normal serum (from the same species as the second layer antibody) or 1 per cent of an 'inert' protein such as albumin may also help to prevent non-specific staining by absorbing any antibodies to albumin that may be present and by competing with the primary antiserum for non-specific binding sites in the tissue. If the primary antibody is being stored in a highly diluted solution (*see* Section 11.1) the presence of 0·1 per cent albumin will also help to prevent adherence of a proportion of the antibody protein to the walls of the storage vessel with consequent lowering of titre (*see also* Section 6.2.2).

5.5.5. *Removal of Endogenous Peroxidase Activity*

With the peroxidase-labelled antibody techniques, the presence in the tissue of enzymically active peroxidase and catalase and haemoprotein, all capable of reacting with hydrogen peroxide and reducing diaminobenzidine, could confuse the final location of the antigen. This, of course, is more of a problem in cryostat sections than in paraffin sections in which much of the native enzyme activity will have been destroyed during the processing, but even here it may be intrusive. Blocking the endogenous 'peroxidase' with hydrogen peroxide at the start of the procedure is usually effective[39–41] (*see also* the Appendix, p. 34).

5.5.6. *Dilution of Primary Antibody*

If a sensitive method such as the PAP method is being used, it should be possible to dilute the primary antibody up to several hundred times the concentration at which it would have to be used for the indirect method for half-an-hour incubation at room temperature. Overnight, or up to 48 hours incubation at 4 °C or at room temperature with the primary antibody is recommended[42] to allow the highest possible dilutions of the primary antibody. This has the major advantage of reducing the concentration of unwanted antibodies in the solution and thus increasing the ratio of specific to background staining. The long incubation time allows the antigen–antibody reaction to reach its maximum, which is probably not achieved with the original short incubation period, even with higher concentrations. In fact, for the PAP method, a high dilution of primary antibody is essential in order to allow space for both antigen-combining sites of the primary antibody to react with tissue antigens. This decreases the possibility of dissociation during washing and indeed, if the concentration is too high, the second, linking, antibody will no longer be in excess; both its combining sites will become attached to the primary antibody and none will be free to combine with the third layer PAP complexes. Consequently no specific staining will be achieved (*see* Section 11.3)

6. TESTS FOR SPECIFICITY OF REACTION

Having achieved a well-fixed and available tissue antigen, a high titre antibody of high affinity and avidity and having eliminated background staining, the question is whether the visible reaction is really specific to the antigen of interest. Specificity testing applies both to the method itself and to the question of what is being stained.

6.1. Reliability of the Method

As antibodies have a limited shelf-life (usually several years) and all kinds of mishaps may occur in the course of a normal laboratory working day and the many stages of a PAP reaction, it is essential to carry through a known positive control slide with every batch of staining to check that all the solutions are in working order.

6.2. Is the Staining Specific, i.e. Due to the Antibody?

6.2.1. *Replacement of Primary Antibody*

A negative control section of the test material should also be carried through with non-immune serum, or preferably, to check for non-specific immunoglobulin attachment, an inappropriate antiserum replacing the primary antiserum and at the same dilution. If the negative control section has the same staining result as the test section, this could be due to inadequately blocked endogenous peroxidase, or to non-specific uptake of immunoglobulins from the serum or from the second antibody. This latter is a particular hazard when dealing with necrotic tissue or with dead or damaged cells in unfixed smears but might be dealt with by extra blocking with non-cross-reacting heterologous antisera, e.g. a sheep antibody when dealing with a rabbit–swine system. An inappropriate tissue should also be stained with the test antibody to make sure that non-antigen-containing structures do not stain.

6.2.2. *Absorption Controls*

An essential control when dealing with a new antiserum, new material or a new tissue localization is to pre-absorb the primary antibody with excess of its specific antigen so that no antibody-combining sites are available to react with the tissue and staining does not occur. If staining does occur (but does not occur on the known antigen-containing control section) then the staining must be due to some cause other than the specific antigen–antibody reaction under investigation. Pre-absorption with an inappropriate antigen should have no effect on staining.

PRECAUTIONS. In order to make the best use of small supplies of expensive pure antigens, absorption should be carried out at the highest dilution of the antibody compatible with consistent clear staining, as the higher the concentration of antibody, the more antigen it will require for saturation. The use of known quantities of antigen (e.g. 1 nmol per vial) makes it possible to compare the reactivities of different antibodies with the same antigen or of one antibody with equivalent molar quantities of several antigens. Parallel staining should be carried out using unabsorbed antibody at the dilution used for the absorption.

LIQUID PHASE ABSORPTION. The incubation of the antigen with the antibody should be carried out at 4 °C for 12–24 h before the solution is used for staining. It must be remembered that antigen–antibody bonds are reversible and over a longer period of time it is possible that some dissociation will take place and free antibody will again be present. A necessary precaution is the addition of a protein such as 0·1 per cent bovine serum albumin (BSA) to the antibody solution prior to absorption and, indeed, prior to storage at high dilution. This large quantity of added protein competes with the very small amount of specific antibody present for unspecific binding to the glass walls of the absorption or storage vessel, and thus prevents removal of antibody from solution. The necessity for this precaution was brought home to us when we successfully 'absorbed' a dilute antibody solution by shaking and incubating it in an empty glass vial overnight. The problem has not recurred since we have routinely diluted our antibodies in a solution containing 0·1 per cent BSA. The addition of albumin will also serve to remove antibodies to albumin in the case of its having been used in immunization as a carrier protein.

SOLID PHASE ABSORPTION. Absorption on antigen-adsorbed Sepharose beads is a more satisfactory way of removing antibody from solution which avoids the difficulties mentioned above. The problem here is that it is difficult to calculate the quantity of antigen-coated beads required or to compare two antibodies.

6.2.3. *Receptor Binding*

One further point to bear in mind is that an 'absorbed' antibody solution containing antigen–antibody complexes may still be able to bind to antigen receptors if they happen to be present in the tissue to be stained, provided that the receptor-binding site on the antigen is different from the antibody-binding site. The antibody part of the receptor-bound complex might then be localized falsely as a positive antigen-producing site. Thus pituitary gonadotropes were stained by absorbed anti-luteinizing hormone-releasing hormone (anti-LHRH) as shown by Sternberger and his colleagues.[43]

6.2.4. *Ionic Binding*

An additional check on the validity of staining has been suggested by Grube.[44] Some non-specific absorption of antibody onto sections may be due to ionic forces. The bonds will be of low affinity and if the rinsing after immunoreaction is carried out under conditions of high salt concentration (0·5 M NaCl) or high pH (buffer at pH 8·6, near the isoelectric point of IgG), non-specific bonds of this nature will be prevented although the conditions will not be so severe as to dissociate the high-affinity specific antigen–antibody bonds. At high dilutions of antibody ionic binding is not a problem.

6.2.5. *Complement Binding*

Another hazard may be due to complement in the antiserum becoming bound to the tissue, and in its turn causing binding of secondary antibody.[45] At the high dilutions of primary antibody that we recommend here, this contingency is not likely to occur.

6.3. The Problem of Cross-Reactivity: What is Being Stained?

When the molecular structure of an antigen is known and its presence in the tissue is a certainty, and when the antibody used for localization has been raised to that particular antigen and staining is prevented by absorption with that antigen but with no other related substance, then its identification should present no problems. However, if the antigen is one of a family of structurally related but different molecules, such as gastrin, cholecystokinin and caerulein, which share C-terminal antigenic sites, then the fact that an antibody to the C-terminal of gastrin stains cells in the duodenum is no guarantee that the substance being stained is truly gastrin, even if absorption with gastrin prevents the staining. This and similar cases call for the use of region-specific antibodies raised to unshared portions of the molecules (*see* Chapter 4). Where such antibodies are not available, no answer to the problem can be obtained by immunocytochemistry alone.

A similar problem is met with in trying to distinguish between the staining of large precursor 'prohormone' molecules and the smaller bioactive molecules derived from them, and possibly stored with them in the same secretory granules. Unless antibodies to the extended precursor part of the molecule are available, as for example for glucagon and proglucagon,[46] no absolute identification can be made. Correlation with radioimmunoassay of chromatographed extracts would have to be used to identify the molecular sizes of the substances present.

It is advisable to use as many antisera as possible when localizing a substance, in the expectation that they may contain antibodies to different antigenic determinants of the immunogen. Absorption with related substances may help to identify the substance being stained. For example, two regulatory peptides, bombesin and substance P, share a C-terminal antigenic site. If staining with an anti-bombesin antibody is not removed by absorption with substance P, the substance being stained is more likely to resemble bombesin than substance P and the antibody being used is likely to be directed to the middle or N-terminal end of the molecule. If staining is removed by absorption with substance P, the antibody is probably directed to the C-terminal end of the molecule and there is no way of establishing by immunocytochemistry whether the localized antigen is substance P or bombesin except by using additional antibodies, specific to other regions of the molecule. A point to bear in mind, in addition, is that an antigen free in solution may differ in its molecular conformation from the same antigen bound in the tissue, and thus its combining properties may also be different.

6.4. Use of Heterologous Antibodies

By this we mean staining tissue of one species with antibodies raised to an antigen from another species of animal. Most of the regulatory peptides, for example, that have been analysed in different animals show differences in amino acid sequences from one species to another, often in the non-bioactive portion of the molecule. There is also frequently more than one form of a peptide in a single species, and these considerations must apply also to other types of antigen. It is obviously impractical to isolate antigens from every species, analyse them and raise and test antibodies to them, unless for a particular piece of research, so comparative endocrinologists, for example, use the available antibodies for immunocytochemical tests, with the proviso that if cross-reactivity is found, as demonstrated by specific (absorbable) staining, the substance localized cannot be said to be the same as the original antigen, but merely to resemble it antigenically. Of course, extraction and analysis may follow the initial localization. Similarly, of course, the function of the antigen-like substance localized in a heterologous species may be completely different from the function of the original antigen in its own species. Nevertheless, cross-immunoreactivity has been valuable in comparative work which has resulted in the current theory that regulatory peptides are at least as old as the first forms of animal life.[47, 48]

7. FINAL RESULT

7.1. Alternatives to Diaminobenzidine

Immunocytochemists have relied on the highly sensitive H_2O_2–diaminobenzidine reaction of Graham and Karnovsky[23] since peroxidase was first introduced as a label.[11] However, recent publicity about the possible carcinogenic properties of diaminobenzidine has led to increased use of alternative couplers for peroxidase development, said not to be carcinogenic, such as 3-amino-9-ethylcarbazole,[49] 4-chloro-1-naphthol[50] and Hanker's reagent[51] (*see* the Appendix, p. 37, for details).* These substances are less satisfactory than diaminobenzidine because, with the exception of Hanker's reagent, the end-product of reaction will not withstand alcohols and solvent-based mounting media so that water-based mounting media have to be used. Hanker's reagent, in our experience, deteriorates on storage and is capricious in action, although when it works well the reaction is quite satisfactory and has the advantage that it is supposed to be specific for plant peroxidase, so that endogenous animal peroxidase does not interfere with the reaction. The end-colour is a purplish brown. The 3-amino-9-ethylcarbazole end-product is red, and that of 4-chloro-1-naphthol, dark-blue. These substances are therefore useful in double staining methods (*see* Section 9).

7.2. Precautions in the Use of Diaminobenzidine

Diaminobenzidine is still the most sensitive coupler and provided that its use is restricted to a fume cupboard and any spills are 'neutralized' with bleach (sodium hypochlorite), and that due care is taken when using it, the advantages probably outweigh the disadvantages (*see* the Appendix, p. 34).

7.3. Background Counterstaining

To provide a fluorescent background contrasting with the green fluorescence of fluorescein, the simplest stain is Evans Blue, originally suggested by Nicholls and

* Nevertheless, it is safer to assume that a substance is dangerous until it is proved otherwise.

McComb.[52] A more specific counterstain, e.g. for mucous cells, is the periodic acid–Schiff reaction, but this must be done before the section is immunostained and may lead to some loss of immunoreactivity. Nuclei may be counterstained for immunofluorescence by methyl green which fluoresces red, or they may be quenched completely by a light stain in haematoxylin. Haematoxylin or methyl green[53] may be used for nuclear counterstaining for immunoperoxidase preparations. Toluidine blue is another useful light microscopical counterstain (*see* the Appendix, p. 34, for details).

8. METHODS OF IMPROVING IMMUNOSTAINING

There is a battery of ways in which one may attempt to increase the signal-to-noise ratio, or contrast, of the immunostain if necessary.

8.1. Increasing the Incubation Time

Petrusz et al.[42] have suggested that long incubation with the first antibody at high dilution can improve the reaction because equilibrium between the antibody and the tissue antigen is more likely to be reached than with shorter incubation times. This also allows for the highest dilution of the antisera, and thus reduction of background staining. If staining is prolonged at a low dilution of the antibody, background staining may increase in prominence. Prolongation of the second and third layer incubation time will also be effective.[54]

8.2. Repetition of Layers

a. By building up layers of antibody, e.g. repeating the second layer after the PAP layer and then repeating the PAP layer again, a complex of antigen–antibody molecules is built up with a concomitant increase in the amount of label over the original site of reaction.[6]

b. Repetition of the primary antibody layer after thorough washing may increase the staining, perhaps because the low affinity antibodies are dislodged during the rinsing process, leaving free binding sites for further high affinity bonds during the next application of the primary antibody.[55]

8.3. Reduction of Background Staining

In addition to the methods described in Section 5.5, this may be achieved, particularly on smears, cultures and cryostat sections, accompanied by improved antibody penetration, by inclusion of a detergent such as Triton X-100, (0·2–1 per cent) in the rinsing buffer.[56] This cuts down non-specific binding of serum proteins to the tissue.[57] Hydrogen peroxide pretreatment, even for the immunofluorescent method where peroxidase blocking is not required, may also be useful.[58] Tris buffer at pH 7·6 was recommended by Graham and Karnovsky for development of peroxidase by the H_2O_2–diaminobenzidine reaction. However, other buffers such as phosphate-buffered saline at pH 7·0–7·2 may also be used during the immunostaining sequence and will also serve as a buffer for the peroxidase reaction. The use of Tris buffer rather than phosphate-buffered saline in the final rinse and in the peroxidase development produces less background staining but may also cut down the intensity of the enzyme reaction. Binding of protein to free aldehyde groups may be prevented by a rinse in sodium or potassium borohydride (*see* Chapters 5 and 15).

8.4. Antibody Penetration

Exposure of antigen to antibody in tissue sections presents no difficulty, but where whole cells are concerned the plasma membrane may prevent large antibody molecules from reaching intracytoplasmic antigen. Antibody penetration into smears, tissue culture preparations, cell suspensions and whole mounts may be improved by making the cell membranes of the tissue permeable with a detergent, as mentioned above, or by repeated freezing and thawing (*see* Chapter 10), or by progressive dehydration through alcohols to xylene, followed by rehydration.[5] For identification of cell surface antigens these procedures should not be used.

8.5. Protease Digestion

As already mentioned (*see* Section 5.4.2) this may improve antigen availability.

8.6. Washing

Very thorough washing of sections from formalin-fixed paraffin-embedded material before immunostaining is begun may also reverse some of the effects of cross-linking by formalin fixation and make more antigen available to the antibody.[5]

9. STAINING SEVERAL ANTIGENS IN ONE SECTION

There are several ways of accomplishing this useful end (*see also* Chapter 7).

9.1. Double (or triple) Immunoperoxidase Method I

One of the first to be described was the double (or triple) immunoperoxidase stain of Nakane[50] in which the initial peroxidase label is developed with diaminobenzidine (DAB): the antigen–antibody complex is then dissociated and the antibody eluted by means of a low pH buffer, leaving the insoluble end-product of the DAB reaction *in situ*. A second antibody is then applied and the reaction is developed to give an end-product of a contrasting colour, say blue with 4-chloro-1-naphthol as the coupler. A further elution and third antibody may then be used with a third developing colour, say red with 3-amino-9-ethylcarbazole (or α-naphthol pyronin, as recommended by Nakane). Provided that the elution is complete (and this varies with the affinity of the antibodies and must be tested in each case) this can be a useful method.

9.2. Double (or triple) Immunoperoxidase Method II

The same kind of sequential staining can be carried out if the first reaction is photographed and the whole complex is removed with acidified potassium permanganate. The section is then bleached and restained with a second system.[59] In practice, it is difficult to achieve the correct balance of potassium permanganate and acid to remove the reaction products without preventing all subsequent reactions.

9.3. Double Immunoenzymatic Method

This method was devised by Mason and Sammons[14] and is described in Chapter 7. Two parallel immunoenzymatic stains are performed simultaneously, the antibodies being raised in different species and not cross-reacting. The disadvantage of this method is that it requires six separate antibodies.

9.4. Double Immunofluorescence

This may be carried out in much the same way as the double immunoenzymatic method, using a mixture or a sequence of differently labelled antibodies in the direct method or non-cross-reacting antibodies in the indirect method, the antibodies being conjugated with fluorescein or rhodamine isothiocyanate. The fluorescence has to be viewed with different filters so the only way to see a simultaneous picture is to photograph the preparation with first one filter and then the other, using a double exposure. Cells expressing both antigens will show as orange, a mixture of the red rhodamine and green fluorescein colours (*see* Chapters 13 and 19). Immunofluorescence may be combined with other labelling methods.

9.5. Gold Labels combined with Other Methods

The gold-labelled antibody or protein A methods may be used at light microscopical level in combination with peroxidase methods. The contrast with the pink of the gold is best if the peroxidase is developed by the 4-chloro-1-naphthol method to give a blue end-product (*see* Plate 3). The gold method must be used last because as the label is attached to the antibody, acid elution will remove it along with the antibody.[60]

9.6. Adjacent Semithin Sections

Adjacent 2–3 µm sections pass through the same structures which may thus be stained with different antibodies (*see* Chapter 11). This is a particularly useful method for resin-embedded material that may be cut easily at 0·5 μm. The gold immunostain is not really intense enough for ordinary light microscopy on such thin sections, but under dark field illumination the scattering of light by the gold particles ensures that the smallest deposit can be seen very clearly (*see* Chapter 6). The visualization of the diaminobenzidine reaction on thin sections is improved by Nomarski differential interference contrast optics which make the stained areas appear in raised relief and greater contrast against the unstained but visible background.

9.7. Radioactive Labels and the Labelled Antigen Methods

These may be combined with most of the above methods.

9.8. Double Staining at Electron Microscopical Level

Gold particles of different sizes may be used to stain two different antigens in the same section, either by a double direct immunostaining method or by the indirect method with gold-labelled second antibodies and primary antibodies raised in different species, e.g. guinea-pig and rabbit (*see* Chapter 6). There are also means of combining immuno-gold staining with immunoperoxidase staining for electron microscopy.

10. IMMUNOCYTOCHEMISTRY FOR THE ELECTRON MICROSCOPE

10.1. Fixation

It used to be thought that conventional glutaraldehyde–osmium fixation was unsuitable for electron microscopical immunocytochemistry, but with the greatly improved sensitivity of today's methods, even these fixatives are not ruled out and, of course, provide much better ultrastructural preservation than substitute fixatives such as very low concentrations of glutaraldehyde or formaldehyde-based fixatives without osmication. Nevertheless, a range of fixatives should be tried for every antigen to be stained, as many may not survive conventional fixation.

Among alternative fixatives are formaldehyde–glutaraldehyde mixtures[61] and a combination of periodate, lysine and paraformaldehyde[62] that is particularly useful for preserving glycoproteins and gives good ultrastructural detail. Larsson[63] has added *p*-benzoquinone to a glutaraldehyde–formaldehyde mixture to preserve peptide antigenicity and another cross-linking fixative, carbodiimide, has also been used in a fixative mixture.[64]

10.2. Tissue Preparation

10.2.1. *Post-Embedding Staining*

Fixed material may be processed to resin blocks in the usual way and immunostaining is then carried out on ultrathin sections mounted on grids (*see* the Appendix, p. 38 and Chapters 5, 6 and 12). If this is unsatisfactory, the cause of the loss of antigenicity must be sought. It may be as much a matter of temperature or the solvents used in the processing as the initial fixation. Polymerization of the embedding resin at room temperature by ultraviolet irradiation has been recommended for preservation of ultrastructure[65] and has been found useful for the localization of vasoactive intestinal polypeptide immunoreactivity which was difficult to demonstrate in heat-cured resin sections.

10.2.2. *Pre-Embedding Staining*

Pre-immunostaining processing can be avoided altogether by carrying out the immunoreaction after fixation on thick (20–50 μm) sections cut on a tissue chopper or vibratome.[67] The immunostained sections are then postfixed in osmium tetroxide, processed and embedded in resin (*see* Appendix, p. 38). The difficulties of this type of method are those of penetration of the antibodies, particularly the PAP which is a rather large complex. Some method of overcoming the permeability problem must be incorporated, for example the inclusion of a detergent in the buffer rinses and even in the antibody solutions, or dehydration and rehydration of the sections before staining. The identification of surface antigens on cells in suspension do not require these precautions.

10.2.3. *Ultrathin Frozen Sections*

A new development, the cryo-ultratome, allows the production of ultrathin frozen sections[68] of fixed material which may be placed directly on grids for immunostaining, thus avoiding the entire embedding procedure. The technique of section-cutting and its uses for light microscopical immunocytochemistry are described in Chapter 5. For producing ultrathin sections, the procedure is similar (*see* Tokuyasu[68] for details).

10.2.4. *Semithin–Thin Method*

It is sometimes easier to avoid the process of immunostaining on grids or free-floating sections, and simply compare photographs of an immunostained 0·05–1 μm resin section with its adjacent ultrathin section in the electron microscope. This method is applicable to peptide antigens in secretory granules in well-defined endocrine cells, for example, but is not adequate for antigens in more delicate intracellular organelles.

10.3. Labels

Because the dense end-product of the peroxidase reaction tends to obscure the morphology of the organelles that are stained, methods using the particulate colloidal gold labels are becoming very popular and have the added advantage that the particles can be made in different sizes, easily distinguishable in the electron microscope, so that double immunostaining procedures can be carried out (*see* Chapter 6).

11. SOME PRACTICAL CONSIDERATIONS

11.1. Storage of Antibodies

Unconjugated primary antisera may be stored frozen, undiluted or at a low dilution (1/10–1/50) in buffer. They should not be repeatedly thawed and refrozen so the solution should be divided into quantities sufficient for a once or twice only dilution. The vials of antibody should be snap-frozen in liquid nitrogen, before storing at −20 to −40 °C.

Alternatively, conjugated or unconjugated antisera may be stored neat at 4 °C with the addition of sodium azide (0·01–0·1 per cent) or merthiolate (0·01 per cent) to prevent bacterial growth. Azide inhibits peroxidase activity but the amount added by commercial suppliers to the neat antibody–peroxidase conjugate or to PAP does not prevent the activity in the diluted working solution. Azide should not be included in the buffers used to dilute peroxidase-conjugated antibodies or PAP, but its presence in the enzyme-developing solution does not seem to be inhibitory (*see* Appendix to Chapter 15). Unconjugated primary antiserum may be kept at its working dilution in buffered saline containing sodium azide or merthiolate and 1 per cent normal serum or albumin to provide a high protein concentration. The inert protein competes with the antibody which is present in very small quantities, and prevents it from sticking to the walls of the vessel (*see also* Section 6.2). Antibody solutions may be kept for many weeks under these conditions. Conjugated antibodies should not be kept diluted as dilution increases the chance of the bond between the label and the antibody becoming dissociated. For the same reason, they should not be repeatedly frozen and thawed. Similar considerations apply to PAP.

11.2. Adherence of Sections to Slides

The immunocytochemical staining process is long and involves many incubation periods and much washing. Unless sections are very firmly attached to slides, they may lift off during the process. Routinely fixed or freeze-dried paraffin sections do not necessarily need an adhesive, provided that they are very well dried at 37 °C before use but, if one is required, ordinary glycerine–albumin solutions are satisfactory. However, if protease digestion is to be used, a stronger adhesive is essential. An almost infallible 'glue' is a polymer of the basic amino acid, lysine, which imparts a positive charge to the slide surface to attract the negatively charged tissue[69] (*see also* the Appendix, p. 36). Paraffin sections should not be baked on a hot plate as this destroys the immunoreactivity of many antigens. Cryostat sections, both fixed and unfixed, may also be picked up on poly(L-lysine)-coated slides, and this is probably more satisfactory than the alternative methods of coating the slides with formol–gelatine or chrome–gelatine. Cryostat sections should be air dried for 1 to 3 h before being stained. Longer drying may result in loss of immunoreactivity of some antigens, e.g. peptides.

Resin sections may be dried onto uncoated slides from a drop of water or 10 per cent acetone, and, provided that they are dried at 37 °C for several hours or overnight, no adhesive is necessary. Epoxy resin is removed before immunostaining by a saturated solution of sodium hydroxide in alcohol.[70] Various other methods have been used, but this one is simple and satisfactory. Glycol methacrylate, extensively used for thin sections in routine histopathology, has so far proved unsatisfactory for immunostaining because of lack of penetration of the antibodies and the impossibility of removing the polymerized plastic from the section. Recently, however, successful immunostaining (for bacteria) was achieved after proteinase digestion, with some other modifications[71] and it may be that the recalcitrance of this embedding medium can be overcome in the same way for other antigens. Other water-soluble resins, such as the London White resin, have potential for immunostaining combined with electron microscopy, but are not yet satisfactory.

11.3 Working Dilution of Antibodies

Every antiserum must be tested on a known positive control to find the 'correct' working dilution for the staining method to be used. The dilution will obviously vary according to whether incubation is to be for half-an-hour at room temperature or overnight at 4 °C. For overnight incubation and the PAP method, for example, an unknown antibody should be tested at doubling dilutions from 1/100 to 1/32 000 to find the initial range, and then further tested if necessary. Even if a dilution of 1/5000 is satisfactory on the positive test tissue, it may have to be altered for another tissue or another fixation method, but the initial test provides a guideline. More concentrated does not usually mean better; for reasons already given, a dilute antibody provides less non-specific background staining and better contrast and, for the PAP method, too concentrated a primary antibody may prevent the PAP layer from attaching. Linking or conjugated antisera and PAP, used for half-an-hour at room temperature, can usually be diluted in the range 1/50 to 1/300. Antisera may also be tested by radioimmunoassay and by the enzyme-linked immunosorbent assay.[13] However, conditions for these *in vitro* techniques are not the same as for antibody exposed to tissue-bound antigen, together with all the other tissue constituents that may be present, and the golden rule for assessing antibodies for immunocytochemistry is to test them by immunocytochemistry (*see* Chapter 3). The other methods have more value, as far as immunocytochemistry is concerned, in assessing potential regional specificity and cross-reactivity of the antibody, since they may expose it to a variety of known antigens.

12. CONCLUSION

The fact that the original method of localizing an antigen with a specific fluorescent antibody, so brilliantly envisaged by Coons, has been developed and expanded over forty years points to its versatility and universal applicability, and ensures that it will survive into the forseeable future. Methods, however, are not ends in themselves, and the reasons for inventing them are amply illustrated in the following chapters.

Appendix

SAMPLE SCHEDULES FOR INDIRECT IMMUNOFLUORESCENCE (IMMUNOPEROXIDASE) AND PEROXIDASE–ANTI-PEROXIDASE (PAP) TECHNIQUES

All primary antisera are diluted in 0·01 M phosphate-buffered normal saline pH 7·0–7·2 (PBS) containing 0·01 per cent sodium azide and 0·1 per cent bovine serum albumin. The second layer antisera and PAP are diluted in PBS alone.

Indirect Method (Immunofluorescence and Immunoperoxidase)

1. Wax sections: remove wax in xylene or other suitable solvent and bring sections to water through graded alcohols. Remove mercury pigment with iodine and bleach in sodium thiosulphate if necessary (peroxidase methods).

Prefixed cryostat sections: mount on poly(L-lysine)-coated slides (*see* the Appendix, p. 36) or on chrome–gelatin or formol–gelatin-coated slides. Allow to dry at room temperature for 30 min to 1 h.

Fresh cryostat sections: allow to dry for a few minutes, then postfix as required.

2. Blocking of endogenous peroxidase activity (peroxidase methods): immerse sections for 30 min in a solution of 0·3 per cent hydrogen peroxide in PBS or in 0·05 M Tris-HCl buffer pH 7·6, or in methanol.

3. Rinse well in PBS: Triton X-100, 0·2–1 per cent may be added to the buffer rinses to decrease background staining. Dry the slides, except for the area of the section, and place in a damp chamber.

4. Background blocking: normal goat serum, 1/5 to 1/30 in PBS, 10–30 min at room temperature. Do not rinse; draw off serum with a pipette or tissue.

5. 1st layer: primary antibody (raised in rabbit), highly diluted, 4 °C overnight or at a lower dilution for 0·5–1 h at room temperature.

6. Rinse three times in PBS, 5 min each rinse. Dry slides except for area of section.

7. 2nd layer: fluorescein-conjugated goat anti-rabbit γ-globulin. 30 min at room temperature at the dilution recommended.

8. (Immunofluorescence) rinse 3 times in PBS and mount section in a mixture of PBS and glycerine 1 : 9. Examine in ultraviolet microscope.

9. (Immunoperoxidase): incubating solution for peroxidase development. Prepare the solution just before use. (Care! Diaminobenzidine (DAB) is carcinogenic.) Dissolve 25–50 mg of DAB* in 100 ml of Tris buffer or PBS. Add 50–100 ml of 100-volume (30 per cent) H_2O_2 (0·015–0·3 per cent final concentration); filter the brownish solution if desired; incubate sections. The development time is rather short—1–5 min. Control by microscopic examination. Counterstain nuclei lightly in haematoxylin, differentiate in acid alcohol, blue, dehydrate, clear and mount. An alternative nuclear stain is 0·1 per cent methyl green, chloroform washed, pH 4 in veronal acetate buffer.[53]

* A convenient and reasonably safe method of dealing with DAB, avoiding the necessity of continually weighing out small samples, is to make a large batch of aqueous concentrate and divide it into vials containing 25–50 mg in solution. The vials are sealed and stored frozen until used.

PAP Method

As for the indirect method up to end of section 6, then as follows:

7. Second layer: unconjugated goat anti-rabbit γ-globulin. 30 min at room temperature at the dilution recommended.
8. Rinse 3 times in PBS. Dry slides as above.
9. 3rd layer: PAP 30 min at room temperature at the dilution recommended. Then continue with rinsing and incubation as for the indirect immunoperoxidase method.

Table. Fixation methods

Method	*Useful for*
Fresh frozen tissue Cryostat sections, unfixed or postfixed in alcohol or acetone	Extracellular antigens, e.g. immunoglobulin in glomerular basement membrane
Smears, impressions, unfixed or fixed in alcohol or acetone Cryostat sections of tissue prefixed by perfusion or immersion in buffered formaldehyde or buffered *p*-benzoquinone (NB-Alteration of temperature or pH may improve fixation)	Cell surface antigens; tissue antigens autoimmune disease Peptides in endocrine cells and nerves Amines, enzymes (as antigens) etc.
Freeze-drying followed by vapour-fixation in formaldehyde, *p*-benzoquinone, diethyl pyrocarbonate Paraffin sections	Intracellular water-soluble antigens; peptides in endocrine cells (not suitable for membrane antigens)
Buffered formalin, formol saline, formol mercury, Bouin's fixative etc. Routine surgical material, paraffin sections (dried at 37 °C) May be treated with a protease before use	Histopathological diagnosis; peptide hormones, immunoglobulins etc.
Periodate–lysine–paraformaldehyde	Glycoproteins
Glutaraldehyde Glutaraldehyde–formaldehyde, periodate–lysine–paraformaldehyde	Electron microscopical immunocytochemistry

TRYPSIN DIGESTION

This can be a useful procedure for revealing over-fixed antigenic sites and is frequently used in staining for immunoglobulins in routinely formaldehyde-fixed tissue. If tissue is fixed in formol sublimate as primary fixative, trypsin digestion is not necessary.

Make a solution of 0·1 per cent trypsin in Tris buffer pH 7·8, or 0·1 per cent calcium chloride, brought to pH 7·8 with 0·1 N NaOH. Warm the solution to 37 °C and check the pH again. After removing wax and mercury pigment, if necessary, immerse the sections in the trypsin solution at 37 °C for 0·5 h. Rinse well in cold distilled water and PBS (TBS) before proceeding with the immunostaining in the usual way.

Note: It is advisable to test a range of times and temperatures for the trypsin digestion to find the optimum for each tissue/antigen. (*See* Chapter 14). Other proteases may be used in the same way.

Poly(L-lysine)-coated Slides

Sections tend to be detached from the slide during trypsin digestion. Make sure they are mounted on slides coated with poly(L-lysine) (from Sigma, mol. wt > 150 000) made up as a 0·01 per cent solution in water and applied to the slide in the same way as a blood film by smearing a small drop over the surface with the end of another slide. The film dries very quickly and the slide is ready to use. A batch of slides may be prepared in advance but should not be kept for more than a week, or they lose their 'stickiness'. It is essential to mark the coated side, because the dried film is invisible. Store the poly(L-lysine) solution frozen in aliquots and refreeze after use.

ABSORPTION CONTROLS

We have found it convenient to divide an antigen solution into vials containing one nanomole of antigen. In general, this is a large enough amount to react completely with the antibody contained in 100 µl of antibody solution at its highest usable dilution for the PAP method with 24 hours of incubation at 4 °C. The nanomole quantities also provide a convenient means of comparing the amounts of different antigens required to saturate an antibody, and avoid the problems of dealing with antigens of different molecular weights. The nanomole aliquots are lyophilized and sealed under vacuum and may then be stored indefinitely in a freezer or at room temperature. An antigenically inert substance such as lactose or albumin may be added to the antigen solution to provide a visual marker for the lyophilized antigen in the sealed vial. When antibody is added to the vial, it should be vigorously shaken to distribute the antibody over the entire surface to make sure that all the antigen is taken up. Incubate the antigen with the antibody for 12–24 h at 4 °C before using the solution for staining. As a control, the antibody should be 'absorbed' with an inappropriate antigen under the same conditions.

ALTERNATIVE METHODS FOR DEVELOPMENT OF PEROXIDASE

None of the methods given below is as satisfactory as the diaminobenzidine method but the reagents used are probably less likely to be carcinogenic and some give coloured end-products different from that of DAB so that they may usefully be used in double immunoperoxidase staining. All incubating solutions must be prepared just before use.

1. Hanker's Method

Hanker–Yates reagent: p-phenylenediamine-HCl—1 part
pyrocatechol—2 parts

Dissolve 75 mg of the mixture in 100 ml 0·1 M Tris buffer pH 7·6 (or PBS) and add 100 µl of 30 per cent H_2O_2. Incubation time is longer than for DAB. Colour of end-product is blue-black. Background is minimal (the reagent is said to be specific for plant peroxidase so endogenous animal peroxidase does not affect the result). Sections may be osmicated if desired and can be dehydrated, cleared and mounted in DPX. A red nuclear counterstain is effective (neutral red or carmalum).

The reagent is available from Polysciences.

2. 4-Chloro-1-naphthol Method

Dissolve 30–40 mg of 4-chloro-1-naphthol in 0·2–0·5 ml absolute alcohol (not more—alcohol reduces peroxidase activity).

Add, with stirring, 100 ml 0·05 M Tris buffer pH 7·6 (or PBS) containing 50–100 µl of 30 per cent H_2O_2. A white precipitate forms. Filter the solution before use. Incubate at room temperature, checking microscopically from time to time. Counterstain with carmalum.

In a modified procedure resulting in a stronger stain (R. Buffa, personal communication), the incubating medium is heated to 50 °C (not more) before use. Filter (coarse filter paper) and use the filtrate while hot.

The reaction-product is dark blue-purple. It is soluble in alcohol and xylene, so cannot be mounted in DPX. A water-soluble mountant such as glycerine jelly is suitable. The reaction-product is said to be more stable if mounted in acid-buffered glycerine pH 2·2.

3. 3-Amino-9-ethylcarbazole Method

Pre-incubate sections in 0·05 M acetate buffer pH 5·0.

Stock Solution (stable at room temperature)

0·4 per cent 3-amino-9-ethylcarbazole in dimethylformamide.

Incubating Solution

3-Amino-9-ethylcarbazole (stock solution)—0·5 ml
0·05 M acetate buffer pH 5·0—9·5 ml
H_2O_2 (30 per cent) —10 µl

Filter on to sections. Reaction time is 5–10 min at room temperature. The reaction product is red. Counterstain lightly with haematoxylin. As the reaction product is alcohol soluble, do not differentiate the haematoxylin in acid-alcohol.

Sections may be thoroughly air dried and mounted in DPX, but it is probably safer to use an aqueous mounting medium.

SAMPLE SCHEDULES FOR ELECTRON MICROSCOPICAL IMMUNOCYTOCHEMISTRY

All buffers are millipore filtered (0·22 μm pore-size) before use for diluting antibodies or washing grids. (Fit filter to syringe for jet washing.) Buffers containing BSA are filtered through a 0·4-μm filter.

Pre-embedding Technique (PAP Method)

Modified from Pickel et al.[67]

1. Fix and wash tissue blocks, e.g. 4 per cent paraformaldehyde + 0·2 per cent glutaraldehyde in 0·01 M phosphate buffer pH 7·2, 4 °C.
2. Cut Vibratome sections (20 μm thickness); put sections into small glass vials.
3. Incubate with 0·25 per cent Triton X-100 in 0·05 M Tris-buffered saline pH 7·3 (TBS) 15 min. Wash 3 × 20 min TBS with agitation or on rollamix.
4. 30 per cent normal goat serum (NGS) in TBS 30 min.
5. Primary (rabbit) antiserum. 24 h 4 °C, diluted in TBS containing 1 per cent NGS. Optimal dilution of primary antiserum is approximately 2 × concentration used for the indirect method at light microscopical level. Wash 3 × 20 min in TBS.
6. Link antibody: goat anti-rabbit IgG 1/100. 2 h at room temperature. Wash 3 × 20 min in TBS.
7. Peroxidase–anti-peroxidase complex (PAP) 1/100. 2 h at room temperature. Wash 3 × 20 min in TBS.
8. Incubate in 0·05 per cent 3,3-diaminobenzidine (DAB) containing 0·01 per cent hydrogen peroxide in 0·1 M Tris-buffered saline pH 7·6, 10 min at room temperature.
9. Fix for 1 h in 1 per cent osmium tetroxide (Millonig's solution). Wash in buffer, dehydrate and embed in Araldite.
10. Cut ultrathin sections from surface to a depth of 1–2 μm.

Post-embedding Techniques

Ultrathin resin or frozen sections are picked up on nickel or gold grids (copper reacts with osmium tetroxide) and are stained directly on the grid. Grids are floated, section down, on drops of antiserum in a microtitre plate or on depressions in parafilm or wax sheets. Grids are individually rinsed by a stream of buffer from a syringe, or communally in a multiple grid holder. All the usual controls are included. In general, dilutions of antibodies for on-grid PAP staining are about the same as for light microscopy and use the same incubation times. Dilutions for the immuno-gold technique lie somewhere between the dilutions used for the indirect method and the PAP method for light microscopy.

PAP Immunostaining on Grid

1. Thin (60–70 nm) silver sections on 300 mesh, uncoated nickel or gold grids.
2. Etch sections in 10 per cent H_2O_2 for 15 min.
3. Jet wash with 0·05 M Tris-buffered saline (TBS) pH 7·3 for 5 min at room temperature, drain buffer from edge of grid on to absorbent paper.
4. Normal goat serum (NGS) 1/30 in TBS for 5 min at room temperature; drain.
5. Primary (rabbit) antibody at dilution in the range used for the PAP method at light microscopical level for 48 h at 4 °C. Re-equilibrate to room temperature for final 1–2 h.
6. Jet wash in TBS. Drain.
7. NGS 1/30 in TBS for 5 min at room temperature; drain.
8. Link antiserum, goat anti-rabbit IgG 1/100 in TBS for 15 min at room temperature.
9. Jet wash in TBS, drain.
10. NGS 1/30 in TBS for 5 min at room temperature, drain.
11. PAP at usual dilution, containing 1 per cent NGS, 20 min at room temperature.
12. Jet wash in TBS, drain.
13. Development of peroxidase in solution of 0·05 per cent DAB + 0·01 per cent H_2O_2 in TBS for 3 min at room temperature.
14. Wash in water and place grids on absorbent paper to dry.
15. Float on 4 per cent (or Millonig's) osmium tetroxide for 25 min.
16. Wash in water.
17. Examine in microscope.

REFERENCES

1. Sternberger L. A. *Immunocytochemistry*, 2nd ed. New York, John Wiley & Sons Inc., 1979.
2. Nairn R. C. *Fluorescent Protein Tracing*, 4th ed. Edinburgh, Churchill Livingstone, 1976.
3. Vandesande F. A critical review of immunocytochemical methods for light microscopy. *J. Neurosci. Methods* 1979, **1**, 3–23.
4. Petrusz P., Ordronneau P. and Finley J. C. W. Criteria of reliability in light microscopic immunocytochemical staining. *Histochem. J.* 1980, **12**, 333–348.
5. Larsson L.-I. Peptide immunocytochemistry. *Prog. Histochem. Cytochem.* 1981, **13**, No. 4.
6. Ordronneau P., Lindstrom P. B. M. and Petrusz P. Four unlabeled antibody bridge techniques: a comparison. *J. Histochem. Cytochem.* 1981, **29**, 1397–1404.
7. Marrack J. Nature of antibodies. *Nature* 1934, **133**, 292–293.
8. Coons A. H., Creech H. J. and Jones R. N. Immunological properties of an antibody containing a fluorescent group. *Proc. Soc. Exp. Biol. Med.* 1941, **47**, 200–202.
9. Coons A. H., Leduc E. H. and Connolly J. M. Studies on antibody production. I. A method for the histochemical demonstration of specific antibody and its application to a study of the hyperimmune rabbit. *J. Exp. Med.* 1955, **102**, 49–60.
10. Riggs J. L., Seiwald R. J., Burckhalter J. H., Downs C. M. and Metcalf T. G. Isothiocyanate compounds as fluorescent labeling agents for immune serum. *Am. J. Pathol.* 1958, **34**, 1081–1097.
11. Nakane P. K. and Pierce G. B. Jr. Enzyme-labeled antibodies: preparation and application for the localization of antigen. *J. Histochem. Cytochem.* 1966, **14**, 929.
12. Massayeff R. and Maiolini R. A sandwich method of enzyme immunoassay. Application to rat and human α-fetoprotein. *J. Immunol. Methods* 1975, **8**, 223–234.
13. Engvall E. and Perlman P. Enzyme-linked immunosorbent assay (ELISA). Quantitative assay of immunoglobulin G. *Immunochemistry* 1971, **8**, 871–874.
14. Mason D. Y. and Sammons R. E. Alkaline phosphatase and peroxidase for double immunoenzymatic labelling of cellular constituents. *J. Clin. Pathol.* 1978, **31**, 454–462.
15. Singer S. J. Preparation of an electron-dense antibody conjugate. *Nature* 1959, **183**, 1523–1524.

16. Larsson L-I and Schwartz T. W. Radioimmunocytochemistry—a novel immunocytochemical principle. *J. Histochem. Cytochem.* 1977, **25**, 1140–1146.
17. Cuello A. C., Priestley J. V. and Milstein C. Immunocytochemistry with internally labeled monoclonal antibodies. *Proc. Natl. Acad. Sci. USA* 1982, **78**, 665–669.
18. Cammisuli S. and Wofsy L. Hapten-sandwich labelling. III. Bifunctional reagents for immunospecific labelling of cell surface antigens. *J. Immunol.* 1976, **117**, 1695–1704.
19. Avrameas S. Indirect immunoenzyme techniques for the intracellular detection of antigen. *Immunochemistry* 1969, **6**, 825–831.
20. Mason T. E., Phifer R. F., Spicer S. S., Swallow R. S. and Dreskin R. D. New intracellular technique for localizing intracellular tissue antigen. *J. Histochem. Cytochem.* 1969, **17**, 190.
21. Mason T. E., Phifer R. F., Spicer S. S., Swallow R. S. and Dreskin R. D. An immunoglobulin–enzyme bridge method for localizing tissue antigens. *J. Histochem. Cytochem.* 1969, **17**, 563–569.
22. Sternberger L. A. and Cuculis J. J. Method for enzymatic intensification of the immunocytochemical reaction without use of labeled antibodies. *J. Histochem. Cytochem.* 1969, **17**, 190.
23. Graham R. C. and Karnovsky M. J. The early stages of absorption of injected horseradish peroxidase in the proximal tubules of mouse kidney. Ultrastructural cytochemistry by a new technique. *J. Histochem. Cytochem.* 1966, **14**, 291–302.
24. Ordronneau P. and Petrusz P. Immunocytochemical demonstration of anterior pituitary hormones in the pars tuberalis of long-term hypophysectomised rats. *Am. J. Anat.* 1980, **18**, 141–158.
25. Sternberger L. A., Hardy P. H. Jr., Cuculis J. J. and Meyer H. G. The unlabeled antibody–enzyme method of immunohistochemistry. Preparation and properties of soluble antigen–antibody complex (horseradish peroxidase–antihorseradish peroxidase) and its use in identification of spirochetes. *J. Histochem. Cytochem.* 1970, **18**, 315–333.
26. Vacca L. L., Rosario S. L., Zimmerman E. A., Tomashefsky P., Ng P.-Y. and Hsu K. C. Application of immunoperoxidase techniques to localize horseradish peroxidase tracer in the central nervous system. *J. Histochem. Cytochem.* 1975, **23**, 208–215.
27. Mason D. Y. and Sammons R. E. The labeled antigen method of immunoenzymatic staining. *J. Histochem. Cytochem.* 1979, **27** 832–840.
28. Larsson L.-I. Simultaneous ultrastructural demonstration of multiple peptides in endocrine cells by a novel immunocytochemical method. *Nature* 1979, **282**, 743–746.
29. Jasani B., Wynford Thomas D. and Williams E. D. Use of monoclonal antibodies for immunolocalisation of tissue antigens. *J. Clin. Pathol.* 1981, **34**, 1000–1002.
30. Guesdon J. L., Ternynck T. and Avrameas S. The use of avidin–biotin interaction in immunoenzymatic techniques. *J. Histochem. Cytochem.* 1979, **27**, 1131–1139.
31. Hsu S.-M., Raine L. and Fanger H. Use of avidin–biotin–peroxidase complex (ABC) in immunoperoxidase techniques. *J. Histochem. Cytochem.* 1981, **29**, 577–580.
32. Wood G. S. and Warnke R. Suppression of endogenous avidin-binding activity in tissues and its relevance to biotin–avidin detection systems. *J. Histochem. Cytochem.* 1981, **29**, 1196–1204.
33. Roth J., Bendayan M. and Orci L. Ultrastructural localization of intracellular antigen by the use of the protein A–gold complex. *J. Histochem. Cytochem.* 1978, **26**, 1074–1081.
34. *Affinity Chromatography*. Publication of Pharmacia Fine Chemicals.
35. Milstein C., Galfre G., Secher D. S. and Springer T. Monoclonal antibodies and cell surface antigens. *Cell Biol. Internatl. Rep.* 1979, **3**, 1–16.
36. Pearse A. G. E. and Polak J. M. Bifunctional reagents as vapour and liquid phase fixatives for immunohistochemistry. *Histochem. J.* 1975, **7**, 179–186.
37. Huang S., Minassian H. and More J. D. Application of immunofluorescent staining in paraffin sections improved by trypsin digestion. *Lab. Invest.* 1976, **35**, 383–391.
38. Fothergill J. E. and Nairn R. C. Purification of fluorescent conjugates: comparison of charcoal and Sephadex. *Nature* 1961, **192**, 1073–1074.
39. Heyderman E. Immunoperoxidase techniques in histopathology: applications, methods and controls. *J. Clin. Pathol.* 1979, **32**, 971–978.
40. Straus W. Inhibition of peroxidase by methanol and by methanol–nitroferricyanide for use in immunoperoxidase procedure. *J. Histochem. Cytochem.* 1971, **19**, 682–688.
41. Straus W. Phenylhydrazine as inhibitor of horseradish peroxidase for use in immunoperoxidase procedures. *J. Histochem. Cytochem.* 1972, **20**, 949–951.
42. Petrusz P., Dimeo P., Ordronneau P., Weaver C. and Keefer D. A. Improved immunoglobulin enzyme bridge method for light microscopic demonstration of hormone-containing cells of rat adenohypophysis. *Histochemie* 1975, **46**, 9–26.

43. Sternberger L. A., Petrali J. P., Joseph S. A., Meyer H. G. and Mills K. R. Specificity of the immunocytochemical LHRH receptor reaction. *Endocrinology* 1978, **102**, 63–73.

44. Grube D. Immunoreactivities of gastrin (G) cells. II. Nonspecific binding of immunoglobulins to G-cells by ionic interactions. *Histochemistry* 1980, **66**, 149–167.

45. Buffa R., Solcia E., Fiocca R., Crivelli O. and Pera A. Complement-mediated binding of immunoglobulins to some endocrine cells of the pancreas and gut. *J. Histochem. Cytochem.* 1979, **27**, 1279–1280.

46. Ravazzola M. and Orci L. Glucagon and glicentin immunoreactivity are topologically segregated in the α granule of the human pancreatic A cell. *Nature* 1980, **284**, 66–67.

47. Van Noorden S. and Falkmer S. Gut-islet endocrinology—some evolutionary aspects. *Invest. Cell Pathol.* 1980, **3**, 21–35.

48. Roth J., Le Roith D., Shiloach J., Rosenzweig J. L., Lesniak M. A. and Havrankova J. The evolutionary origins of hormones, neurotransmitters and other extracellular chemical messengers. Implications for mammalian biology. *N. Engl. J. Med.* 1982, **306**, 523–527.

49. Graham R. C., Jr, Ludholm U. and Karnovsky M. J. Cytochemical demonstration of peroxidase activity with 3-amino-9-ethylcarbazole. *J. Histochem. Cytochem.* 1965, **13**, 150–152.

50. Nakane P. K. Simultaneous localization of multiple tissue antigens using the peroxidase-labeled antibody method: a study in pituitary glands of the rat. *J. Histochem. Cytochem.* 1968, **16**, 557–560.

51. Hanker J. S., Yates P. E., Metz C. B. and Rustioni A. A new specific, sensitive and non-carcinogenic reagent for the demonstration of horseradish peroxidase. *Histochem. J.* 1977, **9**, 789–792.

52. Nichols R. L. and McComb D. E. Immunofluorescent studies with trachoma and related antigens. *J. Immunol.* 1962, **89**, 545–554.

53. Hogg, R. M. Personal communication, 1982.

54. Brandtzaeg P. Prolonged incubation time in immunohistochemistry: effects on fluorescence staining of immunoglobulins and epithelial components in ethanol and formaldehyde-fixed paraffin-embedded tissues. *J. Histochem. Cytochem.* 1981, **11**, 1302–1315.

55. Gu J., Islam K. and Polak J. M. Repeated application of primary antibody improves immunofluorescence staining. *Histochem. J.* 1982 (In press).

56. Hartman B. K. Immunofluorescence of dopamine β-hydroxylase. Application of improved methodology to the localization of the peripheral and central noradrenergic nervous system. *J. Histochem. Cytochem.* 1973, **21**, 312–332.

57. Larsson L.-I. and Mørch-Jørgensen L. Ultrastructural and cytochemical studies on the cytodifferentiation of duodenal endocrine cells. *Cell Tissue Res.* 1978, **194**, 79–102.

58. Islam K. Personal communication, 1982.

59. Tramu G., Pillez A. and Leonardelli J. An efficient method of antibody elution for the successive or simultaneous localization of two antigens by immunocytochemistry. *J. Histochem. Cytochem.* 1978, **26**, 322–324.

60. Gu J., De Mey J., Moeremans M. and Polak J. M. Sequential use of the PAP and immunogold staining methods for the light microscopical double staining of tissue antigens. *Reg. Peptides* 1981, **1**, 365–374.

61. Karnovsky M. J. A formaldehyde–glutaraldehyde fixative of high osmolality for use in electron microscopy. *J. Cell Biol.* 1965, **27**, 137 A.

62. McLean I. and Nakane P. K. Periodate–lysine–paraformaldehyde fixative—a new fixative for immunoelectron microscopy. *J. Histochem. Cytochem.* 1974, **22**, 1077–1083.

63. Larsson L.-I. Ultrastructural demonstration of a new neuronal peptide (VIP). *Histochemistry* 1977, **54**, 173–176.

64. Willingham M. C. and Yamada S. S. Development of a new primary fixative for electron microscopical localization of intracellular antigens in cultured cells. *J. Histochem. Cytochem.* 1979, **27**, 947–960.

65. Shinagawa Y., Yahara S. and Uchida Y. Polymerisation of epoxy resin for electron microscopy in the cold. *J. Electron. Microsc.* 1962, **11**, 133.

66. Probert L., De Mey J. and Polak J. M. Distinct subpopulations of enteric p-type neurons contain substance P and vasoactive intestinal polypeptide. *Nature* 1981, **294**, 470–471.

67. Pickel V. M., Joh T. J. and Reis D. J. Ultrastructural localization of tyrosine hydroxylase in noradrenergic neurons of brain. *Proc. Natl Acad. Sci. USA* 1975, **72**, 659–663.

68. Tokuyasu K. T. Immunochemistry on ultrathin frozen sections. *Histochem. J.* 1980, **12**, 381–403.

69. Husain O. A. N., Millett J. A. and Granger J. M. Use of polylysine-coated slides in the preparation of cell samples for diagnostic cytology with special reference to urine samples. *J. Clin. Pathol.* 1980, **33**, 309–311.

70. Lane B. P. and Europa D. L. Differential staining of ultrathin sections of epon-embedded tissues for light microscopy. *J. Histochem. Cytochem.* 1965, **13**, 579–582.

71. Hermanns W., Liebig K. and Schulz L.-C. Postembedding immunohistochemical demonstration of antigen in experimental polyarthritis using plastic embedded whole joints. *Histochemistry* 1981, **73**, 439–466.

3

Raising and Testing Antibodies for Immunocytochemistry

J. De Mey

1. INTRODUCTION

Immunocytochemical techniques have given us the potential to localize antigens in cells and tissues with strongly improved diversity, selectivity and specificity of staining reactions, as compared to classical staining methods for the demonstration of tissue components.[1–6] They completely rely on the availability of antibodies that will react in a specific way with the antigen *in situ* (which may be very different from *in vitro* conditions). Obtaining such antibodies often presents difficulties, sometimes because immunohistochemistry laboratories lack the necessary equipment for raising and testing antibodies. Nevertheless, it is important that raising and testing of antibodies for immunocytochemistry occur in a coordinated fashion, in order to select those antisera that perform best in a particular localization problem.

I will consider in this chapter general problems related to the preparation of antigens, suitable for immunization, raising and detecting antibodies, and problems related to the antibody specificity, and how one can test and improve this.

2. THE ANTIGEN

An antigen can take many different forms. It can be whole micro-organisms (viruses, bacteria, fungi) or parts thereof,[7, 8] whole eukaryotic cells or subfractions (e.g. membrane preparations). These kinds of antigens have a complex composition and antisera against them will mostly be used for diagnostic purposes to distinguish presence or absence of the group of antigens they are directed to. It is very difficult to obtain monospecific antibodies from this class of antigens, unless extensive absorptions are done. The hybridoma technology[9, 10] clearly has opened new possibilities in this area, and it is recommended to use this approach whenever one wants to obtain monospecific antibodies from mixtures of antigens (*see* Chapter 1).

Often, however, the immunocytochemist wants to raise a specific antibody to one single antigen. Here too, it may be worth considering the hybridoma approach, but it is still very useful to raise serum antibodies as well. In order to obtain monospecific antibodies, it is essential to use highly purified antigens. Often, it will be necessary to use techniques with high resolving power such as preparative isoelectric focusing or affinity chromatography where possible. If the antigen is unsuitable for these techniques, it has often been helpful to first purify it as far as possible and then to separate the mixture by high resolution sodium dodecylsulphate (SDS)–polyacrylamide electrophoresis, cutting out the bands containing the antigen of interest, and using this material to immunize the animal.[11, 12] Using this approach, antibodies have been prepared against proteins that were otherwise very difficult to purify. For testing the purity of protein antigens, the recently introduced, very sensitive silver-staining methods can be used.[13] They are very simple and 20–50 times more sensitive than the normal staining procedures.

Low-molecular-weight antigens such as the peptide hormones are mostly conjugated to a larger carrier protein such as bovine serum albumin, bovine thyroglobulin or keyhole limpet haemocyanin.[14, 15] The possible presence of impurities in the carrier proteins has to be considered. The purity of the peptides must be verified by highly sensitive techniques such as high performance liquid chromatography.

3. RAISING ANTISERA

Rabbits are most often used, but other animals such as goats, sheep or guinea-pigs, or even turkeys can be excellent alternatives. Availability of antibodies made in different species offers the possibility of performing indirect double-labelling experiments without the need to elute antibodies from sections.

It is difficult to make firm statements, based on hard facts, concerning the best possible immunization scheme (e.g. quantity and form of antigen, route, time-schedule). I will, therefore, give as one example the procedure that we have used in our laboratory for the past few years, for antigens such as conjugated peptides and purified soluble proteins (*see also* Chapter 4).

In general, at least two animals are injected per antigen. When enough antigens and animals are available, it is recommended to immunize as many animals as possible. For rabbits and goats, about 1 mg of antigen is solubilized in phosphate-buffered saline (PBS) and homogenized with 1 ml of complete Freund's adjuvant. It does not matter whether the antigen is readily soluble or not. Often, peptide protein conjugates do not dissolve at all.

The homogenate is intradermally and subcutaneously injected at multiple sites along the spinal cord. Usually, 50–200 μl per injection site is given.

Four and eight weeks later, identical booster injections are given, after which monthly boosters of antigen made with incomplete Freund's adjuvant are given.

The first bleeding is taken one week after the second booster, followed by a second bleeding one week later. This is repeated after each booster. For rabbits, blood is taken from the central ear artery with a cannulated needle and a 50-ml plastic syringe. Up to 60 ml blood per bleeding is taken, without damage to the animal. For goats, blood is taken from the jugular vein. Up to 350 ml per bleeding is taken.

Serum is made by incubating the blood (in 12-ml tubes) for 1 h at 37 °C. Serum is centrifuged at low (2500 rev./min) and high speed (3500 rev./min) to remove cells and debris.

Aliquots are stored frozen at −25 °C, or can be lyophilized.

4. DETECTING ANTIBODIES

For most bleedings, it is highly recommended to screen for the presence of antibodies, using an immunocytochemical procedure. This approach presents important advantages.

a. The same principles are used for detection as will be used in future applications. Radioimmunoassay for example uses binding of labelled antigen under *in vitro* conditions. Radioimmunoassay will, indeed, be able to detect the presence of antibodies, but it will not give information about their usefulness in immunocytochemistry.

b. Conditions for tissue or cell preparation which are most appropriate for the localization problem can be selected. The first localization results will give very useful information about the general quality (background and titre) of the antiserum, and its usefulness under staining conditions which should be used to obtain meaningful results. Therefore, direct selection of the animal that gives good quality antiserum, and elimination of the unsatisfactory ones is possible.

c. The particular staining pattern obtained after the first trials will already give an idea of possible specific and non-specific reactions since for many antigens, the localization in the test tissue is known (e.g. restricted to only one cell type or associated with one type of cell organelle).

For the screening of antibodies, a simple technique, such as immunofluorescence is recommended. A second screening can be done with more complicated but more efficient techniques such as immunoperoxidase.

5. TESTING ANTIBODIES: ANTIBODY SPECIFICITY IN IMMUNOCYTOCHEMISTRY

The above described screening will have yielded an antiserum that gives a particular staining pattern.

Now, the specificity of the staining reaction has to be firmly established. The immunocytochemical specificity of an antiserum has different aspects. I will assume that the method specificity of the immunocytochemical procedure used is satisfactory (after performing the usual tests; *see* Vandesande[3]).

5.1. Method Specificity of the Primary Antibody

Grube[16] has shown that antibodies in the primary antiserum can bind to structures such as secretory granules by ionic forces. To avoid the possibility of false positives, he has recommended the use of only high titre antisera at high dilutions (>1500×) and/or raising the salt (NaCl) content of the buffer used as diluent or as rinsing solution to 0·5 M.

5.2. Antibody Specificity

Even a highly purified antigen may contain traces of immunogenic impurities that will induce unwanted antibodies. Such unwanted antibodies may produce a totally normal and specific-looking immunocytochemical staining pattern. This kind of 'specificity', however, is not the desired one. A valuable test for controlling the specificity is the antigen-absorption test. This is often done by adding an excess antigen to the diluted antiserum. A safer way is to use solid phase absorption by which the majority of the antibodies are removed.[3] A negative immunostaining result of such an absorption test indicates that the immunoreactive substances previously shown contain the same antigenic determinants as the ones introduced to the animal by the antigen.

5.3. Affinity Purification of Serum Antibodies

When the antigen absorption test is satisfactorily negative and the serum can be used at high dilutions, giving low background, it can be used as such. The only problems remaining are cross-reactions (*see* Section 5.4) and problems related to the procedure used, such as false negative results because of the use of an inappropriate dilution with the PAP (peroxidase–anti-peroxidase) technique,[3] or differential modification of the immunoreactivity of the antigen, localized at different sites. When the overall staining quality is satisfactory, for example when there is a strong specific component, using the most satisfactory tissue preparation, but there is also a non-specific non-absorbed staining reaction, it will be worth isolating the specific antibodies with antigen-specific affinity chromatography. In our experience, this has practically always been necessary for antibodies against mechanochemical proteins (actin, myosin, α-actinin) and other structural proteins. Fortunately, the yield of such antibodies is often satisfactory. Antisera against peptide hormones, on the other hand, often prove to be of sufficient quality to be used as such. The yield of specific antibodies after affinity purification from such antisera is often also very low.

The following describes the rather simple procedure that is used in our laboratory for purifying antibodies from serum in one step, yielding specific antibodies with >95 per cent immunoglobulin content as judged by analytical SDS–acrylamide electrophoresis.

The antigen is coupled to Sepharose-4B-CNBr (Pharmacia) according to the instructions of the manufacturer. Other carriers and linking groups may be used. Usually, 5–10 mg of the highly purified antigen is coupled per gramme of dry gel to give a total gel volume of 3·5 ml. This amount of gel is added to ±15 ml antiserum and incubated for 2–3 h at room temperature with gentle shaking or rolling. The serum–gel mixture is poured into a column and washed with 10 mM Tris-buffered saline (TBS) at pH 7·6 until the absorbance of the effluent is equal to zero. Then the gel is washed with TBS + 1 M NaCl to elute a small peak of non-specifically absorbed material. The salt is removed from the gel by washing with TBS. To elute the specifically bound antibody, the column is eluted with 0·1 M glycine-HCl buffer pH 2·8 until the baseline is almost reached. The entire trailing peak is collected since the antibodies that are released more slowly are those with the highest affinity. The acid eluant is neutralized with ±10 per cent of 1 M Tris-HCl buffer pH 8·5. This is done during the elution step. Aliquots are stored at −20 °C or −70 °C without further treatment. The gel is removed from the column, washed with 10^{-2} M HCl (pH 2) on a sintered glass filter and re-equilibrated with TBS + 2×10^{-2} M NaN_3, for storage at 4 °C. It can be reused several times. The antibodies thus purified have to be retested to determine the antibody concentration (μg/ml) giving optimal staining reaction and for further characterization of the specificity. An alternative elution method consists of using 4 M $MgCl_2$ in H_2O, pH 6·5–7. The eluted antibody is dialysed against 20 mM Tris-buffered saline pH 8·2. Affinity-purified antibodies give mostly very specific staining patterns with extremely low background levels, comparable to those of monoclonal antibodies. An interesting alternative to the above procedure, especially when only very limited quantities of purified antigens are available, has recently been introduced. It involves electrophoretic transfer of electrophoretically separated proteins (*see* Section 5.5), reaction of antisera with the blots and eluting the antibodies bound to different bands. These antibodies can be retested and used for immunocytochemistry.[17]

5.4. Cross-Reactions and Site Specificity

Many peptides and proteins have homologies in their amino acid sequences and therefore exhibit immunological cross-reactivity. These homologies occur very frequently in secretory peptides found in neurones, and endocrine or paracrine cells, but are also common in related proteins.

An antigenic determinant mostly comprises a sequence of about three to eight amino acid residues. It has therefore been claimed that antibodies should be regarded as site or region-specific detection reagents rather than antigen-specific reagents.[5] With serum antibodies, this important aspect of antibody specificity may be largely hidden by the heteroclonality of the antibodies: various antigenic determinants of the same antigen may be recognized by an even greater variety of antibody molecules. Such antibodies can be made more site-specific by using differential absorption, using either a cross-reacting molecule or parts of the whole antigen. These will remove only certain, cross-reacting antibody populations.

Immunization with smaller fragments of a protein or peptide greatly increases the chances of obtaining region-specific animal sera, and thus helps in avoiding predictable cross-reactions. Such antisera have been very important for specifically localizing different molecular forms belonging to the same family of peptides and proteins.[5,6] Monoclonal antibodies are, by definition, specific for only one antigenic determinant and offer the advantage that their specificity can be precisely determined. By selecting for specificity towards defined sequences, certain predictable cross-reactivities can also be avoided.

The region or site specificity of both animal antisera and monoclonal antibodies must be carefully characterized. This has mostly been achieved through absorption of the antisera or purified antibodies (including monoclonals) against various possible cross-reacting substances such as related proteins or natural and synthetic peptides and peptide fragments. Elegant quantitative cross-reaction studies have been done directly on tissue sections.[18] The test consisted of adding serial dilutions of various peptides and peptide–albumin conjugates to the diluted antisera. The effect of added peptides on labelling intensity was qualitatively observed or quantitatively measured by fluorimetry (depending on the type of section). This approach has the important advantage of using the *in situ* material as substrate for competition reactions. It can give information on the probable degree of identity or similarity of the immunoreactive substances, in different localizations. This study has also given a clear example of the fact that radioimmunological tests do not ensure the specificity of immunohistochemical labelling (*see also* Vandesande[3]). Results of radioimmunological tests reflect the binding properties of a tracer (e.g. a pure preparation of ^{125}I-labelled antigen), to an antiserum, while immunocytochemical procedures reveal the binding properties of an antiserum to an unknown population of immunoreactive tissue substances, as modified by fixation and embedding procedures.[18] The recently introduced gold-labelled antigen detection (GLAD) technique[19] also lends itself very well to this type of antibody characterization. In addition to the above used *in situ* method, immunocytochemical model systems are useful for defining antisera specificity. Such systems include antigens, covalently linked to a solid phase such as Sepharose-4B-CNBr beads.[20] A more elegant model system has recently been introduced by Larsson.[21] It uses submicrogramme quantities of antigens (mostly small peptides) and allows for simultaneous and identical testing of virtually unlimited numbers of peptides under standard immunocytochemical staining conditions.

The test involves immobilization (by formaldehyde vapour) of nanogramme amounts of peptides on filter paper and staining them with the PAP procedure.

A similarly elegant method uses nitrocellulose or diazobenzyloxymethyl (DBM) paper on which various antigens (soluble proteins) have been spotted.[22]

5.5. Positive Identification of Immunoreactive Substances

All the above methods are quite useful for characterizing antisera and antibodies against available molecular forms with more or less predictable cross-reactions. They are, however, dependent on availability of pure test antigens and do not account for unpredictable cross-reactions. Such cross-reactions do occur, certainly when instead of a homologous tissue, a heterologous system is investigated.

The only way to overcome this limitation is to positively identify the immunoreactive substances in the tissue or cells under study. One serious limitation here is that the conditions for testing (especially the state of the antigen that will not be the same as after fixation and tissue embedding) cannot be the same as those of the immunocytochemical procedure.

Three major approaches exist for positively identifying the antigen: biochemical identification, immunoprecipitation and immunoreplica techniques.

5.5.1. *Biochemical Identification*

The tissue containing the immunoreactive material is extracted and the components are separated by high resolution liquid chromatographic techniques such as high performance liquid chromatography. The fractions are screened by a competitive radioimmunoassay and the positive ones are further analysed to identify the immunoreactive substances. The chromatographic characteristics of the immunoreactive substances are compared with those of known immunoreactive standards, mostly the antigen itself or some probable cross-reacting substances. In this way many of the precursor molecules of peptide hormones have been identified, but cross-reactions have also been clarified.

5.5.2. *Immunoprecipitation*

The tissues or cells are extracted. Denaturing conditions, including urea- and detergent-containing buffers are often used, to ensure solubilization of as many components as possible. For cultured cells, metabolic labelling is often used. Tissue extracts are often iodinated.

The extracts are reacted with antibodies and then the antibody–antigen complexes are reacted with either solid phase immunoadsorbent, such as protein A–Sepharose-4B or secondary antibody–Sepharose 4B, or formalin-fixed *Staphylococcus aureus* which has active protein A at its surface.[23] The complexes thus formed are washed, dissolved and analysed by one or two-dimensional electrophoresis techniques, followed by fluorography to detect the tissue components that reacted with the antibodies. Suitable controls, involving pre-immune antiserum or unrelated purified antibodies, have to be included to provide a standard for non-specific absorptions. These techniques are very useful for identification of the antigen. Coprecipitation of components, showing close association with the antigen may give false positives. On the other hand, such coprecipitation may reveal interesting associations of the antigen with other cell components.

5.5.3. *Immunoreplicas*

The development of techniques of transferring proteins, separated by polyacrylamide electrophoresis, to nitrocellulose[24] or diazobenzyloxymethyl (DBM) paper[8] has led to a very elegant and efficient antibody characterization method: the immunoreplica technique.

Initial attempts to characterize antibody binding to separated proteins in acrylamide gels[11,25] were successful but laborious and time consuming. The electroblot techniques offer a method for making exact replicas of separated proteins which are subsequently reacted with antisera[7,8,17] These techniques are extremely sensitive and easy to perform.

With two-dimensional separation techniques, it is possible to identify unequivocally the immunoreactive products. Indirect techniques, using either iodinated protein A or secondary antibodies, followed by fluorography[26] are widely used. The very sensitive silver-staining methods (able to detect less than one nanogramme of protein in a small spot in two-dimensional gels) have made radioactive labelling of the sample unnecessary. Immunoperoxidase techniques are also being used, since their sensitivity is reportedly as high as that of autoradiography. The following describes such an immunoreplica technique as used in our laboratory. As an example, the characterization of an affinity-purified rabbit anti-tubulin is given.

A crude extract of rat brains, homogenized in microtubule polymerizing buffer was separated by one-dimensional SDS electrophoresis, using a 9 per cent acrylamide separating gel. A 2·5-μg sample was dissolved in sample buffer (10 μl of a 0·25 mg/ml solution). Two identical lanes were run. One lane was stained with a silver staining method. The other was electrophoretically transferred to nitrocellulose (NC) paper, as follows. An Electroblot™ apparatus (E–C Apparatus, Florida) was used. The gel was equilibrated for 20 min at room temperature in blotting buffer: 25 mM Tris, 150 mM glycine, 20 per cent methanol.[24] A sheet of NC paper was cut to the size of the acrylamide gel. Two sheets of filter paper were wetted in the buffer and placed on the cathode plate of the apparatus. The acrylamide gel was positioned on the filter paper. The NC paper was wetted with buffer and carefully put on the acrylamide gel. A mark was made to indicate direction of electrophoresis. This mount was covered with two additional sheets of filter paper, also soaked with buffer. Nylon Scotch Brite spacing material was laid on this assembly and the blot unit was closed and positioned in the buffer tank, containing ±6 l of buffer which was recirculated. Transfer was done at room temperature, overnight at 7 V/cm.

After transfer, the acrylamide gel was stained to detect protein bands that were incompletely transferred. This was partly the case (but only to a small degree) for the high-molecular-weight proteins. The NC sheet was subsequently incubated in 5 per cent bovine serum albumin (BSA) buffer: 20 mM Tris, 0·9 per cent NaCl pH 8·2 containing 50 mg/ml BSA type V (Sigma), for 45 min at 37 °C.

For incubation with antibodies, the NC sheet was packed in sealed plastic bags, the dimensions of which closely matched those of the sheet, to reduce reagent consumption. Incubations at room temperature were as follows: rabbit anti-tubulin, 1 μg/ml in 0·1 per cent BSA buffer (same as 5 per cent BSA buffer but containing 1 mg/ml BSA) + 1 per cent normal goat serum (heat inactivated), 2 h; goat anti-rabbit IgG serum (Gibco), diluted 1/40 in 0·1 per cent BSA buffer, 30 min. PAP (prepared in the laboratory), 1/500 in 0·1 per cent BSA buffer, 30 min. Between these incubations, the NC sheet was washed in 3 changes of 0·1 per cent BSA buffer, for 15 min each.

The last wash was in 10 mM Tris-buffered saline pH 7·6. Peroxidase activity absorbed to the sheet was detected by reaction with 0·5 mg/ml diaminobenzidine in 100 mM Tris-HCl buffer pH 7·6 + 0·01 per cent H_2O_2. This reaction mixture was microfiltered before use (0·22 μm). Incubation lasted a couple of minutes.

The results are shown in *Fig.* 3.1. Out of the complex mixture, only one band, corresponding to α and β tubulin was stained, showing specificity of the antibody.

The above procedure could give false negatives since it is not known whether all possible antigens can still react with the antibody after SDS denaturation and absorption to the NC paper.

6. CONCLUSION

Immunocytochemical results can only be validated by the use of high quality, carefully characterized antibodies. Recent developments, briefly described in this chapter, have greatly improved the possibilities for performing this characterization. Nevertheless, a number of pitfalls such as false positives or negatives remain. Specificity of immunocytochemical reactions therefore cannot always be

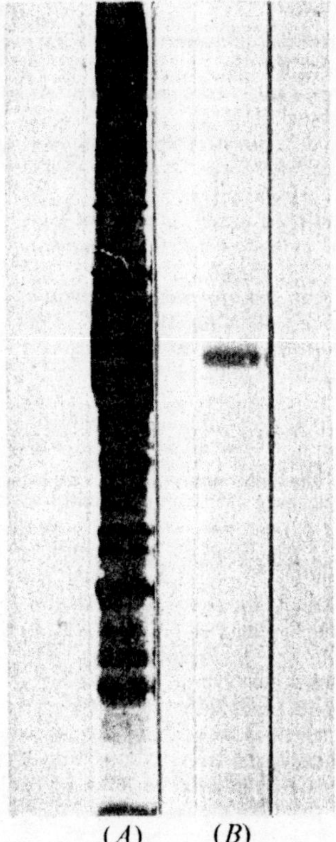

Fig. 3.1. Immunoreplica of rat brain crude homogenate reacted with anti-tubulin and stained with PAP procedure. (*A*) A parallel lane stained with a silver method. (*B*) The immunoreplica of the same proteins after electrophoretic transfer to NC paper: out of the complex mixture, only one band having the mobility of tubulin is stained.

100 per cent proven and this has to be considered when results are interpreted. Nevertheless, with the techniques for antibody characterization at hand today, and the possibilities offered by newly developed procedures for tissue preparation and immunocytochemistry, one can obtain more reliable results than a few years ago.

REFERENCES

1. Coons A. H. Fluorescent antibody methods. In: Danielli, J. F., ed., *General Cytochemical Methods*. New York, Academic Press, 1978; 399–422.
2. Sternberger L. A. *Immunocytochemistry*, 2nd ed. New York, Wiley, 1979.
3. Vandesande F. A critical review of immunocytochemical methods for light microscopy. *J. Neurosci. Methods*. 1979, **1**, 3–23.
4. Forsmann W. G., Pickel V., Reinecke M., Hack D. and Hetz J. Immunohistochemistry and immunocytochemistry of nervous tissue. In: Heym Ch. and Forsmann W. G., ed., *Techniques in Neuranatomical Research*. Berlin, Springer Verlag, 1981: 171–205.
5. Larsson L. I. Peptide immunocytochemistry. *Prog. Histochem. Cytochem*. 1981, **13**, 1–83.
6. Polak J. M., Buchan A. M. J., Probert L., Tapia F., De Mey J. and Bloom S. R. Regulatory peptides in endocrine cells and autonomic nerves. Electron immunocytochemistry. *Scand. J. Gastroenterol*. 1981, **16**, Suppl. 70, 11–23.
7. Legocki R. and Verma D. Multiple immunoreplica technique: screening for specific proteins with a series of different antibodies using one polyacrylamide gel. *Anal. Biochem*. 1981, **111**, 385–392.
8. Symington J., Green M. and Brackmann K. Immunoautoradiographic detection of proteins after electrophoretic transfer from gels to diazo-paper: analysis of adenovirus encoded proteins. *Proc. Natl Acad. Sci. USA* 1981, **78**, 177–181.

9. Köhler G. and Milstein C. Continuous cultures of fused cells secreting antibodies of predefined specificity. *Nature* 1975, **256**, 495–497.
10. Kennett R. H. Monoclonal antibodies. Hybridomas: a new dimension in biological analyses. London, Plenum Press, 1980.
11. Willingham M., Yamada S., Bechtel P., Rutherford A. and Pastan I. Ultrastructural immunocytochemical localization of myosin in cultured fibroblastic cells. *J. Histochem. Cytochem.* 1981, **29**, 1289–1301.
12. Cozzani C. and Hartmann B. Preparation of antibodies specific to choline acetyl transferase from bovine caudate nucleus and immunohistochemical localization of the enzyme. *Proc. Natl Acad. Sci. USA* 1980, **77**, 7453–7457.
13. Merril C., Goldman D., Sedman S. and Ebert M. Ultrasensitive stain for proteins in polyacrylamide gels shows regional variation in cerebrospinal fluid proteins. *Science* 1981, **211**, 1437–1438.
14. Tateishi K., Hamaoka T., Soguira W., Yanaihara C. and Yanaihara N. A novel immunization procedure for production of anti-cholecystokinin-specific antiserum of low cross-reactivity. *J. Immunol. Methods* 1981, **47**, 249–258.
15. Vandesande F. and Dierickx K. Identification of the vasopressin producing and of oxytocin producing neurons of the hypothalamic magnocellular neurosecretory system of the rat. *Cell Tissue Res.* 1975, **164**, 153–162.
16. Grube D. Immunoreactivities of gastrin (G-) cells. II. Non-specific binding of immunoglobulins to G-cells by ionic interactions. *Histochemistry* 1980, **66**, 149–167.
17. Olmsted J. B. Affinity purification of antibodies from diazotized paper blots of heterogeneous protein samples. *J. Biol. Chem.* 1981, **256**, 11, 955–11, 957.
18. Garaud J. C., Eloy R., Moody A. J., Stock C. and Grenier J. F. Glucagon and glicentin-immunoreactive cells in the human digestive tract. *Cell Tissue Res.* 1980, **213**, 121–136.
19. Larsson L.-I. Simultaneous ultrastructural demonstration of multiple peptides in endocrine cells by a novel immunocytochemical method. *Nature* 1979, **282**, 743–746.
20. Streefkerk J., Deelder A., Kors N. and Kornelius D. Antigen-coupled beads adherent to slides: a simplified method for immunological studies. *J. Immunol. Methods* 1975, **8**, 251.
21. Larsson L.-I. A novel immunocytochemical model system for specificity and sensitivity screening of antisera against multiple antigens. *J. Histochem. Cytochem.* 1981, **29**, 408–410.
22. Herbrink P., Van Bussel F. and Warmaar S. The antigen spot test (AST): a highly sensitive assay for the detection of antibodies. *J. Immunol. Methods.* 1982, **48**, 293–298.
23. Oshima R. Identification and immunoprecipitation of cytoskeletal proteins from murine extra-embryonic endodermal cells. *J. Biol. Chem.* 1981, **236**, 8124–8133.
24. Towbin H., Staekelin T. and Gordon J. Electrophoretic transfer of proteins from polyacrylamide gels to nitrocellulose sheets: procedure and some applications. *Proc. Natl Acad. Sci. USA* 1979, **76**, 4350–4354.
25. Adair W., Jurinick D. and Goodenough V. Localization of cellular antigens in sodium dodecylsulfate–polyacrylamide gels. *J. Cell. Biol.* 1978, **79**, 281.
26. Laskey R. and Mills A. Enhanced autoradiographic detection of ^{32}P and ^{125}I using intensifying screens and hypersensitized film. *FEBS Lett.* 1977, **82**, 314–316.

4

Raising Antibodies to Small Peptides

M. Szelke

The main purpose of this chapter is to discuss the chemical aspects of raising antibodies to small peptides that are not immunogenic *per se*. A chemical substance is said to be immunogenic if it is capable of eliciting the production of specific antibodies by lymphocytes. A substance is antigenic if it is capable of binding to specific antibodies.

1. IMMUNOGENS AND THE IMMUNE RESPONSE

1.1 Immunogens

What are the main criteria that determine immunogenicity?

Molecular size.[1–3] Many chemicals (e.g. small peptides, steroids, drugs) of molecular weight less than about 1000 daltons are non-immunogenic, although most of them are antigenic. Large molecules, aggregates and particulate materials are usually good immunogens.

Chemical composition.[1] While most natural proteins with a varied amino acid sequence are strongly immunogenic, homopolymers of amino acids [e.g. poly(L-alanine)] and other macromolecules of monotonous chemical composition (e.g. dextran, polyvinylpyrrolidone) are generally poor immunogens.

Tertiary structure.[1] Immunogenic determinants must be accessible to receptor proteins present on the surface of cells (macrophages, lymphocytes) involved in the immune response. In a complex macromolecule, determinants must be in exposed positions which normally come in contact with the solvent.

How can we raise antibodies to small non-immunogenic molecules? We can render them immunogenic by linking them covalently to large carrier molecule (usually a protein) which is immunogenic in itself[1-3,5]. The small molecule made immunogenic in this way is called a hapten, and the covalent hapten–carrier complex a conjugate.

In addition, we can further stimulate antibody response to such immunogenic conjugates by the use of suitable adjuvants[1-4]. The one most widely used is Freund's complete adjuvant: a mixture of mineral oil, an emulsifying agent and killed mycobacteria. An aqueous solution of the immunogen is intimately mixed with 1–2 volumes of Freund's complete adjuvant to form a stable water-in-oil emulsion, which is used for the primary immunization. Release of immunogen from such a stable emulsion takes place slowly, over a long period. This ensures maximum exposure of the immunogen to cells of the immune system, and also retards its local destruction by phagocytic cells. Lipopolysaccharides present in the killed mycobacteria activate B-lymphocytes and provide a powerful general stimulus to the immune system. Freund's incomplete adjuvant lacks the mycobacteria and should always be used for booster immunizations. When using mineral oil adjuvants, the preparation of intimate and stable water-in-oil emulsions with the antigen is of crucial importance to successful immunization.[4]

1.2 Immune Response

When an antigen is injected into an animal for the first time, antibodies are produced after a time lag of 10–12 days (*Fig.* 4.1). In this primary response antibodies are usually of the high-molecular-weight IgM class; they reach a maximum titre at 15–18 days then disappear from the serum. If the same animal is now re-injected with the same antigen, a more vigorous response follows after a time-lag of only 3–4 days. In addition to IgM antibodies seen in the primary response, much higher levels of IgG antibodies are also produced which persist for long periods.[6] This secondary response to hapten–carrier conjugates depends on cooperation between T- and B-lymphocytes: T-cells recognize the carrier and help the hapten-specific B-cells to respond.

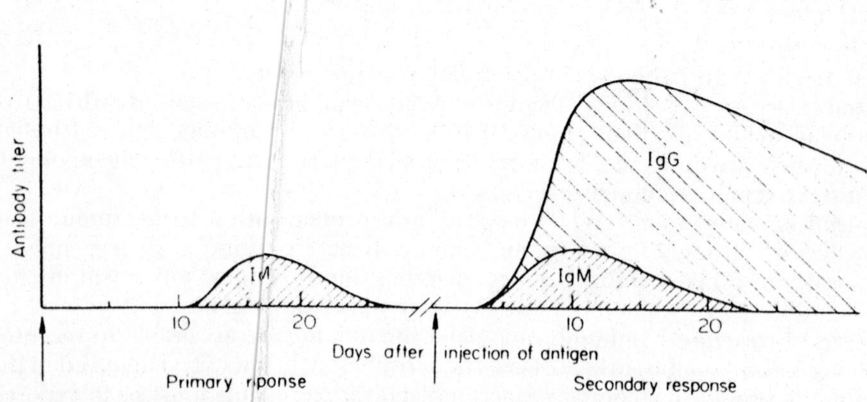

Fig. 4.1. Primary and secondary antibody responses.
This figure is reproduced by kind permission of the publishers from *The Immune System; a Course on the Molecular and Cellular Basis of Immunity* by I. McConnell, A. Munro and H. Waldman. Oxford, Blackwell Scientific Publications Ltd, 1981.

2. PRODUCTION OF ANTISERA

In this section, the various practical aspects of antibody raising will be considered such as the animal species to be used and the route, site and time schedule of the immunization. It must be emphasized that the protocols described here do not necessarily represent the best, or indeed the only ones suitable for the production of antisera. The inherently variable nature of immune response in individual animals makes it difficult to identify significant factors in immunization protocol. Most of the experimental procedures that follow represent a consensus of several laboratories and have been shown to yield antibodies to a variety of small peptides, often in the hands of operators lacking previous experience of immunological techniques.

2.1. Animal Species[1–3]

Ideally, it should be phylogenetically distant from the species against whose peptide antigen one wishes to raise antibodies. The choice is usually limited by practical considerations, such as the availability of accommodation and cost. In general, rabbits, sheep and goats produce more antibody per unit volume of serum than do guinea-pigs and mice. Rabbit is probably the most commonly used species because it is relatively cheap, easy to keep and handle, and affords reasonable quantities (10–30 ml/bleed) of serum. Animals of a mixed breed are generally preferable to inbred strains. It is advisable to use a group of at least 3–6 animals per antigen, in view of the wide individual variation in immune response frequently observed.

2.2. Route of Immunization and Dose of Antigen.[1–4]

The relatively simple subcutaneous or intradermal routes are the ones used most widely and have generally given satisfactory results. They are probably as effective as the technically more difficult intranodal or intraperitoneal methods which also carry a higher risk of mortality. Typically 0·5–1·0 ml of emulsion containing 50–200 μg of the antigen may be divided into 4–6 subcutaneous injections per rabbit, at sites around the neck and shoulders. Alternatively, the multiple-site intradermal technique of Vaitukaitis[7] may be used. The latter involves a greater number of lymph nodes in the primary reaction, is economical of antigen and results in a rapid and vigorous antibody response.

2.3. Immunization Schedule[1–4, 7]

In order to obtain antibodies of high titre and avidity, it is almost always necessary to elicit a secondary immune response (*see Fig* 4.1) by booster immunization which must be given when the antibody titre produced by the primary response begins to fall. Bleeds should be taken at two-week intervals after the primary immunization and antibody titre measured in order to identify the point of decline.

The booster doses should contain a quarter to a half of the amount of antigen used in the primary immunization, and should be emulsified with Freund's incomplete adjuvant to avoid a hypersensitivity reaction.

Further booster injections may be given at intervals of not less than one month. Antiserum should be harvested when antibody titre has reached a plateau, which usually takes place 1–2 weeks after the final booster injection.

3. CONJUGATION OF SMALL PEPTIDES TO A CARRIER PROTEIN[1–3,5]

3.1. Functional Groups Involved

Conjugation is carried out by means of suitable reagents that can form a covalent link between the hapten molecule and the carrier protein. It requires chemically reactive functional groups in both hapten and carrier. In this chapter, we are limiting our discussion to haptens that are small peptides, consequently the functional groups useful for conjugation both in hapten and carrier will be those present in the protein-bound amino acids (*see Fig*. 4.2). In the language of organic chemistry, the chemical reactions leading to conjugation involve nucleophilic groups and, therefore, the practical utility of various functional groups will, by and large, depend on their nucleophilic reactivity.

Fig. 4.2. Functional groups useful in the conjugation of small peptides to carrier proteins.

Amino groups are good nucleophiles and will react with a wide variety of reagents, the ε-amino group of lysine being probably the single most useful functional group for conjugation. N-Terminal α-amino groups of peptides are somewhat less reactive, partly due to their decreased basicity, and partly to steric hindrance.

Carboxyl groups (both those at C-terminal positions and those in the side-chains of aspartate and glutamate residues) can be converted into activated forms (active esters or anhydrides) which will react readily with amine nucleophiles to form amides.

Thiol(-SH) groups of cysteine residues and the phenolic hydroxyl groups of tyrosine residues are also good nucleophiles and will undergo reaction with alkyl or aryl halides. However, one must remember that —SH groups are readily oxidized to form disulphides (—S—S—) which will no longer react with alkylating reagents.

The activated aromatic nuclei of tyrosine and histidine will undergo substitution with diazonium ions generated by diazotization from aromatic amines. The resulting azo linkage forms a stable attachment between carrier and hapten.

Generally, the limited choice of suitable reactive groups on the hapten dictates which type of coupling reagent can be used, and this in turn determines the type of functional groups involved in the attachment to the protein.

3.2. Coupling Reagents

There are two main types of reagents to consider.

a. Those that first react with a functional group (e.g. $—NH_2$ or $—CO_2H$) on the hapten and form a reactive intermediate which sometimes can be isolated. The latter will subsequently enter into a second reaction with a suitable group on the protein to form a covalent link between the two. Carbodiimides and halogenated triazines belong to this type.

b. Bifunctional reagents are molecules bearing two reactive groups (either identical or different) which ideally should be separated by a suitable spacer group. The two active groups may react simultaneously or sequentially with nucleophiles on the hapten and on the carrier protein, cross-linking the two. At the same time two molecules of hapten or two molecules of the protein can be linked together in the same way, resulting in a mixture of the desired conjugate and dimers or sometimes polymers of the hapten and carrier protein, respectively. Glutaraldehyde, diimido esters and bis-diazonium compounds are examples of this second type of coupling reagent.

Table 4.1 summarizes the coupling reagents most frequently used in conjugation, and indicates the type of functional groups with which they will react on a hapten and a carrier protein.

Glutaraldehyde, $OHC—CH_2—CH_2—CH_2—CHO$, is a highly effective coupling reagent for peptides and proteins containing lysine.[8,14,15] As the cross-links formed are stable to acid hydrolysis[10,12] they are unlikely to be bis-aldimines (Schiff's bases) formed by condensation of lysine ε-amino groups and the aldehyde functions of glutaraldehyde. Commercial aqueous glutaraldehyde contains significant amounts of oligomeric α,β-unsaturated aldehydes (formed by dehydration of aldol-condensation products) which will undergo Michael addition with amines to form stable adducts.[10] A spectroscopic study of model reactions between glutaraldehyde and N^{α}-acetyl-L-lysine or 6-aminohexanoic acid also indicates the formation of complex polymeric pyridinium structures.[11] The latter may represent

Table 4.1. Some of the reagents used in the conjugation of small peptides to carrier proteins and the functional groups with which they react.

Functional group on peptide	Reagent	Functional group on protein	Reference
$—NH_2$ (α or ε)	Glutaraldehyde	$—NH_2$ (—SH)	8, 9
	Carbodiimides *4, 5*	$—CO_2H$	13, 16
	Diimido esters *9*	$—NH_2$	9
	Diisocyanates *10, 10a*	$—NH_2$ (—OH)	9
	MeO-Cl_2-triazine *6*	—OH, —SH, $—NH_2$	17–20
	Aryl halides *11*	$—NH_2$, —OH, —SH	9
$—CO_2H$	Carbodiimides *4, 5*	$—NH_2$ (—OH)	13, 16
—SH	MeO-Cl_2-triazine *6*	—OH, —SH, $—NH_2$	17–20
	Alkyl halides *13*	—SH, ($—NH_2$,His)	9
	Iodoacetic acid +carbodiimide	$—NH_2$	13
	Maleimides *12*	—SH	9
—OH (Tyr)	MeO-Cl_2-triazine *6*	—OH, —SH, $—NH_2$	17–20
	Aryl halides *11*	$—NH_2$, —OH, —SH	9
Aromatic (Tyr, His)	Bis-diazonium compounds *8*	Aromatic (Tyr, His); Lys	22, 23

some of the cross-links formed during conjugation. Glutaraldehyde is easy and convenient to use. Excess reagent can be readily inactivated and the cross-linking reaction arrested by the addition of aqueous sodium bisulphite solution.[14]

Carbodiimides, $R^1—N=C=N—R^2$, have been used extensively in forming amide bonds during peptide synthesis[16] and are the reagents of choice when the only reactive functional group of the hapten is carboxyl. Reaction of the peptide carboxyl group with an excess of carbodiimide leads to the formation of *O*-acylisourea *1* (*Fig.* 4.3) which is a highly active acylating agent and reacts rapidly with amino groups on the protein to form stable amide bonds and the urea *3*. A competing side-reaction is the rearrangement of the *O*-acylisourea *1* to the unreactive *N*-acylurea *2*, which is favoured at higher temperatures. In order to suppress it, coupling reactions are usually carried out at 0–5 °C. Carbodiimides can also mediate the formation of symmetrical anhydrides *1a* which are very potent acylating species. They are formed from two molecules of the hapten, one of which is regenerated during the subsequent acylation step. Numerous examples of conjugation with carbodiimides are cited by Bauminger and Wilchek.[13]

Some of the commonly used carbodiimides are shown in *Fig.* 4.4. Since carrier proteins are only soluble in aqueous media, the carbodiimides used for conjugation must be made water soluble. This is achieved by introducing tertiary (e.g. *4*) or quaternary (*5*) ammonium functions which form water-soluble salts. Consequently, the urea *3* (*Fig.* 4.3) formed as a by-product during conjugation is also water soluble and can be separated from the conjugate by dialysis or gel filtration.

Carbodiimides will also react with water to give ureas (*3* in *Fig.* 4.3) and this is another competing side-reaction during conjugation, although much slower than the reaction with carboxyl groups to form *O*-acylisoureas. Salts of water-soluble carbodiimides (e.g. *4* and *5*) are hygroscopic and, therefore, it is essential that they

hapten + carbodiimide or 2 hapten + carbodiimide

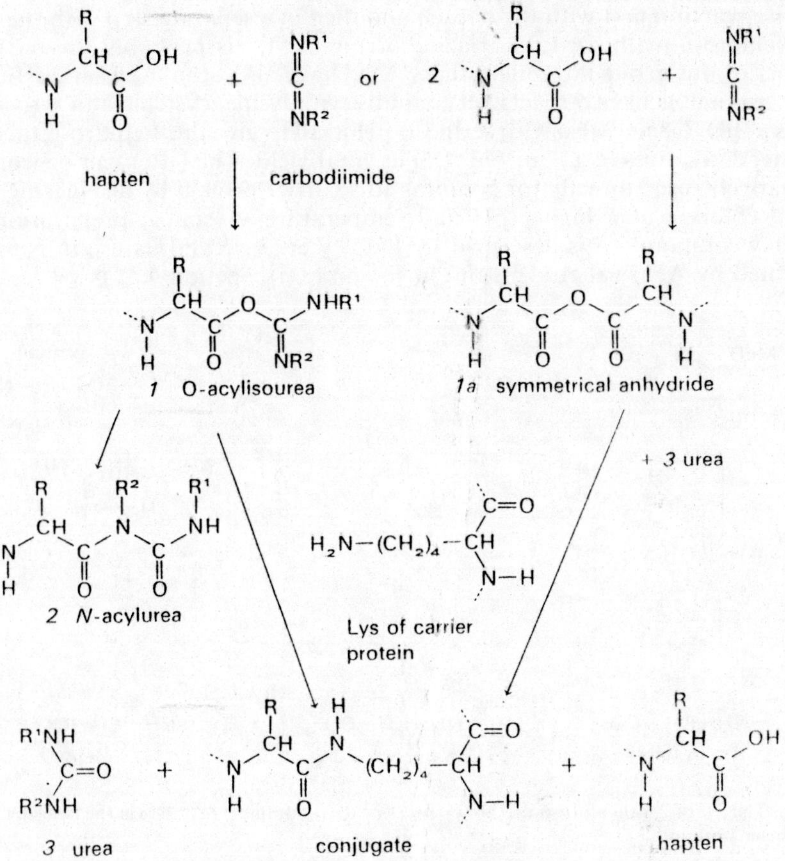

1 O-acylisourea

1a symmetrical anhydride + *3* urea

2 *N*-acylurea

Lys of carrier protein

3 urea + conjugate + hapten

Fig. 4.3. Coupling with carbodiimides.

are kept dry either in sealed containers or in a desiccator in order to avoid inactivation by water during storage.

Chlorinated *s*-triazines deserve a special mention among coupling reagents because of the graded reactivities of their halogen substitutents which enable

4 $CH_3CH_2{-}N{=}C{=}N{-}CH_2CH_2CH_2{-}\overset{+}{N}H(CH_3)_2\ Cl^-$

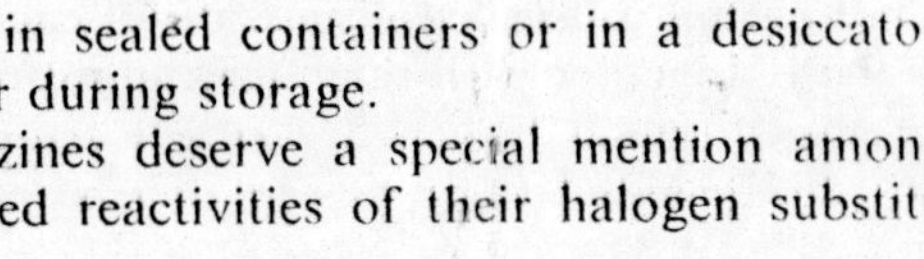

1-Ethyl-3-(3′-dimethylaminopropyl)-carbodiimide hydrochloride

5 $C_6H_{11}{-}N{=}C{=}N{-}CH_2CH_2{-}\overset{+}{N}(CH_3)(C_4H_8O)\ TsO^-$

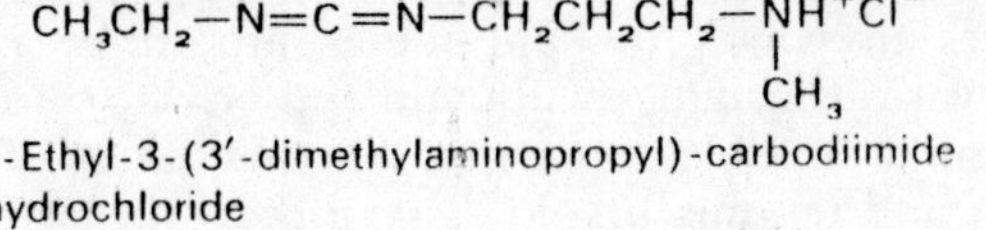

1-Cyclohexyl-3-[2′-(morpholin-4-yl)ethyl]-carbodiimide methyl *p*-toluenesulphonate

Fig. 4.4. Some water-soluble carbodiimide coupling reagents.

stepwise reaction first with the hapten and then in a separate step with the protein. 2,4-Dichloro-6-methoxy-1,3,5-triazine *6* (*Fig.* 4.5)[17] is probably the most useful member of this group although others[18–20] have also been used successfully. One of the two halogens in *6* reacts very rapidly with lysine, cysteine and tyrosine side-chains at pH 7–8 and also with serine at pH 8–10 to give the 4-chloro-6-methoxy-*s*-triazinyl derivatives (e.g. *7* in *Fig.* 4.5) in good yield. The latter can be isolated or, alternatively, used directly for coupling to a carrier protein by displacement of the second chlorine at a higher pH and temperature. A simple preparation of the methoxy compound *6* is described by Dudley et al.,[21] and its use in conjugation illustrated by Agarwal et al.[17] and in Example (i), Section 4.1, p. 62.

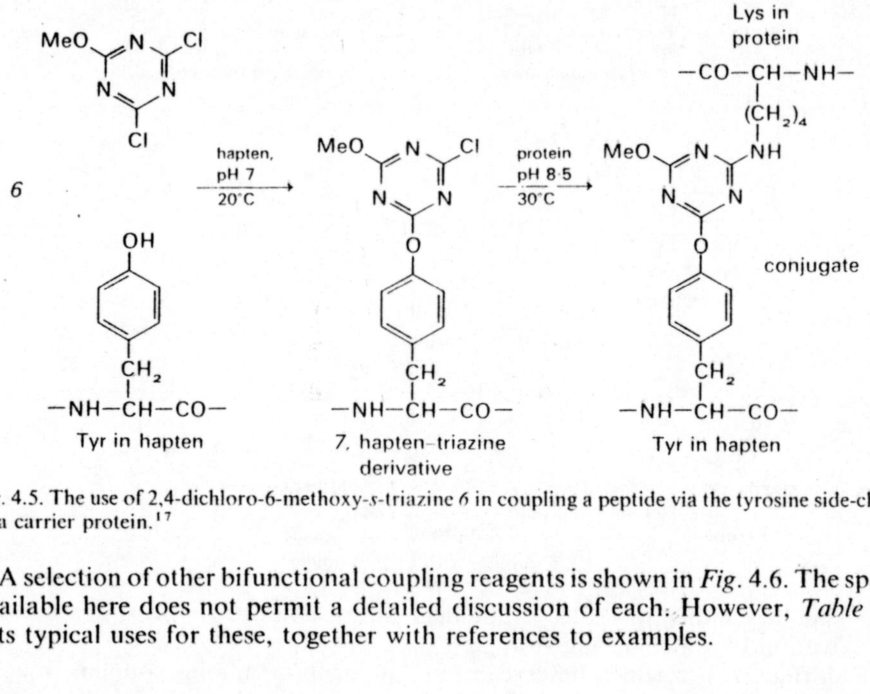

6

hapten, pH 7, 20°C →

protein pH 8·5, 30°C →

Tyr in hapten

7, hapten–triazine derivative

Lys in protein

conjugate

Tyr in hapten

Fig. 4.5. The use of 2,4-dichloro-6-methoxy-*s*-triazine *6* in coupling a peptide via the tyrosine side-chain to a carrier protein.[17]

A selection of other bifunctional coupling reagents is shown in *Fig.* 4.6. The space available here does not permit a detailed discussion of each. However, *Table* 4.1 lists typical uses for these, together with references to examples.

3.3. Carrier Proteins

These must be immunogenic *per se* and, therefore, phylogenetically distant from the animal species in which the antibodies are to be raised. In the rabbit, haemocyanins of various origin, ovalbumin and bovine serum albumin (*see Table* 4.2) have been found suitable as carriers, in that order of decreasing immunogenicity. Flagellin and tetanus toxoid[24] have been suggested as highly antigenic carrier proteins.

Solubility in the media used for conjugation is a useful attribute. It can be influenced by the choice of a suitable buffer, pH and ionic strength. The three proteins listed in *Table* 4.2 all have their isoelectric points at an acid pH and thus they can be dissolved to give a clear solution at pH 7–8, although the solubility of haemocyanin will depend also on its state of aggregation. Bovine serum albumin almost invariably yields soluble conjugates. Chemically, a good carrier protein

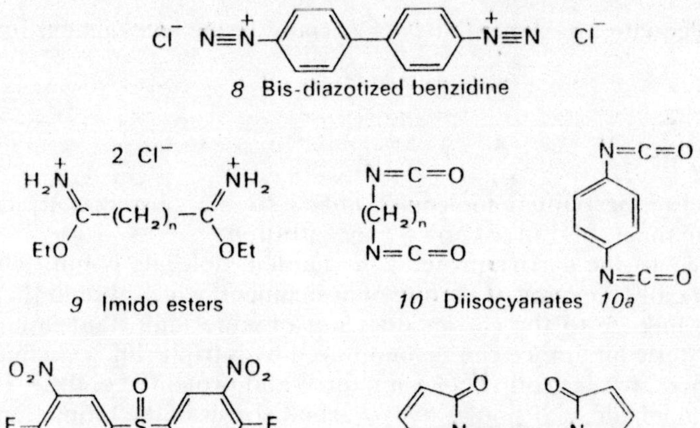

Fig. 4.6. Some bifunctional coupling reagents.

Table 4.2. ***The amino acid composition, molecular weight and isoelectric point of three frequently used carrier proteins***

Amino acid	*Bovine serum albumin*[28]	*Chicken ovalbumin*[29]	M. trunculus *Haemocyanin (submit)*[30]
Lysine	56	20	20
Histidine	17	7	28
Arginine	22	15	20
Aspartic acid	53	32	52
Glutamic acid	75	52	49
Threonine	32	15	22
Serine	26	38	21
Proline	28	15	22
Glycine	16	19	27
Alanine	46	35	29
Valine	35	32	25
Methionine	4	15	19
Isoleucine	13	24	18
Leucine	61	31	38
Tyrosine	19	10	17
Phenylalanine	26	19	27
Tryptophan	2	3	7
Cysteine	1	4	—
Half-cystine	34	2	6
Total	566	387	438
Mol. wt, kdal	65	45	300–9000
pI	4·7	4·6	4·6

should have adequate attachment sites (especially lysine side-chains) for hapten molecules.

3.4. The Conjugate

It should contain the hapten molecules linked to the carrier in a stable and unambiguous manner, and in a known concentration.

When coupled to the carrier protein, the hapten molecule is both chemically altered and sterically hindered at the site of attachment, which should therefore be chosen so that linkage to the carrier does not obscure important immunogenic determinants. Steric hindrance can be minimized by introducing a chemically and antigenically inert spacer group between hapten and protein, e.g. the —$(CH_2)_3$— chain of glutaraldehyde, and similar hydrocarbon chains in the bifunctional imido esters *9* or isocyanates *10* (*Fig.* 4.6).

Although antibodies raised in animals will always be heterogeneous, one can usually increase the specificity of antibodies towards one particular region of the hapten molecule by a judicious choice of attachment site and coupling reagent.

There appears to be no clearcut guideline for an optimum hapten : carrier ratio in the conjugate. A brief survey of published work and experience in the author's laboratory indicates that antibodies to small peptides (comprising 3–15 residues) have been raised successfully with conjugates containing 5–25 molecules of hapten per 50 000 daltons of carrier protein. A low hapten density (and a low dose of immunogen) should favour recognition and binding by immune cells bearing high-affinity binding sites, and therefore may promote the production of high avidity antibodies, although it may take longer to obtain antisera of a high titre.

Analytical procedures for estimating the hapten : carrier ratio in a conjugate will depend on the chemical nature of the hapten. First the conjugate must be separated from uncoupled hapten. This is best done by gel chromatography (e.g. on Sephadex G-50 or Bio-Gel P-30) which enables the estimation of both uncoupled hapten and hapten in the conjugate. Alternatively, uncoupled hapten and reagents can be removed by dialysis (which is conveniently perfomed against isotonic saline so that the dialysed solution is ready for injection) and only the hapten content of the conjugate determined.

Estimation of the hapten : carrier ratio is most readily accomplished by using a radiolabelled hapten. If the hapten molecule contains tyrosine or histidine, a small amount of the ^{125}I-labelled peptide can be prepared as tracer (*see* Example (ii), Section 4.2 p. 64). Comparing the amino acid composition of the conjugate with that of the carrier protein provides another, more generally applicable method.

4. EXAMPLES

4.1. Example (i). Production of Antibodies to Luteinizing Hormone-Releasing Hormone (LHRH)[25]

pGlu—His—Trp—Ser—Tyr—Gly—Leu—Arg—Pro—GlyNH_2 LHRH

1 2 3 4 5 6 7 8 9 10

The decapeptide amide LHRH lacks both a free amino and a carboxyl terminal. Coupling to a carrier protein must therefore be via a side-chain functional group.

As the purpose of the work was to obtain an antibody which was specific for intact LHRH (i.e. for both N- and C-terminal regions), the tyrosine side-chain at the centre of the molecule was chosen as the site of attachment. In the absence of amino and thiol groups, 4,6-dichloro-6-methoxy-*s*-triazine *6* will react specifically with the phenolic hydroxyl of the tyrosine side-chain and was used as the coupling reagent. Haemocyanin isolated from the haemolymph of *Murex trunculus*[26] was selected as the carrier protein because of its high immunogenicity in mammals.[27]

4.1.1. *Reaction of LHRH with the Dichloro-triazine 6*

Under nitrogen atmosphere, LHRH (14 mg) was dissolved in water (0·75 ml) and combined with a solution of $KHCO_3$ (15 mg) in water (0·25 ml). 10 μCi of ^3H-labelled LHRH (New England Nuclear, tritiated in position 3 of the pGlu-residue; specific activity 12 ci/mmol) were added, followed by a solution of 11 mg of the freshly prepared triazine *6*[21] in peroxide-free tetrahydrofuran (0·5 ml). This reaction mixture was stirred at 20 °C for 90 min, when tetrahydrofuran was evaporated *in vacuo* and the residual aqueous solution containing the LHRH–triazine derivative (~1 ml) was extracted with ethyl acetate to remove unreacted triazine.

4.1.2. *Coupling of the LHRH–triazine Derivative to Haemocyanin*

Murex trunculus haemocyanin (75 mg, previously dialysed against isotonic saline and lyophilized) was dissolved in 0·75 ml water, 20 mg $KHCO_3$ in 0·25 ml water were added and this solution was combined under nitrogen with the LHRH–triazine solution obtained as shown above.

The reaction mixture was stirred in a water bath at 30 °C for 24 h, when the resulting cloudy solution was filtered and applied to a column of Bio-Gel P-60 (2·5 × 55 cm) previously equilibrated with 0·4 per cent $NH_4 HCO_3$ buffer. Elution was carried out with the same buffer at 8 ml/h, fractions being collected at 20-min intervals. Absorbance of the effluent was monitored at 280 nm, and a 100-μl aliquot of each fraction was counted for ^3H. Fractions 11–21 contained 61 per cent of the total radioactivity which coincided with strong absorption at 280 nm. The remaining 39 per cent of radioactivity and ultraviolet absorbing material was in fractions 58–68. Fractions 11–21 were pooled and lyophilized yielding 80 mg of LHRH–haemocyanin conjugate containing 8·5 mg of LHRH.

4.1.3. *Immunization*

20 mg of the above conjugate (equivalent to 2·1 mg of LHRH) were dissolved in isotonic saline (1 ml) and this cloudy solution was thoroughly emulsified with Freund's complete adjuvant (1 ml). Three rabbits of mixed breed weighing approximately 3 kg were immunized according to the method of Vaitukaitis[7] using 500 μl of emulsion (equivalent to 500 μg LHRH and divided into 40 intradermal injections) in each animal. Three months later booster injections containing 100 μg LHRH per animal were given, emulsified in Freund's incomplete adjuvant. Two

further booster immunizations were carried out at six-week intervals. Ten days after the last booster immunization all three animals showed high titres of anti-LHRH antibodies. Antisera were harvested and characterized for titre and avidity. The antiserum of highest titre was subsequently used in a radioimmunoassay of LHRH, giving 50 per cent binding at a dilution of 1:80 000. It was found to be highly specific for LHRH and did not cross-react with analogues lacking either N or C-terminal residues.

4.2. Example (ii). Production of Antiserum to the Central (9–20) Segment of Cholecystokinin[15]

The gastrointestinal hormone cholecystokinin (CCK) is a peptide of 33 residues. It shares a C-terminal pentapeptide amide sequence with gastrin, another gastrointestinal hormone of 17 amino acid residues (*see Fig.* 4.7). Antibodies raised to CCK cross-react with gastrin and, vice versa, gastrin antibodies cross-react with CCK. In the past, this made it impossible to differentiate between cells containing these two hormones. Our aim was to solve this problem by synthesizing a central sequence of CCK which is unique to this hormone and to raise antibodies to it. These antibodies will not recognize the C-terminal pentapeptide and therefore will not cross-react with gastrin. The (9–20) decapeptide sequence of CCK was chosen for ease of synthesis and because it contains a lysine residue for conjugation and a histidine residue for labelling with ^{125}I.

Cholecystokinin

1 2 3 4 5 6 7 8

H-Lys-Ala-Pro-Ser-Gly-Arg-Val-Ser-

9 10 11 12 13 14 15 16 17 18 19 20

Met-Ile-Lys-Asn-Leu-Gln-Ser-Leu-Asp-Pro-Ser-His-

21 22 23 24 25 26 SO_3H 28 29 30 31 32 33

Arg-Ile-Ser-Asp-Arg-Asp-Tyr-Met—Gly-Trp-Met-Asp-Phe-NH_2

Gastrin – 17

1 2 3 4 5 6 7 8 9 10 11 SO_3H 13 14 15 16 17

pGlu-Gly-Pro-Tyr-Met-Glu-Glu-Glu-Glu-Glu-Ala-Tyr—Gly-Trp-Met-Asp-Phe-NH_2

Fig. 4.7. Amino acid sequences of porcine cholecystokinin and gastrin-17.

4.2.1. *Synthesis of the CCK (9–20) Fragment*

This was carried out using standard solid phase methodology. After removal from the resin and deprotection, the peptide was purified first by gel chromatography on

Sephadex G-25 SF in 50 per cent acetic acid and then by ion exchange chromatography on carboxymethyl-cellulose. The final product was homogeneous as shown by thin-layer chromatography in several systems and by thin-layer electrophoresis. It contained the correct ratio of amino acids both after acid hydrolysis and after digestion with aminopeptidase M, indicating the absence of racemization.

4.2.2. *Conjugation*

CCK(9–20) (6·8 mg) and approximately 800 000 counts/min of the ^{125}I-labelled peptide were dissolved in 0·1 M phosphate buffer pH 7·5 (0·5 ml) and combined with a solution of chicken ovalbumin (20 mg) dissolved in 1·0 ml of the same buffer. With stirring at 0 °C a 5 per cent solution of glutaraldehyde (300 µl) was added dropwise over 10 min. After 1 h at 0 °C the yellow solution was stirred at 20 °C for 1 h. Excess glutaraldehyde was inactivated with saturated aqueous $NaHSO_3$ (300 µl) and the solution was applied to a column of Bio-Gel P-30 (2·5 × 60 cm) equilibrated in 0·2 M ammonium acetate buffer pH 7. Elution was carried out with the same buffer at 10 ml/h, 5-ml fractions being collected. Transmittance was measured at 280 nm and 100-µl aliquots of each fraction were removed for counting ^{125}I activity. The data are shown in *Fig.* 4.8. There was a single peak of ultraviolet absorption corresponding to the conjugate and excess ovalbumin (peak *A*), and three peaks of radioactivity (peaks *A*, *B* and *C*). Peak *B* corresponded to the elution position of CCK (9–20), peak *C* to $^{125}I^-$. Fractions under peak *A* were combined and lyophilized yielding 22·5 mg of conjugate. The radioactivity in peak *A* represented 42 per cent of the total activity in peaks *A* and *B*, indicating that 2·9 mg (1·8 µmol) of CCK(9–20) had been incorporated into the conjugate. This corresponds to approximately 4 molecules of CCK(9–20) being linked to each molecule of ovalbumin.

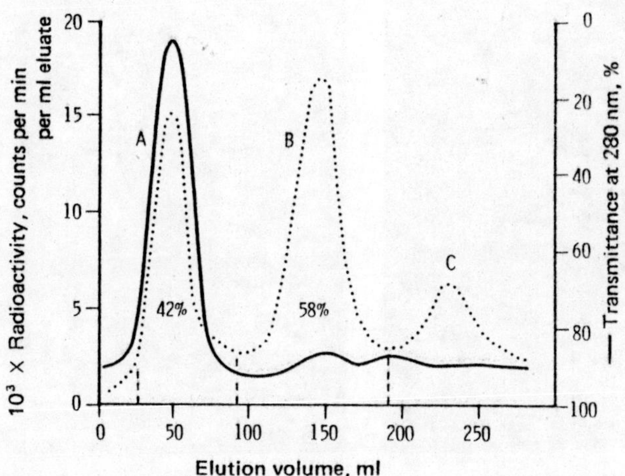

Fig. 4.8. Separation of CCK(9–20) from conjugate on Bio-Gel P-30. Peak *A* contains the conjugate and excess ovalbumin, peak *B*, uncoupled ^{125}I-labelled CCK(9–20) and peak *C*, radio-iodide. (...) Radioactivity; (—) percentage transmission.

4.2.3. *Immunization*

The above conjugate (22·5 mg) was dissolved in physiological saline (1·9 ml). One ml of this solution (corresponding to 1·5 mg of hapten) was thoroughly emulsified with Freund's complete adjuvant (2 ml). Young rabbits (approx. 3 kg) of mixed breed were injected subcutaneously at several sites on the neck and shoulders, using 0·5 ml of the emulsion (equivalent to 250 μg of hapten) per animal. The remaining saline solution (0·9 ml) was emulsified with Freund's incomplete adjuvant (1·8 ml) and 200-μl aliquots of this emulsion (equivalent to 100 μg of hapten) were injected into each animal as boosters three months and six months after the primary immunization. Antisera were harvested two weeks after the second booster injection. Only one of the six rabbits immunized produced a useful antiserum which, however, was entirely specific for CCK and did not cross-react with gastrin. It enabled, for the first time the differential immunostaining of cells containing CCK and gastrin, respectively.[15]

In *Fig.* 4.9 serial thin sections of human small intestine are shown. Cell A is stained both with antibody to gastrin-17 (which cross-reacts with the C-terminus of CCK) and with CCK(9–20) antiserum, demonstrating the presence of CCK. Cells B and C in section *a* reacted with antibody to gastrin-17 but not with CCK(9–20) antiserum, and therefore they must contain gastrin. This example illustrates the value in morphological studies of specific antisera raised to a peptide sequence which is unique to one of a family of related hormones.

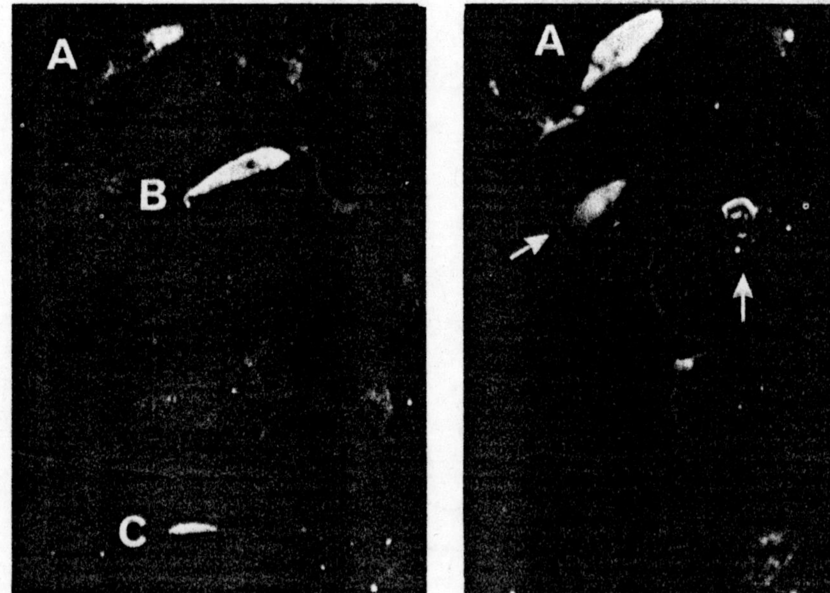

Fig. 4.9. Tangential 1-μm sections through human small intestine fixed in glutaraldehyde and embedded in araldite. Section (*a*) was stained with antibodies to human gastrin-17, section (*b*) with antibodies to CCK(9–20). Only cell A of the three cells stained specifically in section (*a*) reacted with antibodies to CCK(9–20) in section (*b*). Arrows in section (*b*) indicate mucous cells showing non-specific staining.

Acknowledgements

I wish to express my thanks to Dr C. J. Hillyard of the Royal Postgraduate Medical School who first introduced me to the animal techniques involved in raising antibodies, and both to her and to Dr S. I. Girgis for labelling peptides with ^{125}I. I am indebted to Professor J. H. Humphrey for helpful discussions on carrier proteins. The experimental work described above was supported by the Wellcome Trust and the Medical Research Council.

REFERENCES

1. Maurer P. H. and Callahan H. J. Proteins and polypeptides as antigens. *Methods Enzymol.* 1980, **70A**, 49–70.
2. Hurn B. A. L. and Chantler S. M. Production of reagent antibodies. *Methods Enzymol.* 1980, **70A**, 104–142.
3. Chard T. Requirements for a binding assay—the binder. In: *An Introduction to Radioimmunoassay and Related Techniques.* Amsterdam, North Holland, 1978, 377–400.
4. Herbert W. J. Mineral oil adjuvants and the immunization of laboratory animals. In: Weir D. M., ed., *Handbook of Experimental Immunology.* Oxford, Blackwell Scientific Publications, 1978: A.3.1.–A.3.15.
5. Erlanger B. F. The preparation of antigenic hapten–carrier conjugates: a survey. *Methods Enzymol.* 1980, **70A**, 85–104.
6. McConnell I., Munro A. and Waldman H. *The Immune System.* Oxford, Blackwell Scientific Publications, 1981; 109.
7. Vaitukaitis J. L. Production of antisera with small doses of immunogen: multiple intradermal injections. *Methods Enzymol.* 1981, **73B**, 46–52.
8. Reichlin M. Use of glutaraldehyde as a coupling agent for proteins and peptides. *Methods Enzymol.* 1980, **70A**, 159–165.
9. Wold F. Bifunctional reagents. *Methods Enzymol.* 1972, **25B**, 623–651.
10. Richards F. M. and Knowles J. R. Glutaraldehyde as protein cross-linking reagent. *J. Mol. Biol.* 1968 **37**, 231–233.
11. Hardy P. M., Nicholls A. C. and Rydon H. N. The nature of the cross-linking of proteins by glutaraldehyde. *J. Chem. Soc. [Perkin I]* 1976, 958–962.
12. Quiocho F. A. and Richards F. M. The enzymic behaviour of carboxypeptidase-A in the solid state. *Biochemistry* 1966, **5**, 4062–4076.
13. Bauminger S. and Wilchek M. The use of carbodiimides in the preparation of immunizing conjugates. *Methods Enzymol.* 1980, **70A**, 151–159.
14. Sachs D. H. and Winn H. J. The use of glutaraldehyde as a coupling agent for ribonuclease and bovine serum albumin. *Immunochemistry* 1970, **7**, 581–585.
15. Polak J. M., Pearse A. G. E, Szelke M., Bloom S. R., Hudson D., Facer P., Buchan A. M. J., Bryant M. G., Christophides N. and MacIntyre I. Specific immunostaining of CCK cells by use of synthetic fragment antisera. *Experientia* 1977, **33**, 762–763.
16. Rich D. H. and Singh J. The carbodiimide method. In: Gross E. and Meienhofer J., ed., *The Peptides*, Vol. I. New York, Academic Press, 1979: 241–261.
17. Agarwal K. L., Grudzinski D., Kenner G. W., Rogers N. H., Sheppard R. C. and McGuigan J. E. Immunochemical differentiation between gastrin and related peptide hormones through a novel conjugation of peptides to proteins. *Experientia* 1971, **27**, 514–515.
18. Kay G. and Crook E. M. Coupling of enzymes to cellulose using chloro-*s*-triazines. *Nature* 1967, **216**, 514–515.
19. Kay G. and Lilly D. The chemical attachment of chymotrypsin to water-insoluble polymers using 2-amino-4,6-dichloro-*s*-triazine. *Biochim. Biophys. Acta* 1970, **198**, 276–285.
20. Chaudhari A. S. and Bishop C. T. Coupling of amino acids and amino sugars with cyanuric chloride. *Can. J. Biochem.* 1972, **50**, 1987–1991.
21. Dudley J. R., Thurston J. T., Schaefer F. C., Holm-Hansen D., Hull C. J. and Adams P. Cyanuric chloride derivatives. III. Alkoxy-*s*-triazines. *J. Am. Chem. Soc.* 1951, **73**, 2986–2990.
22. Likhite V. and Sehon A. Protein–protein conjugation. In: Williams C. A. and Chase M. W., ed., *Methods in Immunology and Immunochemistry*, Vol. I. New York, Academic Press, 1967: 130–167.

23. Bassiri R. M. and Utiger R. D. The preparation and specificity of antibody to thyrotropin releasing hormone. *Endocrinology* 1972, **90**, 722–727.
24. Humphrey, J. H. Immunology adjuvants. *WHO Tech. Rep. Series*, 1976, No. 595, p. 29.
25. Szelke M. and Fink G. unpublished work, 1974.
26. A generous gift from Professor W. H. Bannister, Department of Physiology and Biochemistry, The Royal University of Malta, 1973.
27. Wood E. J., Salisbury C. M., Formosa N. and Bannister W. H. An electrophoretic and immunologic study of *Murex trunculus* haemocyanin. *Comp. Biochem. Physiol.* 1968, **26**, 345–351.
28. Peters T. and Hawn C. Isolation of two large peptide fragments from the amino and carboxyl-terminal positions of bovine serum albumin. *J. Biol. Chem.* 1967, **242**, 1566–1573.
29. Smith M. B. and Back J. F. Studies on ovalbumin. *Aust. J. Biol. Sci.* 1970, **23**, 1221–7.
30. Ghiretti-Magaldi A., Nuzzolo C. and Ghiretti F. Chemical studies on haemocyanins I. Amino acid composition. *Biochemistry* 1966 **5**, 1943–1951.

5

The Use of Semithin Frozen Sections in Immunocytochemistry

R. C. Richards

INTRODUCTION

For many years, frozen sections have been used to localize molecules by immunocytochemical techniques. Routinely, 5μm thick sections are used and have helped to contribute to our knowledge in many fields of biology. However, with the advent of new technology, it is now possible to purchase attachments for most ultramicrotomes that adapt them to cut ultrathin frozen sections. Such sections are used in immunocytochemistry at the electron microscope level (*see* Chapter 6) or as unfixed material in straightforward structural studies (for a review *see* Tokuyasu[1]). In addition to cutting ultrathin frozen sections, such adapted ultramicrotomes are also able to cut sections in the semithin range (0·1–1 μm thick). Frozen sections of this thickness show increased clarity and resolution when viewed, either under the phase contrast microscope or in stained preparations.[2–4] In addition, when immunocytochemistry is performed on these sections, there is increased sensitivity in detecting antigens and Tokuyasu[1] has pointed out that 'use of these sections allows one to attain the highest resolution available with the light microscope, resulting in a new and powerful tool for immunocytochemical studies of cells and tissues.'

In order to illustrate the marked enhancement in immunocytochemical staining obtainable, I have chosen to take as my example the localization of serum albumin (SA) in the lactating mammary gland of the mouse. The basic technique involved in this study is essentially that reported by Tokuyasu[1,5,6] and is described in the Appendix, p. 75.

LOCALIZATION OF MOUSE SERUM ALBUMIN IN THE MAMMARY GLAND

Milk is a complex fluid derived from material synthesized by the mammary epithelium and also from the ability of this epithelium to transport molecules from the serosa into milk.[7,8] Some of the molecules that are transported into milk include immunoglobulin A(IgA), G(IgG), SA and prolactin (PRL, for a review *see* Koldovsky[9]). Two routes are available for the movement of such molecules into milk: (1) the paracellular route between the epithelial cells and (2) the transcellular route through these cells (*Fig*. 5.1). The paracellular route can be excluded as Pitelka and her coworkers have shown that tight junctions joining the apices of mammary epithelial cells effectively seal off the serosal compartment from the lumina of alveoli.[10] This leaves the transcellular route, and transport of macromolecules across epithelial barriers has been shown to occur via the process of pinocytosis.[11,12] Molecules may either bind to specific receptors on the plasmalemma of cells and be transported in vesicles by a receptor-mediated mechanism or may be carried across the cell in the fluid phase of such vesicles.[13,14] In conjuction with this, receptors for IgG, IgA and PRL, but not for SA, have been identified on isolated mammary epithelial cells.[15–17] Candidates for involvement in this transport process are the many cytoplasmic vesicles (50–150 mm diam.) observed in electron micrographs near the basal plasmalemma of lactating mammary epithelial cells (*Fig*. 5.2). Many of these vesicles are aligned around the periphery of intracellular lipid droplets (*Fig*. 5.2).[18]

In mammary tissue from lactating mice injected with horseradish peroxidase via their tail veins, as an electron opaque marker for the fluid phase of pinocytosis, a large proportion of these cytoplasmic vesicles are positively labelled. Some are 'free' in the cytosol (*Fig*. 5.3) while others are arranged around the periphery of intracellular lipid droplets (*Fig*. 5.4). This suggests that at least some of the vesicles associated with lipid droplets originate from the basal plasmalemma.

In order to look at this transport process in more detail, we decided to use the technique of immunocytochemistry in an attempt to localize serum albumin during its passage through the mammary epithelial cells into milk.

Most previously published work using, in the main, 5 µm frozen sections has purported to show the localization of IgA, IgG, SA and PRL within the mammary epithelial cells during lactation.[19–26] However, with the exception of IgA,[20] where the stain is localized in cytoplasmic vacuoles, the label is shown in a diffuse area in the apical region of the cell. This gross localization does not correspond with the idea that such molecules would be packaged in small cytoplasmic vesicles and certainly there are no large vacuoles or other structures in the apex of the mammary epithelial cell that might contain such macromolecules (*Fig*. 5.1).

In an initial attempt to repeat this earlier work, mammary tissue was fixed in 10 per cent paraformaldehyde for two hours, infused overnight with an 11·6 per cent sucrose solution in 0·1 M cacodylate buffer, before being frozen in isopentane cooled with solid carbon dioxide (Cardice). Sections of 5 µm in thickness were cut on a normal cryostat and transferred to chrome–alum gelatin-coated slides. These slides were then incubated in rabbit antisera raised against mouse serum albumin (rabbit anti-MSA; Nordic Immunological Laboratories) at a dilution of 1 : 1000 for 16 h, washed in phosphate-buffered saline (PBS) for one hour, incubated in goat anti-rabbit serum conjugated to fluorescein or rhodamine (1 : 40) (Miles

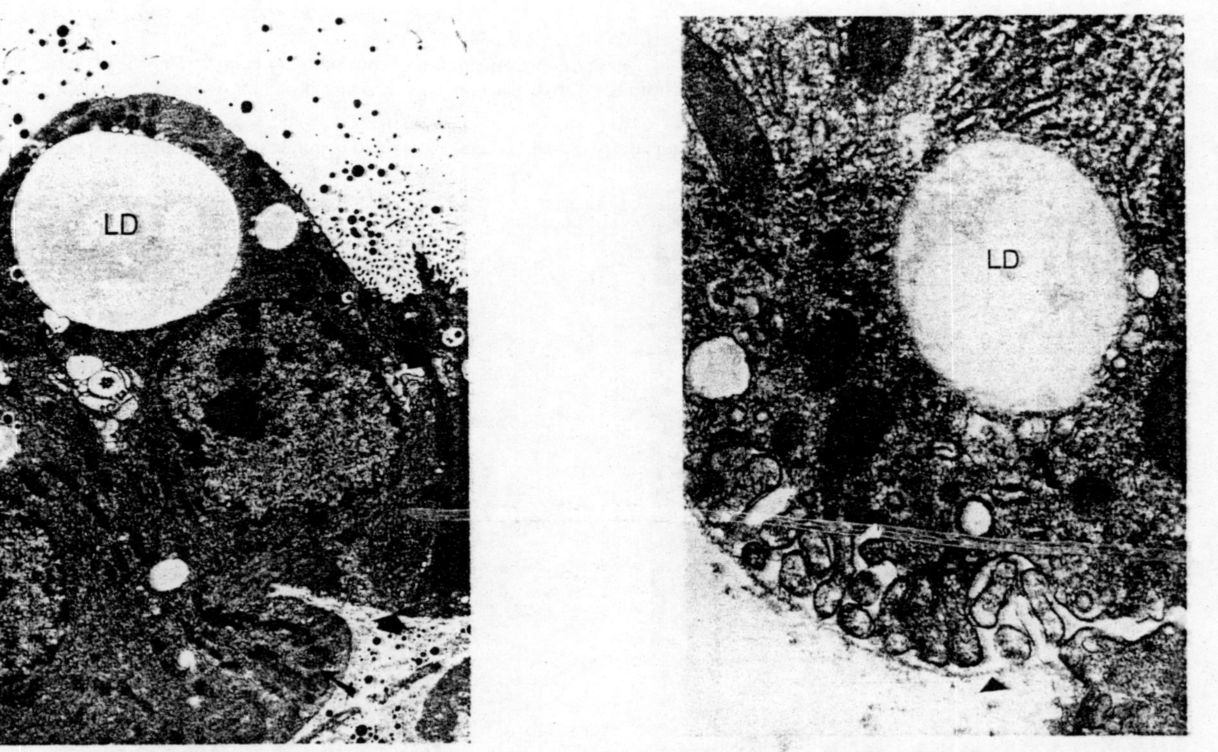

Fig. 5.1. An electron micrograph of part of the mammary epithelium of a mouse during lactation. Note the presence of junctional complexes (arrowhead) at the apices of these cells. Two routes for movement of macromolecules across this epithelial barrier are possible. The paracellular route (short arrow) between adjacent epithelial cells and the transcellular route (long arrow), through the epithelial cells themselves. Lipid droplet, LD. Magnification × 1600.

Fig. 5.2. At high magnification, many small cytoplasmic vesicles (long arrow) can be seen in the basal areas of mammary epithelial cells. Some of these vesicles are arranged around the periphery of lipid droplets (LD). Basal lamina (arrow head). Magnification × 14 300.

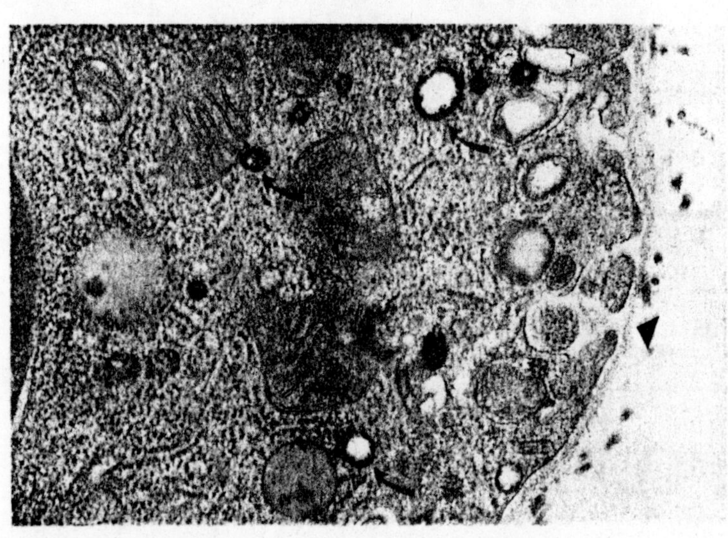

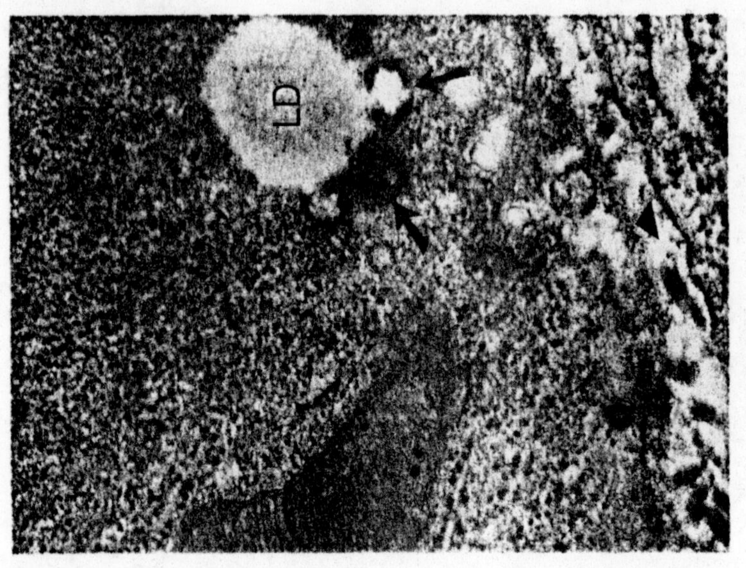

Fig. 5.3. Cytoplasmic vesicles, labelled with horseradish peroxidase (curved arrow), in the basal area of a mammary epithelial cell. Basal lamina (arrow head). Magnification × 14 300.

Fig. 5.4. A mammary epithelial cell showing cytoplasmic vesicles labelled with horseradish peroxidase (long arrow), arranged around the periphery of an intracellular lipid droplet (LD). Basal lamina (arrow head). Magnification × 40 000.

Laboratories) for an hour and finally washed for a further hour in PBS (phosphate-buffered saline). Such preparations when viewed under the fluorescence microscope (*Fig.* 5.5) showed good localization of SA in the connective tissue around the alveoli, but little clear indication of its localization within the mammary epithelial cells was obtained.

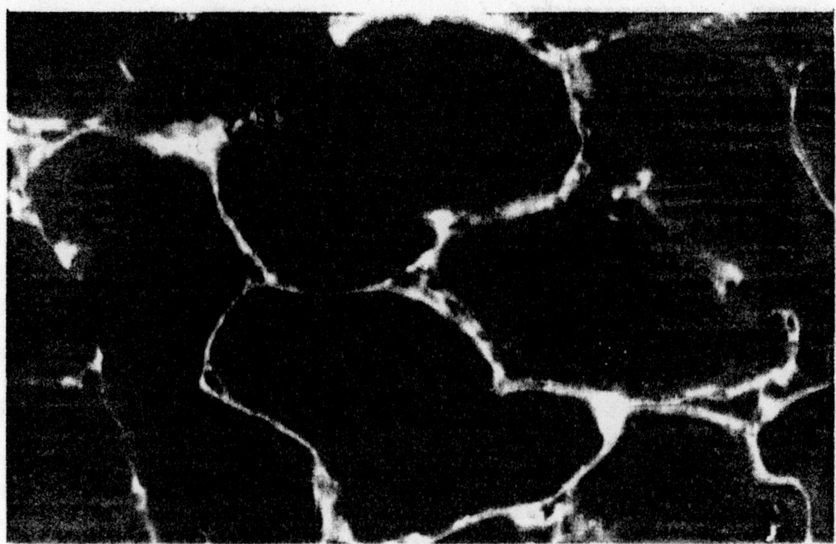

Fig. 5.5. A 5 μm frozen section of the lactating mammary gland stained for mouse serum albumin by an indirect immunofluorescence technique. Note the localization of serum albumin in the connective tissue between the alveoli. Within the mammary epithelial cells there is no clear indication of its distribution.

Controls, where normal rabbit serum (NRS) was substituted for the specific rabbit anti-MSA and where rabbit anti-MSA was pre-absorbed with MSA, showed no fluorescent localization of the protein. However, when IgG or IgA was used for pre-absorbing the antiserum there was no change in the distribution of the MSA localized in the tissue section.

In an attempt to improve on our ability to localize MSA within the epithelial cells, mammary tissue was removed from a lactating mouse, fixed in 10 per cent paraformaldehyde and processed for cutting semithin frozen sections, as outlined in the technology appendix. The semithin sections were again stained for MSA by an indirect immunofluorescence method. Both NRS substitution and pre-absorbed rabbit anti-MSA controls again proved to be negative.

As one can see from *Fig.* 5.6, the distribution of the staining for MSA within the tissue section is much clearer. As in the 5 μm frozen sections, SA can be localized in the connective tissue around the alveoli and, in addition, also shows up clearly in the intercellular spaces between adjacent cells. Again, the milk stored in the alveoli is frequently lost during processing, but when present it can be shown to be positive for MSA. In addition, within the cytoplasm of the mammary epithelial cells one is now able to distinguish a specific localization of MSA to small punctate spots of fluorescence (*Fig.* 5.6). Some of these spots at high magnification can be

Fig. 5.6. A semithin (0·5 μm) frozen section of the mammary gland of a lactating mouse stained for serum albumin. This molecule can be localized in the connective tissue between alveoli and in the lateral spaces between adjacent cells. In addition, staining is also visible within mammary epithelial cells as small punctate spots of fluorescence. Some of these spots of fluorescence are arranged around the periphery of lipid droplets.

seen to form small circles (*Plate* 29*a*) and often surround much larger areas in the apex of these epithelial cells (*Plate* 29*b*).

The circular areas within mammary epithelial cells, surrounded by spots of fluorescence, vary in size but in general are so large that the only structures that might correspond with them are the lipid droplets. This links in quite well with the observation that there are numerous cytoplasmic vesicles around these droplets, with some of them originating from the basal plasmalemma. We would like to suggest that the use of semithin frozen sections has allowed us to visualize MSA within small cytoplasmic vesicles during their transport across the cell.

The resolution achieved by this technique, as illustrated by the localization of MSA, is such that it really does offer a new tool for immunocytochemists, to be used when the distribution of molecules in tissue sections needs to be more precise and when the higher resolution of electron microscopy is not necessarily required.

Appendix

THE PRODUCTION OF SEMITHIN FROZEN SECTIONS

Introduction

The following technique for producing semithin frozen sections (0·1–1 μm thick) is essentially that described by Tokuyasu.[1,5,6] Such sections have been used in immunofluorescence microscopy[2,3,27,28] and have proved to be useful as serial sections in both light and electron microscopy.[29]

Fixation

The conditions of fixation to be used should be selected with a view to optimizing the following four variables: (1) preservation of structure, (2) preservation of antigenicity, (3) retention of accessibility of the antigen to the antibody molecules and (4) immobilization of the antigen. In general, there will be better preservation of antigenicity and accessibility of the antigen when cross-linking by the fixative is low. The best conditions of fixation for an immunocytochemical study will vary with different tissues/cells, different parts of these tissues/cells or with the different proteins under investigation and need to be worked out separately for each antigen/tissue to be studied.

The need for good structural preservation is obviously much less important for light microscopy than for electron microscopy. Conditions of fixation adopted for immunofluorescence microscopy are, therefore, more easily satisfied than for immunoelectron microscopy. Correspondingly, the cutting of 0·1–1 μm sections is much easier than the cutting of sections below 0·1 μm thick.

Formaldehyde is a good fixative for immunocytochemistry as it is a much milder chemical fixative than glutaraldehyde and allows greater retention of antigenicity of sensitive protein/peptide molecules.[30,31] Concentrations of formaldehyde up to 10 per cent have been used in our laboratory to good effect. A characteristic feature of formaldehyde fixation which should be kept in mind and which distinguishes it from glutaraldehyde fixation is its apparent reversibility.[6] In relation to this, formaldehyde, which penetrates tissues very quickly, has been used as a temporary stabilizer before a more permanent fixation by glutaraldehyde is achieved.[6]

Concentrations of glutaraldehyde from 0·1–2 per cent have been used successfully in immunocytochemistry.[28,32,33] However, if glutaraldehyde-fixed tissue is to be used then any free aldehyde groups need to be quenched. This can be accomplished by treating either tissue blocks or sections with a 1 per cent solution of sodium borohydride for 5–10 min.

Sucrose Infusion

In conventional ultramicrotomy, the sectioning quality is determined mainly by the resin embedding medium. However, in cryo-ultramicrotomy the variability in the physicochemical properties of cells or tissues in the frozen state is the decisive factor, there being no embedding medium present. These properties of biological specimens need therefore to be modified in order to obtain the required plasticity which will result in the best sectioning characteristics of the frozen material.

When tissues/cells are frozen, water is thought to be the most brittle component of the block, with any proteins providing a certain amount of plasticity. Experimental protein solutions sectioned in this way show no changes in their plasticity even after fixation with glutaraldehyde.[5] This suggests that areas of tissues/cells with high water content would be more difficult to cut frozen sections from than those areas having a high protein content.

By varying the cutting temperature, it is not possible to compensate completely for this variability in the composition of tissues. In order to control the plasticity of frozen tissue in a more direct manner, Tokuyasu[5] suggested that, before freezing, fixed pieces of tissue/cells be infused with chemically inert hydrophilic substances of low molecular weight. In addition, this treatment also reduces ice crystal damage, most probably by reducing the size of ice crystals formed. Tokuyasu chose sucrose for this purpose and found it to be superior to glucose, dimethylsulphoxide (DMSO) or glycerol.[5] The range of sucrose concentrations for optimum sectioning appears to be 0·6–1·0 M, depending on the tissue under investigation. The hardness of the tissue is inversely related to the sugar concentration within a certain range.[1] We routinely use buffered 1 M sucrose for a variety of cells and tissues.

The infusion is carried out at 4 °C, for at least 30 min depending on the size of the tissue pieces to be used. It has been shown that there is no difference between using graded concentrations of sucrose and the one-step infusion technique.[5]

When cell suspensions are to be sectioned and the cells within the suspension are closely packed then they can be processed as for tissue pieces. However, if the cells are in a relatively dispersed state, the medium surrounding the cells becomes the major factor in influencing sectioning quality. It may thus become necessary to include macromolecular components in the medium in addition to the sucrose. These macromolecules will remain in the medium and will effectively equalize the physical properties of the medium and the cells. Addition of 1–2 per cent gelatin, agarose or dextran to the medium is suggested by Tokuyasu[1] to remedy the situation.

Freezing

Prior to freezing, blocks of tissue may be pretrimmed under a dissecting microscope and mounted on small copper stubs with a small amount of infusion medium. Freezing of the material is accomplished by plunging the copper stub, held with a pair of forceps or an artery forceps, into liquid nitrogen and holding it there until the nitrogen ceases to boil. The copper stub together with the frozen tissue is then quickly transferred to the already cooled head of the microtome and secured firmly with a small screw.

Although liquid nitrogen is notoriously slow at freezing material and slow freezing results in the formation of large ice crystals which disrupt the tissue, tissue sections examined under the light and electron microscope show little evidence of ice-crystal damage when compared with material frozen in isopentane or Freon. As mentioned previously, the sucrose solution is most probably responsible for this protective effect.

Frozen tissue can either be sectioned immediately or may be stored on the copper stubs under liquid nitrogen in a cell-storage unit until required. However, we have become aware of the possibility of the spontaneous occurrence of staining artifacts in tissue incorrectly stored in such units (*see also* Hühn and Nairn[34]).

Trimming

Trimming of frozen tissue after its transfer into the cryobowl of the ultramicrotome can be accomplished using a hand-held precooled scalpel but this risks damaging the specimen. An alternative method, which is similar to that used on the Reichert freezing ultramicrotome, is to use a piece of stainless steel in the position in the cryobowl normally occupied by the diamond knife. This piece of steel after cooling down to the temperature of the cryobowl can be advanced towards the specimen, during its sectioning cycle, enabling it to trim off peripheral areas of the tissue so forming a reasonably shaped block face which will result in greater ease of sectioning and the likelihood of producing a ribbon of sections. Trimming of frozen blocks is not absolutely essential for cutting semithin sections but it is critical for cutting ultrathin frozen sections.

Sectioning

The Sorvall MT2-B ultramicrotome and its LTC-2 cryokit attachment (originally developed by Christensen[35,36]) are used routinely in our laboratory, although other manufacturers produce similar equipment.

Glass knives of good quality perform better in ultracryotomy than in normal ultramicrotomy,[37] but for our studies we routinely use a Dupont diamond knife. It is our experience that 'old' diamond knives, discarded from use in cutting resin sections, give excellent results when cutting semithin or ultrathin sections. This is probably due to the fact that frozen blocks of tissue tend to be softer than resin blocks and in addition any knife marks in the frozen sections seem to disappear when the sections melt on the sucrose pick-up droplet.

After transferring the frozen specimen to the ultramicrotome, a short time (10 min) is allowed for it to equilibrate with the temperature in the cryobowl (usually set at −60 to −70 °C.). The sections are cut dry at the knife edge and sometimes show small areas of interference colours which approximately correspond to the colours of plastic sections.[5] However, normally the average thickness of the sections produced is estimated from the thickness settings dialled on the controls of the microtome. Although sections of 0·1–2 μm can be used for light microscopy we routinely cut sections of 0·5–1 μm thickness.

In order to section tissue blocks or cell pellets smoothly, the correct combination of sucrose infusion concentration and cryobowl temperature is necessary. One normally starts cutting sections by dialling the required temperature and thickness settings on the microtome. However, it is important to remember that if the sections crumble as they come off the knife edge, then one needs to decrease the thickness setting or increase the temperature in the cryobowl. Because of the absence of a fluid on which to float the sections, a very slow speed of sectioning is possible. Although, the automatic mode of operation can be used we normally section extremely slowly by hand. This results in the smooth cutting of thin and flat sections and helps to prevent local heat production at the block face due to the friction of cutting. Faster sectioning also results in the generation of static electricity on the section. It is the production of static electricity which causes a ribbon of sections inconveniently to move back towards the specimen as it approaches the knife edge. At times the interior of the cryobowl becomes so highly charged that, as one attempts to move sections off the knife edge with an eyelash probe, they suddenly jump and disappear. We have found that the use of an antistatic pistol (Zerostat ASP21, Zerostat Instruments Ltd, Huntington) will at times help to reduce static,[37] as does the use of a humidifier in the cutting room. However, this latter procedure may give rise to an additional serious problem encountered in cryo-ultramicrotomy, namely, frosting at the knife edge. This is mainly due to condensation of moisture in the air. It dulls the knife edge and can also prevent the smooth movement of sections away from the knife edge as they are cut. When the knife edge is obviously frosting up one can attempt to clean off the frost using a piece of elder pith. It is also advisable to cover the plastic observation window, which closes off the top of the cryobowl, with a piece of polystyrene, whenever the cryobowl is left idle for any length of time.

A special feature of cryo-ultramicrotomy is that thermal instability is the main source of mechanical error, but this does not play such an important role when cutting semithin sections.

Sections, once cut, may be flattened using an eyelash probe, but it is more important to ensure that the sections are cut smooth and flat in the first place.

One major difficulty in dry cryosectioning is curling of the sections. We have found that this may be remedied by moving to a fresh part of the knife edge, or by cleaning the knife edge with elder pith. However, the sections may be curling because they are being cut too thick at that particular temperature. In this case, the remedy is to either reduce the thickness setting on the microtome or increase the temperature in the cryobowl.

Transfer of the Frozen Sections

Sections, once cut, are moved off the knife edge with the aid of an eyelash probe and are parked down in the boat of the diamond knife. When transfer of these sections onto glass slides is required, a small droplet of a near saturated sucrose solution (2·3 M), suspended on a fine platinum loop (1 mm diameter) at the end of a stick, is introduced into the cryobowl. This droplet remains in a fluid state for sufficient time to collect the frozen sections. The lower surface of the droplet is touched *gently* to the frozen sections, which stick to the sucrose. The correct timing, from insertion of the loop into the cryobowl to contact with the sections, will be apparent after several trials. If the sucrose droplet is too warm it can stick to the plastic of the diamond knife, while if it is too cold then the sections are disrupted in the pick-up process. A sucrose droplet of 2·3 M has a high surface tension and is suitable for structurally stable sections. However, in order to preserve sections of lightly fixed material a sucrose gelatin droplet (2 M sucrose + 0·75 per cent gelatin) is sometimes used. Here, the gelatin serves to reduce the surface tension acting on the section when it melts on the droplet. It should be noted, however, that sections do not flatten as easily on the surface of such a droplet.

When a droplet with sections on its lower surface is brought out of the cryobowl, it immediately becomes opaque as it is frosted by air moisture and then becomes clear again as it warms up to room temperature. The droplet plus attached sections is left for approximately 1–2 min at room temperature to ensure that the sections are completely flattened. The sections are then attached to alcohol-cleaned slides by touching the lower surface of the droplet to the slide. The sections attach firmly to the glass surface and it is only when sections of $>2\ \mu m$ in thickness are used that a subbing solution for the slides, such as chrome–alum gelatin, is required to ensure adhesion of the sections.

Subsequent Treatment of Sections

After transfer to the surface of glass slides, the sections remain covered by a small drop of sucrose from the pick-up loop. The position of the sections on the slide is marked by 'ringing' them using a diamond marker. A further droplet of 2·3 M sucrose is added to the existing droplet covering the sections and a coverslip is placed over them. They are then examined in a phase contrast microscope to check that they have retained their morphology and are properly flattened on the glass surface. Slides are then stored in a moist chamber until required. Prior to staining, the coverslips are soaked off in phosphate-buffered saline.

REFERENCES

1. Tokuyasu K. T. Immunochemistry on ultrathin frozen sections. *Histochem. J.* 1980, **12**, 381–403.
2. Bourguignon L. Y. W., Tokuyasu K. T. and Singer S. J. The capping of lymphocytes and other cells, studied by an improved method of immunofluorescence staining of frozen sections. *J. Cell. Physiol.* 1978, **95**, 239–257.

3. Beyer E. C., Tokuyasu K. T. and Barondes S. H. Localization of an endogenous lectin in chicken liver, intestine and pancreas. *J. Cell Biol.* 1979, **82**, 565–571.
4. Franke W. W., Schmid E., Grund C., Müller H., Engelbrecht I., Moll R., Stadler J. and Jarasch E. D. Antibodies to high molecular weight polypeptides of desmosomes: specific localization of a class of junctional proteins in cells and tissues. *Differentiation* 1981, **20**, 217–241.
5. Tokuyasu K. T. A technique for ultracryotomy of cell suspensions and tissues. *J. Cell Biol.* 1973, **57**, 551–565.
6. Tokuyasu K. T. and Singer S. J. Improved procedures for immunoferritin labeling of ultrathin frozen sections. *J. Cell Biol.* 1976, **71**, 894–906.
7. Richards R. C. Milk secretion: with special reference to its content of hormones, enzymes, and immunoreactive components. In Finn C. A., ed., *Oxford Reviews of Reproductive Biology*, Vol. 1. Oxford, Oxford University Press, 1979: 262–282.
8. Gitlin J. D., Gitlin J. I. and Gitlin D. Selective transport of plasma proteins across mammary gland in lactating mouse. *Am. J. Physiol.* 1976, **230**, 1594–1602.
9. Koldovsky O. Hormones in milk. *Life Sci.* 1980, **26**, 1833–1836.
10. Pitelka D. R. Cell contacts in the mammary gland. In: Larson B. L., ed., *Lactation. A Comprehensive Treatise*, Vol. IV. New York, Academic Press, 1978: 41–66.
11. Rodewald R. Intestinal transport of antibodies in the newborn rat. *J. Cell. Biol.* 1973, **58**, 189–211.
12. Wild A. E. Role of the cell surface in selection during transport of proteins from the mother to foetus and newly born. *Philos. Trans. R. Soc. Lond. Ser. B. Biol Sci.* 1975, **271**, 395–410.
13. Kraehenbuhl J. P. Transport of immunoglobulins across epithelia. In: Silverstein S. C., ed., *Transport of Macromolecules in Cellular Systems*. Berlin, Dahlem Konferenzen, 1978: 213–228.
14. Silverstein S. C., Steinman R. M. and Cohn Z. A. Endocytosis. *Annu. Rev. Biochem.* 1977 **46**, 669–722.
15. Sasaki M., Larson B. L. and Nelson D. R. Kinetic analysis of the binding of immunoglobulin IgG_1 and IgG_2 to bovine mammary cells. *Biochim. Biophys. Acta* 1977, **497**, 160–170.
16. Kuhn L. and Kraehenbuhl J. P. Role of secretory component, a secreted glycoprotein, in the specific uptake of IgA dimer by epithelial cells. *J. Biol. Chem.* 1979, **254**, 11 072–11 081.
17. Shiu R. P. C. and Friesen H. G. Properties of a prolactin receptor from the rabbit mammary gland. *Biochem. J.* 1974, **140**, 301–311.
18. Wooding F. B. P. The mechanism of secretion of the milk fat globule. *J. Cell Sci.* 1971, **9**, 805–821.
19. Dixon F. J., Weigle W. O. and Vazquez J. J. Metabolism and mammary secretion of serum proteins in the cow. *Lab. Invest.* 1961, **10**, 216–237.
20. Kraehenbuhl J. P., Racine L. and Galardy R. E. Localization of secretory IgA, secretory component and J chain in the mammary gland of lactating rabbits by immunoelectron microscopy. *Ann. N.Y. Acad. Sci.* 1975, **254**, 190–202.
21. Brown P. J., Bourne F. J. and Denny H. R. Immunoglobulin-containing cells in pig mammary gland. *J. Anat.* 1975, **120**, 329–335.
22. Nolin J. M. and Witorsch R. J. Detection of endogenous immunoreactive prolactin in rat mammary epithelial cells during lactation. *Endocrinology* 1976, **99**, 949–958.
23. Weisz-Carrington P., Roux M. E. and Lamm M. E. Plasma cells and epithelial immunoglobulins in the mouse mammary gland during pregnancy and lactation. *J. Immunol.* 1977, **119**, 1306–1309.
24. Lee C. G., Ladds P. W. and Watson D. L. Immunocyte populations in the mammary gland of the rat at different stages of pregnancy and lactation. *Res. Vet. Sci.* 1978, **24**, 322–327.
25. Weisz-Carrington P., Roux M. E., McWilliams M., Phillips-Quagliata J. M. and Lamm M. E. Hormonal induction of the secretory immune system in the mammary gland. *Proc. Natl. Acad. Sci. USA* 1978, **75**, 2928–2932.
26. Fujimura M., Chen S. T., Sudo T., Kuyama T. and Saito T. Cellular sites of immunoglobulins. VII. Localization of immunoglobulins in female mammary gland. *Acta Histochem. Cytochem.* 1981, **14**, 163–167.
27. Baumgartner B., Tokuyasu K. T. and Chrispeels M. J. Localization of vicilin peptohydrolase in the cotyledons of mung bean seedlings by immunofluorescence microscopy. *J. Cell Biol.* 1978, **79**, 10–19.
28. Geiger B., Tokuyasu K. T. and Singer S. J. The immunocytochemical localization of α-actinin in intestinal epithelial cells. *Proc. Natl. Acad. Sci. USA* 1979, **76**, 2833–2837.
29. Tokuyasu K. T., Slot J. W. and Singer S. J. Simultaneous observations of immunolabeled frozen sections in LM and EM. *Ninth International Congress of Electron Microscopy* 1978, **2**, 164–165.
30. Kraehenbuhl J. P. and Jamieson J. D. Solid-phase conjugation of ferritin to Fab-fragments of immunoglobulin G for use in antigen localization on thin sections. *Proc. Natl. Acad. Sci. USA* 1972, **69**, 1771–1775.

31. Kyte J. Immunoferritin determination of the distribution of (Na^+, K^+) ATPase over the plasma membranes of renal convoluted tubules. I. Distal segment. *J. Cell Biol.* 1976, **68**, 287–303.
32. Painter R. G., Tokuyasu K. T. and Singer S. J. Immunoferritin localization of intracellular antigens: the use of ultracryotomy to obtain ultrathin sections suitable for direct immunoferritin staining. *Proc. Natl. Acad. Sci. USA* 1973, **70**, 1649–1653.
33. Geuze H. J., Slot J. W. and Tokuyasu K. T. Immunocytochemical localization of amylase and chymotrypsinogen in the exocrine pancreatic cell with special attention to the Golgi complex. *J. Cell Biol.* 1979, **82**, 697–707.
34. Hühn A. and Nairn R. C. A nuclear staining artifact in immunofluorescence. *Clin. Exp. Immunol.* 1967, **2**, 697–700.
35. Christensen A. K. Frozen thin sections of fresh tissue for electron microscopy with a description of pancreas and liver. *J. Cell Biol.* 1971, **51**, 772–804.
36. Bernhard W. and Viron A. Improved techniques for the preparation of ultrathin frozen sections. *J. Cell Biol.* 1971, **49**, 731–746.
37. Frazer T. W. The anti-static pistol as an aid to ultrathin sectioning. *J. Microsc.* 1975, **106**, 97–99.

6 Colloidal Gold Probes in Immunocytochemistry

J. De Mey

1. INTRODUCTION

One decade has passed since the introduction by Faulk and Taylor[1] of their 'immunocolloid method' for studying the distribution of antigens on cell surfaces by electron microscopy (EM). As had previously been shown by Feldherr and Marshall,[2] the electron-dense colloidal gold particles form an excellent marker for transmission EM.

Faulk and Taylor, and subsequently many other investigators, took advantage of the fact that colloidal gold will stably and rapidly adsorb proteins, without a marked change in their biological activities.

This has now been shown to be the case for a number of proteins, such as lectins, enzymes, and tracer molecules for uptake and transport studies.[3–5] For further development in immunocytochemistry, the adsorption to colloidal gold of purified antibodies, protein A and antigens has been of crucial importance. Depending on whether antibodies, protein A or antigens were linked to gold particles, the techniques have respectively been called antibody–gold technique [6–9] or immunogold staining (IGS) method[10] (using IGS reagents), the protein A–gold technique[11,12] or the gold-labelled antigen detection (GLAD) method.[13]

Although, initially, problems were reported with manufacturing stable gold–antibody preparations,[6] much of the basic methodology has now been firmly established and simple working protocols for the reproducible manufacture of gold probes for routine immunocytochemistry have appeared in the literature.[7–10]

Other more recent findings have greatly increased the interest in colloidal gold as an immunocytochemical marker.

Firstly, it has been shown that it is capable of strong emission of secondary electrons, making it a very useful marker for scanning electron microscopy (SEM).[14]

Secondly, accumulations of colloidal gold over antigen-containing sites can be seen in the light microscope.[10,15–19] The typical red colour forms without incubation with an enzyme substrate and its appearance can be monitored continuously. Recently, protein A–colloidal silver has also been used for light microscopical detection of pancreatic hormones and for double staining experiments.[19]

Thirdly, colloidal gold probes can be used to detect antigens in ultrathin plastic[12,13,20–22] or frozen sections.[23] It has also been shown that the smaller sized probes are able to penetrate into whole cultured cells, to label intracellular antigens in the pre-embedding mode.[10]

Fourthly, colloidal gold–antibody probes (IGS reagents) have been used in a novel way to label cell monolayers.[10] The approach consists of growing cells on formvar-coated EM grids, doing a gold-labelling experiment (surface and/or intracellular labelling), and critical point drying the preparations. These can then be viewed by transmission electron microscopy (TEM), even at high acceleration voltages. This procedure bridges the gap between light microscopical observations and thin sections.

Finally, monoclonal antibodies have been shown to provide a first choice material for labelling with colloidal gold.[16] The simplicity and high yield with which monoclonal antibody–gold reagents are manufactured has led to the prediction that in the near future, direct labelling with monoclonal antibody–gold probes will become common practice. The use of various monoclonal antibodies, linked to gold particles of different sizes will increase our possibilities for performing double or multiple labelling experiments.

It thus appears that, starting from a relatively unknown method for labelling cell surface antigens, colloidal gold-containing immune reagents are today becoming widely used in a variety of localization problems.

This chapter aims to review in some detail these recent developments and the principles of the manufacture of stable gold probes for immunocytochemistry. For further literature on colloidal gold as a cytochemical marker, the reader is referred to three other reviews.[3–5]

2. THE PRODUCTION OF GOLD PROBES

2.1. Preparation of Gold Markers

There are three major methods of producing monodisperse gold sols of different sizes. They are all based on the controlled reduction of an aqueous solution of chloroauric acid using phosphorus-saturated ether as reducing agent for 3 nm[24] and ±5 nm gold,[1,3] sodium ascorbate[3] for 10–15 nm gold, or sodium citrate for larger particles (15–150 nm).[25] The sodium citrate technique provides gold particles that are also useful for light microscopy and SEM. Particles produced by the other methods have until now only been used for TEM. These procedures are detailed in the Appendix, p. 101. For a discussion on the stability and characterization of gold sols, *see* Horisberger[4] and Goodman et al.[9]

2.2. Adsorption of Proteins to Colloidal Gold

2.2.1. *General*

Much of the basic methodology for the preparation of colloidal gold–macromolecule conjugates is now established[4,8–10,14] The adsorption of proteins to colloidal gold is influenced by factors such as particle size, ionic concentration (must be kept as low as possible), pH[8] and protein concentration.

The important question of the stability of gold–protein interaction has not yet been completely answered and more work needs to be done. The optimal pH for adsorbing a protein to colloidal gold is related to the pH of its isoelectric point.[8] It follows that proteins which remain soluble at low salt concentrations and at the pH of their isoelectric points will be optimal for coupling. This is the case for protein A, but not for serum or monoclonal antibodies (*see below*).

When the proper size of the particles has been selected, the amount of protein, needed for stabilizing the gold sol against flocculation by salts, has to be determined.

The pH of the gold sol is adjusted to or just basic to the isoelectric point of the protein to be coupled. Proteins have mostly been dialysed against distilled water or 5 mM sodium chloride. It is important to microfilter or centrifuge the protein solution just before use, to eliminate aggregates.

The test involves (1) making a dilution series with small volumes of protein, (2) adding a standard amount of gold sol, (3) letting them react, (4) adding a standard amount of salt to try to destabilize the colloid, and (5) finding the amount of protein that is just enough to protect the colloid against colour change towards the blue.[1,7] This can best be quantitated by measuring the increase in absorbance at 580 nm[8,10] or semiquantitatively by scoring the change in colour (microtitration).[9] It has been noted that the amount of protein determined in this way is not a saturating amount[9] and, therefore, it is better to speak in terms of stabilizing amounts of protein.

For the preparation of a colloidal gold–protein probe, the stabilizing amount is often increased by 10 per cent and is added to the appropriate volume of gold sol.

After a couple of minutes, a stabilizer has to be added. This stabilizer is said to minimize possible aggregation and enhance probe stability. Until now, Carbowax 20 M (polyethylene glycol (PEG) 20 000)[14] is most widely used. When used under proper conditions, bovine serum albumin (BSA), type V (Sigma No. A 4503) gives the same effect.[10] IGS reagents made with BSA perform better in applications involving penetration of the probe into sections or whole cells (*see below*).

Removal of unstabilized marker and free or loosely bound proteins is done by centrifuging the protein–gold preparation. The mobile pool at the bottom of the centrifuge tube is resuspended in a buffered iso-osmotic salt solution, containing a stabilizer (PEG or BSA) and having a pH compatible with optimal biological activity and stability of the bound protein. It is important to have as little free protein in the probes as possible.

Most often the final probe is stored at 4 °C. Addition of glycerol (to 50 per cent) allows it to be stored below 0 °C.[26] This has been reported to increase probe stability.

2.2.2. *Protein A–Gold Probes*

Protein A (pA) from *Staphylococcus aureus* binds to the Fc portion of IgG molecules from several mammalian species[27] and is now widely used as secondary reagent for the localization of antigens by light and electron microscopy.

Romano and Romano[11] were the first to label protein A with colloidal gold. Roth and colleagues[12] modified the method by using PEG as the stabilizer.

serum, and PBS as control). Release of radioactivity from twice washed probes over 3 h at room temperature in the presence or absence of Carbowax 20 M was assayed. Up to 40 per cent of radioactivity could be released in the least favourable instance (conditioned culture medium + calf serum). The presence of PEG had little effect on the antibodies, while for concanavalin A–gold, it reduced the release to 15 per cent. The minimal desorption was still 7 per cent, under routine cell labelling conditions.

In view of these data, Goodman and colleagues[9] advise using the probe immediately after preparation, or washing it before use. These data also show the importance of continuing research on stability of gold probes. It is difficult to estimate the relevance of these data for probes prepared according to other procedures. The relative instability could be due to the presence of the 7 mM buffers in the gold sol before protein adsorption or to other factors during probe preparation. In spite of these apparent drawbacks, antibody–gold probes prepared according to the above method have given satisfactory results in labelling of cell surface antigens.[29] The presence of free antibodies (or protein A) in the gold probes will at most cause a diminishment of the labelling efficiency, when the marker does not need to penetrate into tissues or cells. We have explored the usefulness of IGS reagents for this purpose.[10] It was found that the presence of free antibodies will totally abolish labelling efficiency, because they penetrate much faster than the gold–antibody complex. Gold probes that were only washed once or twice always contained significant amounts of free antibodies (Moeremans and De Mey[68]). Our method for preparing gold probes with negligible free antibody concentrations (*see* the Appendix, p. 102) also derives from that of Horisberger and Rosset[7] and Geoghegan and Ackerman.[8] It contains a number of modifications to account for problems encountered with affinity-purified antibodies (which are less stable than γ-globulin fractions) and to yield probes with very low levels of free antibodies. For detailed evaluation of this method *see* Moeremans and De Mey[68].

The affinity-purified goat, rabbit and mouse antibodies (1 mg/ml) were dialysed against 2 mM borax-HCl buffer pH 9·0. This alkaline buffer increased the solubility of the antibodies in low salt solution. Subsequent trials have shown that probes made by adsorption at pH 9 (pH of the gold sol), have a better stability (defined as the relative appearance of precipitates) during long-term storage than those prepared at more neutral pH values. Both Carbowax 20 M and BSA were tried as stabilizers. The Carbowax was added, dissolved at 1 per cent in H_2O to a final concentration of 0·05 per cent. The BSA was added as a microfiltered solution, 10 per cent in H_2O at pH 9 to a final concentration of 1 per cent. This high pH is far removed from the p*I* of BSA and probably avoids competition phenomena. Radioiodinated immunoglobulins have been used to assess the amount of protein bound and subsequent stability. No significant differences were found between Carbowax and BSA-containing probes except that the latter gave similar or higher yields in gold during manufacturing (depending on gold size) and more satisfactory labelling results in a number of applications. We could confirm that stabilizing amounts of antibodies were not saturating amounts. Washes, spaced in time by several hours, have shown the presence of a population of loosely bound antibodies and non-stabilized gold. Three such washes were needed to eliminate these from the final probes.

The stability of the probes (BSA and Carbowax-containing) was satisfactory:

only 2–6 per cent radioactivity (dependent on the type of probe) was released after 3 days of storage at 4 °C, staying nearly constant for several months. Attempts to scale up the production of IGS reagents have been successful: 1500 ml of gold sol can easily be handled in a preparative ultracentrifuge. IGS reagents prepared in this way pass through microfilters (0·22 μm) without staining the filters and show only nominally changed visible wavelength spectra, fulfilling the criteria of Geoghegan et al.[28] Most importantly, however, 5-nm probes can be used to label intracellular antigens in whole cultured cells with high specificity and efficiency. In such experiments, washing the gold probe before use increases labelling efficiency.

BSA was used in earlier studies, but under different conditions.[6] Most workers, including Horisberger[3] and Geoghegan and Ackerman[8] have repeatedly stated that BSA is to be avoided. This may be true for applications involving lectin labelling where impurities in the BSA may interfere with the lectin binding. BSA, however, is widely used as a complement during incubation of immunocytochemical preparations to block non-specific labelling and has also been used with protein A.[12, 23]

Adsorption of monoclonal antibodies is very efficient because they have much better defined p*I* values.[16] Details of the method, adapted so as to adsorb monoclonals to gold are given in the Appendix, p. 105. Monoclonal antibody–gold IGS reagents are excellent reagents for high resolution, direct immunocytochemistry.

Garaud and colleagues[29] used Tris for buffering gold sols at pH 9 for adsorbing IgG (sheep IgG against rabbit IgG). The probe was washed three times but was resuspended each time in the same low salt solution (0·005 M Tris). The final pellet was resuspended in a small volume of 0·005 M Tris, containing 0·02 per cent sodium azide and 0·04 per cent Tween 20. Tween 20, a wetting agent, was claimed to efficiently stabilize colloidal gold particles. The reagent prepared in this way could be maintained for at least 5 months at 4 °C. No data on probe stability and amounts of free antibodies are as yet available for this interesting new procedure. Satisfactory results were reported for labelling thin plastic sections. The probe was diluted in 0·05 M phosphate buffer, but any isotonic buffer at neutral pH could be used (Garaud, J. C., personal communication, 1982).

3. APPLICATIONS, FINDINGS AND PROSPECTS

3.1. Labelling of Cell Surfaces

3.1.1. *Transmission Electron Microscopy (TEM)*

Most of the initial applications of gold probes in immunocytochemistry involved direct cell surface labelling, using total antiserum[1, 14] or affinity-purified antibody.[7]

Indirect immuno-gold labelling was introduced using horse anti-human IgG to label human red cells coated with human IgG anti-D.[6] To circumvent problems encountered with coating of colloidal gold particles with other immunoglobulins than horse IgG, Romano and Romano[11] subsequently introduced gold-labelled protein A for the indirect marking of cell surface antigens. Applications shown included labelling of lymphocytes, platelets, rat kidney cells (transformed and non-transformed phenotypes), erythrocytes and virus-infected cells, thus establishing the protein A–gold method as widely applicable. Recently, cell-surface galactosyltransferase on HeLa cells was also localized with the protein A–gold method.[31]

With the improvement of methods for manufacturing colloidal gold-labelled immunoglobulin, general applicability of indirect methods using secondary antibody γ-globulin fractions[8,15,32] and affinity-purified antibodies[17] became established as well. Applications included the indirect detection of surface immunoglobulin on B-lymphocytes,[8,15] the major viral glycoprotein gp 70 on surfaces of 3T3 cells and virus in cultures infected with wild-type and mutant ts 29 Rauscher murine leukaemia virus,[32] T-lymphocyte-associated differentiation antigens[17] (*Fig.* 6.1) and platelet antigens[33] using monoclonal antibodies and goat anti-mouse IgG and finally fibronectin[29,34] on cultured cells.

Because of their homogeneity, monoclonal antibodies form first-class material for adsorbing to colloidal gold particles. These can be used in direct labelling experiments.[16] Various monoclonals linked to different sizes of gold particles will greatly increase our potential for performing double or multiple labelling experiments to examine problems such as co-expression of antigens, relative mobility in the membrane and antigen modulation.

3.1.2. *Scanning Electron Microscopy (SEM)*

There exists a large variety of cell surface markers and labelling techniques for scanning electron microscopy (reviewed by Molday and Moher[35]).

In 1975, Horisberger and coworkers introduced colloidal gold as marker for SEM.[14] Advantages of gold markers over other SEM markers are (1) their high secondary electron emission which enables them to be observed on cell surfaces which have not been coated with a heavy metal, (2) the possibility of using them in combined SEM and TEM studies, (3) their detectability by X-ray analysis and back-scattered imaging.[34] Examples of SEM–immuno-gold localization include mannan and non-mannan antigens on *Candida utilis* yeast cells[8,14] and fibronectin on fibroblasts[29] (*Fig.* 6.2) and cultured quail neural crest cells.[34] Horisberger and Rosset showed that 40 and 80 nm gold particles can be used for double SEM-labelling experiments.[7]

3.1.3. *Light Microscopy*

In 1978, Geoghegan and colleagues were the first to report on the applicability of colloidal gold as a marker for light microscopy. B-cells, stained with anti-surface immunoglobulins and gold-labelled secondary antibodies became readily visible in the bright field microscope.[15] This characteristic of colloidal gold could further be confirmed with monoclonal antibodies to differentiation antigens on T-lymphocyte subpopulations using direct[16] or indirect labelling,[17] when staining was done on living cells under conditions that promoted patching and blocked capping and non-specific internalization of the gold markers. Cytospin preparations of labelled cells could subsequently be counterstained with methyl green and monocytes reacted for endogenous peroxidase activity, thus eliminating problems of identifying these cells. Positive lymphocytes were easily recognized and enumerated because they were entirely covered with dark, purplish granules. The combination with peroxidase detection in non-lymphocytic cells, made it possible to adapt the method to a few drops of blood in which red blood cells had been

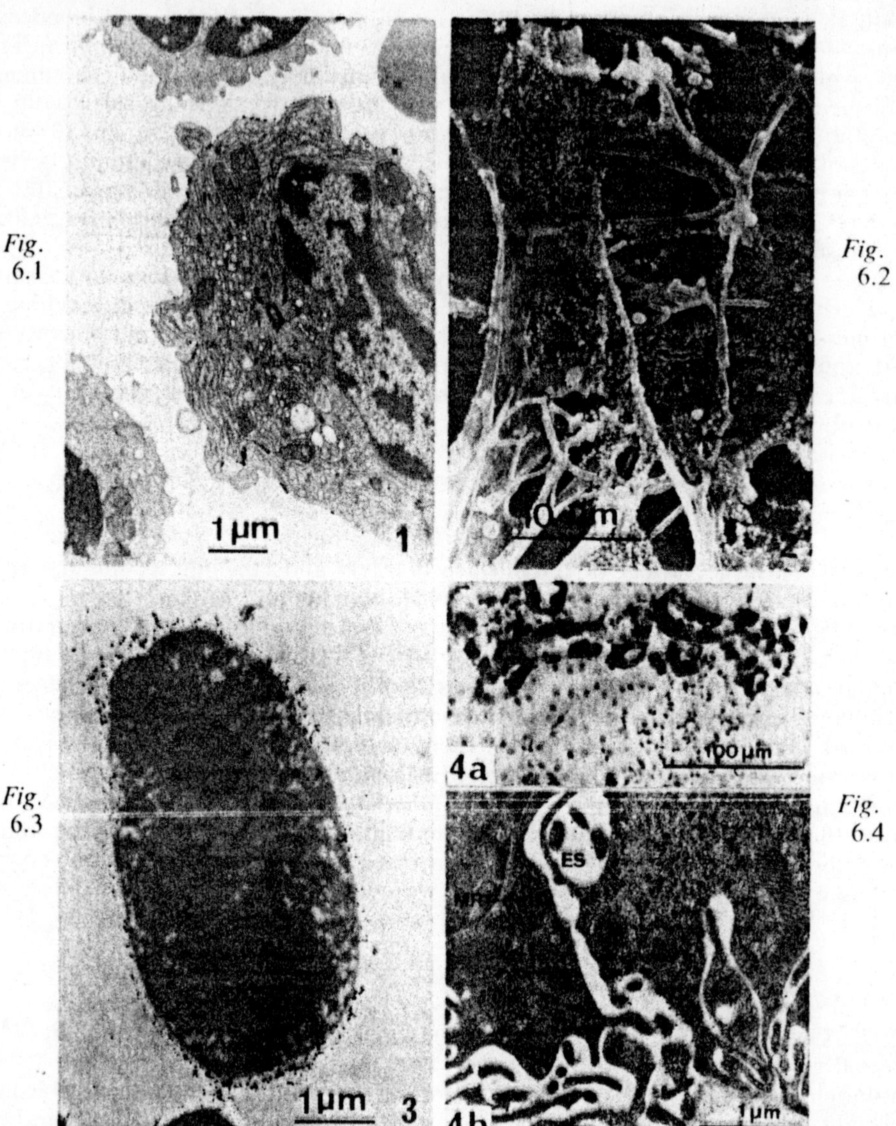

Fig. 6.1

Fig. 6.2

Fig. 6.3

Fig. 6.4

eliminated by lysis. The light microscopical staining of cells with colloidal gold-labelled antibodies was called immuno-gold staining (IGS) method.[10,16,17] A detailed description of this method is given in the Appendix, p. 107.

The IGS method for staining surface antigens on cell suspensions was subsequently combined with a variety of enzyme cytochemical reactions, allowing for simultaneous determination of antigenic and cytochemical profile of the cells. It was found that there is little correlation between the immunoregulatory T-lymphocyte subsets defined by monoclonal antibodies and those defined by Fc receptors.[36]

Recently, we have found that colloidal gold bound at the surface of cells (but also to other antigen-containing structures) can be detected with high sensitivity by polarized light epi-illumination microscopy (De Mey et al.[69]). The dark granules, visible with bright field microscopy strongly back-scatter incident polarized light. Since the polarization of the back-scattered light is lost, it will pass the analyser, while other excitation light will be extinguished. Use of a mercury-arc lamp yields an extremely bright yellow-gold coloured signal, that is still visible against a low background of transmitted non-polarized light, which allows identification of negative or cytochemically stained cells (*Plate* 2). The polarized light epi-illumination system is easily adapated to most epi-fluorescence microscopes and commercially available on a number of them. It has great potential because of the stability of the marker and the selective and bright signal that is emitted. It has the advantage over dark field illumination, which also enhances visibility of colloidal gold, of being entirely specific for colloidal gold or other metals such as silver (autoradiography and immuno-silver staining).[19] As with epi-fluorescence microscopy, this method is most efficient with the stronger immersion-objectives having a large numerical aperture. Details of the microscope equipment that was used for this new detection method are given in the Appendix, p. 109.

Fig. 6.1. Thin section of a plasmacytoma cell, labelled with OKT 10 (Orthomune®) and GAM G40 (Janssen Life Sciences Products). Label associated with the cell membrane is clearly visible. Magnification ×7575. From joint work with Dr M. De Waele (Brussels, Belgium).

Fig. 6.2. SEM of human fibroblasts (He 117) treated with rabbit antihuman fibronectin antibodies followed by 75 nm gold-conjugated second antibody (sheep anti-rabbit).[29] The gold particles are located and aligned along the fibronectin strands and over occasional dense mats of fibronectin. Magnification ×1700. Courtesy of Drs L. Trejdosiewicz, G. Hodges and M. Smolira (London, UK).

Fig. 6.3. *Candida utilis* thin sections were marked simultaneously with gold particles (5 and 26 nm in size) labelled with anti-mannan antibodies and wheat germ agglutinin, respectively. The presence of chitin indicated by the larger particles was mainly restricted to the bud scars and the septum, while the presence of mannan was detected throughout the cell wall by the smaller particles. Magnification ×10 600. Courtesy of Dr M. Horisberger Research Department, Nestlé Products Technical Assistance Co. Ltd, La Tour de Peilz, Switzerland).

Fig. 6.4. Protein A–gold (pAg) technique for light and electron microscopical localization of antigens on sections. Demonstration of the vitamin D-dependent calcium binding protein in rat kidney.[54,55] Immunoreactive sites are found in distal convoluted tubular cells (black: originally the positive cells appear red in the natural colour of colloidal gold) of a paraffin section (*a*).[56] Incubation of thin sections from Epon-embedded kidney reveals the principal cells as the calcium-binding protein-containing cell type whereas the intercalated mitochondria-rich cells are negative (*b*). Gold particle label for calcium-binding protein is restricted to the cytosol. M—mitochondria, ES—extracellular space. Magnification: (*b*) ×23 000. Courtesy of Dr J. Roth (Geneva, Switzerland).

3.2. Localization of Antigens in Thin Sections for Transmission Electron Microscopy

3.2.1. *General*

Post-embedding labelling of intra- and extracellular antigens with gold probes was introduced by Horisberger and Vonlanthen in 1977.[20] Thin sections of plastic embedded *Candida utilis* cells were simultaneously labelled with homologous anti-mannan antibodies linked to 5-nm gold and with WGA linked to 26-nm gold (in order to localize chitin) (*Fig.* 6.3), thus establishing that, as for cell surface labelling, double marking experiments can easily be performed on thin sections. The same approach has been used with anti-soybean agglutinin antibodies adsorbed to 12-nm gold, to localize this antigen in soybeans.[37] For the indirect detection of antibody-binding sites on plastic sections, the pA–gold technique of Romano and Romano[11] was adapted to localize various exocrine pancreatic enzymes,[12,24,38,39] pituitary hormones,[40] other polypeptide hormones,[41,42] renin in kidney and submandibular glands of mice,[43] vitamin D-dependent calcium-binding protein in kidney (*Fig.* 6.4) and duodenum[44–46] and finally galactosyltransferase in HeLa cells[31] and albumin, for studying vascular permeability.[70] The studies of the Geneva group mostly involved labelling of thin sections, embedded in Lowikryl K4M which is claimed to yield enhanced structural preservation and immunocytochemical staining.[47] The study on galactosyltransferase in HeLa cells[31] is most interesting from both scientific and technological viewpoints.* The enzyme is involved in the terminal stages of the biosynthesis of complex heteroglycans which is assumed to occur in the Golgi apparatus: galactosyltransferase activity is often selected as marker for Golgi membrane fractions (for references, *see* Roth and Berger).[31] It is most probably an integral membrane protein, and it was interesting to explore whether such a protein could be localized by a post-embedding procedure. It has been shown that this is indeed the case when Lowikryl K4M is used. The results have provided not only conclusive in situ evidence for the Golgi association of the enzyme, but have also shown that it is present in distinct Golgi subcompartments: two to three transcisternae (secretory side of the organelle). In addition, the immunocytochemical procedure could be performed on material that had been processed for pre-embedding thiamine pyrophosphatase cytochemistry. The data demonstrated that galactosyltransferase immunoreactivity-containing Golgi subcompartments also contain thiamine pyrophosphatase in the cisternae. This is in accordance with a model that proposes a concerted action of both galactosyltransferase and nucleotide diphosphatase in chain elongation of maturing glycoproteins (for references, *see* Roth and Berger).[31]

Geuze and coworkers[23] have shown that the pA–gold technique can be used on ultrathin frozen sections (*see below*). Recent improvements of the imaging of fine structural details[48] have increased the interest in this approach which had previously been shown to be very suitable for indirect immuno-ferritin and immuno-imposil cytochemistry.[49–51]

Colloidal gold-labelled secondary antibodies can substitute for protein A.[21,22,30] It was found that the sensitivity of this approach defined by final

* The mitochondrial proteins, carbamyl phosphate synthetase and a structural membrane protein, OMM.35, were specifically localized in the hepatocyte mitochondrial matrix and inner and outer membranes respectively, using the protein A–gold technique[71]. Both glycol methacrylate and Lowicryl K4M embedding proved to be suitable.

dilution of primary antiserum was between that of the indirect immunofluorescence and the PAP methods (J. M. Polak, personal communication, 1981).

Instead of using indirect labelling techniques, using protein A or secondary antibodies, Larsson[13,52] adsorbed various BSA–peptide conjugates to colloidal gold and used them in the gold-labelled antigen detection (GLAD) technique. This technique has the advantage that the labelled antigens select for the specific high affinity antibodies bound to the sections which can be pretreated with normal serum from the same species as the one producing the primary antiserum. In addition it provides the possibility of testing region specificity of an antiserum by allowing various synthetic peptide fragments to compete with the antigen-coated gold granules for binding to the antibodies.[13,52]

In general, the efficiency with which gold granules can be detected is much higher than that of enzyme reaction products. Gold probes can therefore be used in conjunction with optimally fixed and contrasted thin sections.[12,21] They do not obscure the ultrastructural details of the labelled structures. An example of the latter advantage is the recent proof of the heterogeneity of the peptidergic nerves: antigen (substance P and vasoactive intestinal polypeptide, VIP) defined, separate neurones can be distinguished by the size and appearance of their secondary granules[22] (*Fig.* 6.5). Similar examples[66,67] are provided by immunohistochemical studies on gastrin/CCK-like peptides in the amphibian brain and gastrin in human antrum.[30] Since each single gold granule is easily detected, it is possible to evaluate the degree of non-specific staining versus specific staining by counting gold granules. In most publications, this background has been found to be extremely low.

3.2.2. *Quantitative Immunocytochemistry*

The protein A–gold technique has been used for quantitatively assessing the processing of nine different pancreatic–exocrine enzymes through different intracellular compartments.[38,39] Quantitative evaluation of labelling results are facilitated by the efficiency with which gold granules can be detected and by the very low non-specific background. For such studies, one has to assume that the relative immunoreactivity will be the same for all compartments in which the antigen is detected. Evidence that this may not always be the case is discussed by Swaab et al.[72] A further limitation is that immunocytochemistry detects antigenic epitopes and does not give information on actual biological activity.

3.2.3. *Different Secretory Granule Localization of Molecular Form Variants of the Same or Related Antigens*

The use of gold probes instead of enzymes as markers has considerably improved the resolution of immunocytochemical techniques. Using specific glucagon and glicentin antibodies, it has been shown that glicentin immunoreactivity in the glucagon-containing, specific secretory granules (α granules) of the human pancreatic and gastric A cell, is topologically segregated from glucagon.[41] Glicentin is confined to the electron-lucent periphery, while glucagon is restricted to the core. Slightly different findings (using the same antibody for glicentin) were obtained

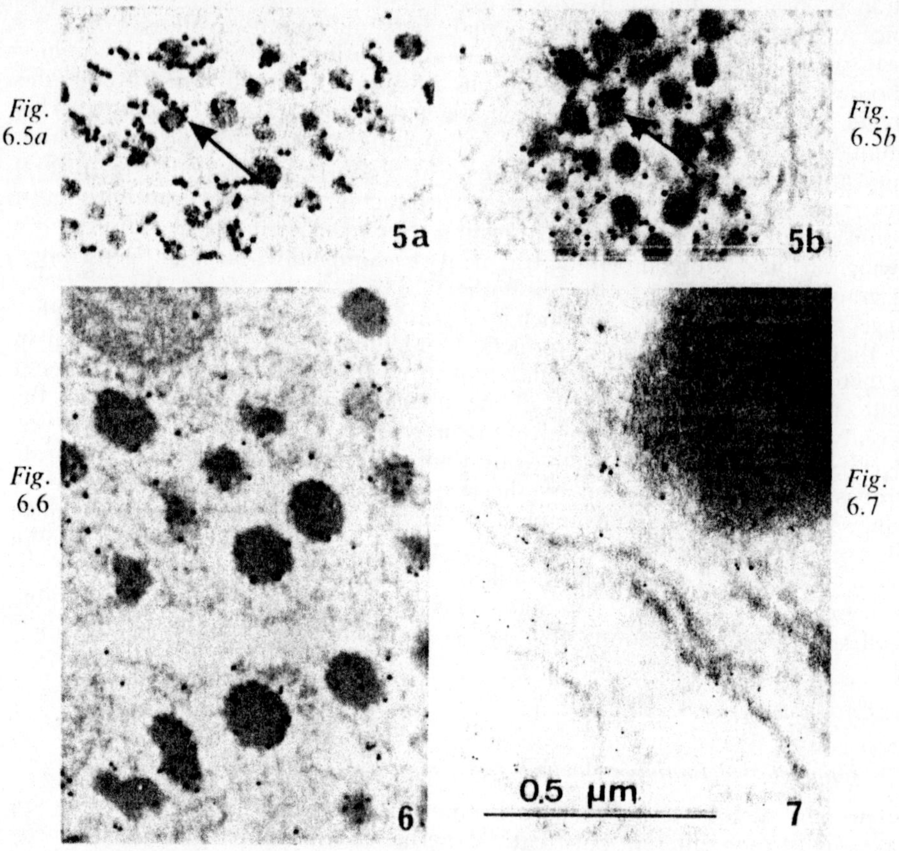

Fig. 6.5a. Substance P immunoreactivity (arrowed) localized to electron-dense vesicles (mean diameter 85±15 nm) in a guinea-pig myenteric nerve terminal. Resin polymerized at 60°C. Immuno-gold staining method with 20-nm gold particles. Uranyl acetate and lead citrate counterstains. Magnification ×50 000.

Fig. 6.5b. Vasoactive intestinal peptide immunoreactivity (arrowed) localized to electron-dense vesicles (mean diameter 98±19 nm) in a guinea-pig myenteric nerve terminal. Resin polymerized by ultraviolet irradiation for 10 days at 18°C. Immuno-gold staining method with 20-nm gold particles. Uranyl acetate and lead citrate counterstains. Magnification ×50 000. From joint work with Dr J. M. Polak and L. Probert (London, UK).

Fig. 6.6. Ultrathin section of human insulin-secreting tumour (insulinoma) immunostained for insulin with IGS method using 20-nm gold particles conjugated to goat anti-guinea-pig immunoglobulins. Uranyl acetate and lead citrate counterstains. Magnification ×40 000. From joint work with Dr J. M. Polak and F. J. Tapia (London, UK).

Fig. 6.7. Ultrathin frozen section of rat liver, perfusion fixed with 20 per cent formaldehyde +0·2 per cent glutaraldehyde. The preparation has been double labelled[19,24,25] with (1) rabbit anti-rat albumin; (2) protein A-G4; (3) rabbit anti-beef catalase; (4) protein A-G 8. Antibodies were affinity-purified and used at 50 μg/ml. Albumin is localized within the RER and catalase within the peroxisome. Magnification ×60 900. Courtesy of Dr J. W. Slot (Utrecht, The Netherlands).

using gold-labelled secondary antibodies. Glicentin is present either throughout the entire granule or only in the halo.[21] Caution in the interpretation of such results is always necessary because presence of one antigen may mask immunoreactivity of another one. Two distinct types of secretory granules are present in normal pancreatic B-cells, one possessing a crystalline core and one with a homogeneous core. Insulin is known to be catabolized from its precursor molecule, pro-insulin, with the liberation of the 29 amino acid C-peptide. Both granule types are immunoreactive to anti-insulin sera but only the granule with a homogeneous core is immunoreactive with anti-C-peptide (pro-insulin) serum.[53] Insulinomas can differ greatly in morphology of their secretory granules and consequently in their labelling with anti-insulin and anti-C-peptide. Some insulinomas have secretory granules that are similar to those containing pancreatic polypeptide which are also very common in many insulinomas. Use of immuno-gold techniques allows a distinction between the two cell types[53] (*Fig.* 6.6). In combination with low temperature tissue processing (embedding in Araldite epoxy resin, cured with ultraviolet irradiation at room temperature),[22] it has been possible to detect antigens that are otherwise difficult to preserve, like VIP in VIP-containing tumours of the pancreas, and define the morphology of their secretory granules[54] (*see also* Roth et al.).[47]

3.2.4. *Localization of Multiple Antigens*

Colloidal gold probes are highly suitable for double or even multiple localization of antigens at the EM level.[34, 37, 73] Double antibody labelling was first introduced with the gold-labelled antigen detection (GLAD) technique.[13] Direct multiple labelling should also be very convenient but has not yet been tried. Labelling gold with monoclonal antibodies[16] should facilitate this. Indirect double labelling techniques have been accomplished using four sequential incubations: antibody one, smallest size pA–gold (5 nm), antibody two, larger sized pA–gold (16 nm).[23] A similar approach was developed by Roth,[24] using 3 and 15-nm gold particles. When the smaller sized gold was used first, no significant interference between the two differently sized gold probes occurred. The elegant technique for sizing protein A–gold,[26] generates very homogenous small (4 and 8 nm) gold probes which perform best to double pA–gold labelling (*Fig.* 6.7) and give the highest labelling density (labelling density sometimes diminishes dramatically with increasing gold particle diameter). Subsequent staining of one of the two exposed faces of plastic thin sections, using the pA–gold technique, also allows for double labelling without interference and further avoids steric hindrance phenomena when antigens are localized in the same organelle.[55] Recently, double indirect labelling has also become possible by using two primary antisera produced in different animal species and species-specific secondary antibodies adsorbed to two gold sizes (*Fig.* 6.8).[56] The advantage of this approach is that primary antibodies and, subsequently, the gold-labelled secondary antibodies can be applied as a mixture. This reduces double labelling to a two step procedure. Combinations of, for example, the latter type of marking with two-face labelling[55] and properly sized gold probes could easily result in multiple demonstration of antigens. The technique used for staining thin sections with IGS reagents is detailed in the Appendix, p. 108.

3.2.5. *Localization of Receptors and Membrane Proteins in Thin Frozen Sections*

The protein A–gold technique on frozen ultrathin sections (*see also* Chapter 5) has recently permitted the visualization of the distribution of the receptor for asialoglycoprotein in rat liver cells with specific anti-receptor antibodies.[57] This receptor mediates the rapid and specific binding and internalization of asialoglycoprotein ligands (galactose-terminal glycoproteins). In a previous study, this receptor had been marked with a desialylated ceruloplasmin-gold probe.[74] So far, direct information of the subcellular localization of the asialoglycoprotein receptor in intact hepatocytes during ligand-induced endocytosis is lacking. Such studies, eventually combined with localization studies on the ligand itself, the desialylated ceruloplasmin-gold probe[74] and other components (such as clathrin, associated with coated vesicles), could give information on the fate of both ligand and receptor after endocytosis. This fate need not necessarily be the same for both. The findings[57] suggested that the receptor normally resides in all surface membranes of the hepatocyte and that during ligand-induced endocytosis, some of the receptor is internalized but remains in the cell periphery. It is not transferred in significant amounts to more central regions of the cell, such as the lysosomes and the Golgi complex. Since data from the literature indicate that, under these conditions, the ligand is transfered further to the Golgi complex–lysosome region, the findings have been interpreted as suggesting that the receptor and ligand are dissociated in the vicinity of the plasma membrane, after which the receptor rapidly returns to the cell surface. Results like these clearly show the new possibilities offered by this immunocytochemical procedure and the gold marker. It must, however, be stressed that even these very interesting results should be considered with caution: immunocytochemical procedures detect immunoreactive antigenic sites and these may become inaccessible after fixation and processing of the tissue for immunocytochemistry, and also by processes such as enzymatic degradation within the cell. For example, it cannot be excluded with certainty that the receptor could be transferred to the Golgi complex–lysosomal region, but the antigenic sites are modified by this transfer in such a way that they are no longer detectable by immunocytochemical procedures. Other examples of localization of membrane proteins in thin frozen sections have been that of GP-2, a glycoprotein characteristic of zymogen granule membrane preparations,[23] and a protein, associated with the sarcoplasmic reticulum of striated muscle, that cross-reacts with an antibody to a 140-kilodalton glycoprotein isolated from intestinal brush border microvilli (*Fig.* 6.9).[65]

3.3. Localization of Antigens in Semithin (0·2–1 μm) and Normal Histological (4–10 μm) Sections for Light Microscopy

Colloidal gold-labelled secondary antibodies have been used for the immuno-gold staining of peptide hormones and neurotransmitters at the light microscopical level.[18,19] Immunoreactive sites typically appear red when viewed by bright field microscopy (*Plate* 3) and the appearance of the staining reaction can be monitored at low magnification. Considerable signal amplification could be obtained by use of the bridge or combined indirect bridge IGS method.[18] The principle of the different IGS methods is illustrated in *Fig.* 6.10. The sequential use of the PAP and the IGS methods has added a valuable complementary method for double staining

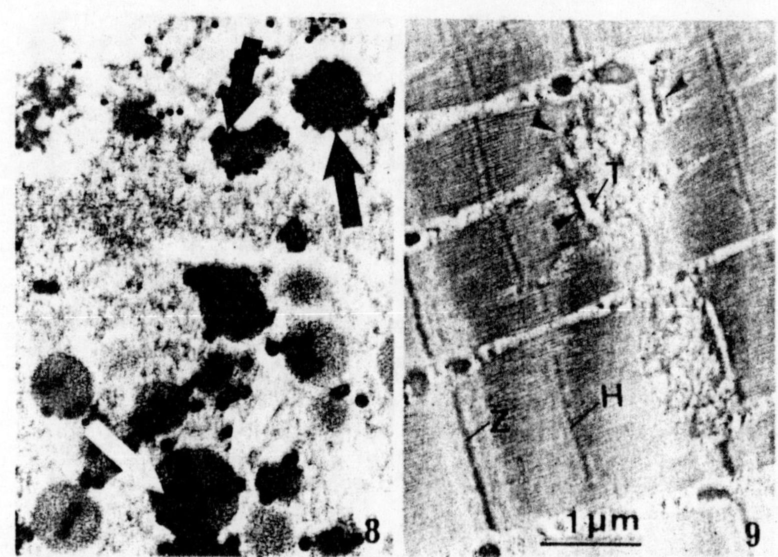

Fig. 6.8. Double immuno-gold staining method applied to ultrathin section of human pancreatic islet. Insulin immunoreactivity (black arrows) is localized to β-granules in the upper half of the micrograph. Glucagon immunoreactivity (white arrow) is localized to α-granules in a neighbouring pancreatic A-cell. The staining procedure was guinea pig anti-insulin mixed with rabbit anti-glucagon followed by goat anti-guinea pig IgG conjugated to 20-nm gold particles mixed with goat anti-rabbit IgG conjugated to 40-nm gold particles. Uranyl acetate and lead citrate counterstains. Magnification × 55 000. From joint work with Dr J. M. Polak and F. J. Tapia (London, UK).

Fig. 6.9. Electron micrograph of a longitudinal thin frozen section from rat skeletal muscle labelled with anti-140 kilodalton glycoprotein, isolated from intestinal brush border microvilli, followed by protein A-gold conjugate. The labelling is found primarily at the periphery of the I band (arrowheads). The Z, A and H bands are not labelled. The labelling is localized to the periphery of the myofibrils and seems to be associated with the membrane of the sarcoplasmic reticulum.[65] Magnification × 13 000. Courtesy of Dr E. Coudrier, H. Peggio and D. Louvard (Heidelberg, GFR).

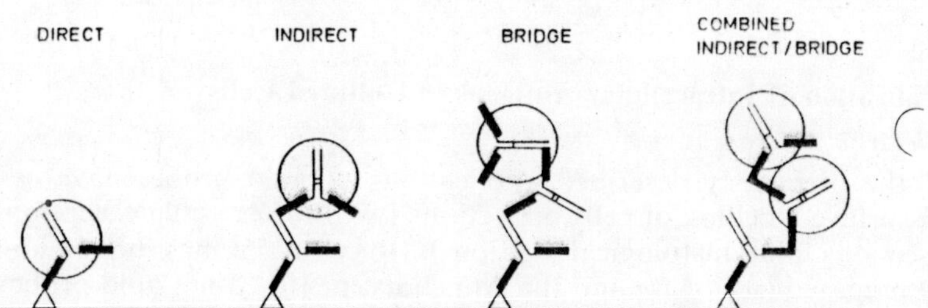

Fig. 6.10. Different ways of using IGS reagents. The direct and indirect method are most useful for EM studies. The bridge and combined indirect and bridge technique give enhanced staining at the light microscopical level. For simplicity, only one antibody molecule per gold granule has been drawn. This is the case for 5-nm gold. With 20-nm gold, at least 20–30 antibody molecules are adsorbed per gold granule (Moeremens and De Mey.[68]) It has however, been noticed that 5-nm gold gives a denser labelling than the 20-nm particles.[7,10,24]

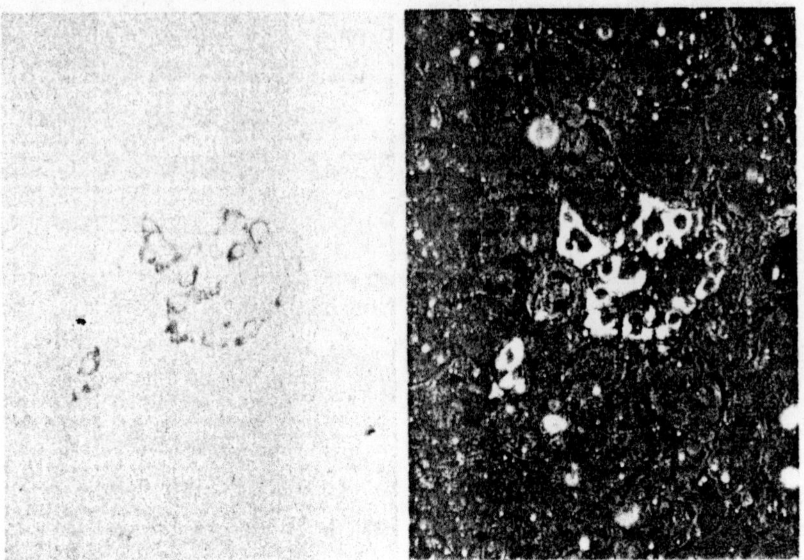

Fig. 6.11. Semithin Epon section of human pancreas, stained with rabbit anti-glucagon (1 : 1000) and GAR G20 (1 : 8) (Janssen Life Sciences Products). Left: normal bright field illumination. Right: dark field illumination (halogen light, 12 v). The positive cells become very bright.

experiments.[18] The same principles can also be applied to semithin sections. Observation of such preparations by dark field illumination (low magnification; *see Fig.* 6.11)[53] or polarized light epi-illumination (high magnification; De Mey et al.[69]) enhances detectability of the gold marker, offering at the same time high resolution and marker stability. Details of these methods are given in the Appendix, p. 109. The recent development of pA–silver staining for light microscopy has enabled Roth[19] to use both pA–gold and pA–silver for double immunocolloid staining of paraffin sections.

3.4. Localization of Intracellular Antigens in Cultured Cells

3.4.1. *Electron Microscopy*

Most of the previously described applications of gold probes have essentially involved surface labelling of cells or sections (with the possible exception of thin frozen sections and histological sections). Recently, it has been shown that properly prepared (*see above* and the Appendix, p. 101) 5-nm gold probes can be used to label whole, adequately fixed, permeabilized cultured PTK_2 cells either in the indirect mode (*Fig.* 6.12),[10] or with monoclonal antibody–gold conjugates in the direct mode (*Fig.* 6.13). This way of labelling can be used in the pre-embedding mode yielding high resolution localization data or, in a novel way, involving preparation of whole cell mounts cultured on electron microscope grids and critical point drying, after labelling (*Fig.* 6.14).[10]

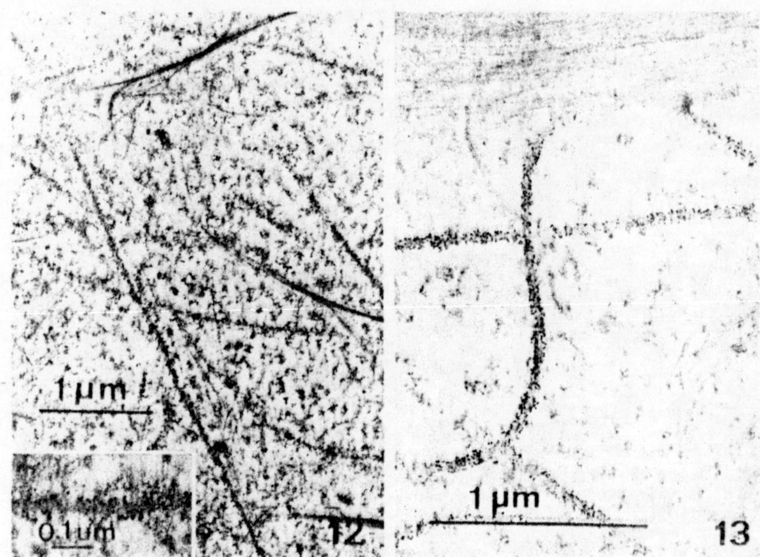

Fig. 6.12. Thin section of a PTK_2 cell stained with anti-tubulin and GAR G5 (Janssen Life Sciences Products). The cells have been fixed in 1 per cent glutaraldehyde and permeabilized with 0·5 per cent Triton X-100. Note the good general preservation and the close association of the gold label with the microtubule wall.[7] Magnification ×15 100. The inset shows a labelled microtubule at higher magnification (×55 400). From joint work with Dr M. De Brabander (Beerse, Belgium).

Fig. 6.13. Thin section of a PTK_2-cell stained with a monoclonal rat anti-tubulin (gift of Dr Kilmartin, MRC, Cambridge, UK) adsorbed to 5-nm gold. This picture illustrates well the high resolution of direct IGS labelling. Magnification ×29 000. From joint work with Dr M. De Brabander (Beerse, Belgium).

A similar approach has also been used for whole mount erythrophore 'cytoskeletons' in which cell residues, left behind after detergent extraction and methanol fixation, were labelled with anti-tubulin and anti-high molecular weight microtubule-associated proteins (anti-HMW-MAPs) and the protein A gold method.[58]

The whole cell mount approach bridges the gap between light microscopical observations and the fractionated picture of thin sections, and allows information to be obtained on three-dimensional antigen distribution using stereo-viewing techniques. This was demonstrated earlier using the PAP procedure.[59] Stereo-viewing is useful for the study of antigens associated with fibrillar elements of the cell (so-called cytoskeleton) such as microtubules, microfilaments and intermediate filaments. It should also allow the simultaneous viewing of extra- and intracellular antigens or lectins by double or multiple labelling techniques.

3.4.2. *Light Microscopy*

Twenty-nm gold probes can be used for the indirect light microscopical visualization of structures as tiny as microtubules, resulting in increased resolution over

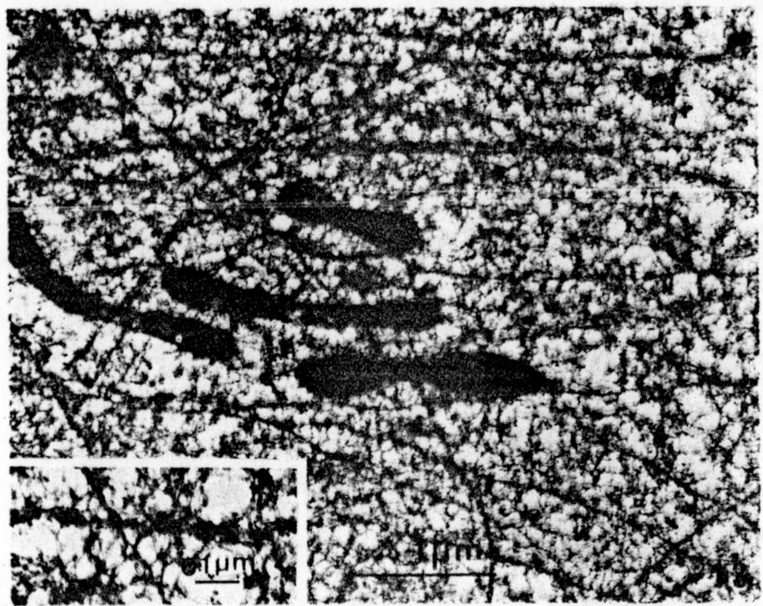

Fig. 6.14. Low power transmission EM view (100 kV) of a whole cell mount of PTK_2 cells, labelled as in *Fig.* 6.12. The three-dimensional distribution of the densely labelled microtubules can be studied relative to unlabelled organelles such as mitochondria, stress fibres and tonofilaments.[7] Magnification × 20 000. The inset shows a high power view of a labelled microtubule. Magnification × 55 400. From joint work with Dr M. De Brabander (Beerse, Belgium).

immunoperoxidase techniques and stable permanent preparations.[10] This characteristic has been used to document microtubule distribution in plant endosperm cells in interphase and during different stages of cell division (*Plate* 4).[60]

4. CONCLUSION

The growing interest in colloidal gold as a versatile, extremely flexible marker for light and electron microscopical immunocytochemical procedures seems fully justified. This chapter has reviewed recent developments and is intended to stimulate other workers to couple their own (monoclonal) antibodies and evaluate their potential. The recent commercialization of a number of gold probes should also help scientists to evaluate these in various unexplored fields. Colloidal gold as a light microscopical marker is still very new. It will probably prove to be very useful in applications such as labelling of cell subpopulations on the basis of specific cell surface antigens and for double staining experiments of histological and semithin sections. Development of other immunocolloid staining methods[19] will increase the flexibility for doing double labelling experiments. There exist of course excellent and extremely sensitive procedures for light microscopy[61,62] and much remains to be learned. However, the exploitation of characteristics of colloidal gold such as its dichroism[63] or its potential of back-scattering incident light (*see* p. 109) may also provide us with further possibilities in the near future for automatic detection and quantitating apparatuses such as cell sorters.

The clearcut potential of colloidal gold in immunocytochemistry lies in its use as a marker for transmission and scanning electron microscopy. Applications reviewed in this chapter clearly illustrate this virtually unlimited potential. The main advantages are the increased resolution, optimal delineation of ultrastructural detail and flexibility in multiple labelling experiments. Whole cell mount pre-embedding immunocytochemistry should permit us to visualize, for example, specific receptors and study their relationship with underlying structures (e.g. coated pits, microfilaments etc.). Staining procedures on thin sections will allow us to study compartmentalization of important proteins.

Although indirect techniques will always be convenient, the expected wave of available monoclonals leads us to speculate that, very soon, direct antibody labelling at the ultrastructural level will become common practice. Immunocytochemistry now offers a wide range of available markers and labelling procedures.[61,62,64] Colloidal gold is only one of many possibilities, and it is important to choose the right marker system to approach a given problem. It must, however, be stressed that no localization method will be better than the immune reagents used, and the tissue preservation achieved. It is salutory that we are now at least aware of these problems, although still far from solving them.

Appendix

I. MAKING MONODISPERSE COLLOIDAL GOLD SOLS

Water absolutely free of organic contaminants and glassware coated with silicone (e.g. Repelcote (Hopkin and Williams), 2% dimethyldichlorosilane in carbon tetrachloride) should be used throughout.

A. 5-nm Particles (slightly modified from Horisberger[3]); *see* Roth[24] for Preparing 3-nm Particles

Make a saturated solution of white phosphorus (Merck) in 100 per cent diethyl ether. The phosphorus reacts rapidly with air. Keep the sticks under water. Cut off a small piece while under water, then transfer it quickly to 10 ml of 100 per cent ether. Close the vial and swirl for at least 2 h to ensure saturation. Centrifuge down remaining solids in a glass centrifuge tube. Mix 1 part saturated solution with 4 parts 100 per cent ether. Meanwhile, make a 1 per cent solution of $HAuCl_4$ (Merck) in H_2O. Microfilter it through a 0·22-μm millipore filter. Add 3 ml of 1 per cent $HAuCl_4$ to 240 ml microfiltered H_2O in a scrupulously cleaned Erlenmeyer flask, fitted with a water-cooled reflux tube. Add 5·4 ml of 0·2 N K_2CO_3 to increase the pH to ±9. Add 2 ml of the diluted phosphorus ether to the $HAuCl_4$ solution, by pipetting it in below the surface to minimize contact with air. Mix this combination slowly at room temperature for 15 min. Then heat and reflux for 20–30 min to remove the ether and to drive the reaction. Cool to 4 °C. Use the colloidal gold solution within 14 days.

A picture of this type of gold particle is shown in *Fig*. 6.15. The mean diameter is 5·5 nm ± 1·1 nm (s.d.).

The visible wavelength absorption spectrum is also shown in *Fig*. 6.15. The maximum wavelength (λ_{max}) is 512 nm.

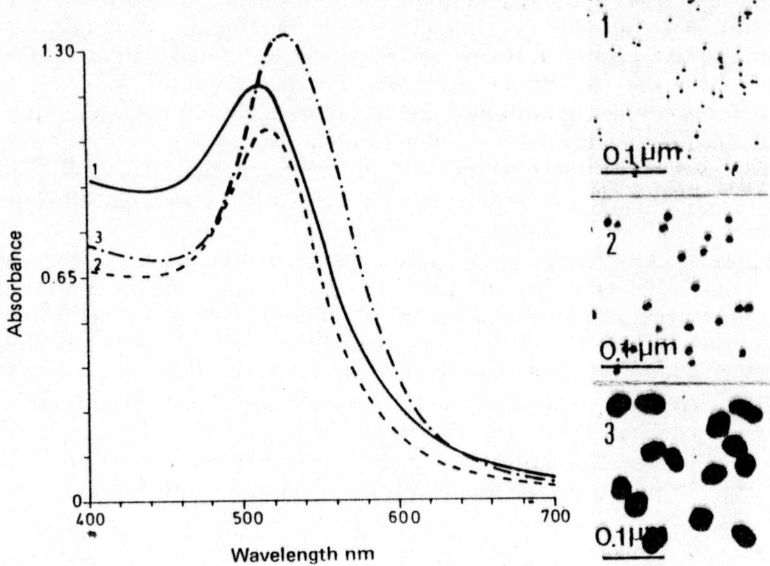

Fig. 6.15. Visible wavelength absorption spectra of three types of gold particles: 5·5, 17 and 40 nm mean size. The inset shows EM pictures. Magnification ×80 750. From joint work with M. Moeremans (Beerse, Belgium).

B. 17–20 nm Particles (modified from Frens[2,3])

Make a 4 per cent stock solution of $HAuCl_4$. Microfilter (0·22 µm). Boil separately 220 ml microfiltered H_2O. Add 5 ml freshly prepared 1 per cent sodium citrate in H_2O to the boiling water. Add 0·5 ml 4 per cent $HAuCl_4$ to this mixture under vigorous mixing. Reflux for 30 min (while continuing swirling). Cool to 4 °C. Use within 14 days. The amount of citrate to be added could differ with laboratory conditions.

A picture of this type of gold particle is shown in *Fig.* 6.15. The mean diameter is 17·3 nm ± 1·7 nm (s.d.).

The visible wavelength absorption spectrum is also shown in *Fig.* 6.15. The maximum wavelength (λ_{max}) is 518·4 nm.

C. 40-nm Particles

Identical, use 2·8 ml of 1 per cent sodium citrate. A picture of this type of gold is shown in *Fig.* 6.15. It is more elongated. The mean diameter is 40 nm ± 5·1 nm (s.d.).

The visible wavelength absorption spectrum is also shown in *Fig.* 6.15. The maximum wavelength (λ_{max}) is 528·5 nm.

II. ADSORBING HETEROGENEOUS IMMUNOGLOBULINS TO COLLOIDAL GOLD

We limit this section to the preparation of immunoglobulin–gold reagents since our procedure has not yet fully been described[10] (Moeremans and De Mey[68]). For preparing protein A–gold, refer to the literature.[12,23,24,26]

A. Preparation of the Antibody for Adsorption

The immunoglobulin concentration is adjusted to about 1 mg/ml. Protein is then dialysed versus 2 mM borax-HCl buffer at pH 9·0. It is recommended to keep the affinity-purified antibodies for as short a period as possible in this low salt buffer. Immediately before use, the protein is centrifuged at 100 000 *g* for 1 h at 0 °C to remove micro-aggregates.

B. Determination of the Optimally Stabilizing Protein Concentration

This is done by constructing a concentration variable adsorption isotherm.[8,15,28] For serum antibodies, the pH of the gold sol is adjusted to 9 with 0·2 N K_2CO_3, just before use. The pH of the gold sol is measured with a gel-filled combination electrode (e.g. Orion).[15] This type of electrode has the advantage of having a low electrolyte flow rate. For example, to work with 20 nm particles, start with a 1 : 3–1 : 5 dilution of the centrifuged, dialysed antibody in 2 mM borax buffer pH 9·0. Make a linear dilution series.

Amount Ab solution	100 µl	90 µl	10 µl	—
Amount borax buffer	—	10 µl	90 µl	100 µl

Centrifuge pH-adjusted gold sol to remove any aggregates: 4800 *g* for 20 min for 5 nm, 250 *g* for 20 min for 20 and 40 nm. Add 1 ml of centrifuged gold to each tube of the dilution series. Vortex, let stand 2 min, then add 100 µl of 10 per cent (w/v) NaCl in H_2O and vortex again. Measure absorbance at 580 nm after approximately 5 min. The optimal amount of antibody is determined by the point where the curve appears asymptotic with the X axis.

For 5-nm sols, a 1 : 2 dilution of the dialysed antibody will be more appropriate since this gold requires more protein to stabilize the colloid. This is because of the greater surface-to-volume ratio of the smaller particles. We have found that pH 9 is compatible with all kinds of serum immunoglobulins (e.g. mouse, goat, rabbit, human etc.).

C. Preparation of an IGS Reagent

The optimally stabilizing amount of antibody is increased by 10 per cent, and any desired volume (up to several litres) of pH-adjusted gold sol is mixed with the appropriate volume of antibody solution (now undiluted). After 2 min of reaction, add a 10 per cent BSA (Sigma, type V) solution to make a final concentration of 1 per cent BSA. The 10 per cent BSA solution is prepared in distilled water, and its pH is adjusted to 9 with NaOH. This stock solution is microfiltered (0·22 µm).

D. Washing the IGS Reagent

Make up 1 per cent BSA in 20 mM Tris-buffered saline pH 8·2 (1 per cent BSA buffer). Microfilter (0·22 µm).

Spin down the IGS reagent. Use 60 000 *g* for 1 h at 4 °C for 5 nm gold; 14 000 g for 1 h at 4 °C for 20 and 40 nm. These are g_{max} values in the tube. As a result, a mobile pool of material at the bottom, not a hard pellet, should be obtained. Most of the supernatant is carefully aspirated and the pool resuspended in the small remaining volume.

Normally, nearly 100 per cent of the gold sol is easily resuspended at this step. Then, the tube is refilled to the original volume with 1 per cent BSA buffer. After overnight equilibration, the centrifugation step is repeated. The supernatant is again aspirated and this washing step is repeated once. Then, the IGS reagent is spun down a final time and resuspended in 1 per cent BSA buffer containing 0·02 M azide to a volume such that, when diluted 1/20 in 1 per cent BSA buffer, the absorbance at 520 nm will be 0·35 for 40-nm gold, 0·5 for 20 nm and 0·25 for 5 nm. One per cent BSA buffer is used as the blank. Of course, these values are arbitrary and only serve for standardization of different batches and for calculating the yield of gold.

If there are visible aggregates in the liquid at any time during the preparation, use a low-speed spin before the high-speed spin and discard the pellet. For 5 nm, use 4800 *g* for 20 min; for 20 and 40 nm, use 250 *g* for 20 and 10 min, respectively. The final product may be sterile filtered through 0·22 μm and stored under sterile conditions, if azide is not desired. During the first washing step in 1 per cent BSA buffer (second centrifugation), sometimes more than 50 per cent of the gold particles may become aggregated irreversibly. The amount lost decreases sharply with subsequent washing steps. This suggests that the washing steps eliminate a fraction of the particles that became destabilized by addition of the 1 per cent BSA buffer. Often, however, a final yield of 70–80 per cent has been obtained. A similar or lower yield was obtained when Carbowax 20 M was used as stabilizer. IGS reagents produced in this way have been shipped to distant places without apparent loss of activity.

E. Elimination of Aggregates

Gold sols prepared as above are reasonably homogeneous, making them suitable for double labelling experiments without a necessity for sizing. A major problem, however, encountered with applications such as intracellular labelling with 5-nm IGS reagents of whole cultured cells has been the presence in the final preparation of micro-aggregates. These do not enter the cells and tend to stick to the cell wall. Since they contribute to the total gold mass, the relative amount of efficient non-aggregated IGS reagent is diminished and labelling quality is compromised. We have recently found that micro-aggregates in 5-nm IGS reagents can efficiently be removed, the first centrifugation being divided in two: for the first step, the settings were 60 000 *g*, 7 min, 4 °C, 38-ml tubes, Kontron TGA50 centrifuge, rotor TFT 50–38. The supernatant at this stage was carefully recovered. Measurement of the absorbance at 520 nm of this supernatant showed that ±20–30 per cent of the gold probe was pelleted. The supernatant was further centrifuged at 60 000 *g*, 60 min, 4 °C, and washed as usual. This procedure yields IGS G5 reagents with much less aggregate, and better labelling efficiency. At the time of writing, it has not been tested whether this procedure would be of any use for larger sized IGS reagents, but there is a good chance that it may for particular applications, such as quantification of gold granules bound to antigens on thin sections.

F. Terminology

We call immunoglobulin–gold conjugates, IGS reagents, as they are used in the immuno-gold staining (IGS) methods.

The first symbols (e.g. GAR, GAM, R, M) designate the kind of immunoglobulin used. For example, GAR is affinity-purified goat anti-rabbit IgG; R and M are protein A–Sepharose-purified rabbit and mouse IgG respectively; G followed by a figure designates the size class of the colloidal gold; R G5 means rabbit IgG adsorbed to colloidal gold with a mean diameter of ± 5 nm.

III. ADSORBING MONOCLONAL ANTIBODIES TO COLLOIDAL GOLD

A. Preparation of the Antibody for Adsorption

Identical as for heterogeneous immunoglobulins.

B. Determination of the Optimally Stabilizing Protein Concentration

For monoclonal antibodies, it is necessary to determine the optimal pH for protein adsorption. This is best done by isoelectric focusing in polyacrylamide or agarose gels, under native conditions. The pH of the gold sol is adjusted to the p*I* of the monoclonal antibody. Use 0·2 M H_3PO_4 for lowering and 0·2 N K_2CO_3 for increasing the pH. Mixture of protein and gold sol will only result in a slight shift of pH that will not influence the adsorption efficiency. If no satisfactory stabilization curve (constructed at the chosen pH, but otherwise identical to the procedure outlined above) is formed, the optimal pH for adsorption can be determined by constructing concentration-variable adsorption isotherms at different pH values. We feel this way is safer than constructing only one pH-variable adsorption isotherm.[8] The curve that will give the lowest baseline (at 580 nm) and stabilizing protein concentration will indicate at the same time optimal pH and stabilizing amount of monoclonal antibody.

C. Preparation and Washing of the Monoclonal IGS Reagent

There is only one difference: before addition of the BSA stock solution, the pH of the gold sol is raised to pH 9·0 with 0·5 M NaOH.

D. Elimination of Micro-aggregates

Same remarks as above.

IV. SOME APPLICATIONS IN WHICH IGS REAGENTS HAVE BEEN USED WITH SUCCESS

A. Light and Electron Microscopical Detection of Leucocyte Cell Surface Antigens with Monoclonal Antibodies[17]

Cell suspensions prepared from any tissue source can be used. The cells should be washed in buffer A (PBS, BSA 1 per cent, 0·2 per cent NaN_3, 1 per cent heat-inactivated human AB serum pH 7·2) and resuspended at a density of 5×10^6 cells/ml in buffer A. 200 µl of this cell suspension is pelleted (300 *g*, 5 min) and resuspended in 25 µl buffer B (PBS, BSA 5 per cent, 4 per cent AB serum, 0·2 per cent NaN_3 pH 7·2).

Add 25 µl of diluted antibody (in buffer A) or buffer A (to serve as a method control for the GAM G40). The optimal dilution should be determined with a dilution series.

Mix well and incubate the cell mixture at room temperature for 20–30 min.

Agitate every 10 min.

Add 2 ml buffer A and wash cells by centrifugation at 300 *g* for 5 min. Remove the supernatant and repeat this twice.

Resuspend the last pellet of washed cells in 25 µl of buffer B.

Add 25µl of GAM G40 (absorbance at 520 nm = 0·35 when diluted 1/20 with 1 per cent BSA buffer). This GAM G40 can often be diluted up to five times with 1 per cent BSA buffer (*see* Preparation of an IGS reagent).

Mix well and incubate the cell mixture at room temperature for 30–60 min.

Agitate every 10 min.

Repeat washing procedure.

Resuspend the last pellet in fixative. For electron microscopy, use 1 per cent glutaraldehyde in 0·1 M cacodylate buffer pH 7·2, for 30 min. For light microscopy, use 0·01 per cent glutaraldehyde for 10 min. Mix well to avoid formation of cell clumps.

For electron microscopy, pellet the cells, treat with osmium tetroxide, prestain with uranyl acetate, dehydrate and embed in an epoxy resin according to standard techniques.

For light microscopy, repeat washing procedure.

Resuspend the last pellet of washed cells in an appropriate volume of PBS, BSA 5 per cent. Cytospin the preparations or make smears from this cell suspension.

Further Characterization of Immunoreactive Cells by Enzyme Cytochemistry

In order to reveal cells containing endogenous peroxidase activities, the preparations are treated as follows:

The freshly made preparations are fixed for 2 min in 4 per cent formaldehyde in absolute ethanol.

Rinse in distilled water. Air dry.

Freshly prepare the peroxidase reaction mixture:

1. 1 ml H_2O_2 30 per cent + 9 ml distilled water.
2. Add 0·1 ml of (1) to 10 ml distilled water.
3. Dissolve 5 mg diaminobenzidine in 10 ml 100 mM Tris buffer, pH 7·6.

4. Add 0·2 ml of (2) to (3).
5. Microfilter through millipore (0·22 μm). Avoid contact with skin of this solution.
6. Incubate preparations for 20 min in this reaction mixture.
7. As an alternative, the Hanker Yates reagent (Polysciences) may be used:

   Tris-HCl, 0·1 M, pH 7·6
   Hanker Yates 1 mg/ml
   H_2O_2 final concentration 0·006 per cent
   Incubate 10 min

Rinse with water.

Counterstain for 3 min with 1 per cent methyl green in distilled water. For obtaining best results, impurities are extracted from the stock solution with chloroform. A 1 per cent methyl green solution is added to an equal volume of chloroform in a separation funnel. The solution is shaken and the two phases are separated. Repeat this twice.

Rinse with distilled water, dehydrate through an ethanol series and xylene.

Mount in a permanent resin.

Antigen-bearing cells can easily be detected in the normal bright field light microscope, since they are surrounded by a large number (> 10) of clearly visible dark granules. The antigen-negative lymphocytes can be distinguished by both their morphological characteristics and the absence of peroxidase reaction. The usual cytochemical reactions such as acid alpha-naphthyl acetate esterase, acid phosphatase and beta-glucuronidase can also be performed on the cell preparations, but, with the above methods, the incubation times have to be shortened to avoid overstaining. This allows the study of both the antigenic and the enzymatic profiles of cell suspensions and will be of considerable help for characterizing leukaemias.

The polarized light epi-illumination technique is very helpful for detecting the positive cells. The signal is strong enough to combine polarized light epi-illumination with moderate transmitted light.

B. Light Microscopical Demonstration of Antigens in Histological (4–10 μm) and Semithin (0·2–2 μ) Sections

For light microscopical use of the IGS method, the indirect, bridge or combined indirect and bridge method can be used.[18] The latter gives a particularly strong signal amplification, because it increases the number of gold particles accumulating near the antigen-containing sites.

Preparation of tissues will differ from application to application. The most suitable method will not differ very much from standard methods that give satisfactory results with other labelling methods. I refer to different chapters of this book for specific examples of tissue preparation.

Incubate the preparation with 5 per cent normal serum (same species as secondary antibody) made up in 0·1 per cent BSA buffer (20 mM Tris, 150 mM NaCl, 1 mg/ml BSA pH 8·2).

Remove excess normal serum and replace by appropriately diluted primary antiserum made up in 0·1 per cent BSA buffer supplemented with 1 per cent normal serum. When the optimal dilution is unknown, it should be determined by making up a series of serial dilutions. Incubate from a couple of hours to overnight, at room temperature in a moist atmosphere.

Wash the preparations with 0·1 per cent BSA buffer. Use 3 changes of ±10 min each.

For the indirect and combined indirect and bridge IGS method, incubate with the appropriate secondary antibody–gold IGS reagent diluted in 1 per cent BSA buffer. G20 is recommended. Gold particles smaller than 10 nm cannot be used for observation of gold by colour. The optimal dilution and incubation time should be determined. Centrifuge the diluted reagent, for 10 min, 250 *g*). In under the microscope. For the bridge method, unlabelled secondary antibody diluted in 0·1 per cent BSA buffer is used at this stage.

Wash the preparations with 0·1 per cent BSA buffer. Use 3 changes of ±10 min each.

For the bridge and combined indirect and bridge method, incubate with the immunoglobulin–gold IGS reagent (G20; same animal species as first antibody), diluted in 1 per cent BSA buffer. The optimal dilution and incubation time should be determined. Centrifuge the diluted reagent for 10 min, as above.

Wash the preparations with 0·1 per cent BSA buffer as above.

Rinse in distilled water (all methods).

Fix with 1 per cent glutaraldehyde in fixing buffer (10 min).

Rinse in water. At this stage, a suitable light counterstaining may be used.

Dehydrate and mount in a permanent resin.

Antigen-containing cells are stained red by these procedures. For thin sections, dark field illumination enhances detectability of the label. Polarized light epi-illumination can also be used with the help of lenses with high numerical aperture and oil immersion, when high resolution and sensitivity is required. In many cases however, normal bright field microscopy will give satisfaction. As has been shown,[18] these procedures lend themselves extremely well to sequential staining of two antigens in the same section, following peroxidase staining of the first antigen, and elution of antibodies.

C. Indirect Labelling of Antigens in Thin Plastic Sections for Transmission Electron Microscopy

This method is similar to immunostaining of ultrathin sections, previously described[12,13,40] and almost identical to Probert et al.[22]

Epoxy resin embedded sections are etched with 10 per cent H_2O_2 for 15 min. Etching is used for tissues that have been osmicated, although it has been reported to be unnecessary.[21] A large number of antigens resist osmication.[12,21,22] This results in excellent ultrastructural detail.

Sections mounted on EM grids are etched by floating the grids, or immersing them (two face staining) in drops (20 µl) of 10 per cent H_2O_2, which are placed in wells of microtest plates.

The grids are rinsed by dipping them in a beaker with water and are immersed in drops (20µl), 5 per cent normal serum (same species as second antibody) in 0·1 per cent BSA buffer. For transfer of grids away from the microtest plates, two fine forceps are used. One (left) is used for tilting the grid, the other for holding the grid. The grids are transferred to and immersed in 20 µl drops (in wells) of diluted specific antibody in 1 per cent normal serum (0·1 per cent BSA buffer) and incubated. Optimal dilution and incubation time should be determined. The best is to choose a fixed time and adapt the dilution to it.

For washing, the grids are transferred to a special grid holder (similar to Polysciences Cat. No. 7332) bathing in 0·1 per cent BSA buffer in a petri dish. The holder has a perforated bottom and also the cover is perforated over each space. This allows for free passage of fluids when the holder is closed. The closed grid holder with grids is moved to a beaker with 150 ml 0·1 per cent BSA buffer which is slowly stirred for ± 10 min. This washing procedure is efficient but very gentle and particularly useful when Formvar-coated slot grids are used.

The grid holder is again moved to a petri dish with 0·1 per cent BSA buffer and opened, taking care that all grids are placed at the bottom of each space.

These are then transferred to appropriately diluted secondary antibody–gold IGS reagent in 1 per cent BSA buffer in wells and incubated for about one hour.

Washing as above, the last wash in buffer without BSA.

The grids are postfixed with glutaraldehyde, rinsed with H_2O and the sections are contrasted with uranyl acetate and lead citrate, rinsed and dried as usual.

The above procedure can be adapted for double or multiple labelling experiments in different ways (*see* p. 95).

V. POLARIZED LIGHT EPI-ILLUMINATION MICROSCOPY

The microscope used was a Reichert Polyvar®, equipped with a 200-W HBO mercury-vapour lamp, a 100-W halogen lamp, an epi-illuminator (cat. No. 901.137) and an epi-polarized light module (Pol.) (cat. No. 300.644). The latter contains a polarizer, for the excitation light, a dichroic half-mirror and an analyser that will extinguish reflected light having the same polarization as the excitation light. Light, back-scattered by the gold particles, is not extinguished by the polarizer. The objectives used are a planapochromat 40 x, I.K., iris, maximum N.A. 1.00 (cat. No. 242.752) and a planapochromat 100 x, I.K., iris, maximum N.A. 1.32 (cat. No. 245.052). The epi-light can be adjusted with the diaphragms in the epi-illuminator. The microscope is further equipped with a universal bright field–dark field condensor (cat. No. 283.902).

Acknowledgements

My colleagues of the Laboratory of Oncology, M. De Brabander, M. Moeremans, G. Geuens and R. Nuydens are gratefully acknowledged for their contribution to this work. Thanks are also expressed to the group of the Histochemistry Department of Hammersmith Hospital, London, especially Drs J. Polak, L. Probert, F. Tapia, I. Varndell and J. Gu and to Drs M. De Waele and B. Van Camp from the Haematology Laboratory of the University of Brussels, Medical School (Vrije Universiteit Brussel) for the fruitful collaborations and the pictures they kindly provided. I also am indebted to all persons or groups of persons for their generous gifts of illustrative material and preprints of unpublished work. I finally thank Mrs B. Wouters for typing this manuscript. This work has been supported by a grant from the IWONL (*Instituut ter bevodering van Wetenschappelijk Onderzoek in Landbouw en Nijverheid, Brussels*).

REFERENCES

1. Faulk W. P. and Taylor G. M. An immunocolloid method for the electron microscope. *Immunochemistry* 1971, **8**, 1081–1083.

2. Feldherr C. M. and Marshall J. M. The use of colloidal gold for studies of intracellular exchanges in the ameba *Chaos chaos*. *J. Cell Biol.* 1962, **12**, 640–645.
3. Horisberger M. Evaluation of colloidal gold as a cytochemical marker for transmission and scanning electron microscopy. *Biol. Cellul.* 1979, **36**, 253–258.
4. Horisberger M. Colloidal gold: a cytochemical marker for light and fluorescent microscopy and for transmission and scanning electron microscopy. In: Johari O., ed., *Scanning Electron Microscopy II*. AMF O'Hare, Illinois, SEM Inc., 1981: 9–31.
5. Goodman S. L., Hodges G. M. and Livingston D. C. A review of the colloidal gold marker system. In: Johari O., ed., *Scanning Electron Microscopy II*. AMF O'Hare, Illinois, SEM Inc., 1981: 133–145.
6. Romano E. L., Stolinski C. and Hughes-Jones N. C. An antiglobulin reagent labelled with colloidal gold for use in electron microscopy. *Immunochemistry* 1974, **11**, 521–522.
7. Horisberger M. and Rosset J. Colloidal gold, a useful marker for transmission and scanning electron microscopy. *J. Histochem. Cytochem.* 1977, **25**, 295–305.
8. Geoghegan W. D. and Ackerman G. A. Adsorption of horseradish peroxidase, ovomucoid and anti-immuno-globulin to colloidal gold for the indirect detection of concanavalin A, wheat germ agglutination and goat anti-human immuno-globulin G on cell surfaces at the electron microscopic level: a new method, theory and application. *J. Histochem. Cytochem.* 1977, **25**, 1187–1200.
9. Goodman S. L., Hodges G. M., Trejdosiewicz L. and Livingston D. C. Colloidal gold markers and probes for routine application in microscopy. *J. Microsc.* 1981, **123**, 201–213.
10. De Mey J., Moeremans M., Geuens G., Nuydens R. and De Brabander M. High resolution light and electron microscopic localization of tubulin with the IGS (immuno-gold staining) method. *Cell Biol. Int. Rep.* 1981, **5**, 889–899.
11. Romano E. L., Romano M. *Staphylococcus* protein A bound to colloidal gold: a useful reagent to label antigen–antibody sites in electron microscopy. *Immunochemistry* 1977, **14**, 711–715.
12. Roth J., Bendayan M. and Orci L. Ultrastructural localization of intracellular antigens by the use of protein A–gold complex. *J. Histochem. Cytochem.* 1978, **26**, 1074–1081.
13. Larsson L.-I. Simultaneous ultrastructural demonstration of multiple peptides in endocrine cells by a novel immunocytochemical method. *Nature* 1979, **282**, 743–746.
14. Horisberger M., Rosset J. and Bauer H. Colloidal gold granules as markers for cell surface receptors in the scanning electron microscope. *Experientia* 1975, **31**, 1147–1149.
15. Geoghegan W. D., Scillian J. J. and Ackerman G. A. The detection of human B-lymphocytes by both light and electron microscopy utilizing colloidal gold-labelled anti-immuno-globulin. *Immunol. Commun.* 1978, **7**, 1–12.
16. De Mey J., Moeremans M., De Waele M., Geuens G. and De Brabander M. The IGS (Immuno-Gold Staining) method used with monoclonal antibodies. *Proceedings of Colloquium on the Protides of the Biological Fluids*. Peeters M., ed., Oxford, Pergamon Press, 1981; 943–947.
17. De Waele M., De Mey J., Moeremans M. and Van Camp B. The immuno-gold staining method: an immunocytochemical procedure for leukocyte characterization by monoclonal antibodies. In: Knapp W., ed., *Leukemia Markers*. London, Academic Press, 1981: 173–176.
18. Gu J., De Mey J., Moeremans M. and Polak J. Sequential use of the PAP and immuno-gold staining methods for the light microscopical double staining of tissue antigens. Its application to the study of regulatory peptides in the gut. *Regulat. Pept.* 1981, **1**, 365–374.
19. Roth J. Applications of immunocolloids in light microscopy. *J. Histochem. Cytochem.* 1982, **30**, 691–696.
20. Horisberger M. and Vonlanthen M. Localization of mannan and chitin on thin sections of budding yeasts with gold markers. *Arch. Microbiol.* 1977, **115**, 1–7.
21. Garaud J. C., Eloy R., Moody A. J., Stock C. and Grenier J. F. Glucagon and glicentin-immunoreactive cells in the human digestive tract. *Cell Tissue Res.* 1980, **213**, 121–136.
22. Probert L., De Mey J. and Polak J. Distinct subpopulations of enteric p-type neurons containing substance P and vasoactive intestinal polypeptide. *Nature* 1981, **294**, 470–471.
23. Geuze H., Slot J., Van der Ley P., Schuffer R. and Griffith J. Use of colloidal gold particles in double labelling immuno-electron microscopy of ultrathin frozen tissue sections. *J. Cell Biol.* 1981, **89**, 653–665.
24. Roth J. Evaluation of the protein A–gold (pAg) technique for labelling of multiple antigens: preparation of pAg complexes with 3 nm and 15 nm gold particles. *Histochem. J.* 1982, in press.
25. Frens G. Controlled nucleation for the regulation of particle size in monodisperse gold suspensions. *Nature. Phys. Sci.* 1973, **241**, 20–22.
26. Slot J. and Geuze H. Sizing of protein A–colloidal gold probes for immuno-electron microscopy. *J. Cell Biol* 1981, **90**, 533–536.

27. Forsgren A. and Sjöquist J. 'Protein A' from *S. aureus*. I. Pseudoimmune reaction with human γ-globulin. *J. Immunol.* 1966, **97**, 822.

28. Geoghegan W. D., Ambegaonkar S. and Calvanico N. Passive gold agglutination. An alternative to passive haemagglutination. *J. Immunol. Methods* 1980, **34**, 11–21.

29. Trejdosiewicz L. K., Smolira M. A., Hodges G. M., Goodman S. L. and Livingston D. C. Cell surface distribution of fibronectin in cultures of fibroblasts and bladder derived epithelium: SEM-immunogold localization compared to immunoperoxidase and immunofluorescence. *J. Microsc.* 1981, **123**, 227–236.

30. Garaud J. C., Doffoel M., Stock C. and Grenier J. F. Are 'G' cells the only source of gastrin in the human antrum? *Biol. Cell.* 1982 in press.

31. Roth J. and Berger E. Immunocytochemical localization of galactosyltransferase in HeLa cells: codistribution with thiamine pyrophosphatase in *trans*-Golgi cisternae. *J. Cell Biol.* 1982, **93**, 223–229.

32. Yeger H. and Kalnins V. Immunocytochemical localization of Gp 70 over virus-related submembranous densities in ts mutant Rauscher murine leukemia virus-infected cells at the non-permissive temperature. *Virology* 1978, **91**, 489–492.

33. Bain B., Catovsky D., O'Brien M., Prentice H., Lawlor E., Kumaran T., McCann S., Matutes E. and Galton D. Megakaryoblastic leukemia presenting as active myelofibrosis. A study of four cases with the platelet-peroxidase reaction. *Blood* 1981, **58**, 206–213.

34. Sieber-Blum M., Sieber F. and Yamada K. Cellular fibronectin promotes adrenergic differentiation of quail neural crest cells *in vitro*. *Exp. Cell Res.* 1981, **193**, 285–295.

35. Molday R. and Moher P. A review of cell surface markers and labelling techniques for scanning electron microscopy. *Histochem. J.* 1980, **12**, 273–315.

36. De Waele M., De Mey J., Moeremans M., Smet L., Broodtaerts L. and Van Camp B. Cytochemical profile of immunoregulatory T-lymphocyte subsets defined by monoclonal antibodies. 1982, Submitted for publication.

37. Horisberger M. and Vonlanthen M. Ultrastructural localization of soybean agglutinin on thin sections of *Glycine max* (Soybean) Var. Altona by the gold method. *Histochem.* 1980, **65**, 181–186.

38. Bendayan M., Roth J., Perrelet A. and Orci L. Quantitative immunocytochemical localization of pancreatic secretory proteins in subcellular compartments of the rat acinar cell. *J. Histochem. Cytochem.* 1980, **28**, 149–160.

39. Bendayan M. and Ørstavik T. Immunocytochemical localization of kallikrein in the rat exocrine pancreas. *J. Histochem. Cytochem.* 1982, **30**, 58–66.

40. Batten T. F. C. and Hopkins C. R. Use of protein A-coated colloidal gold particles for immunoelectron microscopic localization of ACTH on ultrathin sections. *Histochemistry* 1979, **60**, 317–320.

41. Ravazzola M. and Orci L. Glucagon and glicentin immunoreactivity are topologically segregated in the α granule of the human pancreatic A cell. *Nature* 1980, **284**, 66–67.

42. Roth J., Ravazzola M., Bendayan M. and Orci L. Application of the protein A–gold technique for electron microscopic demonstration of polypeptide hormones. *Endocrinology* 1981, **108**, 247–253.

43. Tanaka T., Gresik E. W., Michelakis A. M. and Barka T. Immunocytochemical localization of renin in kidneys and submandibular glands of SWR/J and C57 B2/6J mice. *J. Histochem. Cytochem.* 1980, **28**, 1113–1118.

44. Roth J., Thorens B., Hunziker W., Norman A. W. and Orci L. Vitamin D-dependent calcium binding protein: immunocytochemical localization in chick kidney. *Science* 1981, **214** 197–200.

45. Roth J., Brown D., Norman A. and Orci L. Localization of the vitamin D-dependent calcium binding protein in mammalian kidney. *Am. J. Physiol.* 1982, in press.

46. Thorens B., Roth J., Norman A., Perrelet A. and Orci L. Immunocytochemical localization of the vitamin D-dependent calcium binding protein in chick duodenum. *J. Cell Biol.* 1982, **94**, 115–122.

47. Roth J., Bendayan M., Carlemalm E., Villiger W. and Garavito M. Enhancement of structural preservation and immunocytochemical staining in low temperature embedded pancreatic tissue. *J. Histochem. Cytochem.* 1981, **29**, 663–671.

48. Tokuyasu K. T. A study of positive staining of ultrathin frozen sections. *J. Ultrastr. Res.* 1978, **63**, 287–307.

49. Painter R. G., Tokuyasu K. T. and Singer S. J. Immunoferritin localization of intracellular antigens: the use of ultracryotomy to obtain ultrathin sections suitable for direct immunoferritin staining. *Proc. Natl Acad. Sci. USA* 1973, **70**, 1649–1653.

50. Tokuyasu K. T. and Singer S. J. Improved procedures for immunoferritin labelling of ultrathin frozen sections. *J. Cell Biol.* 1976, **71**, 894–906.
51. Dutton A., Tokuyasu K. T. and Singer S. J. Iron–dextran antibody conjugates: general method for simultaneous staining of two components in high-resolution immuno-electron microscopy. *Proc. Natl Acad. Sci. USA* 1979, **76**, 3392–3396.
52. Larsson L.-I. Peptide immunocytochemistry. *Proc. Histochem. Cytochem.* 1981, **13**, 1–85.
53. Varndell I. M., Tapia F. J., Probert L., Buchan A. M. J., Gu J., De Mey J., Bloom S. R. and Polak J. M. Immunogold staining method for the localization of regulatory peptides. *Peptides* 1982, in press.
54. Capella C., Polak J. M., Buffa R., Tapia F. J., Heitz Ph., Usellini L., Bloom S. R. and Solcia E. Morphological patterns and diagnostic criteria of VIP-producing endocrine tumors of the pancreas. A histological, histochemical, ultrastructural and biochemical study of 32 cases. *Cancer* 1982, in press.
55. Bendayan M. Double immunocytochemical labelling applying the protein A–gold technique. *J. Histochem. Cytochem.* 1982, **30**, 81–85.
56. Tapia F. J., Varndell I. M., Probert L., Gosselin E. J., De Mey J. and Polak J. M. Immunogold staining methods (IGS) for the simultaneous ultrastructural localization of molecular form variants of pancreatic hormones. *Proc. R. Microsc. Soc.* 1982, **17**, 133 (Abstract).
57. Geuze H. J., Slot J. W., Strous G. J. A. M., Codish H. F. and Schwartz A. L. Immunocytochemical localization of the receptor for asialoglycoprotein in rat liver cells. *J. Cell Biol.* 1982, **92**, 865—870.
58. Ochs R. and Stearns M. Colloidal gold immunolabelling of whole-mount erythrophore cytoskeletons: localization of tubulin and HMW-MAPs. *Biol. Cell* 1981, **42**, 19–28.
59. De Mey J., Wolosewick J. J., De Brabander M., Geuens G., Joniau M. and Porter K. R. Tubulin localization in whole glutaraldehyde fixed cells, viewed with stereo high-voltage electron microscopy. In: De Brabander M., Marcel M. and De Ridder L., ed., *Cell Movement and Neoplasia*. Oxford, Pergamon Press, 1980: 21–28.
60. De Mey J., Lambert A. M., Bajer A. S., Moeremans M. and De Brabander M. Visualization of microtubules in interphase and mitotic plant cells of *Haemanthus* endosperm with the immunogold staining (IGS) method. *Proc. Natl Acad. Sci. USA* 1981, **79**, 1848–1902.
61. Sternberger L. A. *Immunocytochemistry* 2nd ed. New York, John Wiley and Sons, 1979.
62. Hsu S., Raine L. and Fanger M. Use of avidin–biotin–peroxidase complex (ABC) in immunoperoxidase techniques: a comparison between ABC and unlabelled antibody (PAP) procedures. *J. Histochem. Cytochem.* 1981, **29**, 577–580.
63. Inoué S., Bajer A., Molé-Bajer J. Dichroism of mitotic microtubules displayed by gold-conjugated anti-tubulin. *J. Cell Biol.* 1981, **91**, 321a.
64. De Mey J. A critical review of light and electron microscopic immunocytochemical techniques used in neurobiology. *J. Neurosci. Methods*, 1982, in press.
65. Coudrier E., Reggio H., Coward D. The cytoskeleton of intestinal microvilli contains two polypeptides immunologically related to proteins of striated muscle. *Cold Spring Harbor Symp. Quant. Biol.* 1982, **46**, 881–892.
66. Doerr-Schott J., Garaud J.-C. Ultrastructural identification of gastrin-like immunoreactive nerve fibers in the brain of *Xenopus laevis* by means of colloidal gold or ferritin immunocytochemical methods. *Cell Tissue Res.* 1981, **216**, 581–589.
67. Doerr-Schott J. and Garaud J.-C. Gastrin-like peptides in the amphibian brain: an immunohistochemical study. *Peptides* 1981, **2**, Suppl. 2, 99–107.
68. Moeremans M. and De Mey J. (1982). in preparation.
69. De Mey J. et al. (1982). In preparation.
70. Bendayan M. Use of the protein A-gold technique for the morphological study of vascular permeability. *J. Histochem. Cytochem.* 1980, **28**, 1251–1254.
71. Bendayan M. and Shore G. Immunocytochemical localization of mitochondrial proteins in the rat hepatocyte. *J. Histochem. Cytochem.* 1982, **30**, 139–147.
72. Swaab D. F., Pool C. W. and Nijveldt F. Influence of vasopressin and oxytocin in the rat hypothalamo-neurohypophyseal system. *J. Neurotransm.* 1975, **36**, 195–215.
73. Horisberger M. and Vonlanthen M. Multiple marking of cell surface receptors by gold granules: simultaneous localization of three lectin receptors on human erythrocytes. *J. Microsc.* 1979, **115**, 97–102.
74. Horisberger M. and Vonlanthen M. Simultaneous localization of an hepatic binding protein specific for galactose and of galactose-containing receptors on rat hepatocytes. *J. Histochem. Cytochem.* 1978, **26**, 960–966.

7 Double Immunoenzymatic Labelling

D. Y. Mason, B. Falini and
Z. Abdulaziz, H. Stein

1. INTRODUCTION

The purpose of this chapter is to describe the methods which may be used for labelling pairs of antigens in tissue sections by double immunoenzymatic procedures. The number of antigens which can be detected in human tissue is increasing steadily year by year, principally as a result of the production of new monoclonal antibodies, and this has led to a growing need for methods which will enable the relative distribution patterns of pairs of antigens to be visualized simultaneously in tissue sections. Immunofluorescent techniques have been widely used in the past for double labelling of antigens. However immunoenzymatic procedures offer several advantages which are likely to result in their being used on an increasingly wide scale in the future.

2. METHODS OF DOUBLE IMMUNOENZYMATIC STAINING

Double immunoenzymatic labelling may be achieved by two fundamentally different approaches, which may be referred to as single-enzyme and double-enzyme methods. In the former type of procedure both antigens are labelled using an immunoperoxidase technique, two different enzyme substrates yielding distinctively coloured reaction products being used to reveal each antigen. In the two-enzyme approach on the other hand two unrelated enzymes are used to label the two antigens.

2.1. Development of Peroxidase with Two Different Substrates

This approach was first used as long ago as 1968 by Nakane, at the time when immunoperoxidase techniques were in their infancy, to label different cell populations in the rat pituitary.[1] Nakane's achievement was all the more noteworthy for the fact that he succeeded in demonstrating not merely two but three different antigens simultaneously (growth hormone, thyrotropic hormone and luteinizing hormone). His technique involved first incubating sections sequentially with rabbit anti-growth hormone, peroxidase-conjugated sheep anti-rabbit Ig and then diaminobenzidine and H_2O_2 (yielding a brown reaction product). The section was then incubated in a low pH buffer in order to elute the first two antibodies, since without this step there was a risk that, firstly, the primary rabbit anti-growth hormone antibody would be recognised subsequently by the anti-rabbit Ig in the second immunoperoxidase sandwich, and, secondly, that residual peroxidase conjugate might participate in the substrate reaction for the second immunoperoxidase reaction and thus cause spurious double labelling.

Following the acid elution step a second indirect immunoperoxidase sandwich procedure was applied, using rabbit anti-thyrotropic hormone as primary antibody, followed by the same peroxidase conjugate as before and then development with alpha-naphthol pyronin yielding a pink-coloured reaction product. A further acid elution step was used to remove these antibodies and the third and final immunoperoxidase sandwich applied using rabbit anti-luteinizing hormone as the primary antibody and 4-chloro-1-naphthol as the substrate. This resulted in cells containing luteinizing hormone being labelled with a greyish-blue coloured reaction product.

Nearly a decade passed before further reports appeared of multiple antigen labelling by Nakane's method. Antigens labelled in this way included gastrin and somatostatin in human pancreas,[2] growth hormone and prolactin in rat and human pituitary,[3] kappa and lambda light chains in Reed–Sternberg cells,[4] ACTH and luteinizing hormone in guinea-pig pituitary,[5] hypothalamic peptides in rat[6,7] and neurophysins in frog and mammalian hypothalamus.[8]

Although these investigators all used labelling procedures which were in principle identical to that of Nakane, a number of minor modifications were introduced. Tramu et al.[5] reported that elution with acidic buffer did not remove all traces of the first antibody sandwich and recommended the use of potassium permanganate oxidation as an alternative and more efficient procedure. However, in the following year Sternberger and Joseph reported that the development of the first antibody label with diaminobenzidine effectively masks antigenic sites on the primary antibody, and also blocks any residual enzyme activity, so that the subsequent elution step, rather than needing to be made more efficient as Tramu et al. had suggested, could be omitted without risk of spurious double labelling.[7] It remains to be determined how generally valid this observation is when different antibodies and tissues are studied. However, Sternberger and Joseph's report is of relevance when considering ways in which pairs of monoclonal antibodies may be labelled by double immunoenzymatic procedures (*see* Section 2.5.4 *below*).

2.2. Use of Two Unrelated Enzymes as Antibody Labels

The risk that reagents in the first immunoperoxidase sandwich will remain on the section and give rise to spurious staining when the second immunoperoxidase sandwich is applied may be avoided by using two unrelated enzymes as antibody labels. Since the late 1960s, Avrameas and his colleagues have explored the use of a variety of different enzymes as alternatives to peroxidase for labelling antibodies.[9] One of these enzymes (glucose oxidase) was used by Campbell and Bhatnagar[10] in conjunction with peroxidase for double immunohistological labelling of luteinizing hormone and growth hormone in rat pituitary. The use of two different enzymes, together with the fact that primary antibodies from different species were used (thereby avoiding the risk that the second stage antibodies would recognize the inappropriate primary antibody), meant that it was not necessary to carry out an elution step between the application of the first and second immunoenzymatic sandwiches.

Shortly after this report an alternative method was developed in the authors' laboratory in which alkaline phosphatase was used in conjunction with peroxidase.[11–13] Apart from establishing these two enzymes as a suitable pair of labels for double staining, this report also demonstrated that both sandwiches could be applied simultaneously without cross-reactivity, a finding which considerably shortens the time required for double labelling.

The same pair of enzymes (peroxidase and alkaline phosphatase) was used by Falini et al.[14] and by Malik and Daymon[15] in studies of Ig chain expression in human lymphoid tissue. The former authors used the highly specific 'labelled antigen' procedure for each immunoenzyme sandwich. A further report of double immunoenzymatic staining with peroxidase and alkaline phosphatase is to be found in the study by Klareskog et al. of HLA-DR expression in human tissue samples.[16]

Recently there has been renewed interest in the use of glucose oxidase as an antibody label,[17] and Clark et al. have reported its use in combination with peroxidase for double labelling.[18]

2.3. Choice of Optimal Double Immunoenzymatic Procedure

In selecting the optimal technique to be used for double immunoenzymatic labelling it is necessary to consider the following factors:

a. Whether to use one enzyme or two.
b. If two enzymes are to be used which second enzyme is optimal.
c. Which method should be used for linking enzyme to antibody.
d. Whether antibodies (or antigens) are available in sufficient quantities for direct conjugation.
e. Whether the two primary antibodies are raised in the same or different species.

2.4. One Enzyme or Two

The major disadvantage of using the same enzyme (i.e. peroxidase) for labelling both antigens is that this inevitably entails performing the procedure as a sequential technique, in which one sandwich has to be applied and developed before the second sandwich can be applied. In consequence the technique will take twice as long as a single immunoenzymatic labelling procedure. In addition, the risk that the second sandwich will reveal residual elements from the first sandwich (*see above*) should be considered, although this risk may have been exaggerated in the past.

When using two different enzymes as antibody labels, it is often possible to apply the reagents of each sandwich simultaneously so that the only extra time required to complete the procedure is that needed for the development of the second enzyme label. However, it should be noted that application of both immunoenzyme sandwiches simultaneously presupposes that the two primary antibodies are raised in different species (*see below under* Section 2.5.3) or alternatively that some means (e.g. hapten labelling or a biotin–avidin system—*see below*) can be used to avoid non-specific cross-reactivity).

2.5. Choice of Enzymes

If two different enzymes are used the question is raised as to which is the most suitable choice of second enzyme (peroxidase being universally employed as the first enzyme). As noted above alkaline phosphatase has been used in the authors' and other laboratories as second enzyme, a choice dictated by the fact that the enzyme is relatively inexpensive, by its stability and by the fact that it gives strong labelling with several different substrates, thus offering a choice of differently coloured reaction products. When revealed with a blue reaction product (using naphthol-AS-MX and Fast Blue), excellent contrast is obtained with the peroxidase diaminobenzidine reaction product (*Plates* 5–11), so that cells showing a mixed reaction product can be clearly visualized. Alternatively, if the enzyme is developed with Fast Red in place of Fast Blue an excellent contrast with haematoxylin is obtained (*see Plates* 8 and 9), although this reaction product does not lend itself as well as the blue product to the demonstration of cells containing two antigens.

As noted above, glucose oxidase has also been used for double immunohistological labelling.[17,18] This enzyme has the theoretical advantage over alkaline phosphatase (and indeed over peroxidase itself) that no endogenous enzyme activity exists in mammalian tissues. However, in practice, endogenous peroxidase activity can either be blocked, e.g. by methanol–H_2O_2 or by periodate oxidation, or else ignored, whilst endogenous alkaline phosphatase activity is relatively easy to inhibit, e.g. with levamisole,[19] acetic acid[20] or periodate oxidation.[21] Hence there is little reason at present for using glucose oxidase or other enzymes in preference to alkaline phosphatase, and the combination of this enzyme with peroxidase seems likely to be the enzyme pair of choice in the future for double immunoenzymatic labelling.

2.5.1. *Methods of Labelling Antibody with Enzyme*

There are two principal methods whereby antibodies may be labelled with enzymes. Firstly, covalent complexes of antibody and enzyme may be prepared (usually employing glutaraldehyde or periodate for conjugation). Alternatively, an unlabelled antibody method may be employed in which a second stage anti-Ig antiserum acts as a bridge between the primary antiserum and a third stage reagent consisting of immune complexes of enzyme and anti-enzyme antibody.

The authors' laboratory relies primarily upon the use of two unlabelled antibody techniques (*Fig.* 7.1), since this method is sensitive and does not involve the

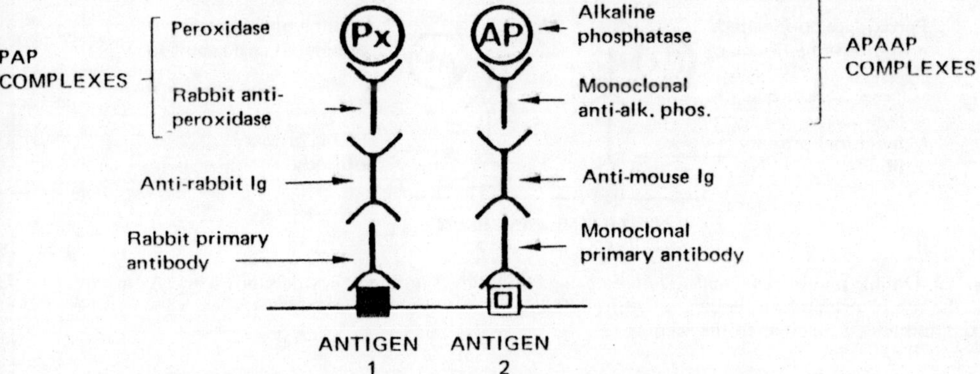

Fig. 7.1. Diagrammatic representation of the way in which two antigens may be labelled simultaneously by applying two unlabelled antibody sandwiches. Since the two primary antibodies are raised in different species the reagents at each level may be mixed and applied simultaneously to the section.

necessity for covalent conjugation of enzyme to antibody. However, we have recently started to use peroxidase-conjugated antibody for double labelling pairs of monoclonal antibodies (*see* Section 2.5.4 *below*).

Recently a number of alternative systems for linking enzymes to antibodies have been introduced. One such technique is based upon the use of avidin and biotin.[22,23] This approach has yet to be exploited on a wide scale but may prove of value when labelling pairs of monoclonal antibodies. A second newly introduced technique, which may also be of value in this context, is based upon the use of hapten-labelled antibodies.[24–26] These procedures are discussed in further detail below (*see* Section 2.5.4).

2.5.2 *Availability of Purified Antibodies or Antigens*

If antibodies can be obtained in sufficient quantity in a purified state the possibilities for labelling with enzymes are greatly enhanced. In particular, direct conjugation with enzymes, haptens or with biotin becomes possible and may therefore facilitate double staining when using two antibodies raised in the same species. When using commercial polyclonal antisera it is often possible to obtain sufficient antibody for such conjugation. Monoclonal antibodies on the other hand are usually obtainable in only small amounts from commercial sources. In these circumstances double labelling is only feasible by using either subclass specific techniques (*see below*) or sequential labelling procedures (*see below*).

As noted above, Falini et al.[14] reported double labelling by using two labelled antigen sandwiches. However, this procedure is only feasible if antigen is available in substantial amounts. Since many antigens detected by monoclonal antibodies have yet to be isolated, let alone purified in substantial yield, this highly specific technique is of limited application.

2.5.3. *Species in which Primary Antibodies are Raised*

When embarking on double immunoenzymatic labelling it is important to consider whether the two primary antibodies are raised in the same or in different species. In

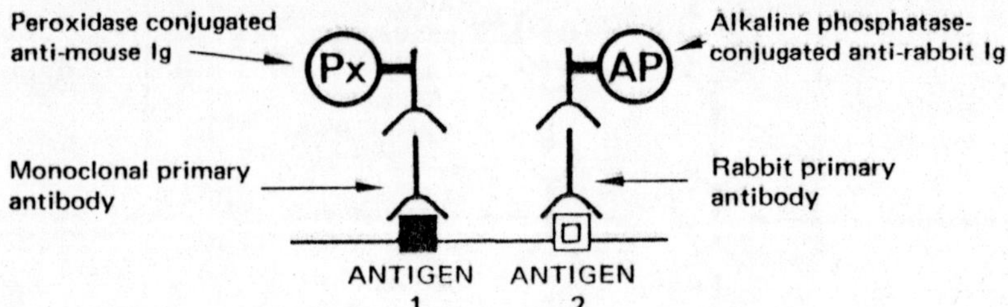

Fig. 7.2. Double immunoenzymatic staining using two indirect immunoenzyme sandwiches. As in *Fig*. 7.1, the use of primary antibodies raised in different species enables the reagents at each level to be mixed and added together to the section.

the latter instance double labelling is relatively easily achieved, since two different anti-Ig reagents, specific for the appropriate species, can be used as second stage reagents (*Figs*. 7.1 and 7.2). If both primary antibodies are raised in the same species, however, the question arises as to how the second stage reagents can distinguish between the two primary antibodies. A variety of techniques are available for overcoming this problem and they are discussed briefly below. With the increasing use of monoclonal antibodies the development of such systems will become steadily more important.

2.5.4. *Double Labelling of Two Primary Antibodies Raised in the Same Species*

a. USE OF DIRECTLY LABELLED PRIMARY ANTIBODIES. The disadvantage of this approach (*Fig*. 7.3) is that for each antigen it is necessary to isolate and directly conjugate the antibody with an enzyme. Furthermore, direct immunoenzymatic techniques are inherently less sensitive than indirect procedures and labelling of the antibody with enzyme may result in loss of antibody activity.

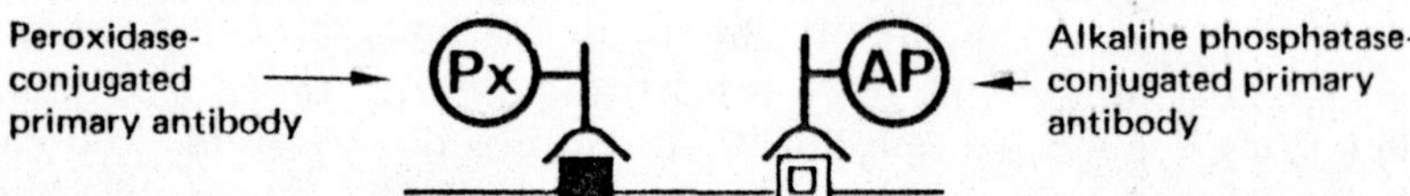

Fig. 7.3. Double immunoenzymatic labelling using two enzyme-conjugated primary antibodies. As noted in the text, this avoids the problem encountered when the two primary antibodies are raised in the same species. However, it is frequently impracticable to directly conjugate two antibodies in this way.

b. DOUBLE LABELLED ANTIGEN TECHNIQUE. This procedure (*Fig*. 7.4) has been used to label kappa and lambda light chains and haemoglobin and albumin in tissue sections[14] but suffers, as noted above, from the limitation that it is only applicable to antigens available in substantial quantity.

c. SEQUENTIAL APPLICATION OF TWO IMMUNOENZYMATIC SANDWICHES. This technique (*Fig*. 7.5) has already been discussed above and its potential drawbacks outlined. However it has the advantage that it is relatively easy to perform and

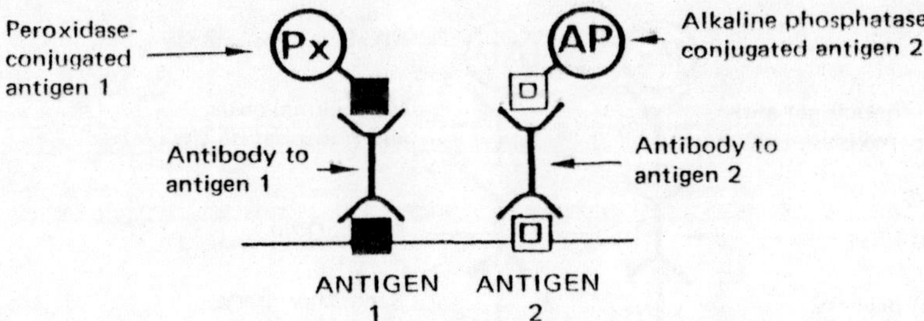

Fig. 7.4. Double immunoenzymatic labelling using two labelled antigen sandwiches.[14]

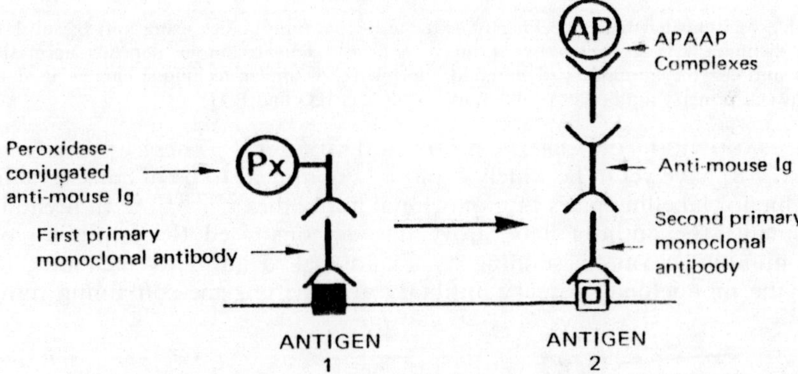

Fig. 7.5. Double immunoenzymatic labelling of monoclonal antibodies. Since the second stage reagent in each sandwich is an antibody against mouse Ig it is necessary to perform the two labelling procedures sequentially. The potential risk that the bridging anti-mouse Ig in the second sandwich may detect the primary antibody in the first sandwich is reduced by the development of the peroxidase reaction which tends to mask antigenic sites on the first monoclonal antibody.[7]

the authors' laboratory has used it to label a variety of pairs of monoclonal antibodies. This method is of particular use when only limited amounts of monoclonal antibodies are available.

d. USE OF CLASS OR SUBCLASS-SPECIFIC SECOND ANTIBODIES. This approach (*Fig.* 7.6) is of particular relevance in the context of monoclonal antibodies, since these reagents are always of a single class or subclass.[27] It has the disadvantage, however, that the appropriate second antibodies (e.g. anti-mouse IgG1, anti-mouse IgG2a etc.) are not widely available and are relatively expensive. Further problems include the fact that such antibodies may not be truly specific and that the number of subclass-specific antigenic sites on the primary antibody may be few in number, with the consequence that only weak staining can be achieved. It is also not possible to perform double labelling by this method if the two primary antibodies are of the same subclass. Since many monoclonal antibodies are of the IgG1 subclass this represents an important potential limitation.

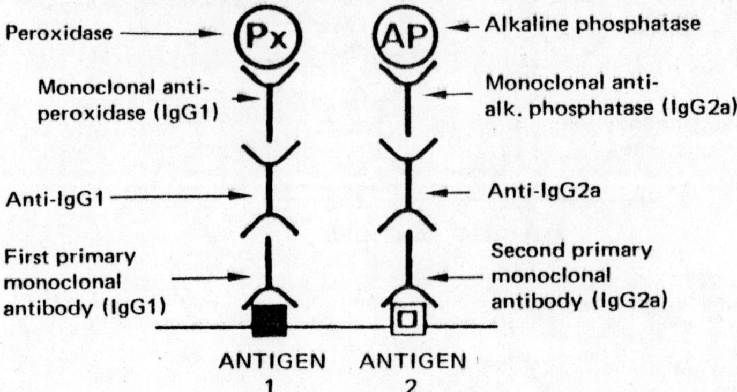

Fig. 7.6. Double immunoenzymatic labelling of two monoclonal antibodies using anti-Ig subclass antibodies to distinguish between the two primary reagents. This technique depends upon the availability of anti-enzyme antibodies of different subclasses. A similar technique can be used to distinguish between primary antibodies of different classes (i.e. IgG and IgM).

e. HAPTEN SANDWICH OR BIOTIN–AVIDIN TECHNIQUES. These procedures (*Figs*. 7.7 and 7.8) have yet to be widely explored but appear to be of considerable potential value for labelling pairs of monoclonal antibodies.[22,23,26,28] In preliminary experiments the authors' laboratory has demonstrated the feasibility of performing immunoenzymatic staining by an unlabelled antibody technique in which both the monoclonal primary antibody and the enzyme-containing third

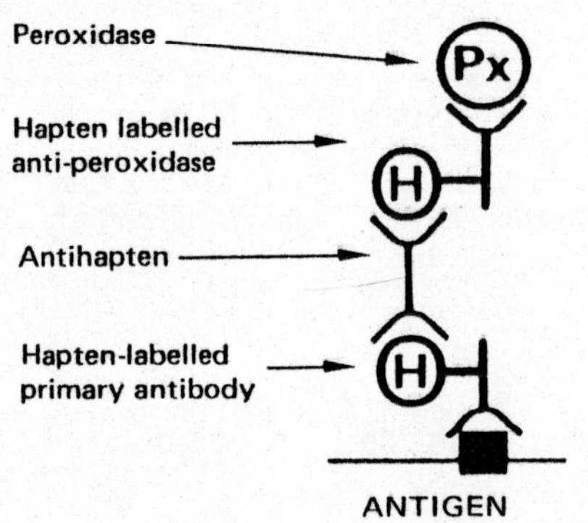

Fig. 7.7. Immunoenzymatic staining of a monoclonal antibody using a hapten-sandwich procedure. With the techniques illustrated in *Figs*. 7.3–7.6 and in *Fig*. 7.8, this procedure allows two primary antibodies raised in the same species to be labelled.

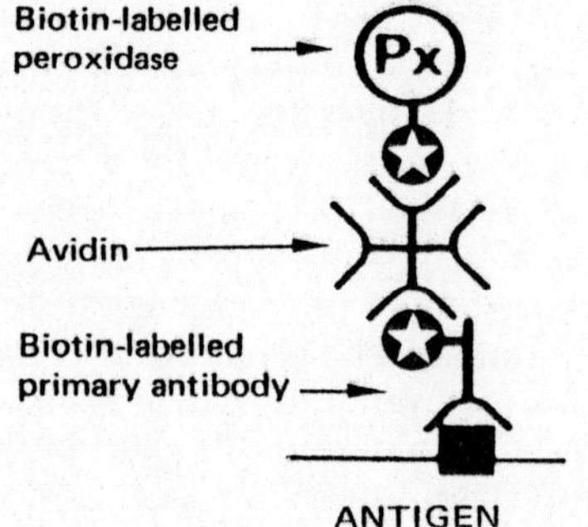

Fig. 7.8. Example of a biotin–avidin labelling system. Use of biotinylated primary antibodies may provide an alternative solution to the problem of labelling two antibodies raised in the same species (*see text* and *Figs*. 7.3–7.7).

stage are haptenated with the same hapten (arsanilic acid), the two stages being linked by anti-hapten bridging antibody. Similar results have been reported by other authors using dinitrophenyl-labelled heterologous antisera.[24, 25] In order to achieve double immunoenzymatic labelling a hapten sandwich of this sort may be combined with a second hapten sandwich, based upon a non-cross-reacting hapten such as *p*-aminobenzoylglycine or *p*-aminobenzoylglutamic acid.[28] Alternatively, a biotin–avidin system may be used in conjunction with a hapten sandwich.[29]

3. COMPARATIVE PROPERTIES OF DOUBLE IMMUNOENZYMATIC AND DOUBLE IMMUNOFLUORESCENT LABELLING PROCEDURES

Although a few non-fluorescent labels have been used in the past for the purposes of double labelling tissue antigens (e.g. radioisotopes or colloidal gold) fluorochromes have been most widely used as alternatives to enzymes. It is therefore pertinent to outline briefly the relative advantages and disadvantages of these two techniques (*Table* 7.1).

The most important aspect of this comparison is that double immunoenzymatic staining is particularly suitable for detecting a minor population of cells in a tissue

Table 7.1. ***Comparison between immunoenzymatic and immunofluorescent double labelling techniques***

Technique	*Advantages*	*Disadvantages*
Double immunofluorescent labelling	Relatively rapid to perform Each label can be visualized without interference from the other, making it possible to detect small amounts of one antigen even in the presence of large amounts of the other	Fluorescent labels fade on excitation and storage When examining double labelled preparations each field must be studied in turn with two different filter systems, making it tedious to pick out small numbers of cells which differ in their labelling pattern from that of the majority of cells Requires fluorescent microscope
Double immunoenzymatic labelling	Labels are permanent and can be visualized using a standard microscope There is no loss of the label intensity when examining sections at low magnifications, and labels may easily be visualized without a microscope Both labels are visualized simultaneously, so that minor cell populations, differing in their reactions from the major populations, can rapidly be picked out	Labelling takes longer to perform than immunofluorescent staining One label may mask the presence of trace amounts of the other label

section (or cell smear) which differs in its reactivity from the reactions of the major populations. An example in this context is provided by Reed–Sternberg cells in cases of Hodgkin's disease. By double staining for kappa and lambda chains (*Plate* 10) it is readily possible to identify these cells since their mixed reaction pattern differs strikingly from the single labelling reactions of other Ig-containing cells (plasma cells). Even when such cells are present at low frequency (for example in Hodgkin's disease of the nodular paragranuloma type) or are of the small mononuclear variety they can rapidly be identified by double immunoenzymatic labelling because of the clarity with which their labelling pattern contrasts with that of plasma cells. Double immunofluorescent staining on the other hand would be a laborious means of identifying such cells since it would necessitate examining large numbers of cells alternately with two different filter systems.

When it is necessary, however, to assess whether small amounts of one antigen are present in a cell together with larger amounts of a second antigen, double immunofluorescent labelling is clearly superior to double immunoenzymatic procedures. Each fluorescent label is visualized separately using selected filters without interference from the other label. In contrast, when examining sections stained by double immunoenzymatic techniques it may be difficult to be certain whether staining is single or mixed. This is particularly true when examining cell surface antigen staining in cryostat sections in which labelling is limited to a thin rim of staining around individual cells.

Generally speaking the search for mixed cells by double immunoenzymatic methods is most appropriately used when antigens are intracytoplasmic in location (rather than on the surface membrane) and when it is known that substantial amounts of each antigen are present in all positive cells.

From these considerations it will be apparent that double immunoenzymatic labelling of monoclonal antibodies on cryostat sections is particularly suited to analysing pairs of antigens which are already known to be present in different cell populations (e.g. helper and suppressor T-cell antigens, T- and B-cell antigens). When studying non-overlapping populations of this sort, the ability of the immunoenzymatic procedure to reveal both antigens at the same time lends itself particularly well to assessing at a glance the codistribution of two cell populations.

4. PRACTICAL APPLICATIONS OF DOUBLE IMMUNOENZYMATIC LABELLING

The concluding section of this chapter describes some of the ways in which double immunoenzymatic labelling has been applied in the authors' laboratory to the analysis of human tissue biopsies. The majority of this work has been based upon the study of lymphoid tissue samples, initially in paraffin sections and more recently in cryostat sections.

4.1. Double Staining of Kappa and Lambda Light Chains

Histologists concerned with the immunoenzymatic analysis of human lymphoid tissue biopsies frequently place great weight on the results of staining for kappa and lambda light chains, since the detection of a population of cells containing only one type of light chain provides strong evidence for its neoplastic nature. Difficulties of interpretation may arise, however, when positive cells are found in sections stained for each light chain class, and it is then of great value to know if these antigens are present in the same or different cell populations. If the latter is true it indicates that each Ig-positive cell population consists of cells which have synthesized Ig themselves, whereas if cells showing mixed staining are revealed it indicates that the Ig has been acquired from the cells' environment (since it is rare, if not unknown, for both classes of light chain to be synthesized by the same cell).

Examples of the results of staining human lymphoid tissue in this way are illustrated in *Plates* 5 and 10. One of the findings to emerge from such studies was the frequency with which cells in routine paraffin-embedded biopsies of human lymphoid tissue contain serum immunoglobulin which has been absorbed from their environment.[12,30,31] This artefact, which probably takes place during the processing of the tissue, has been partly overlooked in the past (principally because of the fact that cells of this sort are often scattered against a negative background) and may account for reports of lymphoma expressing cytoplasmic Ig of both kappa and lambda light chain type. As illustrated in *Plate* 11, double labelling makes genuine endogenous Ig-producing plasma cells stand out dramatically against a background of non-specifically labelled cells.

In addition to cells which contain both kappa and lambda chains as a result of passive diffusion of serum IgG into their cytoplasm it is also possible to identify cells which appear to have selectively taken up IgG.[12,31] These cells, of which Reed–Sternberg cells and the giant cells sometimes found in cases of non-Hodgkin's lymphoma represent the most obvious examples, may be distinguished from cells which have acquired their Ig as a result of passive diffusion by staining them for serum albumin, since this constituent (along with other serum proteins) is absent from Reed–Sternberg cells.[31]

4.2. Double Staining for Ig and J Chain

In a recent study of a series of biopsies of reactive lymphoid tissue by double staining for Ig and J chain,[32] a population of cells expressing the latter constituent alone was detected (*Plate* 7). Although Ig-positive J chain-negative cells are frequently encountered in reactive lymphoid tissue, there have been no previous reports of cells in human lymphoid tissue biopsies showing the reverse pattern. The explanation of this dissociation between the expression of Ig and J chain is not known, but its discovery led to a search for lymphoid neoplasms showing the same pattern of reactivity. As a result of screening more than sixty high grade lymphoma biopsies three cases of this sort were identified.[32]

4.3. Staining of Cryostat Sections with Monoclonal Antibodies

In the past two years the authors' laboratory has made increasing use of cryostat sections (in preference to sections of paraffin-embedded tissue), since many antigens are preserved in frozen sections which do not survive paraffin embedding. Greater use of cryostat material has been stimulated by the availability of an increasing range of monoclonal antibodies, many of which detect such labile antigens.

For example, *Plate* 6 shows a lymph node containing metastatic carcinoma double stained for HLA-DR and for an epithelial antigen. This allowed the metastatic deposits to be visualized simultaneously with the lymphoid tissue areas, and also revealed the fact that a proportion of carcinoma cells expressed the HLA-DR antigen. Until recently HLA-DR has been considered to be restricted principally to cells of the lymphoreticular system. It is now realized, however, that a wide range of non-lymphoid cells may express this antigen, both in the normal state and following neoplastic transformation. Double labelling of the sort illustrated in *Plate* 6 is clearly of value in analysing this phenomenon.

In another case, a cryostat section of a fibrosarcoma was double stained for HLA-DR and for a cytokeratin antigen (detected by antibody LE61). Single staining for each of these antigens had revealed numerous positively staining, elongated or spindle-shaped cells. This raised the question of whether these two antigens were expressed on the same or different cells, the similarity in morphological appearances suggesting that the former possibility would be true. Double labelling, however, revealed that these antigens were present in different cell populations. These findings were interpreted as indicating that the neoplastic cells (i.e. fibroblasts) were HLA-DR negative, and that the HLA-DR positive cells were 'dendritic' cells of the sort which are found ubiquitously in all human tissues.

5. CONCLUSIONS

The preceding review of double immunoenzymatic labelling techniques summarizes the current state of this art. However, the subject is evolving rapidly and the availability of an apparently inexhaustible supply of new monoclonal antibodies means that the need for such labelling methods becomes increasingly pressing. Exploitation of new techniques such as hapten sandwich labelling (*see above*) should lead to simplification of these techniques and hence to their wider application.

Appendix

MATERIALS

Tris-Buffered Saline (TBS)

Stock solution. 0·5 M Tris pH 7·6 [6·05 g Trizma (Sigma) base in 100 ml distilled water]. Adjust to pH 7·6 with 0·5 M hydrochloric acid. For use dilute Tris 1 : 10 with 0·9 per cent saline.

Enzyme Substrates

Alkaline phosphatase substrate. Dissolve 2 mg naphthol AS-MX phosphate in 0·2 ml dimethylformamide in a glass tube. Add 9·8 ml 0·1 M Tris (pH 8·2). This buffer is stable at 4 °C for several weeks.

Peroxidase substrate. Diaminobenzidine tetrahydrochloride is dissolved in TBS at 0·6 mg/ml. This solution may be stored for several weeks at 4 °C. Hydrogen peroxide is added just before use (*see below*).

Immunoenzyme Reagents

Peroxidase–anti-peroxidase (PAP) and peroxidase-conjugated anti-mouse Ig. These reagents may be obtained commercially (Dakopatts).

Alkaline phosphatase–anti-alkaline phosphatase (APAAP) complexes. This reagent is not at present commercially available. It has been prepared in the authors' laboratory by adding calf intestinal alkaline phosphatase to tissue culture supernatant containing monoclonal mouse anti-alkaline phosphatase.

METHODS

Double Labelling using Primary Antibodies Raised in Different Species

The technique outlined below to illustrate this type of double immunoenzymatic labelling is the method used for staining kappa and lambda light chains in formol sublimate-fixed paraffin-embedded human tonsil. The dilutions are those which are found to be optimal in the authors' laboratory.

i. *Dewaxing*
   Xylene 60 s
   Xylene 60 s
   Ethanol 60 s
   Ethanol 60 s
   70 per cent alcohol 60 s
   Wash in tap water

ii. *Removal of mercury pigment*
   0·5 per cent iodine in 70 per cent alcohol for 5 min
   Rinse in tap water
   2·5 per cent sodium thiosulphate in distilled water for 1 min
   Wash in tap water

iii. *Blocking of endogenous peroxidase**
   0·5 per cent H_2O_2 for 15 min
   Wash in tap water
   Rinse in distilled water
   Transfer to Tris-buffered saline (TBS)

iv. *Reduction of non-specific staining**
   Apply normal sheep serum (1/10) for 10 min
   Drain off but do not wash
   Dry slides around sections

v. *First layer antibodies* (monoclonal mouse anti-lambda chains 1/16 + rabbit anti-kappa chains 1/500)
   Allow reagent to remain on section for at least 30 min
   Wash slides in TBS for 5 min
   Dry slides around sections

vi. *Second layer antibodies* (sheep anti-mouse immunoglobulin 1/25 + sheep anti-rabbit immunoglobulin 1/25)†
   Leave this reagent on slides for at least 30 min

* Blocking of endogenous peroxidase and reduction of non-specific staining are not essential steps and may be omitted in many cases without affecting the final quality of the labelling reaction.

† These reagents on occasion contain cross-reactive antibodies (i.e. sheep anti-rabbit Ig may detect determinants on mouse Ig) which can give rise to spurious mixed staining. This problem can be circumvented when necessary by pre-absorbing these antisera with a solid phase immunoabsorbent of immunoglobulin from the appropriate species.

Wash slides in fresh TBS for 2 min
Dry slides around sections

vii. *Third layer enzyme–antibody complexes* (monoclonal mouse APAAP 1/4 + rabbit PAP 1/100)
Leave this reagent on the sections for at least 30 min
Wash slides in fresh TBS for 2 min
Dry slides around section

viii. *Development of peroxidase reaction*
Immediately before staining add sufficient concentrated H_2O_2 to the stock diaminobenzidine solution (*see above*) to produce a final concentration of 0·01 per cent. Remove slides from the TBS in which they have been washed, tip off excess TBS and flood the slides with enzyme substrate. Allow reaction to develop for 7–10 min. The development of the reaction may be followed under the microscope. When staining reactive lymphoid tissue, clusters of plasma cells (recognizable by their eccentrically placed nuclei) will begin to become visible after a few minutes. Terminate the reaction before generalized background staining appears. Wash slides in fresh TBS for 2 min

ix. *Development of alkaline phosphatase reaction*
Immediately before staining dissolve Fast Blue BB salt at 1 mg/ml in naphthol AS-MX phosphate buffer (*see above*). Remove slides from the TBS in which they are being washed, tip off all excess buffer, and pipette the alkaline phosphatase substrate directly onto the slide. If the staining solution appears cloudy at this stage it may be filtered onto the slide. Allow incubation to continue for 10–15 min. The development of the reaction may be observed under the microscope. Plasma cells should become visible at this stage because of their blue reaction product. They will be found in the vicinity of the peroxidase-stained plasma cells. Terminate the reaction by washing in TBS. Wash extensively in tap water

x. *Mounting of slides*
Sections stained as described above are mounted in an aqueous material such as Apathy's mountant

Double Labelling of Two Monoclonal Antibodies

The reaction described below in order to illustrate this type of labelling is the procedure used to stain T- and B-cells in cryostat sections of human tissue.

1. *Fixation*
Air-dried cryostat sections are fixed in acetone at room temperature for 10 min and then air dried
2. *First antibody* (monoclonal mouse anti-T-cell)
Leave this reagent on the sections for at least 30 min
Wash slides in fresh TBS 2 min
Dry slides around section
3. *Immunoperoxidase conjugate* (rabbit anti-mouse Ig 1/50 + normal human serum 1/25)
Leave this reagent on the section for at least 30 min
Wash slides in fresh TBS 2 min
Dry slides around section
4. *Development of peroxidase reaction*
This is carried out as described above
Wash slides in fresh TBS 2 min after completion of reaction
5. *Application of second antibody* (monoclonal anti-B-cells)
Leave this reagent on the sections for at least 30 min
Wash slides in fresh TBS 2 min
Dry slides around section

6. *Bridging antibody* (sheep anti-mouse immunoglobulin 1/25 + normal human serum 1/25)
   Leave this reagent on the slides for at least 30 min
   Wash slides in fresh TBS 2 min
   Dry slides around sections
7. *APAAP complexes*
   Leave this reagent on the section for at least 30 min
   Wash slides in fresh TBS 2 min
   Dry slides around sections
8. *Development of alkaline phosphatase reaction*
   This is carried out as described above
9. *Mounting of slides*
   Sections are mounted as outlined above

REFERENCES

1. Nakane P. K. Simultaneous localization of multiple tissue antigens using the peroxidase-labeled antibody method: a study on pituitary glands of the rat. *J. Histochem. Cytochem.* 1968, **16**, 557–558.
2. Erlandsen S. L., Hegre O. D., Parsons J. A., McEvoy R. C. and Elde R. P. Pancreatic islet cell hormones. Distribution of cell types in the islet and evidence for the presence of somatostatin and gastrin within the D cell. *J. Histochem. Cytochem.* 1976, **24**, 883–897.
3. Martin-Comin J. and Robyn C. A comparative immunoenzymatic localization of prolactin and growth hormone in human and rat pituitaries. *J. Histochem. Cytochem.* 1976, **24**, 1012–1016.
4. Poppema S., Elema J. D. and Halie M. R. The significance of intracytoplasmic proteins in Reed Sternberg cells. *Cancer* 1978, **42**, 1793–1803.
5. Tramu G., Pillez A. and Leonardelli J. An efficient method of antibody elution for the successful simultaneous localization of two antigens by immunocytochemistry. *J. Histochem. Cytochem.* 1978, **26**, 322–324.
6. Joseph F. A. and Sternberger L. A. The unlabeled antibody method. Contrasting color staining of β-lipotropin and ACTH-associated hypothalamic peptides without antibody removal. *J. Histochem. Cytochem.* 1979, **27**, 1430–1437.
7. Sternberger L. A. and Joseph F. A. The unlabeled antibody method. Contrasting color staining of paired pituitary hormones without antibody removal. *J. Histochem. Cytochem.* 1979, **27**, 1424–1429.
8. Vandesande F. and Dierickx K. Immunocytochemical localization of somatostatin-containing neurons in the brain of *Rana temporaria*. *Cell Tissue Res.* 1980, **205**, 43–53.
9. Avrameas S. Enzyme markers: their linkage with proteins and use in immuno-histochemistry. *Histochem. J.* 1972, **4**, 321–330.
10. Campbell G. T. and Bhatnagar A. S. Simultaneous visualization by light microscopy of two pituitary hormones in a single tissue section using a combination of indirect immunohistochemical methods. *J. Histochem. Cytochem.* 1976, **24**, 448–452.
11. Mason D. Y. and Sammons R. E. Alkaline phosphatase and peroxidase for double immunoenzymatic labelling of cellular constituents. *J. Clin. Pathol.* 1978, **31**, 454–462.
12. Mason D. Y., Stein H., Naiem M. and Abdulaziz Z. Immunohistological analysis of human lyphoid tissue by double immunoenzymatic labelling. *J. Cancer Res. Clin. Oncol.* 1981, **101**, 13–22.
13. Mason D. Y. and Woolston R.-E. Double immunoenzymatic labelling. In: Bullock G. and Petrusz P., ed., *Techniques in Immunocytochemistry*, Vol. 1. London, Academic Press, 1982: 135–152.
14. Falini B., deSolas I., Halverson C., Parker J. W. and Taylor C. R. Double labeled-antigen method for demonstration of intracellular antigens in paraffin-embedded tissues. *J. Histochem. Cytochem.* 1982, **30**, 21–26.
15. Malik N. J. and Daymon N. E. Improved double immunoenzyme labelling using alkaline phosphatase and horseradish peroxidase. *J. Clin. Pathol.* 1982, in press.
16. Klareskog L., Forsum U., Wigren A. and Wigzell H. Spatial relationships between HLA-DR expressing cells and T-lymphocytes of different subsets in the rheumatoid sinovial tissue. *Scand. J. Immunol.* 1982, **15**, 501–507.

17. Rathlev T., Hocko J. M., Franks G. F., Suffin S. C., O'Donnell C. M. and Porter D. D. Glucose oxidase immunoenzyme methodology as a substitute for fluorescence microscopy in the clinical laboratory. *Clin. Chem.* 1981, **27**, 1513–1515.
18. Clark C. A., Downs E. C. and Primus F. J. An unlabelled antibody method using glucose oxidase–antiglucose oxidase complexes (GAG): a sensitive alternative to immunoperoxidase for the detection of tissue antigens. *J. Histochem. Cytochem.* 1982, **30**, 27–34.
19. Ponder B. A. and Wilkinson M. M. Inhibition of endogenous tissue alkaline phosphatase with the use of alkaline phosphatase conjugates in immunohistochemistry. *J. Histochem. Cytochem.* 1981, **29**, 981–984.
20. Dearnaley D. P., Sloane J. P., Ormerod M. G., Steele K., Coombes R. C., Clink H. McD., Powles T. J., Ford H. T., Gazet J.-C. and Neville A. M. Increased detetection of mammary carcinoma cells in marrow smears using antisera to epithelial membrane antigen. *Br. J. Cancer* 1981, **44**, 85–90.
21. Bulman A. S. and Heyderman E. Alkaline phosphatase for immunocytochemical labelling: problems with endogenous enzyme activity. *J. Clin. Pathol.* 1981, **34**, 1349–1351.
22. Guesdon J-L., Ternynck T. and Avrameas S. The use of avidin–biotin interaction in immunoenzymatic techniques. *J. Histochem. Cytochem.* 1979, **27**, 1131–1139.
23. Warnke R. and Levy R. Detection of T- and B-cell antigens with hybridoma monoclonal antibodies: a biotin–avidin–horseradish method. *J. Histochem. Cytochem.* 1980, **28**, 771–776.
24. Farr A. G. and Nakane P. K. Use of anti-hapten in immunohistochemistry and immunoassay. *J. Histochem. Cytochem.* 1981 **29**, 891.
25. Jasani B., Wynford-Thomas D. and Williams E. D. Use of monoclonal anti-hapten antibodies for immunolocalisation of tissue antigens. *J. Clin. Pathol.* 1981, **34**, 1000–1002.
26. Wofsy L., Henry C. and Cammisuli S. Hapten-sandwich labelling of cell-surface antigens. *Contemp. Top. Mol. Immunol.* 1978, **7**, 215–237.
27. Tidman N., Janossy G., Bodger M., Granger S., Kung P. C. and Goldstein G. Delineation of human thymocyte differentiation pathways utilising double staining techniques with monoclonal antibodies. *Clin. Exp. Immunol.* 1981, **45**, 457–467.
28. Wallace E. F. and Wofsy L. Hapten-sandwich labelling. IV. Improved procedures and non-cross-reacting hapten reagents for double-labelling cell surface antigens. *J. Immunol. Methods* 1979, **25**, 283–289.
29. Caligaris-Cappio F., Gobbi M., Bofill M. and Janossy G. Infrequent normal B lymphocytes express features of B-chronic lymphocytic leukaemia. *J. Exp. Med.* 1982, **155**, 623–628.
30. Coruh G. and Mason D. Y. Serum proteins in human squamous epithelium. *Br. J. Dermatol.* 1980, **102**, 497–505.
31. Mason D. Y., Bell J. I., Christensson B. and Biberfeld, P. An immunohistological study of human lymphoma. *Clin. Exp. Immunol.* 1980, **40**, 235–248.
32. Mason D. Y. and Stein H. Reactive and neoplastic human lymphoid cells producing J chain in the absence of immunoglobulin: Evidence for the existence of J chain disease? *Clin. Exp. Immunol.* 1981, **46**, 305–312.

8 Lectin Histochemistry

B. A. J. Ponder*

1. INTRODUCTION

Lectins are sugar-binding proteins or glycoproteins of non-immune origin.[1] The basis of their use as reagents in pathology and cell biology is their ability to recognize complex carbohydrate structures in glycoproteins and glycopeptides, in particular those of the cell membrane. Because lectins are of non-immune origin, it is incorrect to speak of 'lectin immunocytochemistry'. I shall use instead 'lectin histochemistry'. This chapter will illustrate some of the ways in which lectin histochemistry may provide an alternative to immunocytochemistry with antibodies.

Many hundreds of lectins of plant and animal origin have been described, and the list continues to increase. A recent comprehensive review is provided by Goldstein and Hayes.[2] Some of the lectins more commonly used in biological research (and which are commercially available) are shown in *Table* 8.1. The specificity of a lectin is generally described in terms of binding to free monosaccharide, as in this Table, but it is important to realize that this is an oversimplification. Lectins with a nominally identical monosaccharide specificity have the ability to recognize fine differences in more complex structures. A detailed analysis of the specificity of twelve lectins, which illustrates this point, has been made by Debray et al.[3] Translated to histochemistry, this means that lectins with identical monosaccharide specificity may give quite different staining patterns on tissue sections, which are related to the distribution of more complex oligosaccharides or glycopeptides.

There is increasing evidence that the carbohydrate components of cell-membrane molecules are important in defining cell types, and that changes in these

Table 8.1. Some commonly used lectins

Lectin	*Source*	*Sugar specificity*	$\varepsilon_{280}^{0.1\%}$
Abrus precatorius haemagglutinin	Jequirity bean	β-D-Gal	1·56
Bandeiraea simplicifolia lectin	*Bandeiraea simplicifolia*	α-D-Gal >α-D-Gal *N*Ac	1·40
Concanavalin agglutinin	Jack bean	α-D-man >α-D-Glc	1·20
Dolichos biflorus agglutinin	Horse gram	α-D-Gal *N*Ac	1·38
Lens culinaris agglutinin	Lentil	α-D-Man >α-D-Glc	1·25
Lotus tetragonolobus agglutinin	Asparagus pea	α-L-Fuc	1·78
Peanut agglutinin	*Arachis hypogaea* (peanut)	β-D-Gal β-D-Gal (1→3)-D-Gal *N*Ac	0·89
Phaseolus lunatus limensis agglutinin	Lima bean	α-D-Gal *N*Ac	1·16
Ricinus communis agglutinin I	Castor bean	β-D-Gal >α-D-Gal	1·17
Soybean agglutinin	*Glycine max* (soybean)	α-D-Gal *N*Ac >β-D--Gal *N*Ac	1·28
Ulex europaeus agglutinin	Gorse	α-L-Fuc	1·30
Wheat germ agglutinin	*Triticum vulgaris* (wheat germ)	$(\beta\text{-D-Glc }N\text{Ac}(1\rightarrow4)_3)>$ $(\beta\text{-D-Glc }N\text{Ac}(1\rightarrow4)_2)$	1·46

Data from Goldstein and Hayes[2] and from BDH-Vector Product Information Booklet.

components are associated with cellular differentiation, maturation and neoplastic transformation.[4–8] Lectins therefore provide a range of readily available and well-defined potential reagents for the identification of cell types and for the study of these processes. Much of the work in this area so far has involved the study of cells in suspension, in particular subsets of lymphocytes,[9–12] but the examples in this chapter will show that lectins are also simple and convenient reagents for use on tissue sections.

The finding that a particular lectin will mark a particular phenotype is almost always an empirical one, made in practice by screening the tissues of interest against a panel of different lectins. It is empirical because sufficiently detailed knowledge of the cellular glycoproteins or glycolipids is rarely available. In some special cases, the lectin-binding characteristics of the cells can be predicted. One example is the study of blood group substances. It is already known that certain lectins exhibit blood-group specificities (for example, *Dolichos biflorus* agglutinin has anti-A_1 specificity, and *Ulex europaeus* anti-O). In general, however, no detailed knowledge of the carbohydrate determinants recognized by the lectin, or of the nature or function of the molecules which bear them, is available. The position is similar to that with monoclonal antibodies where the antigen has not been identified. Lectins are nevertheless perfectly valid reagents as markers of cellular phenotype provided two points are borne in mind: first, that the determinants recognized by the lectin may be common to several different glycoproteins or glycolipids and second, that without more detailed information one will not usually know precisely how changes in the lectin-binding characteristics of cells are related to observed differences in cell behaviour. As with monoclonal antibodies, it may in some cases be worthwhile to use the lectin to purify the molecules to which it binds, in order to characterize them more fully.

2. HISTOCHEMICAL APPLICATION OF LECTINS

Lectins have potential uses as histochemical markers of different cell types, of differentiation and maturation, and of neoplastic change. Examples of these uses will be taken from our own unpublished work and from the literature. A new application of lectins, to mark cellular genotypes in tissue sections of genetically mosaic (chimeric) mice, will also be presented to illustrate the process of empirical screening of lectins to find reagents to tackle a particular problem. Technical details of the preparation of lectin conjugates and of staining methods are given in the Appendix.

2.1. Lectins as Markers of Differentiation

Numerous studies[9–12] have shown that murine and human lymphocytes can be fractionated into populations with different properties by binding to different lectins, using either agglutination or lectin bound to a solid support. Peanut (*Arachis hypogaea*) agglutinin (PNA), for example, has been reported to be a marker of immature T-lymphocytes in the cortex of the mouse thymus.[9] A smaller proportion of PNA-positive cells was also found in other organs.[13] Studies with peanut lectin–peroxidase conjugate on tissue sections[14] have shown the feasibility of extending these studies of lymphocyte subpopulations to their anatomical localization in different lymphoid tissues. In these studies, Rose et al.[14] found that about 20 per cent of lymphocytes from Peyer's patches in the mouse intestine bound PNA and that these PNA-positive lymphocytes were localized to the germinal centre of the Peyer's patches. The origin of the lymphoid cells in the germinal centres is still unclear. The demonstration that PNA binds to cells in this anatomical site suggests its use as a reagent to separate these cells so that their properties can be studied further. The finding of PNA positivity in both cortical thymocytes and germinal centre cells incidentally illustrates the difficulty of interpretation of lectin binding in biological terms. It is not clear whether the positivity at both sites merely reflects the coincidental presence of similar sugar moieties, or whether it indicates a common biological characteristic, such as immaturity, in the two populations.[14]

In breast epithelium, PNA has been reported to be a marker to discriminate between epithelial and myoepithelial lineages arising from rat mammary stem cells in culture and in tissue sections.[15]

2.2. Lectins as Markers of Cell Maturation

Maturation in an epithelium occurs as the progeny of the basal cells acquire the characteristics of the fully mature or differentiated cell type. This process may be accompanied by changes in cell membrane glycoproteins or glycolipids which can be recognized by alterations in the binding of specific lectins. For example, three different lectins from *Bandeiraea simplicifolia* and *Ulex europaeus* were found to show specificity for the surface of cells in different layers of the epidermis in the new-born rat.[16] Expression of binding sites for peanut agglutinin (PNA) was found to alter with maturation in two cell lineages in mouse tissues: spermatocytes and squamous epithelium of the oesophagus.[17] In normal human bladder epithelium, binding sites for concanavalin A are predominantly located in the plasma membranes of the basal cells.[18] A marker of cells of different stages of maturation might seem to be potentially of most interest to investigate the pathogenesis of abnormal states of the epithelium, such as neoplasia. A caution should be entered here on the lines of that made in Section 1: that incidental changes in cell-surface oligosaccharide groupings may accompany the induction of the abnormal state and these may alter the pattern of lectin binding which is seen. Without detailed information about the specific molecules which are involved, the interpretation of changes in lectin-binding patterns in this situation is unlikely to be secure.

2.3. Lectins as Markers of Specific Cell Type

Lectins may sometimes be useful as markers of a particular cell type or structure that would otherwise be difficult to distinguish in a tissue section. Thus the B4 lectin from *Bandeiraea simplicifolia*, which has specificity for α-D-galactopyranosyl groups, binds to basement membranes of a variety of mouse and rabbit tissues, in particular the basement membranes of capillaries.[19] In our screening of mouse tissues for polymorphic differences in lectin binding between strains, described below, we have found that *Dolichos biflorus* agglutinin (DBA) binds to the cell membrane of capillary endothelial cells in a variety of tissues of RIII and DDK mice but not to other components of these tissues (*Fig*. 8.1*a*). Electron microscopy (*Fig*. 8.1*b*) confirms that the lectin is binding to the surface of the endothelial cells. In both these instances, the lectin–peroxidase conjugates provide histochemical stains for capillary blood vessels. Of possibly greater interest is our preliminary finding that in certain tissues of RIII mice the vascular endothelium is consistently DBA negative. This demonstrates that, in this respect at least, there are tissue-specific differences in vascular endothelium. The nature of these differences and their possible significance has yet to be explored.

Studies with DBA in other mouse strains have shown a restricted distribution of binding sites on tissue sections, including collecting tubules of the kidney, bile and pancreatic duct epithelium, spermatozoa and oocytes, and gastrointestinal epithelial cells and secreted mucin.[17] A study of lectins as histological markers for tissues within the human kidney showed a consistent and restricted binding pattern for PNA and DBA to distal tubules and collecting ducts, for *Ulex europaeus* agglutinin (UEA) to vascular endothelium (independent of the blood group of the tissue donor), and predominant localization of the binding of wheat germ agglutinin to glomeruli.[20] A panel of seven lectins has also been used to demonstrate different components of human skeletal muscle in tissue sections.[21] Alterations in the patterns of staining may be of value in the investigation of muscle disorders.

Lectins used in this way are valuable as histological markers in relation to a particular problem. When such a problem arises, it may well be worth the relatively small effort involved to screen a panel of lectins against the appropriate tissue sections. Several commercial kits of conjugates of a set of representative lectins are available for this purpose.

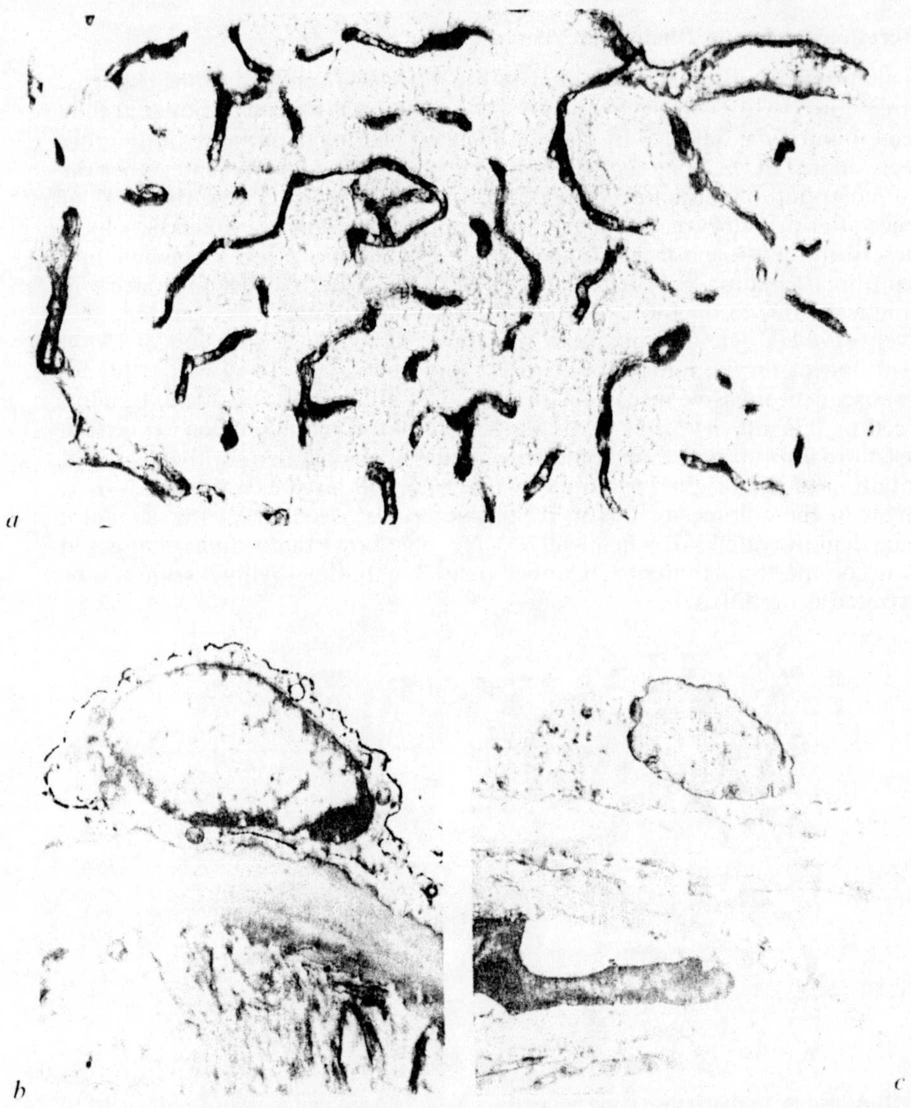

Fig. 8.1. *a.* Demonstration of capillaries in salivary gland of RIII mouse, using *Dolichos biflorus* agglutinin (DBA) peroxidase conjugate. 4-μm paraffin section of tissue fixed in 10 per cent formol–saline. Uncounterstained. For method, *see* the Appendix.

b, c. Localization of DBA-binding sites to surface of endothelial cells in RIII mouse aorta. Magnification: (*b*) × 7800, (*c*) sugar control × 4500. 3-mm square pieces of aorta were incubated for 1 h in DBA–peroxidase in PBS with or without 0·1 M *N*-acetylgalactosamine, washed thoroughly in PBS, postfixed by immersion in 1 per cent OsO_4 (phosphate-buffered) for 2 h at room temperature, dehydrated in ethanol and embedded via propylene oxide in Epon/Araldite resin. 2-μm sections were cut with a glass knife, stained with toluidine blue for light microscopy, and selected areas cut on the diamond knife for electron microscopy. The sections were examined without further staining at 60 kV in a Phillips EM 400 electron microscope. (Method and pictures provided by Dr P. A. Monaghan, Ludwig Institute for Cancer Research (London Branch)).

2.4. Alterations in Lectin Binding in Neoplasia

Neoplastic transformation is accompanied by a variety of cell membrane changes. Many are known to involve glycoproteins[7,8,22,23] and it is not surprising that they have been found to be reflected by changes in lectin binding. A number of tumours have been studied in the hope that changes in lectin binding will indicate either the invasive potential of tumours (superficial bladder cancer),[24] or the risk of recurrence after primary treatment (breast cancer).[25] In addition, differences have been described in the pattern of binding of PNA conjugates to benign and malignant breast lesions.[15,26] It remains to be shown whether this approach will have practical value in diagnosis or prognosis. Bladder cancer has been the most extensively studied. The aim has been to show that loss of expression of blood group substances on the initial non-invasive tumours is correlated with a higher risk of subsequent invasive disease. Although most studies have shown a trend in this direction, it is not clear that the assessment of blood group antigen expression provides more information than could be gained from assessment of the degree of differentiation of the original tumours by conventional histology alone.

Changes in the colonic mucins in dysplastic crypts associated with neoplasia have been demonstrated histochemically.[27] *Fig.* 8.2 shows that similar changes in dysplastic colonic epithelium of CBA mice treated with dimethylhydrazine can be demonstrated using DBA.

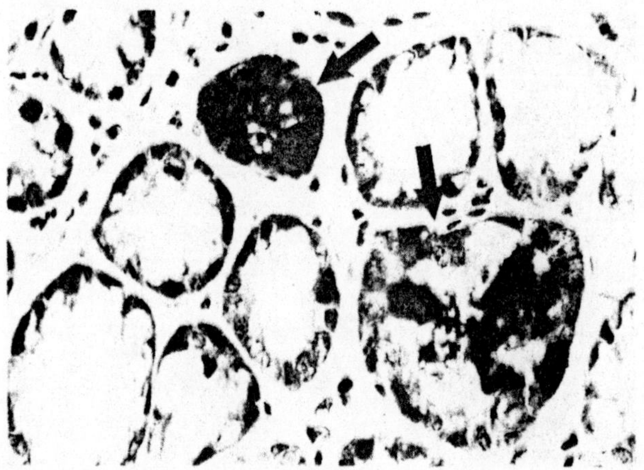

Fig. 8.2. DBA binding to dysplastic crypts in distal colon of CBA/Ca mouse treated with dimethylhydrazine. 4-μm paraffin section, formol–saline fixed tissue, counterstained with haemalum. DBA–peroxidase staining appears black.

2.5. Polymorphism for Lectin Binding Sites Between Inbred Strains of Mice, and its Application to the Study of Mouse Chimeras

Genetically mosaic (or chimeric) mice are made by aggregation of four- to eight-cell embryos of mice of different strains.[28] The aggregated embryos are placed into the uterus of a pseudopregnant 'incubator mother' and there develop normally. The tissues of the resulting mice are a mosaic of cells derived from the component embryos. These mice are of great potential interest for the study of cell lineages, of cellular organization during embryogenesis and in adult tissues, and for studies of the pathogenesis of a variety of abnormal conditions, including neoplasia (for review *see* McLaren).[29] A major problem has been the lack of cellular markers of genotype which could be used to demonstrate the mosaic cell populations in tissue sections. As part of a programme to find such markers, we screened a panel of lectins against a range of tissue sections from mice of ten different inbred strains, looking for polymorphic differences in lectin binding between strains.

We used the lectins shown in *Table* 8.1, conjugated to horseradish peroxidase by the method given in the Appendix. The mouse strains were chosen from those readily available to us to include strains with as different ancestry as possible, and comprised the following: CBA/Ca, C57BL/6, BALB/c, DDK, RIII, HTG, IF, 101H, SEA and GE. The tissues screened included skin, stomach, small intestine, colon, bronchus, liver, kidney, pancreas, salivary gland, uterus, bladder and brain. Screening was done on 4-μm frozen sections which were fixed briefly, after cutting, in 10 per cent formol–saline. Composite blocks were made each containing two tissues, and sections from up to four similar blocks from mice of different strains were mounted on the same slide. We used frozen sections because we had no way of knowing whether fixation and embedding would destroy or modify lectin binding sites. (For a screening exercise such as this, the usual strategy is to start with frozen sections, try a variety of fixatives on these and then to proceed with the more promising lectins or antibodies to a trial of other fixatives and methods of embedding. At each stage, the results are compared with simultaneously stained frozen sections.)

Considering the degree of polymorphism in the outbred human population it was perhaps surprising that we emerged from the screening with only one polymorphic difference between the mouse strains tested: for binding of *Dolichos biflorus* agglutinin (DBA) to gut epithelium and vascular endothelium. The polymorphic differences are illustrated in *Fig*. 8.3, and summarized in *Table* 8.2. DBA has a monosaccharide specificity for *N*-acetylgalactosamine, and addition of 0·1 M sugar to the incubation of the lectin with the tissue section completely abolished immunocytochemical staining. Other lectins, such as *Bandeiraea simplicifolia* agglutinin (BSA), although they were also blocked by *N*-acetylgalactosamine did not show the polymorphism which was seen with DBA. This emphasizes the point made at the start of this chapter, that the specificities of lectin binding are more complex than the quoted monosaccharide specificities would suggest.

The use of the DBA polymorphism to demonstrate the mosaic cell populations in chimeric tissues is shown in *Fig*. 8.4. and *Plate* 12. There are three technical points to note. First, although the illustrations are of frozen sections, fixed with formol–saline after cutting, we have found that the determinant (or determinants) recognized by DBA is stable to a wide range of fixatives, and to conventional embedding in paraffin and Epon–Araldite resin. Electron microscopy is possible from resin-embedded sections as shown in *Fig*. 8.1. Whether this would be true for the determinants recognized by other lectins, or by DBA in another situation (where the carrier molecules might be different) could only be established by experiment. With DBA, we have preliminary evidence that the intensity of staining diminishes with storage of paraffin blocks or sections, so that it is probably wise always to evaluate freshly prepared material in the first instance.

Second, we have no knowledge of whether DBA is binding to one or several different molecules in these tissues, of what the molecules are or their biological

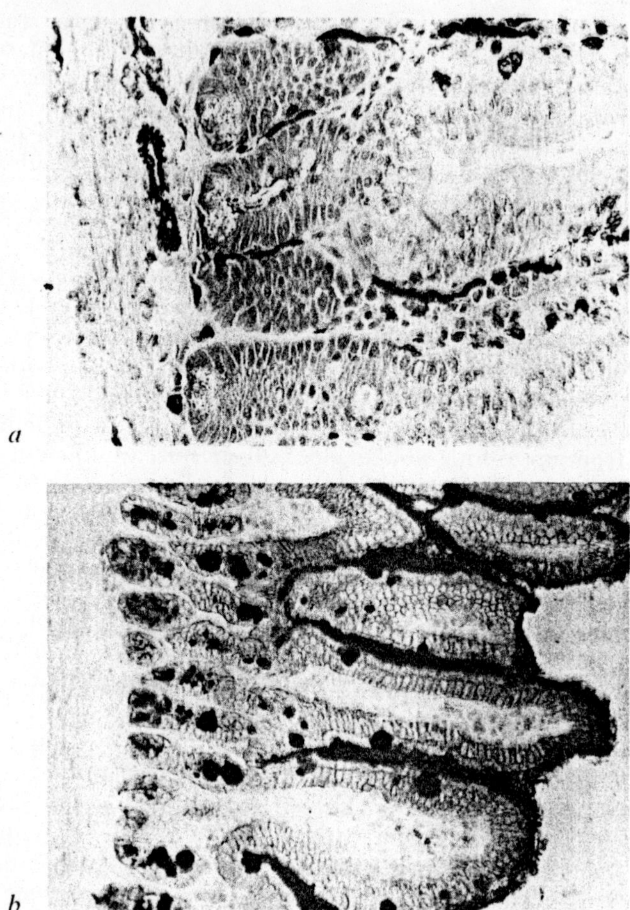

Fig. 8.3. Polymorphism for DBA–peroxidase binding.
a. RIII small intestine. Epithelium negative; endothelium positive.
b. C57BL/6 small intestine. Epithelium positive; endothelium negative.
5-μm frozen sections. Counterstained with haemalum.

Table 8.2. ***Polymorphism between inbred mouse strains for binding of*** **Dolichos biflorus** ***agglutinin***

Strains	*DDK*	*RIII*	*C57BL/6*	*CBA/Ca*
Gut epithelium				
Small intestine	– [a]	–	+	+
Caecum	+	+	+	+
Proximal colon	–	–	+	+
Distal colon	–	–	+	–
Vascular endothelium	+ [b]	+ [b]	– [c]	– [c]

[a] One of six mice jejunal epithelium positive.
[b] Tissue-specific differences in DBA-binding sites in arteries, veins and capillaries (*see text*).
[c] Small vessels in brain weakly positive.

function. So long as DBA is being used simply as a marker of cellular genotype, this is irrelevant. What is important, is to be sure that the DBA binding sites are indeed expressed strictly according to genotype: in other words, cells of the DBA-negative component of the chimera must not have been induced to become DBA positive by outside influences, including the presence of DBA-positive cells. We have compared the cell populations revealed by DBA and by immunocytochemistry for H2 haplotypes in adjacent sections from chimeras between strains polymorphic for both DBA binding and H2 haplotype. In every case so far studied, they coincide. This suggests that in normal tissues of these chimeras, the expression of DBA-binding sites is determined by cellular genotype, and that DBA is a satisfactory cellular marker. This may not, however, prove to be so in abnormal states, such as neoplasia, which we have already seen to be associated with changes in cellular expression of lectin binding sites.

Third, in any such system in which only one component is positively identified, there is a risk of misinterpretation. For example, in the arteriole in *Fig.* 8.4 is it safe to assume that the DBA-negative areas are indeed C57BL in origin? An ideal cellular marker system would overcome this by providing for positive identification of both components of the chimera.

Finally, a comment about the chimeras themselves. In 14 chimeras we have examined using DBA and H2 antigens as cellular markers, we have not so far found an intestinal crypt which is of 'mixed' composition (*see Fig.* 8.4 and *Plate* 12). This implies that developmentally each crypt arises from a single progenitor cell, a finding which we had not expected. The DBA-binding polymorphism for gut epithelium is present in RIII and C57BL/6 embryos at 12 days gestation, so it should be possible to study the histogenesis of intestinal crypts in relation to the mosaic cell population in the embryonic gut epithelium.

3. CONCLUSIONS

In this chapter lectins have been considered solely as potential alternative reagents for immunocytochemistry, and not at all in terms of their possible biological functions.

The parallel development of monoclonal antibodies and of immunocytochemical techniques in recent years has created a surge of interest in immunohistological investigation. It has yet to be seen how lectins will fit into this picture: their potential has still to be fully evaluated.

The chief advantage of lectins over antibodies as immunocytochemical reagents is their ready availability. Their binding specificity is more or less well defined, and reproducible, and enzyme or fluorescein conjugates are either commercially available or fairly easily prepared (*see* the Appendix). Addition of the appropriate sugar to the incubation provides in most cases (*see* the Appendix) a simple control for the specificity of the lectin histochemical stain.

In each of the examples quoted earlier in this chapter, the requirement was for a reagent which would distinguish one population of cells from another, or mark a single tissue component, on an empirical basis. No doubt it would have been possible to raise antibodies to do the same job. If further investigations were planned of the biological differences between the cell populations thus distinguished, a panel of antibodies would probably be a worthwhile investment. But as

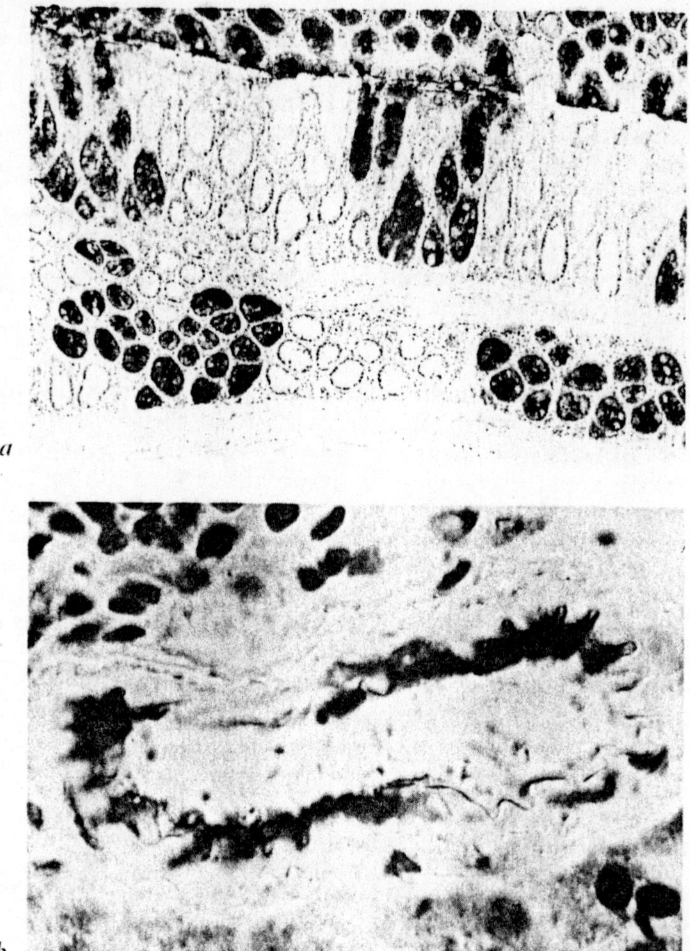

Fig. 8.4. Demonstration of mosaic cell populations in chimeric mice.
a. CBA↔C57BL/6 distal colon. 5-μm frozen section, counterstained with haemalum. The C57BL/6 crypts bind DBA–peroxidase and appear black.
b. DDK↔C57BL/6 arteriole in pancreas. 5-μm frozen section, counterstained wuth haemalum. The DDK endothelium binds DBA–peroxidase and appears black. For methods, *see* the Appendix.

a first step, screening with a panel of lectins may be a relatively cheap and simple way of obtaining both histological markers which will demonstrate that distinct cell populations exist, and reagents which will help in the separation of the populations for further study where appropriate.

Acknowledgement

Work from the author's laboratory described in this chapter was funded by the Cancer Research Campaign.

Appendix

CONJUGATION OF LECTIN TO PEROXIDASE

Commercial conjugates are useful for screening, where small amounts of a variety of different lectins are required. For more extensive studies with a particular lectin, it may be worthwhile to prepare your own conjugates. The method we use is given below. It is not particularly difficult or time consuming. For large scale use, preparation of the lectin from the raw material may be considered. The methods are beyond the scope of this chapter (*see* Colowick and Kaplan[31]).

Procedures for Conjugation[30]

Activation

1. Dissolve 10 mg of Sigma Type VI horseradish peroxidase in 1 ml of 0·3 M sodium bicarbonate solution (1·25 g/50 ml).
2. Add 50 µl of 1 per cent fluorodinitrobenzene in absolute alcohol and mix gently for 1 h at room temperature.
3. Add 1 ml of 0·06 M sodium periodate in distilled water (0·62 g/50 ml) and mix gently for 30 min at room temperature.
4. Add 2 drops ethylene glycol and mix gently for 1 h at room temperature.
5. Desalt peroxidase on small Sephadex G-25 column equilibrated with bicarbonate.

Conjugation

6. Normally an equal amount of lectin is conjugated to the peroxidase but the ratio can be altered according to requirements. Thus 10 mg of lectin are dissolved in 25 ml of bicarbonate buffer and added to the 'activated' peroxidase. This gives a lectin concentration of about 0·3 mg/ml. The appropriate specific sugar (approximately 0·1 M) is added to protect the lectin binding site during conjugation. This sugar is subsequently removed by dialysis or during chromatographic purification (*see below*).
7. Conjugated 3 h with gentle stirring at room temperature.
8. 1 mg sodium borohydride added 1 h (room temperature).
9. pH dropped to 6·4 using 1 N HCL and left overnight (room temperature).

Purification

10. Conjugates can be separated from free lectin, free peroxidase and high-molecular-weight components by chromatography on suitable gels such as Sephadex G-200, Sephacryl S-300 in PBS etc. If chromatography is to be carried out the conjugate has to be concentrated to a suitable volume for loading using, for example, an Amicon 52 ultrafiltration cell with a PM-30 membrane. It should be

noted however that 'crude' (unpurified) conjugates can sometimes be used with adequate results. (Remember to dialyse out the inhibitory sugar in this case.)

11. The absorbance of the purified conjugate is measured at 280 and 403 nm and then the solution is filtered through a 0·22-μm millipore filter into a sterile bottle. Before carrying out this step it should be ascertained whether the lectin being conjugated absorbs onto millipore. 'Conditioning' the filter using 1 per cent BSA may be useful.

12. The sodium azide concentration is increased to 0·1 per cent and the conjugate is stored at 4 °C.

Approximation of concentrations:

$$\text{HRP concn (mg/ml)} = \frac{A_{403}}{1\cdot5}$$

where A = absorbance

$$\text{Protein concn (mg/ml)} = \frac{A_{280} - (A_{403}/2)}{\varepsilon^{1\%}_{280}\ \text{of lectin}}$$

$$\text{Molar ratio} = \frac{\text{HRP concn}}{\text{Lectin concn}} \times \frac{M_r\ \text{lectin}}{M_r\ \text{HRP}}$$

II. HISTOCHEMICAL STAINING OF TISSUE SECTIONS WITH LECTIN–PEROXIDASE CONJUGATES

Frozen sections on gelatin-subbed slides (try a range of fixatives applied to the dried sections for a few minutes to see which gives the best combination of specific staining and tissue morphology), or, if found by experiment to be possible, paraffin or resin sections.

Follow standard immunocytochemical procedure:

i. Wash in phosphate-buffered saline pH 7·5 containing 0·5 per cent bovine serum albumin (PBS–BSA)
ii. Blot away excess moisture
iii. Add lectin conjugate diluted in PBS–BSA (generally 100–200 μl depending on the size and number of sections on the slide). Incubate 1 h at room temperature in a moist chamber
   For control slides, make the incubation mixture in a solution (about 0·1 M) of the appropriate sugar (*see below*)
   The optimum dilution of the lectin conjugate is first found by experiment: a range of three or four three-fold increasing dilutions is usually appropriate for the first trial. In our hands, lectin–peroxidase conjugates prepared by the method given above and not concentrated after final chromatographic purification (i.e. 10 ml final volume from 10 mg starting material) generally are used at a titre of about 1 in 10.
iv. Wash three times by flooding the slide with PBS and finally immerse the slides in a rack in a bath of PBS.
v. Demonstrate peroxidase using appropriate substrate (a choice of methods is available, yielding different coloured products). For most purposes we use the standard diaminobenzidine (DAB) technique, which gives a brown product insoluble in alcohol or xylene. *See* p. 34.
vi. Counterstain in Meyer's haemalum (should be freshly made up every few days for best results) and mount in XAM mountant.

III. CONTROLS FOR THE SPECIFICITY OF LECTIN BINDING

The monosaccharide specificities of some commonly used lectins are given in *Table* 8.1. Addition of these sugars at 0·1 M concentration will generally inhibit specific lectin binding to the tissue section. In some cases (e.g. in our experience with concanavalin A, *Ricinus communis*, *Abrus precatorius* and *Phaseolus limensis*), complete inhibition may not be obtained even with higher concentrations of the sugar. In this case, more complex oligosaccharides for which the lectin has higher affinity may be used if they are available. An alternative (which we have not used ourselves) is to pretreat the sections with the appropriate glycosidase to remove the specific terminal sugars.[17]

REFERENCES

1. Kocourch J. and Horejsi V. Defining a lectin. *Nature* 1981, **290**, 188.
2. Goldstein I. J. and Hayes C. E. The lectins: carbohydrate-binding proteins of plants and animals. *Adv. Carbohydr. Chem. Biochem.* 1978, **35**, 127–340.
3. Debray H., Decont D., Strecker G., Spik G. and Montreuil J. Specificity of twelve lectins towards oligosaccharides and glycopeptides related to *N*-glycosylproteins. *Eur. J. Biochem.* 1981, **117**, 41–55.
4. Edelman, G. M. Surface modulation in cell recognition and cell growth. *Science* 1976, **192**, 218–226.
5. Salik S. E. and Cook G. M. W. Comparison of early embryonic and differentiating cell surfaces. *Biochim. Biophys. Acta* 1976, **419**, 119–136.
6. Gooi H. C., Feizi T., Kapadia A., Knowles B. B., Solter D. and Evans, M. J. Stage-specific embryonic antigen involves $\alpha 1 \rightarrow 3$ fucosylated type 2 blood group chains. *Nature* 1981, **292**, 156–158.
7. Burridge K. Changes in cellular glycoproteins after transformation: identification of specific glycoproteins and antigens in sodium dodecylsulphate gels. *Proc. Natl Acad. Sci. USA* 1976, **73**, 4457–4461.
8. Rapin A. M. C. and Burger M. M. Tumour cell surfaces: general alterations detected by agglutinins. *Adv. Cancer Res.* 1974, **20**, 1–94.
9. Reisner Y., Linker-Israfli M. and Sharon N. Separation of mouse thymocytes into two sub-populations by the use of peanut agglutinin. *Cell Immunol.* 1976, **25**, 129–134.
10. Boldt D. H. and Lynis R. D. Fractionation of human lymphocytes with plant lectins. *Cell Immunol.* 1979, **43**, 82–93.
11. Kimura A., Orn A., Holmquist G., Wizzell H. and Ersson B. Unique lectin-binding characteristics of cytotoxic T lymphocytes allowing their distinction from natural killer cells and 'K' cells. *Eur. J. Immunol.* 1979, **9**, 575–578.
12. Mattes M. J. and Holden H. T. The distribution of *Helix pomatia* lectin receptors on mouse lymphoid cells and other tissues. *Eur. J. Immunol.* 1981, **11**, 358–365.
13. London J., Berrih S. and Bach J. F. Peanut agglutinin T. A new tool for studying T-lymphocyte subpopulations. *J. Immunol.* 1978, **121**, 438–446.
14. Rose M. L., Birbeck M. S. C., Wallis V. J., Forrester J. A. and Davies A. J. S. Peanut lectin binding properties of germinal centres of mouse lymphoid tissue. *Nature* 1980, **284**, 364–366.
15. Newman R. A., Klein P. J. and Rudland P. S. Binding of peanut lectin to breast epithelium, human carcinomas and a cultured rat mammary stem cell: use of the lectin as a marker of mammary differentiation. *J. Natl Cancer Inst.* 1979, **63**, 1339–1346.
16. Brabec R. K., Peters B. P., Bernstein I. A., Gray R. H. and Goldstein I. J. Differential lectin binding to cellular membranes in the epidermis of the newborn rat. *Proc. Natl Acad. Sci. USA* 1981, **77**, 477–479.
17. Watanalte M., Muramatsur T., Shirana H. and Ugai K. Discrete distribution of binding sites for *Dolichos biflorus* agglutinin and for peanut agglutinin (PNA) in mouse organ tissues. *J. Histochem. Cytochem.* 1981, **29**, 779–790.
18. Kjoergaard J., Starklint H. and Bierring F. Ultrastructural visualisation of concanavalin A binding receptor sites in neoplastic and non-neoplastic transitional cell epithelium of the human urinary bladder. *Urol. Int.* 1980, **35**, 271–280.

19. Peters B. P. and Goldstein I. J. The use of fluorescein-conjugated *Bandeiraea simplicifolia* B4-isolectin as a histochemical reagent for the detection of α-D-glactopyranosyl groups. *Exp. Cell Res.* 1979, **120**, 321–334.
20. Holthofer H., Virtanen I., Pettersson E., Tornmoth, T., Alfthen O., Linder E. and Miettinen A. Lectins as fluorescence microscopic markers for saccharides in the human kidney. *Lab. Invest.* 1981, **45**, 391–399.
21. Pena S. D. J., Gordon B. B., Karpat G. and Carpenter S. Lectin histochemistry of human skeletal muscle. *J. Histochem. Cytochem.* 1981, **29**, 592–546.
22. Kim Y. S., Isaacs R. and Perdomo J. M. Alterations of membrane glycopeptides in human colonic adenocarcinoma. *Proc. Natl Acad. Sci. USA* 1974, **71**, 4859–4873.
23. Bramwell M. E. and Harris H. An abnormal membrane glycoprotein associated with malignancy in a wide range of different tumours. *Proc. R. Soc. Lond. B. Biol. Sci.* 1978, **201**, 87–106.
24. Cummings K. B. Carcinoma of the bladder: predictors. *Cancer* 1980, **45**, 1849–1855.
25. Furmanski P., Kirkland W. L., Gargala T., Rich M. A. and the Breast Cancer Prognostic Study Clinical Associates. Prognostic value of concanavalin A reactivity of primary human breast cancer cells. *Cancer Res.* 1981, **41**, 4087–4092.
26. Howard D. R., Ferguson P. and Batsakis J. G. Carcinoma-associated cytostructural antigenic alterations: detection by lectin binding. *Cancer* 1981, **47**, 2872–2877.
27. Filipe M. I. Mucous secretion in rat colonic mucosa during carcinogenesis induced by dimethylhydrazine. A morphological and histochemical study. *Br. J. Cancer* 1975, **32**, 60–77.
28. Mintz B. Allophenic mice of multi-embryo origin. In: Daniel J. C. Jr, ed., *Methods in Mammalian Embryology*. San Francisco, Freeman, 1971: 182–214.
29. McLaren A. *Mammalian Chimeras*. Cambridge, Cambridge University Press, 1976.
30. Nakane P. K. and Kawaoi A. Peroxidase labelled antibodies: a new method of conjugation. *J. Histochem. Cytochem.* 1974, **22**, 1084–1091.
31. Colowick S. P. and Kaplan N. O. Complex carbohydrates, part B. In: V. Ginsburg (ed.), *Methods in Enzymology* vol. 28. London, Academic Press, 1972.

9

Immunocytochemical Localization of Noradrenaline, Adrenaline and Serotonin

A. A. J. Verhofstad, H. W. M. Steinbusch, H. W. J. Joosten, B. Penke, J. Varga and M. Goldstein

I. INTRODUCTION

Noradrenaline, adrenaline and serotonin are found in many mammalian as well as non-mammalian species. In order to assess their biological significance a brief introductory review will be given of the distribution of cells containing these compounds in mammals.

Noradrenaline has been found in the central and peripheral nervous systems. From the work of Dahlström and Fuxe, and others it is known that in the central nervous system noradrenaline is localized in neurones.[1–3] The cell bodies of these noradrenergic neurones are concentrated in a limited number of areas of the brain stem. From these cell bodies, nerve processes are distributed over the brain and spinal cord. In the peripheral nervous system noradrenaline is present in the sympathetic postganglionic neurones.[4,5] The cell bodies of these neurones are found in the ganglia of the sympathetic trunk and in the ganglia of the prevertebral autonomic nerve plexuses. Axons of these cell bodies are distributed to organs controlled by the sympathetic nervous system. Recently, a limited number of small ganglionic cells containing noradrenaline have been described in the autonomic ganglia.[6] It has been suggested that these cells belong to the so-called small intensely fluorescent (SIF) cells. Outside the nervous system high concentrations of noradrenaline have been found in the adrenal medulla of many species (in some species, e.g. rabbit and guinea-pig, concentrations are extremely low) as well as in the extra-adrenal medullary tissues (extra-adrenal chromaffin tissues or chromaffin paraganglia).[7] The carotid body and related structures (non-chromaffin paraganglia) also seem to contain noradrenaline.[8,9]

Adrenaline is mainly detected in the adrenal medulla. Thus the adrenal medulla is able to store adrenaline as well as noradrenaline. Remarkable species differences with regard to the adrenaline : noradrenaline ratio have been reported.[7,10] It was shown that separate adrenaline and noradrenaline-medullary cells can be distinguished by several methods. For different species the ratio of the two storing cell-types seems to correlate quite well with the adrenaline : noradrenaline ratio.[10,11] Recently, adrenaline-containing neurones have been demonstrated in the central nervous system. The cell bodies of these adrenergic neurones are located in the lower brain stem.[12]

Serotonin is also found in several organs. In the central nervous system serotoninergic cell bodies have been described in the brain stem.[1,2,13,14] Processes emanating from these cell bodies are widely spread over the brain and spinal cord. A limited number of serotonin-containing neurones are supposed to be involved in the innervation of the gastrointestinal tract.[15–18] However, the major portion of serotonin is stored in mostly isolated cells of the gastrointestinal epithelium, viz. the enterochromaffin cells.[19–21] Evidence has also been obtained indicating the presence of serotonin in the so-called neuroepithelial bodies of the lung[22–24] and in endocrine organs, e.g. in the thyroid gland.[25] Finally, in certain species (rat, mouse), mast cells contain serotonin.[26]

From this brief review it will be evident that noradrenaline, adrenaline, and serotonin are distributed throughout the organism, stored in neuronal as well as in epithelial cells. Consequently, one might expect these compounds to be involved in numerous regulatory functions, either as neurotransmitters or as hormones or hormone-like substances.

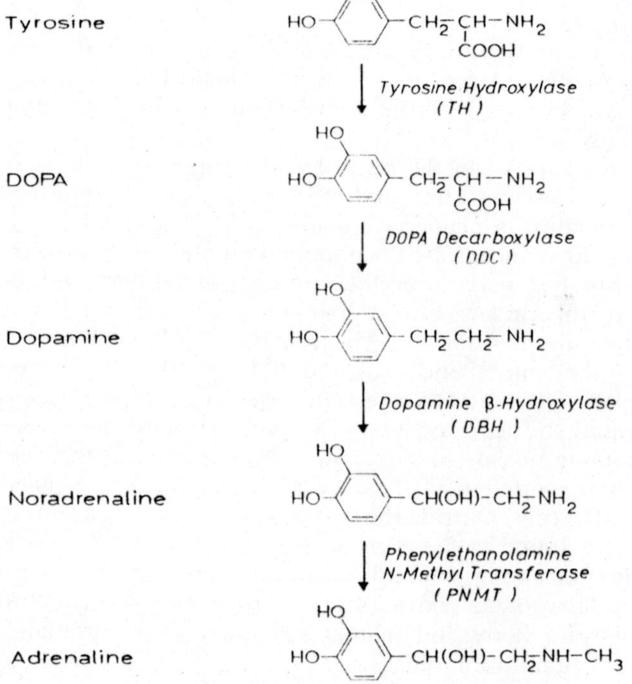

Fig. 9.1. Biosynthesis of noradrenaline and adrenaline.

In addition, tumours have neen described that contain either serotonin[27] (carcinoids) or noradrenaline and/or adrenaline[28] (phaeochromocytomas). These tumours are supposedly derived from the enterochromaffin cells or from the adrenal medulla and extra-adrenal medullary tissues, respectively.

It is now generally accepted that *noradrenaline* and *adrenaline* have common precursors and are derived from the same amino acid, L-tyrosine[29,30] (*Fig.* 9.1). Consequently, cells synthesizing noradrenaline contain tyrosine, dopa (3,4-dihydroxyphenylalanine) and dopamine (3,4-dihydroxyphenylethylamine), as well as the biosynthetic enzymes tyrosine hydroxylase, dopa decarboxylase and dopamine β-hydroxylase, while adrenaline and the corresponding enzyme, phenylethanolamine *N*-methyltransferase, are absent. On the other hand, cells synthesizing adrenaline contain all components of the biosynthetic pathway. Thus, noradrenaline is present in the noradrenaline as well as in the adrenaline-synthesizing cells. In the first case it forms the end-product, in the second an intermediate step to adrenaline. Therefore, one might expect the noradrenaline concentration in the noradrenaline-synthesizing cells to be much higher than in the adrenaline-synthesizing cells.

The biosynthetic pathway of *serotonin* is indicated in *Fig.* 9.2. Serotonin is the end-product of a pathway starting with the conversion of the amino acid L-tryptophan into 5-hydroxytryptophan.[31] Finally, 5-hydroxytryptophan is converted to serotonin (5-hydroxytryptamine).

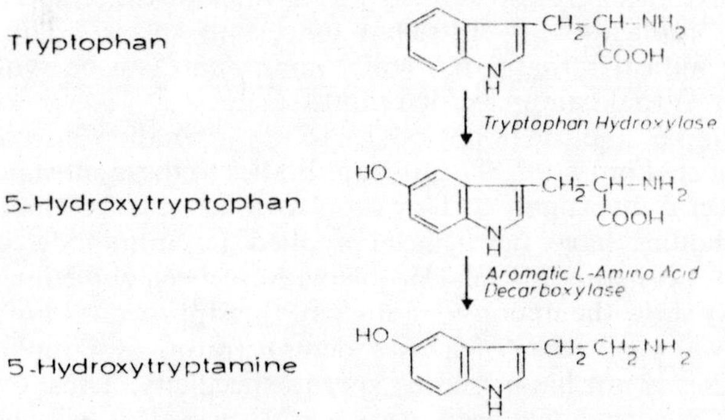

Tryptophan

Tryptophan Hydroxylase

5-Hydroxytryptophan

Aromatic L-Amino Acid Decarboxylase

5-Hydroxytryptamine

Fig. 9.2. Biosynthesis of serotonin.

It has been claimed that both pathways (*Figs.* 9.1 and 9.2) have one enzyme in common,[32] i.e. aromatic L-amino acid decarboxylase (dopa decarboxylase). However, there are other reports indicating that two different types of decarboxylase exist.[33,34]

2. IDENTIFICATION OF NORADRENALINE, ADRENALINE AND SEROTONIN-SYNTHESIZING CELLS

Cells synthesizing noradrenaline, adrenaline, or serotonin can now be identified by two different immunocytochemical procedures; firstly, by applying antibodies to the biosynthesizing enzymes or, secondly, by employing antibodies to the end-products, noradrenaline, adrenaline or serotonin.

The *first approach* (antibodies to the synthesizing enzymes) was introduced in the late sixties and early seventies.[35–40] Since then many studies based upon this principle have been published.[41–43] Keeping in mind the biosynthetic pathway, it will be evident that at least two types of antibodies are required to determine whether a particular cell or cell group contains *noradrenaline or adrenaline*. In practice, two adjacent serial sections are stained with an antibody to dopamine β-hydroxylase which converts dopamine to noradrenaline and an antibody to phenylethanolamine *N*-methyltransferase which converts noradrenaline to adrenaline, respectively.[39] If both reactions are positive, one is dealing with an adrenaline-synthesizing cell. Alternatively, cells only immunoreactive to dopamine β-hydroxylase supposedly synthesize noradrenaline. For the *serotonin-synthesizing cells* antibodies to both tryptophan hydroxylase[43–45] and aromatic L-amino acid decarboxylase[38] have been used. The latter antibody also recognizes the decarboxylating enzyme of the noradrenaline and adrenaline pathways. Therefore, antibodies to aromatic L-amino acid decarboxylase can only be used as a marker for serotonin synthesis if other indicators are also available. Application of antibodies to the synthesizing enzymes has proved to be a great achievement. However, there are several practical and fundamental limitations. First, in order to raise antibodies, enzymes have to be isolated and purified as far as possible without disturbing their immunogenic properties. Secondly, antibodies to the biosynthesizing enzymes might have little cross-reactivity to enzymes of other species.[46] As a consequence, their application is restricted to certain species. Thirdly, immunoreactivity solely indicates that a particular compound can be synthesized. Whether it is really produced cannot be determined.

The *second approach*, i.e.. the use of antibodies to noradrenaline, adrenaline, or serotonin was first reported in 1978.[47] To date, antibodies to these substances have been produced by several investigators. However, as far as can be traced in the literature, these antibodies have never been applied in immunocytochemical studies. Although not all methodological problems have been solved, so far the results obtained clearly show the major advantage of this approach. Immunoreactions have been observed in a variety of species, demonstrating that application of these types of antibodies is not hampered by species specificity. Thus, one of the drawbacks of the first approach has been overcome. However, it will be obvious that antibodies to noradrenaline, adrenaline and serotonin can only be used to reveal the presence or absence of these compounds. Whether these compounds are really synthesized at the site or merely stored cannot be determined, unless, in addition, antibodies to the biosynthesizing enzymes are used. Thus, the first and second approaches are complementary to each other. Ideally, they should be used simultaneously.

3. IMMUNOCYTOCHEMISTRY USING ANTIBODIES TO NORADRENALINE, ADRENALINE AND SEROTONIN

The detection of compounds of low molecular weight by antibodies shares problems generally encountered in immunocytochemistry. However, there are two problems to be mastered in particular, viz. the production of specific antibodies, d the proper preservation of the substances to be demonstrated. This ... plain why efforts in this field were reported only quite recently. This section is med at summarizing the results obtained so far.

.1. Preparation of Antisera

Since noradrenaline, adrenaline, and serotonin are regular constituents of many species, they are not regarded as foreign materials by the immune apparatus. In addition, the molecules are too small to be immunogenic. However, as was shown by Landsteiner,[48] such small molecules might present immunogenic determinants after linkage to a foreign carrier protein. This principle has been adopted widely to raise antibodies to non- or weakly immunogenic substances, like small peptides, peptide fragments, steroid hormones, drugs etc. Likewise, several investigators were able to raise antibodies to noradrenaline, adrenaline and serotonin. The specificities of these antibodies were tested by bioassay, complement fixation tests, immunoprecipitation, immunoelectrophoresis and/or radioimmunoassay.[49–61] The first reports on the applicability of these antibodies in immunocytochemistry appeared in 1978 (serotonin)[47] and 1980 (noradrenaline and adrenaline).[62] Since then, other reports dealing with the development of an immunocytochemical technique for serotonin were published.[63–67] Obviously, conjugates prepared by linkage of haptens to a carrier protein carry different types of antigenic determinants, related partly to the hapten and partly to the original carrier protein. Consequently, antisera raised will contain a variety of antibodies with different specificity, titre and avidity. As far as the antibodies recognizing the small molecular compounds are concerned, one might expect that their characteristics would be determined by the number of hapten molecules bound, the type of linkage used for attachment and the steric configuration of the resulting immunogen. It should also be considered whether the hapten molecules are altered during coupling, and whether just one or a number of different coupling products are formed. In some papers the proposed structure of the immunogen synthesized was indicated.[49,51,52,54–57,59–61,68,69] However, a detailed characterization of the immunogens currently used for the raising of antibodies is lacking.

In the present study different coupling procedures were tried. So far, the best immunocytochemical results were obtained when immunogens were prepared using formaldehyde as a coupling agent. Noradrenaline, adrenaline and serotonin were coupled to bovine serum albumin following the descriptions of Grota and Brown[56,57] (noradrenaline and serotonin; for serotonin *see also* Steinbusch[14] and Miwa et al.[59,60] (adrenaline). Usually rabbits were used for immunization. Each animal was injected intramuscularly with an emulsion of 0·5 ml of the antigen-containing solution (2 mg/ml) and 0·5 ml complete Freund's adjuvant. Booster injections (incomplete Freund's adjuvant, otherwise the same mixture) were given subcutaneously at biweekly intervals. Serum samples were tested by immunodiffusion according to Ouchterlony, immunoelectrophoresis and immunofluorescence microscopy. Immunoelectrophoresis revealed the presence of precipitating antibodies to the carrier protein as well as to the hapten-carrier complex. Antibodies to the carrier protein could be removed by liquid phase absorption with bovine serum albumin (5 mg bovine serum albumin per ml serum; incubated for 1 h at

37 °C, then kept for 18 h at 4 °C; centrifugation at 20 000 *g* for 20 min at 4 °C). This purification led to a somewhat higher contrast in immunofluorescence microscopy. Moreover, unwanted staining of collagen-rich tissues (e.g. blood vessel walls, fibrous capsule of the adrenal) could also be eliminated.

3.2. Preparation of the Tissues

It will be evident that a high titre of well-defined antibodies with high avidity is a prerequisite for successful immunostaining. However, negative or unacceptable results might be caused by other factors as well, including tissue preparation. Ideally, during tissue preparation noradrenaline, adrenaline and serotonin should be preserved without any loss of concentration. In addition, their antigenicity should be unaltered. In this respect one has to be aware that the chemical structure of these compounds might be changed during tissue preparation, e.g. during fixation. As a consequence of such changes the antibody might not recognize the antigens (haptens) any more. Conversely, if the structure of the haptens is changed during the preparation of the immunogen, a corresponding alteration introduced to the tissues, e.g. during fixation, might be a prerequisite to obtaining a positive immunocytochemical reaction. Thirdly, unwanted absorption of immunoglobulins by tissue components should be as low as possible. Finally, the preservation of the morphology of the tissue should be acceptable. Some of the above factors have been studied, whilst others have not yet been investigated thoroughly. So far, two preparatory steps seem to be crucial, viz. *fixation* and *embedding*. The latter is not relevant where isolated cells are concerned. Acceptable results were obtained with both the indirect immunofluorescence technique of Coons and collaborators,[70] and with the immunoperoxidase technique of Sternberger et al.[71,72] (peroxidase–anti-peroxidase technique). Unless otherwise stated, all experimental animals remained pharmacologically untreated.

Fixation appeared to be very essential. Unfixed tissues (cryostat sections) did not reveal any immune reactions. Antibodies to different types of immunogens were tested on tissues fixed by a variety of fixatives. As indicated above, the best immunostaining was achieved by antibodies to immunogens which were prepared utilizing formaldehyde as a coupling agent. Accordingly, the following data refer to these types of antibodies only. Fixation with 4 per cent paraformaldehyde in 0·1 M sodium phosphate buffer pH 7·3, seemed to be the fixative of choice. The best results were obtained if tissues were fixed by perfusion. However, immersion fixation of small tissue pieces gave satisfactory results as well. Using 4 per cent paraformaldehyde and other conditions mentioned below, intense and distinct serotonin immunoreactivity was observed in cryostat sections of tissues containing serotonin. Likewise, antisera to noradrenaline revealed intense cytoplasmic staining of the noradrenaline-storing cells present in the adrenal medulla and the extra-adrenal medullary tissues. Adrenaline-storing cells of the adrenal medulla were stained by antisera to adrenaline. The noradrenergic neurones of the central and peripheral nervous system stained either weakly or were negative. However, they became positive if 0·5 per cent glutaraldehyde was added to the fixative. Adrenergic neurones were not stained either upon fixation with 4 per cent paraformaldehyde or by fixation with 4 per cent paraformaldehyde plus 0·5 per cent glutaraldehyde. Other fixatives (e.g. 2 per cent glutaraldehyde, alone or combined with 4 per cent paraformaldehyde, Bouin's solution, 0·4 per cent benzoquinone or fixatives containing dichromate) gave negative or unacceptable results in the brain (antisera to noradrenaline, adrenaline, and serotonin), in tissues innervated by the sympathetic postganglionic noradrenergic neurones (antiserum to noradrenaline), as well as in the adrenal medulla (antisera to noradrenaline and adrenaline). However, intense immunostaining was observed in the enterochromaffin cells of the gastrointestinal epithelium following liquid or vapour fixation [73] with benzoquinone or fixation in Bouin's solution.

Thus, as a rule, the best results were obtained with 4 per cent paraformaldehyde or with 4 per cent paraformaldehyde plus 0·5 per cent glutaraldehyde (noradrenergic neurones). This might be explained by assuming that noradrenaline, adrenaline and serotonin have to be linked to tissue proteins, similarly to the linking of the hapten to the carrier protein. However, as mentioned above, enterochromaffin cells can be stained by serotonin antibodies even after fixation with benzoquinone. This observation does not seem to be in agreement with the above hypothesis, unless one assumes that benzoquinone links serotonin to tissue proteins in the same way as does formaldehyde. Secondly, previous investigations showed that antibodies raised by a serotonin–bovine serum albumin conjugate, synthesized according to the mixed anhydride method, also stained formaldehyde-fixed tissues.[69]

Apparently, fixation requirements of neuronal cells and epithelial cells storing noradrenaline, adrenaline or serotonin are different. This conclusion is based upon the following observations. Firstly, under conditions appropriate for the noradrenaline-storing cells of the adrenal medulla and the extra-adrenal medullary tissues (4 per cent paraformaldehyde), the noradrenergic neurones were weakly stained or were negative. In fact, 0·5 per cent glutaraldehyde had to be added to the formaldehyde solution to obtain a positive reaction. Secondly, adrenaline-storing cells of the adrenal medulla fixed by 4 per cent paraformaldehyde are stained by antibodies to adrenaline, while adrenergic neurones are not. Thirdly, while both serotoninergic neurones and enterochromaffin cells can be stained following fixation with 4 per cent paraformaldehyde, only the latter are positive if tissues are fixed by 0·4 per cent benzoquinone.

It will be evident that the matter of fixation is not yet fully understood and requires further investigation. In addition, fixation of noradrenaline and adrenaline within the neuronal cells needs to be improved. Nevertheless, the present fixation by 4 per cent paraformaldehyde allows studies combining antibodies to the end-products, serotonin (all tissues), noradrenaline and adrenaline (adrenal medulla and extra-adrenal medullary tissues) with others, to, for example, the synthesizing enzymes and peptides.

Following fixation and rinsing (in the present study a buffer containing 5 per cent sucrose was used) a number of methods can be used to produce *tissue sections*, viz. frozen sections, paraffin sections, plastic sections and sections cut with a vibrating knife microtome.

3.2.1. *Frozen Sections*

Most of the present observations are based upon study of cryostat sections (*Figs.* 9.3, 9.4, 9.5, 9.6*a*, *b* and 9.7). Tissues can be frozen on cryostat chucks either with

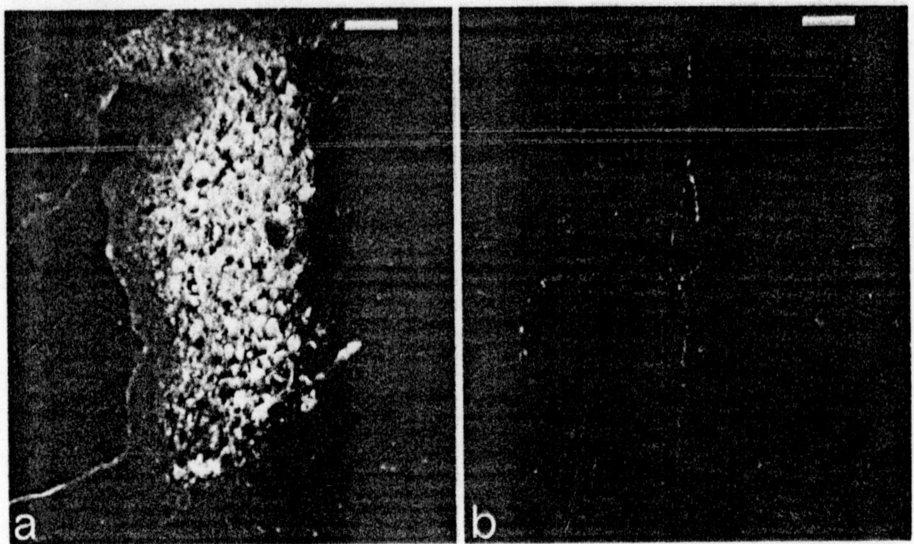

Fig. 9.3. Photomicrographs of rat brain fixed by perfusion with 4 per cent paraformaldehyde–0·5 per cent glutaraldehyde. Cryostat sections (thickness 7 μm) stained by indirect immunofluorescence with an antiserum to noradrenaline. Cell bodies immunoreactive to noradrenaline in the locus coeruleus (*a*) and a nerve fibre with varicosities in the molecular layer of the cerebellum (*b*). Bars: (*a*) 70 μm and (*b*) 14 μm.

powdered carbon dioxide or with isopentane cooled by liquid nitrogen (used at a temperature of −150 °C). Pending use they can be stored at −70 °C or lower.

Serial sections of 7–10 μm in thickness, with reasonably good structural preservation can be obtained easily. During freezing and sectioning cellular membranes are disrupted. This improves penetration by antibodies during staining. However, one must also consider the possible loss of antigenic substances during the staining procedure. Up till now, it is not known whether this loss is significant. Frozen sections cut on a freezing microtome can also be used. Both cryostat sections and sections cut on a freezing microtome should be as thin as possible, firstly because penetration of antibodies might not be satisfactory on thick sections and, secondly, because under high power, thick sections appear less brilliant and distinct than thin sections, due to the relatively low depth of the field. Consequently, detailed high power observations of thick sections are hardly possible.

3.2.2. *Paraffin Sections*

While frozen sections, especially cryostat sections, are extremely useful for studying the cellular distribution of noradrenaline, adrenaline and serotonin, it would be most rewarding if tissues could be embedded in paraffin without loss of immunostaining properties. Serial cryostat sections of small and fragile tissues are hard to achieve. Moreover, in general, cellular details are badly preserved in frozen sections. Hence, if different cell types occur in a particular area, it might be impossible to estimate which particular cells contain the immunoreactive material.

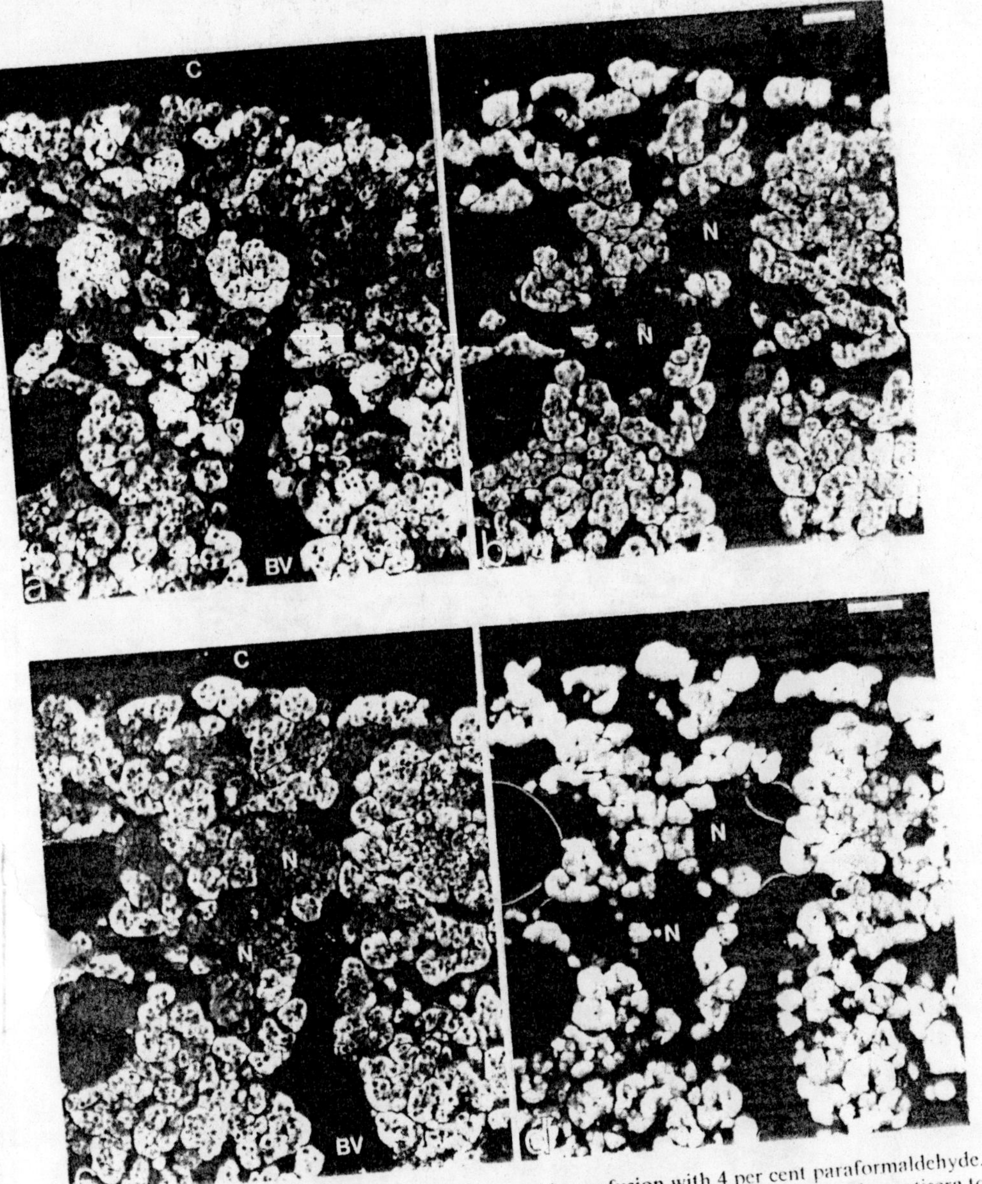

Fig. 9.4. Photomicrographs of rat adrenal gland fixed by perfusion with 4 per cent paraformaldehyde. Consecutive cryostat sections (thickness 7 μm) stained by indirect immunofluorescence using antisera to noradrenaline (*a*), adrenaline (*b*), bovine dopamine *β*-hydroxylase (*c*) and bovine phenylethanolamine *N*-methyltransferase (*d*). The adrenal medulla shows noradrenaline-storing cells (positive in *a* and *c*, negative in *b* and *d*; two groups are indicated by N), and adrenaline-storing cells (positive in *a–d*). C, cortex; BV, blood vessel. Bars: 35 μm.

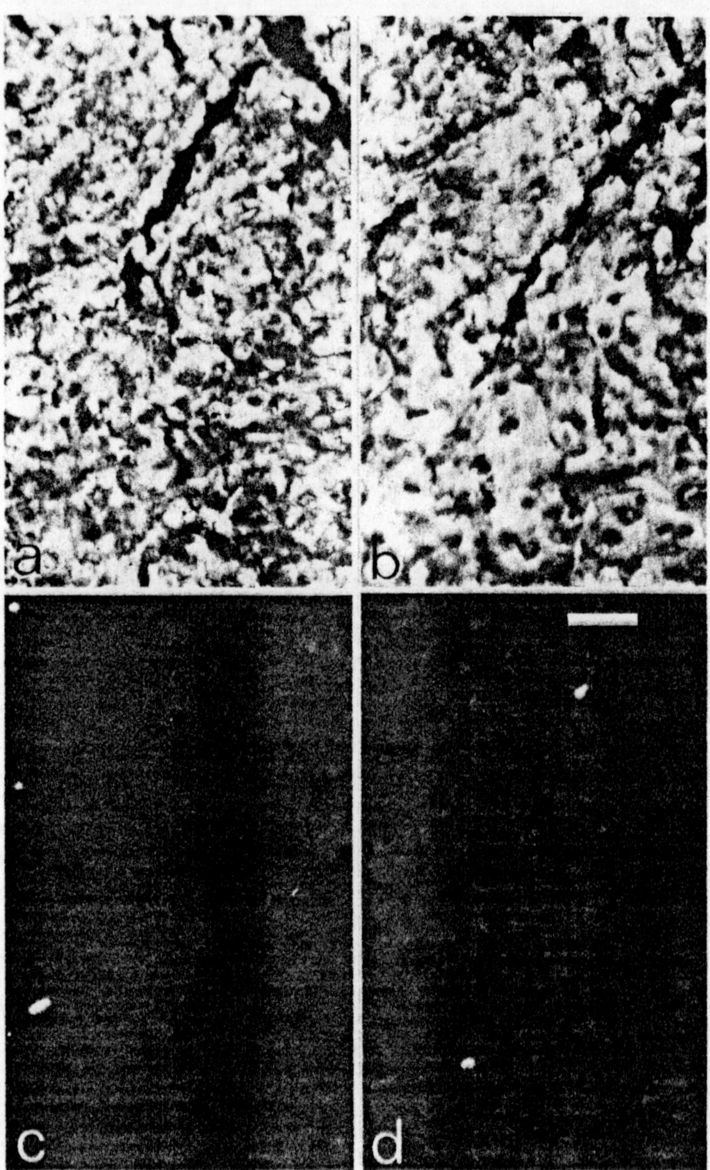

Fig. 9.5. Photomicrographs of an extra-adrenal phaeochromocytoma (presumably a tumour of the organs of Zuckerkandl) fixed by immersion with 4 per cent paraformaldehyde. Consecutive cryostat sections (thickness 7 µm) stained by indirect immunofluorescence with antisera to bovine dopamine β-hydroxylase (*a*), noradrenaline (*b*), bovine phenylethanolamine *N*-methyltransferase (*c*) and adrenaline (*d*). Tumour cells seem to store only noradrenaline (only *a* and *b* are positive). Bar, 43 µm.

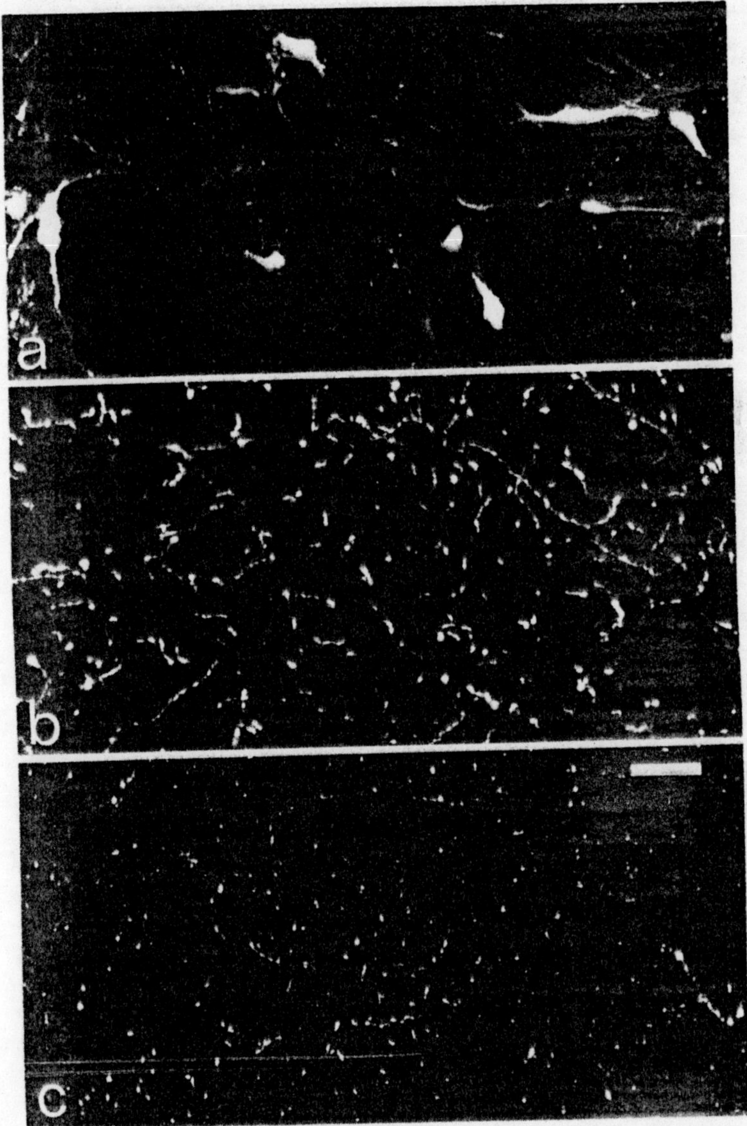

Fig. 9.6. Photomicrographs of rat brain fixed by perfusion with 4 per cent paraformaldehyde. Cryostat sections (thickness 7 μm; *a* and *b*) or Epon sections (thickness 1 μm; *c*) stained by indirect immunofluorescence using an antiserum to serotonin. Immunoreactive cell bodies in the brain stem (*a*) and nerve fibres (*b*, *c*). Bars: (*a*) 35 μm and (*b*, *c*) 14 μm.

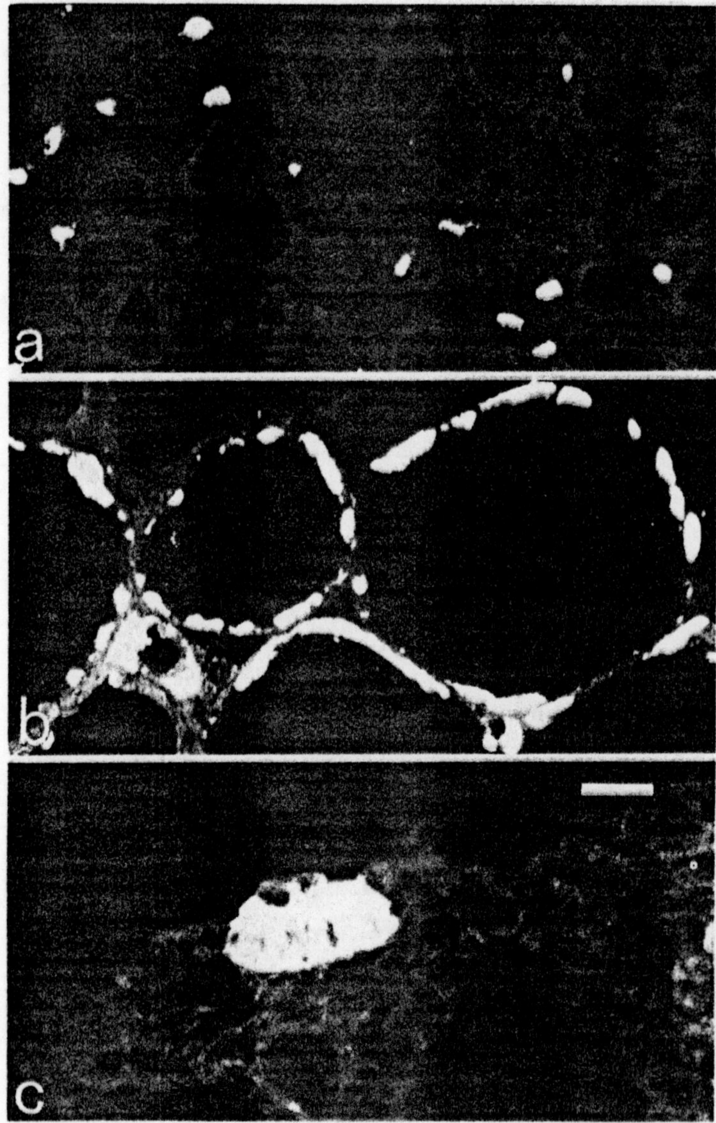

Fig. 9.7. Photomicrographs of rat thyroid (*a*), sheep thyroid (*b*) and neonatal rabbit lung (*c*). Tissues fixed by perfusion (*a*, *c*) or immersion (*b*) with 4 per cent paraformaldehyde. Cryostat sections (thickness 7 μm) stained by indirect immunofluorescence with an antiserum to serotonin. Note that in the rat thyroid (*a*) only mast cells are stained, while in the thyroid gland of the sheep (*b*) parafollicular cells are serotonin-immunoreactive. A neuroepithelial body is shown in the bronchiolar epithelium of the lung (*c*). Bars: (*a*, *b*) 43 μm and (*c*). 28 μm.

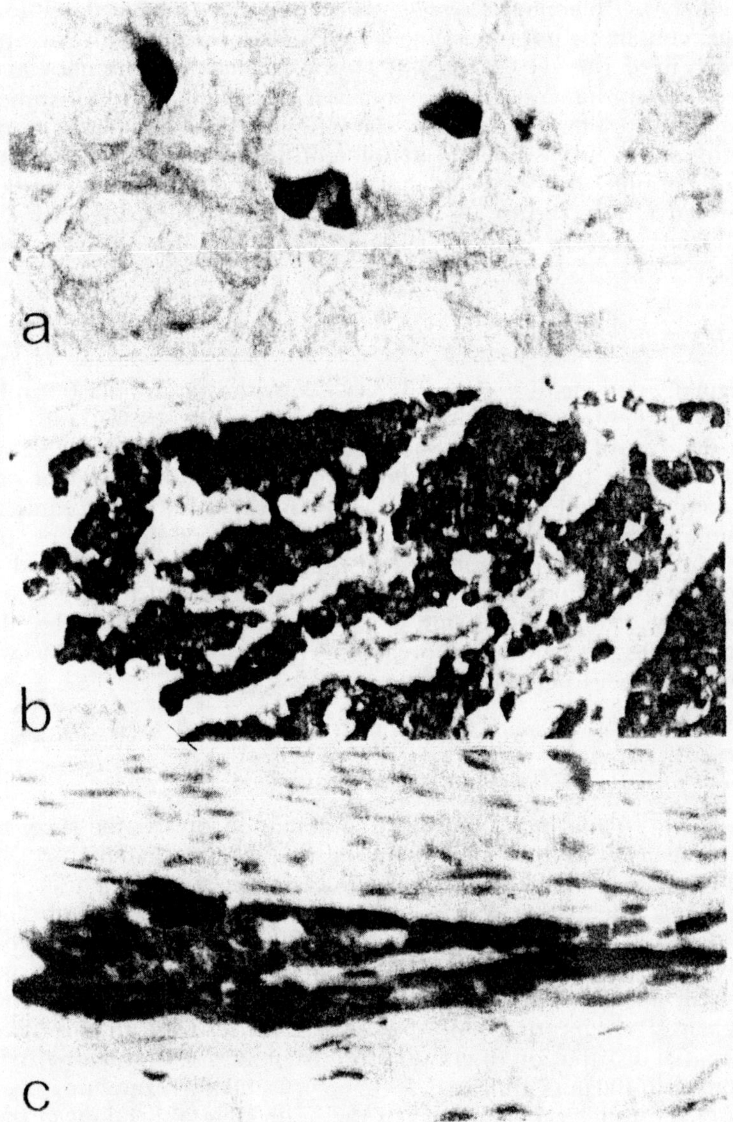

Fig. 9.8. Photomicrographs of human colon (*a*, *b*) and urinary bladder (*c*) fixed by immersion with 4 per cent formaldehyde. Paraffin sections (thickness 10 μm) stained according to the peroxidase anti-peroxidase technique using an antiserum to serotonin. Section (*c*) counterstained with haematoxylin. Typical serotonin-immunoreactive (enterochromaffin) cells are present in the mucosa. Immunoreactivity in carcinoid tissue invading the external muscular layer (*b*). Tumour cells in the muscular layer of the urinary bladder (*c*). These tumour cells are only partly serotonin-immunoreactive. Bars: (*a*, *c*) 17 μm and (*b*) 44 μm.

Finally, in view of possible applications in pathology, it would be extremely helpful if previously collected material (e.g. tumours) could be studied as well ('retrospective pathology'). Preliminary results show that deparaffinized sections of epithelial tissues containing noradrenaline, adrenaline or serotonin give acceptable results if properly fixed, rinsed etc. (*Fig*. 9.8). However, much less immunoreactivity can be observed in noradrenergic or serotoninergic neurones in paraffin-embedded tissues. These initial results indicate that the binding characteristics of noradrenaline and serotonin might be different in different tissues. Apparently, during the preparation of paraffin sections (dehydration, clearing, embedding, dewaxing) antigenic material is lost and/or its antigenicity might be impaired. It will be evident that further exploration of the possible use of paraffin sections is needed.

3.2.3. *Plastic Sections*

Although originally designed for electron microscopy, nowadays plastic embedding is used also for light microscopical purposes. Thin sections (1–2 μm) can be stained upon removal of the plastic. Such sections show structural preservation and resolution superior to that of paraffin sections. In addition, under certain conditions a combined light and electron microscopical study on the same tissue block is possible. If the antigen survives the different preparatory steps, plastic embedding might be useful for immunocytochemistry as well. In the present study paraformaldehyde-fixed, Epon-embedded, tissues were stained following the removal of Epon by sodium methoxide as described by Mayor et al.[74] Initial experiments revealed serotonin immunoreactivity in neuronal (*Fig*. 9.6*c*) as well as in epithelial cells known to contain serotonin.

3.2.4. *Sections Cut with a Vibrating Knife Microtome*

Actually, a vibrating knife microtome is a combination of a tissue slicer and a microtome. By means of a vibrating knife (usually a disposable safety razor blade) serial sections of unfrozen and unembedded tissues can be cut along a desired plane of sectioning. Thus, this procedure avoids disruption of cellular membranes by freezing and thawing, and the loss of antigenic material during dehydration etc. Unfortunately, producing sections is time consuming. Moreover, not all tissues are equally easily cut and thin sections are hard to produce. As a consequence, in immunocytochemistry, vibrating knife microtome sections are mainly used for studying the spatial distribution of immunoreactive material. Immunoperoxidase-stained sections (50–100 μm) appeared extremely useful for examining the dendritic pattern of serotoninergic neurones of the brain stem.[75] Sections cut with a vibrating knife microtome also proved to be suitable for the demonstration of serotoninergic nerve processes by immuno-electron microscopy (pre-embedding technique; Léránth Cs., personal communication, 1982).

3.3. Staining Methods

In the present study both the indirect immunofluorescence technique of Coons et al.[70] and the immunoperoxidase (peroxidase–anti-peroxidase) technique of Sternberger et al.[71,72] have been employed (*see* the Appendix for details).

3.4. Specificity

Immunocytochemistry is based upon the binding of an antibody to an antigenic determinant present in the tissue. One should realize that antigenic determinants usually represent only a certain portion of proteins and peptides. As long as these portions are specific for the proteins or peptides to be demonstrated, the antibody can be used to reveal their cellular localization. However, antisera raised by conventional immunization of animals with, for example, proteins, will contain antibodies with different specificities recognizing different parts (antigenic determinants) of the immunogen or tissue components. Therefore, an antiserum may only be used as a specific marker substance, if cross-reactivity to tissue components other than the ones to be studied has been ruled out. Consequently, besides immunological controls, tissues either containing or devoid of a particular substance should be examined. Since fixation, embedding etc. might change the antigenicity of the tissue, the specificity of an antiserum is only relevant under the conditions employed during tissue preparation.

In the present study, antisera were raised by complexes obtained by linkage of haptens to a carrier protein. Inevitably, these antisera are heterogeneous. Only those antibodies recognizing the haptens or hapten-derived compounds are relevant. As was indicated earlier, routinely, all sera were treated with the carrier protein (5 mg bovine serum albumin per/ml antiserum) to absorb antibodies immunoreactive to the carrier substance. Several *immunological control* tests were performed. *Firstly*, pre-immune (non-immune) instead of immune serum was used. *Secondly*, serum samples were absorbed with different concentrations of conjugates of noradrenaline, adrenaline or serotonin. *Thirdly*, in the case of serotonin antisera, the binding capacity of different haptens was examined by blocking (inhibition) experiments. Both absorption and blocking experiments were done by incubating serum samples with different concentrations of the conjugates or haptens for 1 h at 37 °C, followed by incubation for 18 h at 4 °C. Before use, all sera were centrifuged for 20 min at 20 000 *g* and 4 °C to remove any precipitate. All three controls were tested by indirect immunofluorescence on appropriately fixed rat tissues containing either noradrenaline (brain stem, gastrointestinal tract, adrenal medulla), adrenaline (adrenal medulla), or serotonin (brain stem, duodenum). The results clearly indicate that the presented staining results are due to antibody binding (no staining with pre-immune (non-immune) sera). From the absorption experiments (*Table* 9.1) one may conclude that antibodies raised by a noradrenaline-containing conjugate do not recognize adrenaline or serotonin. In addition, antibodies to the adrenaline conjugate do not bind noradrenaline or serotonin. Likewise, antibodies to the serotonin conjugate do not cross-react with noradrenaline or adrenaline. Thus, the antisera raised seem to be highly specific to the hapten portion of the immunogen. For serotonin, additional support was obtained from the blocking (inhibition) experiments. As shown in *Table* 9.2, serotonin immunoreactivity could only be blocked by low concentrations of serotonin and, at much higher doses, also with 5-methoxytryptamine and dopamine. However, it is questionable whether 5-methoxytryptamine is a real

*Table 9.1. **Absorption experiments***

NORADRENALINE-antisera, tested on			☐ Rat medulla oblongata ☐ Rat ductus deferens ☐ Rat adrenal medulla
Absorbent*	BSA-NA	BSA-A	BSA-5-HT
Fluorescence	negative	normal	normal
ADRENALINE-antisera, tested on			☐ Rat adrenal medulla
Absorbent*	BSA-A	BSA-NA	BSA-5-HT
Fluorescence	negative	normal	normal
SEROTONIN-antisera, tested on			☐ Rat medulla oblongata ☐ Rat duodenum
Absorbent*	BSA-5-HT	BSA-NA	BSA-A
Fluorescence	negative	normal	normal

*2 mg/ml antiserum; BSA bovine serum albumin; NA noradrenaline; A adrenaline; 5-HT 5-hydroxytryptamine (serotonin).

*Table 9.2. **Blocking (inhibition) experiments***

Concn, μM	2	4	8	16	32	64	128	256	512	1024	2048	4096	8192
5-HT	++	+	+	+	–	–	–	–	–	–	–	–	–
5-HTP							+++	+++	+++	+++	+++	+++	+++
L-Trp							+++	+++	+++	+++	+++	+++	+++
DA							++	++	++	+	+	+	+
NA							+++	+++	+++	+++	+++	+++	+++
A							+++	+++	+++	+++	+++	+++	+++
5-MT					++	++	+	+	+	–	–	–	–
Oct							+++	+++	+++	+++	+++	+++	+++
Syn							+++	+++	+++	+++	+++	+++	+++
Hist							+++	+++	+++	++	++	++	++

+++ normal fluorescence ++ some reduction + strong reduction – no fluorescence. 5-HT 5-hydroxytryptamine (serotonin); 5-HTP 5-hydroxytryptophan; L-Trp L-tryptophan; DA dopamine; NA noradrenaline; A adrenaline; 5-MT 5-methoxytryptamine; Oct octopamine; Syn synephrine; Hist histamine.

Fluorescence intensity in comparable sections of *rat medulla oblongata* after adding different concentrations of several substances to samples of serotonin antiserum, previously absorbed with bovine serum albumin (dilution 1 : 500).

In *rat duodenum* higher concentrations are required (data not shown).

constituent of the brain.[77] Since no staining was observed in areas containing dopaminergic neurones (*see below*), cross-reactivity to dopamine does not seem to be of importance in immunocytochemistry, at least with the present methods of fixation etc. No blocking was obtained by the precursors of serotonin (5-hydroxytryptophan and L-tryptophan), noradrenaline, adrenaline and related haptens.

In addition to these immunological controls, *tissue experiments* were performed to prove that the antisera do actually demonstrate noradrenaline, adrenaline or serotonin. *First*, guided by previous histochemical and biochemical observations, tissues were selected containing cells storing one or the other of these compounds. Thus, for noradrenaline, sections through the rat brain stem (containing noradrenergic cell bodies and noradrenergic nerve processes), sections of the duodenum and ductus deferens of the rat (containing noradrenergic sympathetic nerve terminals) and sections of the adrenal medulla of rat, hamster and guinea-pig were examined. The adrenal medulla was studied in three species, taking advantage of well-known species differences. Rat and hamster adrenals possess separate noradrenaline-storing cells, while in the guinea-pig such cells do not seem to occur, i.e. in the adrenal medulla only adrenaline-storing cells can be found.[10] Moreover, in the rat the noradrenaline-storing cells are without a preferential position, whilst in the hamster they are located at the cortical–medullary boundary.[78] For the same reasons, antisera raised by the adrenaline-containing conjugate were tested on adrenals of the same species. The specificity of the antisera to serotonin were tested on sections of the rat brain stem (containing serotoninergic cell bodies and nerve processes) and on rat duodenum (enterochromaffin cells). In this case, too, advantage was taken of remarkable species differences. Thus, thyroid glands of rats and sheep were studied. Both organs contain serotonin. However, in the rat thyroid serotonin occurs in mast cells, while in the sheep serotonin is stored in the parafollicular cells.[79] Rat parafollicular cells do not seem to contain serotonin. All experiments showed that the immunocytochemical findings are in good agreement with previously obtained histochemical and biochemical data. Thus, cells of the locus coeruleus, known to contain noradrenaline, were only stained by the antiserum to noradrenaline (*Fig.* 9.3*a*). These cell bodies are innervated by serotoninergic nerve terminals.[80] This could easily be demonstrated by application of antiserum to serotonin. Raphe nuclei containing serotoninergic perikarya (*Fig.* 9.6*a*) were only stained by serotonin antisera. No staining was observed in the substantia nigra known to contain dopaminergic cell bodies. Duodenum and ductus deferens showed noradrenaline-immunoreactive nerve fibres with a distribution that conforms to that of the sympathetic postganglionic axons. The adrenal medulla of rat and hamster revealed two cell types. The majority of the parenchymal cells were immunoreactive to both noradrenaline and adrenaline antisera (adrenaline-storing cells). A few groups of limited numbers of cells were only immunoreactive to noradrenaline antisera (noradrenaline-storing cells). In the rat (*Fig.* 9.4*a*, *b*) these groups were not limited to certain sites in the adrenal medulla, whilst in the hamster they were located at the boundary between cortex and medulla. In the guinea-pig adrenal medulla no cells were found that were only immunoreactive for noradrenaline, i.e. all parenchymal cells were immunoreactive to both noradrenaline and adrenaline antisera. Enterochromaffin cells of the rat duodenum were only stained by antisera to serotonin (*Fig.* 9.8*a*). Thyroids of rats and sheep showed patterns of immunoreactivity for serotonin as expected, viz. mast cells in the rat (*Fig.* 9.7*a*) and parafollicular cells in the sheep (*Fig.* 9.7*b*). In a *second* series of tissue experiments both antibodies to the synthesizing enzymes and antibodies to noradrenaline, adrenaline and serotonin were applied. It might be assumed that adjacent sections stained with the proper counterpart antibodies will reveal a similar distribution, if the latter are indeed specific for noradrenaline, adrenaline or serotonin. For practical reasons, in the case of antibodies to noradrenaline and adrenaline this was only studied in the brain stem of the rat (noradrenergic neurones) and the adrenal medulla of the rat, hamster and guinea-pig (noradrenalin and adrenaline-storing cells). It could be demonstrated that cells immunoreactive for dopamine β-hydroxylase but negative for phenylethanolamine *N*-methyltransferase, i.e. synthesizing noradrenaline, were only immunoreactive to noradrenaline antisera (*Fig*. 9.4*a–d*). In the adrenal medulla of rat and hamster these cells were distributed as described above. In addition, cells immunoreactive for both enzymes, i.e. synthesizing adrenaline, were immunoreactive for noradrenaline as well as for adrenaline (*Fig*. 9.4*a–d*). In the guinea-pig adrenal medulla, all parenchymal cells were stained by all four antisera. Likewise, in the rat brain stem serotoninergic perikarya were immunoreactive for serotonin as well as for the synthesizing enzyme dopa decarboxylase.[81] *Thirdly*, serotonin immunoreactivity in the rat brain and spinal cord was studied following treatment with neurotoxins 5,6- and 5,7-dihydroxytryptamine, known to cause a depletion of serotonin. Under these conditions serotonin immunoreactivity was strongly reduced.[81–83] A similar approach might give additional information about the specificity of antisera to noradrenaline and adrenaline. However, such experiments have not yet been performed.

Both immunological and 'tissue controls' clearly indicate that antisera raised by immunogens prepared by linkage of noradrenaline, adrenaline and serotonin to bovine serum albumin can be used as specific markers for cells storing these compounds. However, in the rat brain, cell bodies supposed to contain dopamine (substantia nigra) showed some weak staining with the antiserum raised by the noradrenaline conjugate. This might indicate that these antibodies cross-react with dopamine to a certain extent. It will be evident that this observation deserves further investigation.

3.5. Sensitivity

The sensitivity of an immunocytochemical technique is determined by many factors, e.g. the characteristics of the antisera, tissue preparation, immunostaining procedures. In order to assay the sensitivity of the staining methods based upon the application of antibodies to noradrenaline, adrenaline and serotonin, the immunofluorescent staining results on cryostat sections were carefully compared with known histochemical and biochemical data. For noradrenaline and adrenaline present in neuronal cells final conclusions cannot be drawn, because, as was mentioned earlier, fixation needs to be improved. Nevertheless, even under the currently used conditions noradrenergic cell bodies and fibres can be demonstrated quite well. The antibody techniques seem to be highly sensitive for noradrenaline and adrenaline stored in epithelial cells. Thus, studying the development of the noradrenaline and adrenaline-storing cells of the rat adrenal medulla, their earliest appearance as estimated immunocytochemically corresponds quite well with biochemical data.[84] Serotonin immunoreactivity was observed in neuronal and epithelial cells positively stained by existing histochemical techniques for serotonin. However, additionally, highly immunoreactive nerve terminals were revealed in previously negative-seeming brain areas, such as the cerebral cortex, known to contain substantial amounts of serotonin.[14,85] These findings indicate a considerable increase in sensitivity for the visualization of serotonin compared with the older techniques. Preliminary experiments showed that in the peroxidase–anti-peroxidase technique higher dilutions of the first antisera can be used. However, it might be questioned whether this really means an increased sensitivity, i.e. an increase in the number of immunoreactive cell bodies and fibres identified, or merely better staining of those already recognized. Initial studies also show that embedding procedures reduce the sensitivity of the immunostaining, especially in neuronal cells storing noradrenaline or serotonin.

3.6. Applications in Biology and Pathology to Date

Compared to the existing histochemical techniques (e.g. fluorescence induced by formaldehyde, glyoxylic acid) the presently described procedures allow the demonstration of noradrenaline, adrenaline and serotonin separately with equal (noradrenaline, adrenaline) or much higher sensitivity (serotonin).

New histochemical techniques render it possible to solve questions left by previous ones. In spite of being developed quite recently, some reports have already appeared illustrating this statement. Thus, detailed studies of the distribution of the serotoninergic neurone system in the rat brain and spinal cord have been published.[14,85] The number of serotoninergic fibres and terminals in the rat central nervous system appears to be much higher than was hitherto known. The presence of a serotoninergic innervation of the gastrointestinal tract could be demonstrated convincingly.[17,18] Likewise, in the lung it could be proved that the so-called neuro-epithelial bodies do indeed contain serotonin (*Fig.* 9.7*c*).[86]

In addition, these new techniques have introduced new subjects, e.g. the precise distribution of serotoninergic axons by the combined use of immunocytochemistry and retrograde axonal tracing techniques,[75,87,88] the presence of serotonin and related neuroactive principles in the same neurone,[81–83,89,90] and the ultrastructural localization of serotonin.[91] Following these lines serotonin and noradrenaline-immunoreactive small ganglionic cells were described in the superior cervical ganglion of the rat.[6] Preliminary observations showed that the above antibody techniques might be useful tools in the diagnosis of tumours supposedly derived from adrenal or extra-adrenal medullary tissues (phaeochromocytomas; *Fig.* 9.5*a–d*) or from enterochromaffin cells (carcinoids; *Fig.* 9.8*b*, *see also Fig.* 9.8*c*). Antibodies to noradrenaline and adrenaline are currently being used for a detailed study on the development of the noradrenaline and adrenaline-storing cells of the adrenal medulla of the rat.[84]

4. CONCLUDING REMARKS

1. Two immunocytochemical procedures can be used to detect cells containing noradrenaline, adrenaline, or serotonin, i.e. antibodies to the synthesizing enzymes and those to the end-products. Ideally, both approaches should be used simultaneously.

2. By conventional immunization with immunogens consisting of noradrenaline, adrenaline, or serotonin linked to the carrier protein bovine serum albumin, antibodies can be raised which are specific to the respective hapten. These antibodies can be used to study the cellular distribution of these compounds, provided proper tissue preparation ensures specific reactions.

3. Further study is required to define more precisely the structure of the immunogens, to improve fixation, particularly that of neuronal noradrenaline and adrenaline, and to make feasible the application of immunocytochemical staining of embedded tissues, in particular following plastic embedding (allowing studies on

thin sections as well as the combined use of light and electron microscopy). Last but not least, it would be worthwhile to try to raise more homogeneous antisera, possibly monoclonal antibodies.

Appendix

The indirect immunofluorescence technique has been used for cryostat, paraffin, and Epon sections. All sections were mounted on glass slides coated with chrome–alum gelatin (0·5 g chrome–alum and 0·5 g gelatin in 100 ml distilled water). Before staining, paraffin sections were dewaxed in xylene (2 × 15 min, room temperature). Epon was removed by sodium methoxide (3–5 min, room temperature). The primary and secondary antisera used are indicated in *Table* 9.3. All sera were diluted with 0·1 M phosphate-buffered saline pH 7·3 containing 0·1 per cent Triton X-100 to improve the penetration by the antibodies.[40] In order to prevent drying, sections were incubated according to the descriptions of Tung.[76] The sections were rinsed with 0·1 M phosphate-buffered saline pH 7·3 at room temperature for 2 × 15 min before the first incubation, for 30 min between the first and second antiserum and following the second incubation. The sections were mounted in a mixture of glycerin and 0·1 M phosphate-buffered saline (3 : 1 v/v) and examined in a Zeiss Universal microscope equipped for epifluorescence. Photomicrographs were taken with Kodak Tri-X or Agfapan Vario-XL film. The sections were stored at −20 °C.

Table 9.3. **Indirect immunofluorescence technique**

Primary antisera
Sheep anti-bovine serum albumin–noradrenaline 1 : 500–1000
Rabbit anti-bovine serum albumin–adrenaline 1 : 50–100
Rabbit anti-bovine serum albumin–serotonin 1 : 400–500
18 h (overnight), 4 °C
Secondary antisera
Rabbit anti-sheep immunoglobulins, labelled with fluorescein isothiocyanate (Nordic, Tilburg, The Netherlands) 1 : 16
Sheep anti-rabbit immunoglobulins, labelled with fluorescein isothiocyanate (Statens Bakteriologiska Laboratorium, Stockholm, Sweden) 1 : 16
30 min, room temperature

The peroxidase–anti-peroxidase technique was applied to the same types of sections as well as to sections cut with a vibrating knife microtome. Paraffin and Epon sections were pretreated as described above. In the case of immersion-fixed tissues, sections were treated before staining for 30 min at room temperature with 0·3 per cent H_2O_2 in absolute methanol to block the endogenous peroxidase activity of the erythrocytes. The three types of antisera utilized are indicated in *Table* 9.4. The first and second antisera were diluted with 0·1 M phosphate-buffered saline containing 0·1 per cent Triton X-100. The third antiserum (peroxidase–anti-peroxidase complex) was diluted in buffered saline without Triton X-100. Incubation was performed as mentioned above (sections upside down[76]). The sections were rinsed at room temperature with 0·1 M phosphate-buffered saline, before the first (2 × 15 min) as well as following the second and third incubation (30 min). Next, they were exposed to a peroxidase substrate solution containing 0·075 per cent 3,3′-diaminobenzidinetetrahydrochloride (Sigma) and 0·01 per cent H_2O_2 in 0·05 M Tris-HCl buffer pH 7·6 for 15 min at room temperature. Finally, the sections were dehydrated and mounted in Entellan (Merck). Agfapan-25 or Agfapan Vario-XL films were used for photomicrography. The sections were stored at room temperature.

Table 9.4. ***Peroxidase–anti-peroxidase technique***

Primary antisera
Sheep anti-bovine serum albumin–noradrenaline 1 : 1000–1500
Rabbit anti-bovine serum albumin–adrenaline 1 : 100–200
Rabbit anti-bovine serum albumin–serotonin 1 : 1000–1500
18 h (overnight), 4 °C

Secondary antisera
Swine anti-sheep immunoglobulins (Cappel Laboratories, Cochranville, USA) 1 : 30
Goat anti-rabbit immunoglobulins (Fc-specific) (Nordic, Tilburg, The Netherlands) 1 : 30
2 h, room temperature

Tertiary antisera
Sheep peroxidase–anti-peroxidase complex (Cappel Laboratories, Cochranville, USA) 1 : 90
Rabbit peroxidase–anti-peroxidase complex (Dakopatts, Copenhagen, Denmark) 1 : 90
1 h, room temperature

Notes added in proof

1. Additional absorption experiments on tissues fixed by 4 per cent paraformaldehyde showed that serotonin immunoreactivity could *not* be reduced by conjugates synthesized by linkage of dopamine or histamine to bovine serum albumin using formaldehyde as a coupling reagent (concentrations of the conjugates up to 2 mg/ml; cf. Table 9.1). Thus, under the conditions described in this chapter, serotonin antibodies do not seem to cross-react with dopamine or histamine.

2. Recently,[92] important methodological aspects of the immunocytochemical procedure for serotonin have been reported. Gelatin gels containing serotonin or related compounds were fixed by 4 per cent paraformaldehyde and rinsed as described in section 3.2 of this chapter. Cryostat sections were stained according to the indirect immunofluorescence technique employing the same antiserum as used in the present communication. Blocking (inhibition) experiments revealed that serotonin immunoreactivity could be reduced by serotonin, 5-methoxytryptamine, and at concentrations less than needed for serotonin also with 6-hydroxy-1,2,3,4-tetrahydroxy-β-carboline and 6-methoxy-1,2,3,4-tetrahydro-β-carboline.

In addition, it was found that approximately 46 per cent of serotonin and 59 per cent of 6-hydroxy-1,2,3,4-tetrahydro-β-carboline were retained in the gels following fixation and rinsing. On the contrary, 5-methoxy-tryptamine and 6-methoxy-1,2,3,4-tetrahydro-β-carboline were eluted almost completely. The authors suggest that the high specificity of the presently described immunocytochemical procedure for serotonin is based upon the conversion of serotonin into 6-hydroxy-1,2,3,4-

tetrahydro-β-carboline during the synthesis of the immunogen as well as during tissue fixation, and the high fixation efficiency of this substance.

3. Additional blocking (inhibition) experiments (cf. Table 9.2) showed that 6-hydroxy-1,2,3,4-tetrahydro-β-carboline is able to reduce serotonin immunoreactivity in tissues as well (rat brain and duodenum fixed by 4 per cent paraformaldehyde). Fluorescence was abolished completely at concentrations 8–16 times less than needed for serotonin. Hence, if fixation in the tissue is as effective as found in the above non-biological model, staining results obtained in tissues containing 6-hydroxy-1,2,3,4-tetrahydro-β-carboline should be interpreted with caution.

REFERENCES

1. Dahlström A. and Fuxe K. Evidence for the existence of monoamine-containing neurons in the central nervous system. I. Demonstration of monoamines in the cell bodies of brain stem neurons. *Acta Physiol. Scand. Suppl. 232*, 1964, **62**.
2. Fuxe K. Evidence for the existence of monamine neurons in the central nervous system. IV. Distribution of monoamine nerve terminals in the central nervous system. *Acta Physiol. Scand. Suppl. 247*, 1965, **64**, 39–85.
3. Lindvall O. and Björklund A. Organization of catecholamine neurons in the rat central nervous system. In: Iversen L. L. and Iversen S. D., ed., *Chemical Pathways in the Brain: Handbook of Psychopharmacology*, Vol. 9. New York, Plenum Press, 1978, pp. 139–231.
4. Norberg K. A. Transmitter histochemistry of the sympathetic adrenergic nervous system. *Brain Res.* 1967, **5**, 125–170.
5. Jacobowitz D. M. The peripheral autonomic system. In: Hubbard J. I., ed., *The Peripheral Nervous System*. New York, Plenum Press, 1974, pp. 87–110.
6. Verhofstad A. A. J., Steinbusch H. W. M., Penke B., Varga J. and Joosten H. W. J. Serotonin-immunoreactive cells in the superior cervical ganglion of the rat: evidence for the existence of separate serotinin- and catecholamine-containing small ganglionic cells. *Brain Res.* 1981, **212**, 39–49.
7. Holzbauer M. and Sharman D. F. The distribution of catecholamines in vertebrates. In: Blaschko H. and Muscholl E., ed., *Catecholamines: Handbook of Experimental Pharmacology*, Vol. 33. Berlin, Springer, 1972: 110–185.
8. Hansen J. T. and Christie D. S. Rat carotid body catecholamines determined by high performance liquid chromatography with electrochemical detection. *Life Sci.* 1981, **29**, 1791–1795.
9. Böck P. The paraganglia. In: Oksche A. and Vollrath L., ed., *Handbuch der mikroskopischen Anatomie des Menschen*, Band VI, Teil 8. Berlin, Springer, 1982.
10. Coupland R. E. The adrenal medulla. In: Beck F. and Lloyd J. B., ed., *The Cell in Medical Science*, Vol. 3: Cellular specialization. London, Academic Press, 1975: 193–242.
11. Coupland R. E. Observations on the form and size distribution of chromaffin granules and on the identity of adrenaline and noradrenaline-storing chromaffin cells in vertebrates and man. *Mem. Soc. Endocrinol.* 1971, **19**, 611–635.
12. Hökfelt T., Fuxe K., Goldstein M. and Johansson O. Immunohistochemical evidence for the existence of adrenaline neurons in the rat brain. *Brain Res.* 1974, **66**, 235–251.
13. Fuxe K. and Jonsson G. Further mapping of central 5-hydroxytryptamine neurons: studies with the neurotoxic dihydroxytryptamines. In: Costa E., Gessa G. L. and Sandler M., ed., *Serotonin: New Vistas, Histochemistry and Pharmacology: Advances in Biochemical Psychopharmacology*, Vol. 10. New York, Raven Press, 1974: 1–12.
14. Steinbusch H. W. M. Distribution of serotonin-immunoreactivity in the central nervous system of the rat—cell bodies and terminals. *Neuroscience* 1981, **6**, 557–618.
15. Gershon M. D. Properties and development of peripheral serotonergic neurons. *J. Physiol. (Paris)* 1981, **77**, 257–265.
16. Gershon M. D. and Tamir H. Release of endogenous 5-hydroxytryptamine from resting and stimulated enteric neurons. *Neuroscience* 1981, **6**, 2277–2286.
17. Furness J. B. and Costa M. Neurons with 5-hydroxytryptamine-like immunoreactivity in the enteric nervous system: their projections in the guinea pig small intestine. *Neuroscience* 1982, **7**, 341–349.

18. Costa M., Furness J. B., Cuello A. C., Verhofstad A. A. J., Steinbusch H. W. M. and Elde R. P. Neurons with 5-hydroxytryptamine-like immunoreactivity in the enteric nervous system: their visualization and reactions to drug treatment. *Neuroscience* 1982, **7**, 351–363.

19. Barter R. and Pearse A. G. E. Mammalian enterochromaffin cells as the source of serotonin (5-hydroxytryptamine). *J. Pathol. Bacteriol.* 1955, **69**, 25–31.

20. Vialli M. Histology of the enterochromaffin cells. In: Erspamer, V., ed., *5-Hydroxytryptamine and Related Indolealkylamines: Handbook of Experimental Pharmacology*, Vol. 19. Berlin, Springer, 1966: 1–65.

21. Penttilä A. Histochemical reactions of the enterochromaffin cells and the 5-hydroxytryptamine content of the mammalian duodenum. *Acta Physiol. Scand. Suppl. 281*, 1966, **69**.

22. Lauweryns J. M., Cokelaere M. and Theunynck P. Neuro-epithelial bodies in the respiratory mucosa of various animals. A light optical, histochemical and ultrastructural investigation. *Z. Zellforsch. Mikrosk. Anat.* 1972, **135**, 569–592.

23. Lauweryns J. M. and Peuskens J. C. Neuroepithelial bodies (neuroreceptor or secretory organs?) in human infant bronchial and bronchiolar epithelium. *Anat. Rec.* 1972, **172**, 471–482.

24. Lauweryns J. M., Cokelaere M. and Theunynck P. Serotonin producing neuroepithelial bodies in rabbit respiratory mucosa. *Science* 1973, **180**, 410–413.

25. Paasonen M. K. 5-Hydroxytryptamine in mammalian thyroid gland. *Experientia* 1958, **14**, 95–96.

26. Parrat J. R. and West G. B. 5-Hydroxytryptamine and tissue mast cells. *J. Physiol. (Lond.)* 1957, **137**, 169–178.

27. Kaplan E. L. The carcinoid syndromes. In: Friesen S. R., ed, *Surgical Endocrinology: Clinical Syndromes*. Philadelphia, Lippincott, 1978: 120–147.

28. Manger W. M. and Gifford R. W. *Pheochromocytoma*. New York, Springer, 1977.

29. Von Euler U. S. Synthesis, uptake and storage of catecholamines in adrenergic nerves. In: Blaschko H. and Muscholl E., ed., *Catecholamines: Handbook of Experimental Pharmacology*, Vol. 33. Berlin, Springer, 1972: 186–230.

30. Stjärne L. The synthesis, uptake and storage of catecholamines in the adrenal medulla. The effect of drugs. In: Blaschko H. and Muscholl E., ed., *Catecholamines: Handbook of Experimental Pharmacology*, Vol. 33. Berlin, Springer, 1972: 231–269.

31. Hagen P. B. and Cohen L. H. Biosynthesis of indolealkylamines. Physiological release and transport of 5-hydroxytryptamine. In: Erspamer V., ed., *5-Hydroxytryptamine and Related Indolealkylamines: Handbook of Experimental Pharmacology*, Vol. 19. Berlin, Springer, 1966: 182–211.

32. Christenson J. G., Dairman W. and Udenfriend S. On the identity of dopa decarboxylase and 5-hydroxytryptophan decarboxylase. *Proc. Natl Acad. Sci. USA* 1972, **69**, 343–347.

33. Sims K. L., Davis G. A. and Bloom F. E. Activities of 3,4 dihydroxy-L-phenylalanine and 5-hydroxy-L-tryptophan decarboxylases in rat brain: assay characteristics and distribution. *J. Neurochem.* 1973, **20**, 449–464.

34. Rahman M. K., Nagatsu T. and Kato T. Aromatic L-amino acid decarboxylase activity in central and peripheral tissues and serum of rats with L-DOPA and L-5-hydroxytryptophan as substrates. *Biochem. Pharmacol.* 1981, **30**, 645–649.

35. Geffen L. B., Livett D. G. and Rush R. A. Immunohistochemical localization of protein components of catecholamine storage vesicles. *J. Physiol. (Lond.)* 1969, **204**, 593–605.

36. Fuxe K., Goldstein M., Hökfelt T. and Joh T. H. Immunohistochemical localization of dopamine β-hydroxylase in the peripheral and central nervous system. *Res. Commun. Chem. Pathol. Pharmacol.* 1970, **1**, 627–636.

37. Goldstein M., Fuxe K., Hökfelt T. and Joh T. H. Immunohistochemical studies on phenylethanolamine *N*-methyltransferase, dopadecarboxylase and dopamine β-hydroxylase. *Experientia* 1971, **27**, 951–952.

38. Hökfelt T., Fuxe K. and Goldstein M. Immunohistochemical localization of aromatic L-amino acid decarboxylase (DOPA decarboxylase) in central dopamine and 5-hydroxytryptamine nerve cell bodies of the rat. *Brain Res.* 1973, **53**, 175–180.

39. Hökfelt T., Fuxe K., Goldstein M. and Joh T. H. Immunohistochemical studies of three catecholamine synthesizing enzymes: aspects on methodology. *Histochemie* 1973, **33**, 231–254.

40. Hartman B. K. Immunofluorescence of dopamine β-hydroxylase. Application of improved methodology to the localization of the peripheral and central noradrenergic nervous system. *J. Histochem. Cytochem.* 1973, **21**, 312–332.

41. Swanson L. W. and Hartman B. K. The central adrenergic system. An immunofluorescence study of the localization of cell bodies and their efferent connections in the rat utilizing dopamine β-hydroxylase as a marker. *J. Comp. Neurol.* 1975, **163**, 467–506.

42. Fuxe K., Hökfelt T., Agnati L. F., Johansson O., Goldstein M., Perez de la Mora M., Possani L., Tapia R., Teran L. and Palacios R. Mapping out central catecholamine neurons: immunohistochemical studies on catecholamine-synthesizing enzymes. In: Lipton M. A., DiMasco A. and Killam K. F., ed., *Psychopharmacology. A generation of Progress*. New York, Raven Press, 1978: 67–94.

43. Pickel V. M., Joh T. H. and Reis D. J. Monoamine-synthesizing enzymes in central dopaminergic, noradrenergic and serotoninergic neurons. Immunocytochemical localization by light and electron microscopy. *J. Histochem. Cytochem.* 1976, **24**, 792–806.

44. Joh T. H., Shikimi T., Pickel V. M. and Reis D. J. Brain tryptophan hydroxylase: purification of, production of, antibodies to, and cellular and ultrastructural localization in serotonergic neurons of the rat midbrain. *Proc. Natl Acad. Sci. USA* 1975, **72**, 3575–3579.

45. Gershon M. D., Dreyfus C. F., Pickel V. M., Joh T. H. and Reis D. J. Serotonergic neurons in the peripheral nervous system: identification in gut by immunohistochemical localization of tryptophan hydroxylase. *Proc. Natl Acad. Sci. USA* 1977, **74**, 3086–3089.

46. Grzanna R. and Coyle J. T. Rat adrenal dopamine β-hydroxylase: purification, and immunologic characteristics. *J. Neurochem.* 1976, **27**, 1091–1096.

47. Steinbusch H. W. M., Verhofstad A. A. J. and Joosten H. W. J. Localization of serotonin in the central nervous system by immunohistochemistry: description of a specific and sensitive technique and some applications. *Neuroscience* 1978, **3**, 811–819.

48. Landsteiner K. *The Specificity of Serological Reactions*. 1947. Cambridge, Harvard University Press, 1947.

49. Went S. and Kesztyüs L. Das Antiadrenalin. Synthetische Herstellung eines Adrenalinantigens. Aminoadrenalin und Adrenalin-azoproteine. *Naunyn-Schmiedebergs Arch. Pharmakol. Exp. Pathol.* 1939, **193**, 609–614.

50. Went S., Kesztyüs L. and Szilágyi T. Weitere Untersuchungen über die physiologische Wirkung des Adrenalylazoprotein-Antikörper. *Naunyn-Schmiedeberg's Arch. Pharmakol. Exp. Pathol.* 1943, **201**, 143–149.

51. Fillipp G. and Schneider H. Die Synthese des Serotoninazoproteins. Experimente zur Frage der Serotoninimmunität. *Acta Allergol.* 1964, **19**, 216–228.

52. Ranadive N. S. and Sehon A. H. Antibodies to serotonin. *Can. J. Biochem.* 1967, **45**, 1701–1710.

53. Spector S. Application of immunology to neurochemistry. In: Costa E. and Greengard P., ed., *Advances in Biochemical Psychopharmacology*, Vol. 1. New York, Raven Press, 1969: 181–190.

54. Peskar B. and Spector S. Serotonin: radioimmunoassay. *Science* 1973, **179**, 1340–1341.

55. Spector S., Berkowitz B., Flynn E. J. and Peskar B. Antibodies to morphine, barbiturates and serotonin. *Pharmacol. Rev.* 1973, **25**, 281–291.

56. Grota L. J. and Brown G. M. Antibodies to indolealkylamines: serotonin and melatonin. *Can. J. Biochem.* 1974, **52**, 196–202.

57. Grota L. J. and Brown G. M. Antibodies to catecholamines. *Endocrinology* 1976, **98**, 615–622.

58. Kellum J. M. and Jaffe B. M. Validation and application of radioimmunoassay for serotonin. *Gastroenterology* 1976, **70**, 516–522.

59. Miwa A., Yoshioka M., Shirahata A. and Tamura Z. Preparation of specific antibodies to catecholamines and L-3,4 dihydroxyphenylalanine. I. Preparation of the conjugates. *Chem. Pharm. Bull. (Tokyo)* 1977, **25**, 1904–1910.

60. Miwa A., Yoshioka M. and Tamura Z. Preparation of specific antibodies to catecholamines and L-3,4 dihydroxyphenylalanine. III. Preparation of antibody to epinephrine for radioimmunoassay. *Chem. Pharm. Bull. (Tokyo)* 1978, **26**, 3347–3352.

61. Delaage M. A. and Puizillout J. J. Radioimmunoassays for serotonin and 5-hydroxyindole acetic acid. *J. Physiol. (Paris)* 1981, **77**, 339–347.

62. Verhofstad A. A. J., Steinbusch H. W. M., Penke B., Varga J. and Joosten H. W. J. Use of antibodies to norepinephrine and epinephrine. In: Eränkö O., Soinila S. and Päivärinta H., ed., *Histochemistry and Cell Biology of Autonomic Neurons, SIF cells and Paraneurons: Advances in Biochemical Psychopharmacology*, Vol. 25. New York, Raven Press, 1980: 185–193.

63. Facer P., Polak J. M., Jaffe B. M. and Pearse A. G. E. Immunocytochemical demonstration of 5-hydroxytryptamine in gastrointestinal endocrine cells. *Histochem. J.* 1979, **11**, 117–121.

64. Nemoto N., Kawaoi A. and Shikata T. An attempt of immunohistochemical detection of indoleamines. *Acta Histochem. Cytochem.* 1979, **12**, 575.

65. Sherman S. P., Li C. Y. and Carney J. A. Microproliferation of enterochromaffin cells and the origin of carcinoid tumours of the ileum. *Arch. Pathol. Lab. Med.* 1979, **103**, 639–641.

66. Buffa R., Crivelli O., Lavarini C., Sessa F., Verme G. and Solcia E. Immunohistochemistry of brain 5-hydroxytryptamine. *Histochemistry* 1980, **68**, 9–15.

67. Consolazione A., Milstein C., Wright B. and Cuello A. C. Immunocytochemical detection of serotonin with monoclonal antibodies. *J. Histochem. Cytochem.* 1981, **29**, 1425–1430.

68. Miwa A., Yoshioka M. and Tamura Z. Preparation of specific antibodies to catecholamines and L-3,4-dihydroxyphenylalanine. II. The site of attachment on catechol moiety in the conjugates. *Chem. Pharm. Bull. (Tokyo)* 1978, **26**, 2903–2905.

69. Steinbusch H. W. M., Verhofstad A. A. J. and Joosten H. W. J. Antibodies to serotonin for neuroimmunocytochemical studies on the central nervous system. Methodological aspects and applications. In: Cuello A. C., ed., *IBRO Handbook Series: Methods in the Neurosciences.* Volume: Neuroimmunocytochemistry. New York, Wiley & Son. In press.

70. Coons A. H. Fluorescent antibody methods. In: Danielli J. F., ed., *General Cytochemical Methods.* New York, Academic Press, 1958: 399–422.

71. Sternberger L. A., Hardy P. H., Cuculis J. J. and Meyer H. G. The unlabeled antibody enzyme method of immunohistochemistry. Preparation and properties of soluble antigen-antibody complex (horseradish peroxidase–antihorseradish peroxidase) and its use in identification of spirochetes. *J. Histochem. Cytochem.* 1970, **18**, 315–333.

72. Sternberger L. A. *Immunocytochemistry.* New York, Wiley & Sons, 1979.

73. Kishimoto S., Polak J. M., Buchan A. M. J., Verhofstad A. A. J., Steinbusch H. W. M. Yanaihara N., Bloom S. R. and Pearse A. G. E. Motilin cells investigated by the use of region-specific antisera. *Virchows Arch. Abt. B Zellpathol. (Cell Pathol.)* 1981, **36**, 207–218.

74. Mayor H. D., Hampton J. C. and Rosario B. A simple method for removing the resin from epoxy-embedded tissue. *J. Biophys. Biochem. Cytol.* 1961, **9**, 909–910.

75. Steinbusch H. W. M., Nieuwenhuys R., Verhofstad A. A. J. and Van der Kooy D. The nucleus raphe dorsalis of the rat and its projection upon the caudatoputamen. A combined cytoarchitectonic, immunohistochemical and retrograde transport study. *J. Physiol. (Paris)* 1981, **77**, 157–174.

76. Tung K. S. K. A useful technique applicable to the immunofluorescence test. *J. Immunol. Methods* 1977, **18**, 391–392.

77. Bosin T. R., Jonsson G. and Beck O. On the occurrence of 5-methoxytryptamine in brain. *Brain Res.* 1979, **173**, 79–88.

78. Eränkö O. Histochemical demonstration of noradrenaline in the adrenal medulla of the hamster. *J. Histochem. Cytochem.* 1956, **4**, 11–13,

79. Falck B., Larson B. V. Mecklenburg C., Rosengren E. and Svenaeus K. On the presence of a second specific cell system in mammalian thyroid gland *Acta Physiol. Scand.* 1964, **62**, 491–492.

80. Pickel V. M., Joh T. H. and Reis D. J. A serotoninergic innervation of noradrenergic neurons in locus coeruleus: demonstration by immunocytochemical localization of the transmitter specific enzymes tyrosine and tryptophan hydroxylase. *Brain Res.* 1977, **131**, 197–214.

81. Hökfelt T., Ljungdahl A., Steinbusch H., Verhofstad A., Nilsson G., Brodin E., Pernow B. and Goldstein M. Immunohistochemical evidence of substance P-like immunoreactivity in some 5-hydroxytryptamine-containing neurons in the rat central nervous system. *Neuroscience* 1978, **3**, 517–538.

82. Johansson O., Hökfelt T., Pernow B., Jeffcoate S. L., White N., Steinbusch H. W. M., Verhofstad A. A. J., Emson P. C. and Spindel E. Immunohistochemical support for three putative transmitters in one neuron: coexistence of 5-hydroxytryptamine, substance P- and thyrotropin releasing hormone-like immunoreactivity in medullary neurons projecting to the spinal cord. *Neuroscience* 1981, **6**, 1857–1881.

83. Gilbert R. F. T., Emson P. C., Hunt S. P., Bennet G. W., Marsden C. A., Sandberg B. E. B., Steinbusch H. W. M. and Verhofstad A. A. J. The effects of monoamine neurotoxins on peptides in the rat spinal cord. *Neuroscience* 1982, **7**, 69–87.

84. Verhofstad A. A. J., Steinbusch H. W. M., Joosten H. W. J., Coupland R. E., Colenbrander B. and MacDonald A. A. Development of the noradrenaline- and adrenaline-storing cells of the adrenal medulla and the possible influence of the adrenal cortex. *Acta Endocrin. (Kbh.)* 1981, **97**, Suppl. 243, 285.

85. Lidov H. G. W., Grzanna R. and Molliver M. E. The serotonin innervation of the cerebral cortex in the rat—an immunohistochemical analysis. *Neuroscience* 1980, **5**, 207–227.

86. Lauweryns J. M., de Bock V., Verhofstad A. A. J. and Steinbusch H. W. M. Immunohistochemical localization of serotonin in intrapulmonary neuroepithelial bodies. *Cell Tissue Res.* 1982, **226**, 215–223.

87. Steinbusch H. W. M., Van der Kooy D., Verhofstad A. A. J. and Pellegrino A. Serotonergic and non-serotonergic projections from the nucleus raphe dorsalis to the caudate–putamen complex in the rat, studied by a combined immunofluorescence and fluorescent retrograde axonal labeling technique. *Neurosci. Lett.* 1980, **19**, 137–142.

88. Van der Kooy D., Hunt S. P., Steinbusch H. W. M. and Verhofstad A. A. J. Separate populations of cholecystokinin and 5-hydroxytryptamine-containing neuronal cells in the rat dorsal raphe, and their contribution to the ascending raphe projections. *Neurosci. Lett.* 1981, **26**, 25–30.

89. Glazer E. J., Steinbusch H., Verhofstad A. and Basbaum A. I. Serotonin neurons in nucleus raphe dorsalis and paragigantocellularis of the cat contain enkephalin. *J. Physiol. (Paris)* 1981, **77**, 241–245.

90. Belin M. F., Weisman-Nanopoulos D., Steinbusch H., Verhofstad A., Maître M., Jouvet M. and Pujol J. F. Mise en évidence de glutamate décarboxylase et de sérotonine dans un même neurone au niveau du noyau du raphé dorsalis du rat par des méthodes de double marquage immunocytochimique. *C. R. Acad. Sci. [D] (Paris)* 1981, **293**, 337–341.

91. Pelletier G., Steinbusch H. W. M. and Verhofstad A. A. J. Immunoreactive substance P and serotonin present in the same dense-core vesicles. *Nature* 1981, **293**, 71–72.

92. Schipper J. and Tilders F. J. H. A new technique for studying specificity of immunocytochemical procedures. Specificity of serotonin immunostaining. *J. Histochem. Cytochem.* 1982. In press.

10

Immunocytochemistry of Cell and Tissue Cultures

K. R. Jessen

This chapter will deal exclusively with some technical and biological aspects of the combined use of cultured cells and immunohistochemical techniques. The discussion is based chiefly on experience with nervous tissue in culture and *in situ* using indirect immunofluorescence methods. The essential points made, however, should be generally applicable to other cell systems and to peroxidase as well as fluorescence methods.

1. CULTURE TYPES AND GENERAL PRINCIPLES

Cells can be maintained alive in a variety of ways after removal from the body. In the context of immunohistochemical studies, it is essential to distinguish between two situations. Firstly, parts of organs, constituting relatively large pieces of tissue, can be maintained as explants (organ culture or organotypic tissue culture). Immunohistochemical studies on such systems are usually carried out in the same manner as on freshly dissected tissue, i.e. by fixation and/or freezing of the cultured organ fragment followed by sectioning and application of antibodies to free floating or mounted sections. For example, antibodies have been used in this way to visualize intrinsic peptide-containing neurones in the gut, using hemisections of the gut wall maintained in culture for 2–3 weeks to achieve extrinsic denervation.[1] Since such studies are, as far as the immunohistochemical method is concerned, identical to those on freshly dissected tissues, they will not be dealt with further in this chapter. Secondly, very small tissue fragments or single cells may be cultured (tissue or cell culture, respectively). For immunohistochemistry these preparations differ in one important respect from those mentioned above, since in this case the antibodies are applied to intact cells, living, fixed or frozen, sitting on a glass or plastic surface; no sectioning is therefore involved in the procedure. The localization of antigens in this type of preparation is in general straightforward and does not pose any major new technical problems. They are particularly suitable for the visualization of cell surface molecules, in which case the antibodies are applied directly to the living cells, while excellent results have also been obtained on the localization of intracellular antigens. For such studies, the cell membrane must be made permeable to large molecules by fixation, lipid extraction, or freezing, prior to the application of antibodies. The next two sections of this chapter deal with methodological aspects of antigen localization in this type of culture system. In the last section the advantages of the combined use of immunohistochemical and tissue and cell culture techniques for solving biological problems will be illustrated by two examples from neurobiology.

Prior to a more detailed discussion, however, it is useful to raise two general cautionary points. Firstly, although cells often retain their differentiation and specific properties to a remarkable degree in culture, as the voluminous literature on the subject demonstrates (*see* Federoff and Hertz[2] and Fischbach and Nelson[3]), it is important to remember that there are many examples to the contrary. An example from our own studies is provided by a surface antigen found on glial cells from the peripheral nervous system, termed rat neural antigen-1 (RAN-1).[4] This antigen is defined by a mouse antiserum raised against a rat neural tumour. It has proved useful for cell-type identification in culture, since it appears to be restricted to peripheral glia, and the staining obtained in culture is bright and unambiguous.[5,6] We have found, however, that this antigen is barely detectable on Schwann cells prior to culturing. This result was obtained by immunostaining of a preparation of the rat sciatic nerve, in which the nerve is teased immediately after removal from the body, exposing single Schwann cells as they lie arranged along individual axons (unpublished data). Thus, Schwann cells appear to express much higher levels of RAN-1 in culture than they do in situ. This is illustrated in *Figs*. 10.1 and 10.2. Our observations on another glial surface antigen, rat neural antigen-2 (RAN-2)[7] suggest that in contrast to RAN-1, those peripheral glia that express this antigen in situ lose it during maintenance in culture (unpublished data). These two examples show that in some instances, care is needed in extrapolating from immunohistochemical findings obtained in culture, to mature cells *in vivo*.

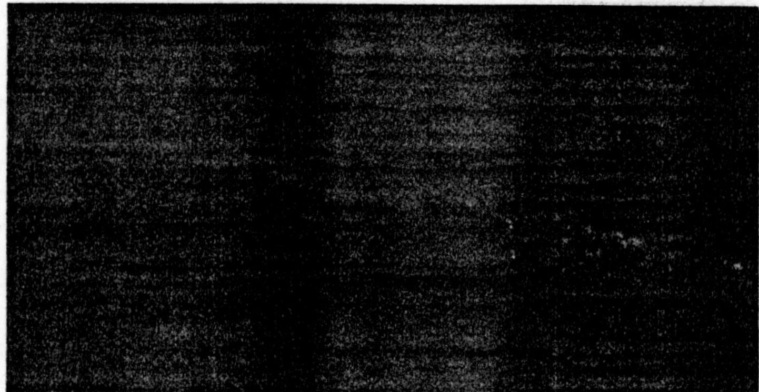

Fig. 10.1. Weak RAN-1 immunoreactivity in a small axon bundle from the teased rat sciatic nerve.

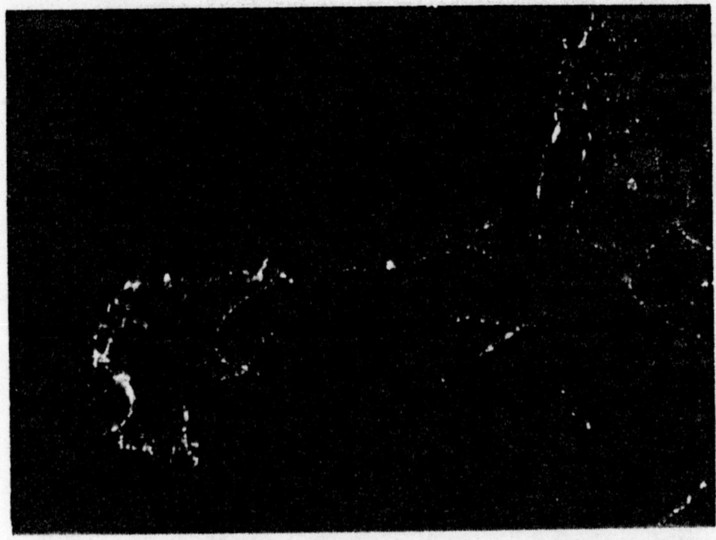

Fig. 10.2. Unambiguous speckled RAN-1 immunostaining on the surface of three cultured enteric glial cells lying at the border of the outgrowth zone. Rat myenteric plexus, 7 days in culture.

The second point is technical. Although variation exists, both between different antigens and cell types, it holds true in general that the quality and reproducibility of immunostaining increases as the cell density in the culture decreases. To obtain immunostaining of high quality in cultures of dissociated cells, the cells should be relatively sparsely seeded and preferably not allowed to grow to confluence before being used for antigen localization. Similarly, in tissue cultures, clearest localization of antigens is obtained at the borders of the outgrowth zone around the explant, while the immunostaining near, and especially inside, the explant is generally unsatisfactory. The importance of cell density on the quality of immunostaining is illustrated in *Figs*. 10.2 and 10.3.

2. LOCALIZATION OF INTRACELLULAR ANTIGENS

As mentioned above, antibody will not penetrate into cells if applied directly to living cultures. For the localization of intracellular antigens, therefore, the first consideration is to find a suitable method for rendering cell membranes permeable to the antibodies, thus allowing access to the antigen. This can be achieved either by chemical agents or by freezing.

2.1. Chemical Permeabilization

In principle, any method can be used which (*a*) effectively permeabilizes the cell membrane; (*b*) does not interfere with the configuration of the antigen in question to the extent of reducing antibody binding to unworkable levels; (*c*) does not lead to high non-specific background fluorescence, e.g. by reacting with substances such

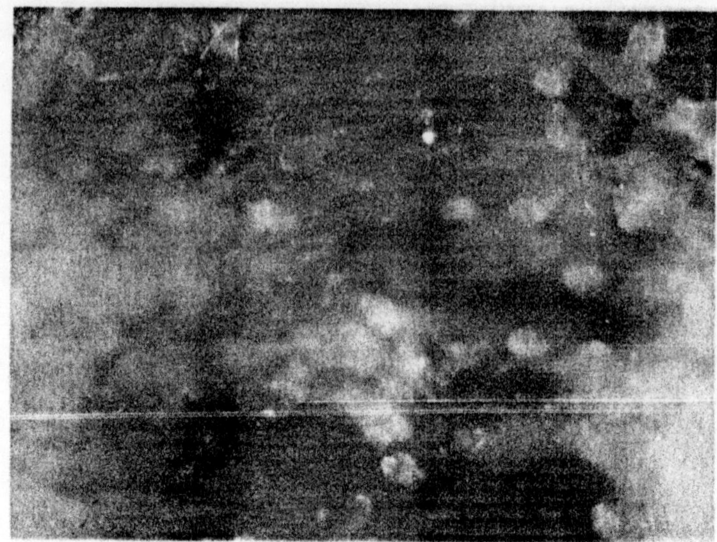

Fig. 10.3. Unsatisfactory RAN-1 immunostaining in the same culture as illustrated in *Fig.* 10.2, showing the staining pattern observed in culture areas where cell density is high.

as polylysine or collagen, that in some culture systems are used to coat culture surfaces, or by being autofluorescent itself and (*d*) preserves cell morphology adequately. Chemical agents in common use for immunohistochemical studies on cultured cells are discussed below, while *Table* 10.1 shows various possible combinations of these treatments. It is worth stressing that the procedure that gives optimal results depends on the antigen, cell type and culture system in question, and is best determined empirically for each new situation. The information provided below and in *Table* 10.1 should provide a useful starting point for efforts in that direction (*see* Osborn[8] for useful discussion and Pearse[9] concerning the chemistry of fixatives and precipitants).

Table 10.1. ***Chemical permeabilization***

Treatment	*Detergent extraction*	*Cross-linking*	*Precipitation/ extraction*	*First antibody*
1	———	———	→ Acid–alcohol or methanol or acetone	→
2	———	→ Formaldehyde or benzoquinone	———	→
3	———	→ Formaldehyde or benzoquinone	→ Acid–alcohol or methanol or acetone	→
4	→ Triton X-100 or NP40	→ Formaldehyde or benzoquinone	———	→
5	→ Triton X-100 or NP40	→ Formaldehyde or benzoquinone	→ Acid–alcohol or methanol or acetone	→

2.1.1. *Cross-Linking Reagents*

Formaldehyde. This frequently used fixative reacts in a complex manner with tissue proteins resulting in the formation of links between adjacent protein chains. Weak formaldehyde fixation in an aqueous solution effectively permeabilizes cell membranes to antibody molecules, while a more thorough fixation, e.g. in higher concentrations of formaldehyde or in glutaraldehyde, may be considerably less successful in achieving high permeability. Most frequently 1–3 per cent formaldehyde is made up in 0·1 M phosphate-buffered saline (PBS) at neutral pH, and fixation is carried out for 5–20 min at room temperature. Formaledhyde fixation is commonly used, in combination with a subsequent acetone or alcohol treatment, in the localization of actin, myosin, and intermediate filament proteins (Treatment 3, *Table* 10.1).[10] It can also be employed alone for the localization of neuropeptides (Treatment 2, *Table* 10.1), although in our hands benzoquinone gives better results in this case, and it gives good preservation of cell morphology.

Benzoquinone. As with formaldehyde, this compound exhibits a variety of reactions with proteins, and appears to act as a weak bifunctional cross-linking reagent. In concentrations of 0·4–0·5 per cent in PBS it has proved superior to a number of other fixatives for immunostaining of peptide antigens in endocrine tissue.[11] Fixation in this solution for 30 min at room temperature (Treatment 2, *Table* 10.1) has been successfully used in the visualization of neuropeptides[12] and other small molecules such as serotonin[13] in cultured neurones.

2.1.2. *Protein Precipitants/Liquid Extractants*

Methanol, ethanol and acetone. These agents remove lipid from tissues and severely distort the tertiary structure of proteins, causing protein precipitation. They are, therefore sometimes referred to, together with other compounds having similar effects, as 'precipitant fixatives'. They are most commonly used as 100 per cent methanol, 5 per cent acetic acid in ethanol (acid–alcohol) and 100 per cent acetone respectively, at −10 to −20 °C for 10 to 20 min. It is important to wash the cultures thoroughly between this treatment and application of the first antibody. This can be achieved by dipping the coverslip carrying the cultures repeatedly in several beakers of washing solution, e.g. tissue culture medium such as Eagle's minimum essential medium buffered with HEPES buffer (MEM-HEPES). When using acid–alcohol, the reaction between the acid and the colour pH indicator in the washing solution provides a useful way of judging the effectiveness of the washes. These agents are often used in combination with cross-linking reagents (Treatment 3, *Table* 10.1). For instance, in the localization of intracellular proteins such as cytoskeletal intermediate filaments and myelin proteins, both methanol and acid–alcohol also yield excellent results when used on their own (Treatment 1, Table 10.1).[5,6] In this case, the preservation of morphology tends to vary considerably from area to area in the culture, although it is often very good and background fluorescence is low.

2.1.3. *Detergents*

Triton X-100 and NP40. The use of these gentle non-ionic detergents in conjunction with immunohistochemistry falls into two categories. Firstly, they can be used prior to fixation, for extraction of cell membranes, an approach primarily adopted for studies on cytoskeletal antigens. This can result in the total removal of intracellular membranes as well as the plasma membrane, although the degree to which this takes place can to some extent be controlled by varying the detergent concentration and incubation time. Since cytoskeletal proteins are often sensitive to their ionic environment and also because of the presence of proteolytic enzymes in unfixed cells, the type of buffer employed during the extraction is an important consideration, and may have to be determined empirically for each new type of experiment. A fairly safe general procedure is to use 0·1–0·5 per cent detergent made up in PIPES buffer containing 1 mM EGTA for the removal of Ca^{2+}, and 2 mM Mg^{2+}; incubation time most often varies from 1 to 3 min. This is generally followed by cross-linkage (Treatment 4, *Table* 10.1) or cross-linkage and precipitant fixation (Treatment 5, *Table* 10.1).[13] Secondly, detergents are often used subsequent to fixation in order to remove remaining lipids (not shown in *Table* 10.1). [12] Since the proteins are now fixed, the choice of buffer is not critical. A common procedure is an overnight wash in 0·3 per cent detergent made up in PBS.

2.2. PERMEABILIZATION BY FREEZING

Sometimes it is important to avoid any loss in the level of antibody–antigen binding, which often occurs during chemical permeabilization. In these instances, freezing may be the method of choice. This is achieved by placing the coverslip carrying the culture, with the cell side facing upwards, on a metal plate cooled with solid CO_2. The cells freeze in a few seconds, judged by the surface of the coverslip turning white, due to the freezing of the thin film of culture medium covering the cells. The coverslip is now removed from the plate and allowed to thaw completely during a few minutes. This procedure is repeated two or three times before application of the first antibody. While leading to optimal preservation of some antigens, a modification of this method may be necessary if the antigen is freely soluble. In the absence of fixation for anchoring or precipitating such molecules, they may diffuse out of the cells during the freeze–thaw cycles. This can be effectively prevented by covering the culture with the first layer of antibodies before the start of the freeze–thaw procedure. At the end of the last thaw, some more antibody solution may be added; the coverslip soon reaches room temperature and incubation with first antibody is now continued in the same way as following chemical permeabilization.

At the end of permeabilization by chemical agents or freezing, the immunostaining and subsequent mounting does not differ in principle from the procedures used for other material, e.g. freeze-mounted tissue sections. It will often be found, however, that washing time after first and second antibody layers can be shortened considerably. Provided that 5 or 6 changes in volumes of about 25 ml each or larger, are employed, 30–45 s are often adequate. An example of intracellular immunostaining is provided in *Fig.* 10.4, showing the localization of substance P to nerve fibres in explant cultures of the myenteric plexus from the guinea-pig (*Fig.* 10.4).

Localization of intracellular antigens in culture offers, in some respects, technical advantages over immunostaining of in situ material, e.g. tissue sections, and can be a very useful adjunct to in situ studies. For instance, cultures can be

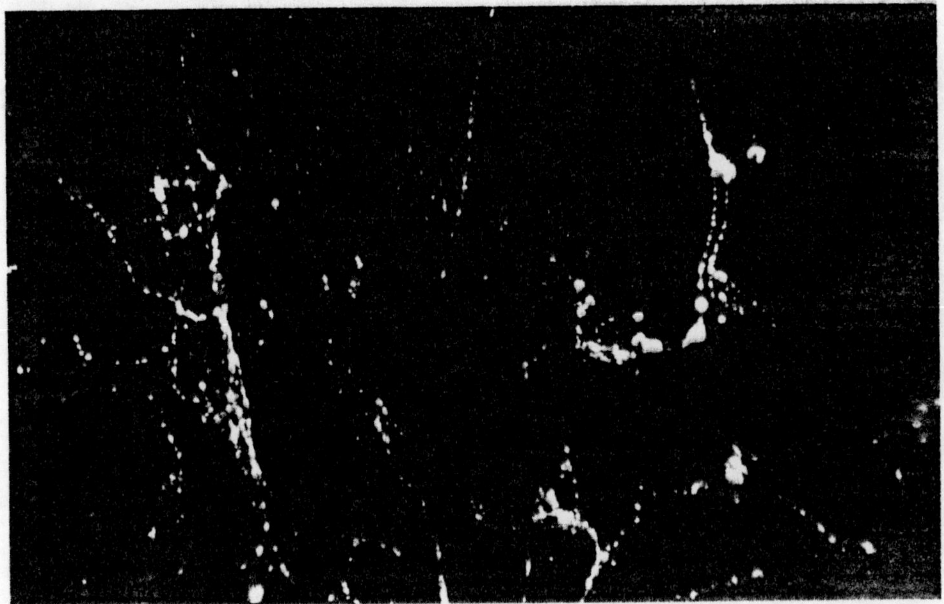

Fig. 10.4. Substance P-like immunoreactivity inside varicose nerve fibres in an explant culture from the enteric nervous system. Guinea-pig myenteric plexus, 10 days in culture.

used to answer unambiguously questions about which tissue elements are responsible for immunostaining observed in sections, otherwise not answerable except by electron immunohistochemistry. Two examples are provided in the localization of glial fibrillary acidic protein (GFAP) and glutamine synthetase in the gastrointestinal tract.[15,16] When frozen sections of the gut wall are treated with antibodies against GFAP or glutamine synthetase, and processed according to the indirect immunofluorescent technique, high quantities of both antigens are revealed inside the small ganglia of the myenteric plexus, as judged by the intense fluorescence (*Figs.* 10.5*a* and 10.6*a*). However, the question as to which cellular elements inside the ganglion carry these antigens cannot be answered with any certainty from this type of observation. In the absence of other information, they could be glial, neuronal or both, and it is not clear from the staining pattern, whether the antigens are intracellular or reside on cell surface membranes. To answer these questions, immunostaining was carried out on explant cultures of the rat myenteric plexus (*Figs.* 10.5*b* and 10.6*b*). In these cultures, individual neurones, glial cells and their processes can be clearly observed, both with phase contrast and fluorescence optics. It was found that both of these antigens were localized in the glial cells, where they were present both in the cell bodies and in the processes. Further, neither antigen could be visualized unless the cells had previously been permeabilized. This shows that both of these antigens are exclusively intracellular in this cell type (*see below*).

3. LOCALIZATION OF CELL SURFACE ANTIGENS

In the case of cell surface antigens, permeabilization of the cell membrane prior to the application of antibodies is unnecessary. It is best avoided, unless there are

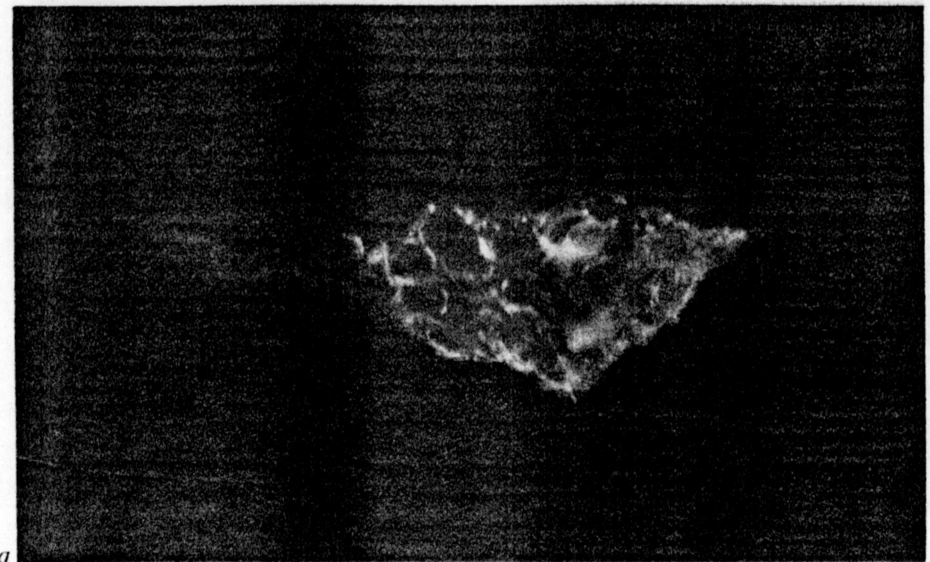

a

b

Fig. 10.5. GFAP-like immunoreactivity. (*a*) Frozen section from the rat colon, showing one ganglion of the myenteric plexus; the localization within the ganglion, of the intense immunoreactivity is difficult to determine. (*b*) Four immunoreactive enteric glial cells in culture; rat myenteric plexus, 15 days in culture. (Reproduced by permission from *Nature*, **286**, no. 5744, 736–737. Copyright (c) 1980, Macmillan Journals Limited.

specific reasons to the contrary, since it will in most cases lead to a weaker immunostaining and may cause higher background fluorescence, through non-specific antibody binding to intracellular components. The first layer of antibody can be applied to the cells immediately after removal from the tissue culture medium or following a brief wash in medium without serum or in PBS. Since the cells are unfixed, incubation time in the first antibody is usually short, 20–60 min at room temperature, and overnight incubation, commonly used for intracellular antigens, is not recommended. Washes following first and second antibody layers proceed in a similar manner to that described for intracellular localization. Unless the cells are to be observed and photographed immediately following the

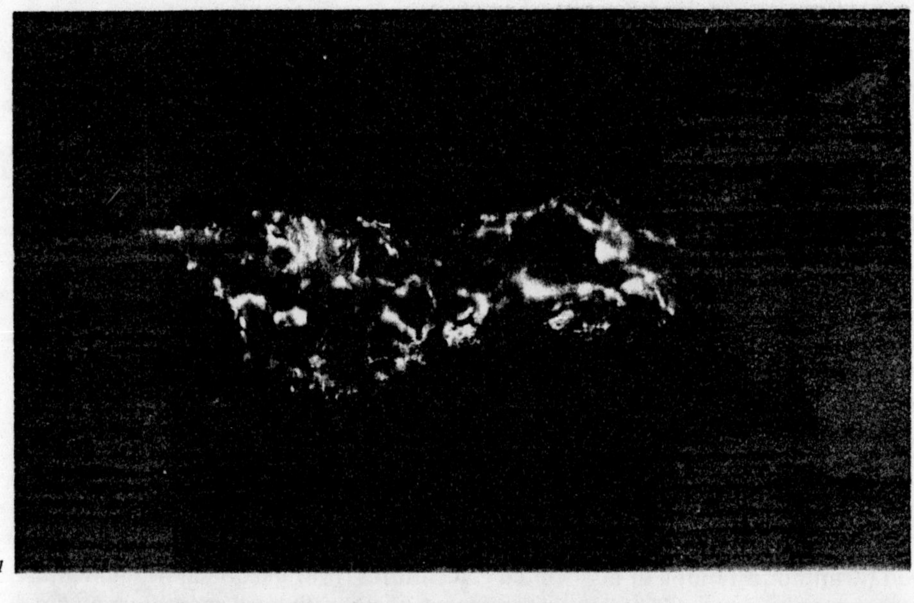

a

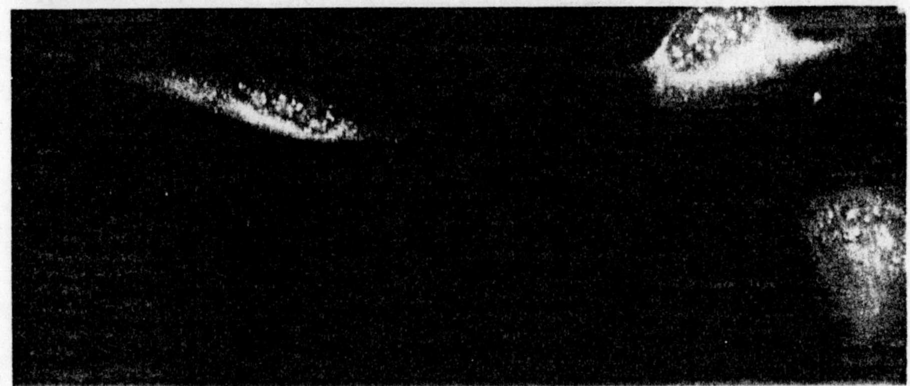

b

Fig. 10.6. Glutamine synthetase-like immunoreactivity. (*a*) Frozen section from the rat colon showing one myenteric ganglion embedded between two muscle layers. Bright immunoreactivity can be seen inside the ganglion, although its cellular localization can not be clearly defined. (*b*) Three immunopositive enteric glial cells in culture. Rat myenteric plexus 10 days in culture.

immunostaining, treatment with formaldehyde or acid–alcohol is recommended to preserve cell morphology. This is most commonly done after the wash which follows the second layer of antibody. Alternatively, it can be carried out between the first and second layers, immediately after the wash following the first layer of antibody. An example of surface immunostaining is shown in *Plate* 13. This Figure, as well as *Fig.* 10.2, illustrates clearly the characteristically uneven or speckled appearance of the fluorochrome. In most cases this will be caused by patching of membrane molecules, which can take place in unfixed membranes during the staining procedure.[17] This phenomenon may serve to enhance the sensitivity of the immunostaining, since aggregates of a fluorochrome are more easily visualized and distinguished from background fluorescence, than the same quantity of the fluorochrome evenly distributed over the cell membrane.

As with intracellular antigens, the immunohistochemical localization of cell surface antigens in culture often provides clear-cut answers to questions about antigen localization. This applies both to the determination of which cell type carries the antigen and to extracellular versus intracellular localization: an antigen which can be visualized without permeabilization must reside on the cell surface, while intracellular localization is indicated if permeabilization is a prerequisite for antigen detection. These features are illustrated in studies on an antigen defined by a monoclonal antibody designated 38D7.[18] This antibody, raised by immunizing with dorsal root sensory ganglia, produced a staining pattern in peripheral ganglia which could not have been interpreted with any degree of certainty without electron immunohistochemistry or tissue culture studies. The antigen distribution in a frozen section of the rat dorsal root ganglion is shown in *Fig.* 10.7. The precipitate, indicating the presence of the antigen, is found surrounding, or associated with the periphery, of neuronal cell bodies. On the basis of this observation, the antigen could reside in at least four different locations: inner or outer aspect of neuronal membranes, or inner or outer aspect of the membrane of the satellite cells that surround the neuronal perikaya. This problem was easily solved by immunostaining of dissociated cell cultures from the dorsal root ganglia. The antigen was restricted to the outer aspect of neuronal cell membranes (*Fig.* 10.8). The results of similar experiments on a variety of cell and explant cultures from the central and peripheral nervous system suggest that the antigen 38D7 is restricted to the surface membranes of peripheral neurones while being absent

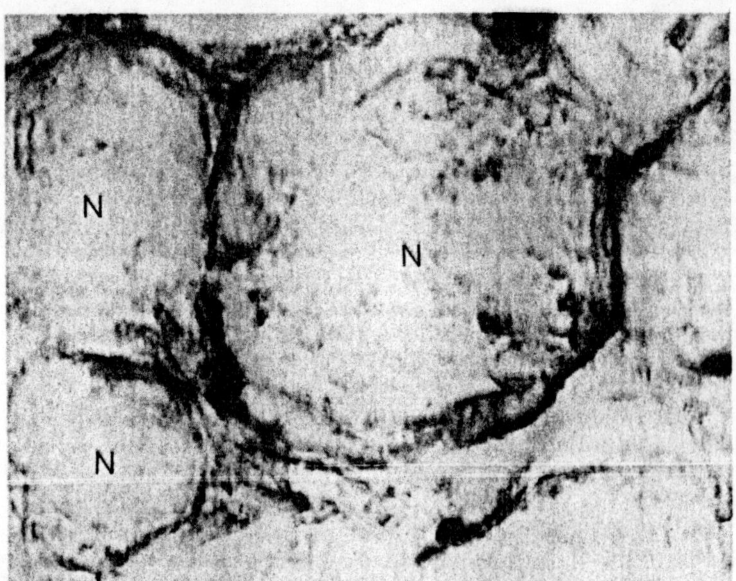

Fig. 10.7. 38D7 immunoperoxidase labelling associated with the peripheral areas of sensory neurones (N). Frozen section from a rat dosal root sensory ganglion. Photograph kindly supplied by Dr. Rhona Mirsky, University College, London, and reproduced by permission from *Nature*, **291**, no. 5814, 418–420. Copyright (c) 1981 Macmillan Journals Limited.

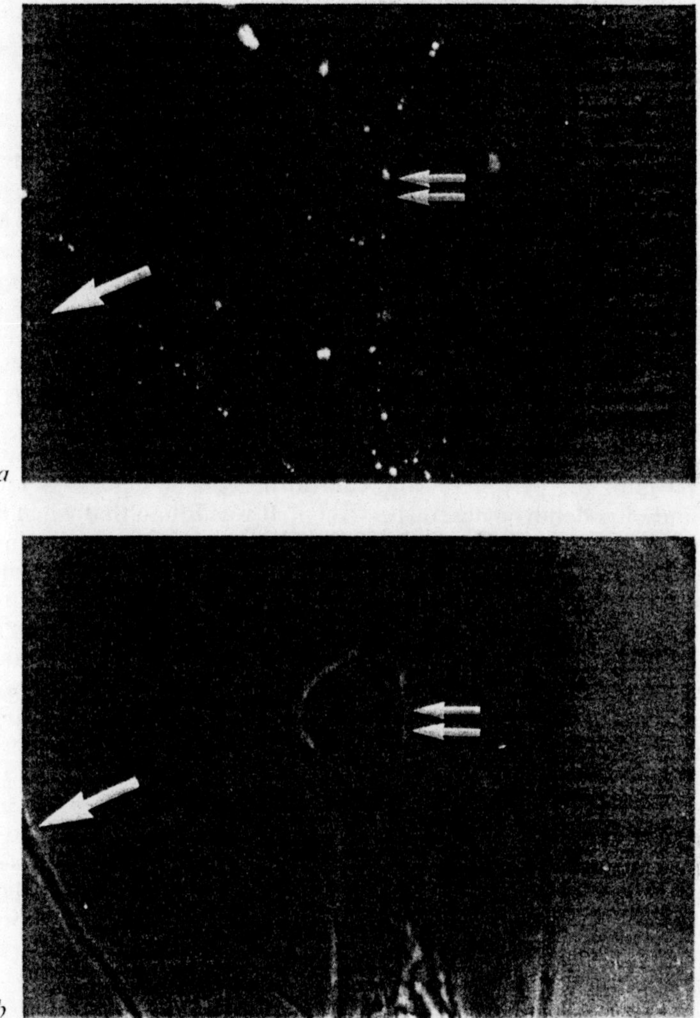

Fig. 10.8. 38D7 immunofluorescence staining of a culture containing sensory neurones and glia. Note immunopositive neuronal cell body (double arrow) and fibres, while a glial cell (single arrow) is negative. (*a*) and (*b*) represent fluorescence and phase contrast view respectively, of the same field. (Photographs kindly supplied by Dr Rhona Mirsky, University College London.)

from central neurones and nearly all non-neuronal cells. It would thus represent the first chemical marker to comprehensively distinguish between peripheral and central neurones.

4. APPLICATION OF IMMUNOHISTOCHEMISTRY AND CULTURE TECHNIQUES TO PROBLEMS IN NEUROBIOLOGY

The combination of tissue culture methods and immunological technology represents a novel approach of considerable promise in studies on the nervous system. Below, two recent examples are described briefly. The first deals with neuro-glia interactions in the peripheral nervous system, while the second addresses questions concerning brain development.

4.1. Neuronal Control of Myelin Synthesis

Schwann cells and oligodendrocytes elaborate multilayered myelin sheaths around axons in the peripheral and central nervous systems, respectively. Since some serious diseases involve the failure of these cells to form or maintain the myelin sheath, it is of great importance to understand the factors that control myelin synthesis and maintenance. One of the central questions in this context concerns the control of synthesis, within the Schwann cell or oligodendrocyte cytoplasm, of the major biochemical constituents of myelin, such as galactocerebroside (GC), sulphatide (S), and myelin basic protein (BP). This problem was studied by Mirsky and her colleagues, using antibodies against GC, S and BP, and dissociated cell cultures from parts of the peripheral and central nervous system as a source of Schwann cells and oligodendrocytes, respectively.[6] It was found that when myelin-forming Schwann cells were dissociated from the axons they normally surround and put into culture, GC, S and BP were all immunohistochemically demonstrable in the Schwann cells for periods of up to 16–20 h after they had been put into culture (*Fig.* 10.9). Interestingly, however, the number of positive cells fell rapidly after longer culture periods, so that after 5–6 days none of the Schwann cells contained a detectable quantity of these substances. Mere contact with axons in Schwann cell cultures of this type was not enough to prevent this disappearance of myelin molecules, as shown by coculturing the Schwann cells with peripheral neurones. Oligodendrocytes on the other hand behaved differently. GC, S and BP, all of which were readily detectable by immunohistochemistry from the onset of culturing, were fully maintained in the oligodendrocytes when grown in dissociated cultures without neurones for periods of up to several weeks. These results suggest that, surprisingly, the organization of myelin synthesis differs between these two cell types, in that 'rat Schwann cells require a continuing signal from appropriate axons to make detectable amounts of myelin-specific glycolipids and proteins, while oligodendrocytes do not.[6]

4.2 Biological Clocks Versus Positional Information in Brain Development

In a study of brain development, Abney et al.[19] used cell type-specific immunohistochemical markers to define and identify the major cell types: neurones, defined by tetanus toxin binding; astrocytes, defined by the filament protein GFAP; oligodendrocytes defined by their content of galactocerebroside (GC). By immunostaining freshly prepared cell suspensions from embryonic and neonatal rat brain, a regular sequence of appearance of these markers in normal development was observed: tetanus-positive neurones were present already in 10-day-old embryos, (the earliest time tested), GFAP-positive astrocytes appeared at 15–16 days in utero, while galactocerebroside-positive oligodendrocytes were first detected 2–3 days after birth. These authors now addressed the question, as to whether this regular timing had been programmed into the respective precursor

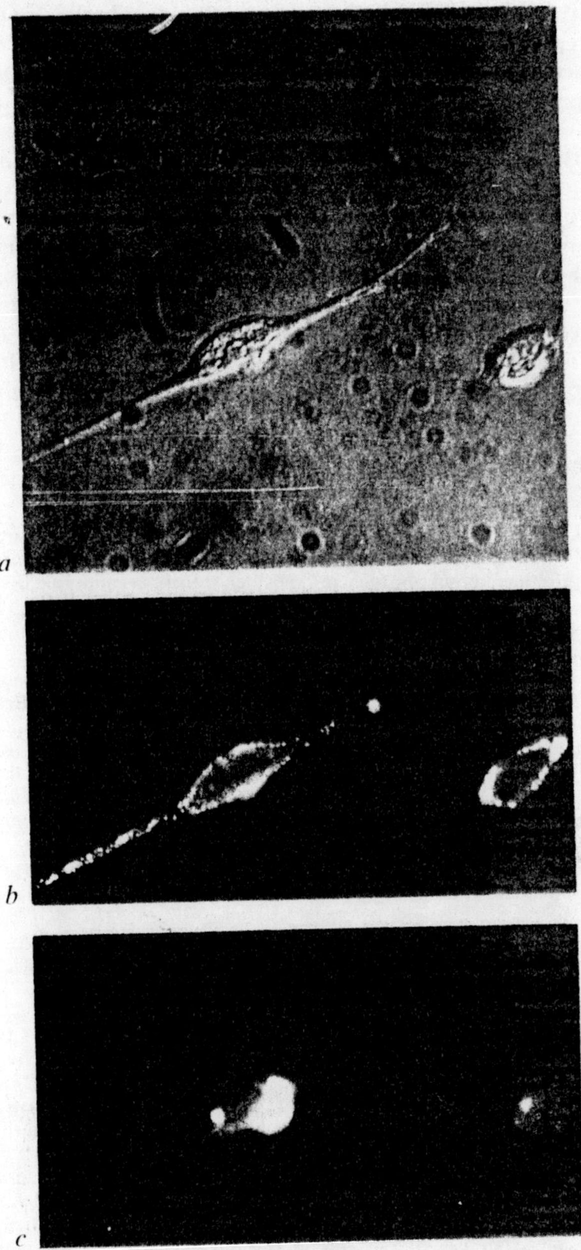

Fig. 10.9. Two Schwann cells in a 16–20 h cell culture of the newborn rat sciatic nerve. The culture was incubated with mouse anti-RAN-l antibodies followed by goat anti-mouse Ig conjugated to rhodamine and, after fixation, with rabbit-anti-BP followed by goat anti-rabbit Ig conjugated to fluorescein. The cells were viewed by Nomarski optics. (*a*), rhodamine (*b*) and (*c*) fluorescein. Both Schwann cells are RAN-l positive and one of them contains readily demonstrable basic protein. Photographs kindly supplied by Dr. Rhona Mirsky, University College, London, and reproduced from *The Journal of Cell Biology* 1980, **84**, 483–494, by copyright permission of the Rockefeller University Press.

cells already by the tenth day of embryonic life, or whether extrinsic signals, determined by the position of the precursor cells within the developing tissue, were required as triggers (positional information, *see* Wolpert[20]). Dissociated cell cultures were prepared from brains of 10-day-old embryos and maintained *in vitro* for up to several weeks. The appearance of the marker antigens was monitored by immunostaining the cells after varying periods *in vitro*. This revealed a pattern, which exactly matched the normal developmental timing. Thus, neurones binding tetanus toxin were found from the onset of culturing; GFAP-immunofluorescent astrocytes first appeared after 5–6 days in culture (10-day-old embryo +5–6 days in culture equalling 15–16 days *in vitro*); and galactocerebroside-containing oligodendrocytes were first seen after 13–14 days of culturing (10-day-old embryo+3–14 days in culture equalling second and third day after birth, since the gestation period is 21 days). It was concluded 'that biological clocks are more important than positional information in gliogenesis between 10 days of gestation and 1 week after birth in the rat'.[19]

REFERENCES

1. Schultzberg M., Dreyfus C. F., Gershon M. D., Hökfelt T., Elde R., Nilsson F., Said S. and Goldstein M. VIP, enkephalin, substance P and somatostatin-like immunoreactivity in neurons intrinsic to the intestine: immunohistochemical evidence from organotypic tissue cultures. *Brain Res.* 1978, **155**, 239–248.
2. Federoff S. and Hertz L. (eds) *Cell Tissue and Organ Cultures in Neurobiology*. New York, Academic Press, 1977.
3. Fischbach G. D. and Nelson P. G. Cell culture in neurobiology. In: Brookhart J. M. and Mountcastle V. B., ed., *Handbook of Physiology: The Nervous System*. Bethesda, American Physiological Society, 1977, pp. 719–774.
4. Fields K. L., Gosling C., Megson M. and Stern P. L. New cell surface antigens in rat defined by tumours of the nervous system. *Proc. Natl Acad. Sci. USA* 1975, **72**, 1286–1300.
5. Raff M. C., Fields K. L., Hakomori S., Mirsky R., Pruss R. M. and Winter J. Cell-type-specific markers for distinguishing and studying neurons and the major classes of glial cells in culture. *Brain Res.* 1979, **174**, 283–308.
6. Mirsky R., Winter J., Abney E. R., Pruss R. C., Gavrilovic J. and Raff M. C. Myelin-specific proteins and glycolipids in rat Schwann cells and oligodendrocytes in culture. *J. Cell Biol.* 1980, **84**, 483–494.
7. Bartlett P. F., Noble M. D., Pruss R. M., Raff M. C., Rattray S. and Williams C. A. Rat neural antigen-2 (RAN-2): a cell surface antigen on astrocytes, ependymal cells, Müller cells and leptomeninges defined by a monoclonal antibody. *Brain Res.* 1981, **204**, 339–352.
8. Osborn M. Localization of proteins by immunofluorescent techniques. *Techniques in Cellular Physiology P107*. Amsterdam, Elsevier/North Holland Scientific Publishers Ltd, 1981: 1–28.
9. Pearse A. G. E. *Histochemistry: Theoretical and Applied*, Vol. 1. Edinburgh, Churchill Livingstone, 1980.
10. Gröschel-Stewart U., Chamley J. H., McConnell J. and Burnstock G. Comparison of the reaction of cultured smooth and cardiac muscle cells and fibroblasts to specific antibodies to myosin. *Histochemistry* 1975, **43**, 215–224.
11. Polak J. M. and Pearse A. E. G. Bifunctional reagents as vapour and liquid-phase fixatives. *Histochem. J.* 1975, **7**, 179–186.
12. Jessen K. R., Saffrey M. J., Van Noorden S., Bloom S. R., Polak J. M. and Burnstock G. Immunohistochemical studies of the enteric nervous system in tissue culture and in situ: localization of vasoactive intestinal polypeptide (VIP), substance-P and enkephalin immunoreactive nerves in the guinea-pig gut. *Neuroscience* 1980, **5**, 1717–1735.
13. Saffrey M. J. Personal communication, 1982.
14. Osborn M. and Weber K. The detergent-resistant cytoskeleton of tissue culture cells includes the nucleus and the microfilament bundles. *Exp. Cell Res.* 1977, **106**, 339–349.
15. Jessen K. R. and Mirksy R. Glial cells in the enteric nervous system contain glial fibrillary acidic protein. *Nature* 1980, **286**, 736–737.

16. Jessen K. R. and Mirksy R. Immunohistochemical methods define a new type of glial cell. *Neurosci. Lett.* 1981, Suppl. 7, 396.
17. de Petris S. and Raff M. C. Normal distribution, patching and capping of lymphocyte surface immunoglobulin studied by electron microscopy. *Nat. New Biol.* 1973, **241**, 257.
18. Vulliamy T., Rattray S. and Mirsky R. Cell surface antigen distinguishes sensory and autonomic peripheral neurons from central neurons. *Nature* 1981, **291**, 418–420.
19. Abney E. R., Bartlett P. P. and Raff M. C. Astrocytes ependymal cells and oligodendrocytes develop on schedule in dissociated cell cultures of embryonic rat brain. *Dev. Biol.* 1981, **83**, 301–310.
20. Wolpert L. Positional information and the spatial pattern of differentiation. *J. Theoret. Biol.* 1969, **25**, 1–47.

11

Immunocytochemistry of Regulatory Peptides

J. M. Polak and
S. R. Bloom

Brown-Sequard, with his self-imposed experiments, involving injection of testis in the hope of recovering his youth, initiated the idea of 'chemical messengers acting at a distance'. This concept was further validated by Bayliss and Starling[1] during the discovery of the very first hormone, secretin. The concept of blood-borne messengers predominated for more than half a century and seriously challenged Pavlov's Nobel prize-winning concept of nervism.

Indeed, the first idea that endocrinology dealt exclusively with hormones produced in ductless glands, later including scattered endocrine cells, and acting via the circulation, stayed with us for very many years. Immunocytochemistry was instrumental in revolutionizing this concept of endocrinology.[2] Using specific antibodies to substance P we,[3] and others,[4] showed the simultaneous localization of substance P in some endocrine argentaffin cells of the gut mucosa and in enteric nerves. A year or two later, some other peptides, vasoactive intestinal polypeptide (VIP)[5,6], somatostatin[7] and the enkephalins, [8] were found in peripheral autonomic nerves, the latter two being also localized by immunocytochemistry to typical endocrine cells of the gut,[9–11] pancreas,[9] thyroid[12,13] and adrenal.[14,15] It is now widely accepted that a considerable number of active peptides, all accommodated under the broad term 'regulatory peptides', are present both in the typical endocrine cells of Feyrter's diffuse endocrine[16] or Pearse's APUD (*a*mine or amine *p*recursor *u*ptake and *d*ecarboxylation) system[17] and in central[18] and peripheral nerves.[19]

A new aspect of endocrinology has thus emerged. Unlike classical glandular endocrinology, dealing with circulating hormones and their excess or deficiency following malfunction of particular endocrine glands, this newly recognized endocrine system is characterized by the production and release of a number of regulatory peptides, only some of which will be acting as circulating hormones. Others will act only in the area of their release from endocrine cells or nerves, as potent modulators or as neurotransmitters. In addition, it is possible that these regulatory peptides are released simultaneously and act in an orchestrated (agonistic or synergistic) manner.

1. THE DIFFUSE NEUROENDOCRINE SYSTEM

Feyrter first introduced the idea of a 'diffuse endocrine system' when he described the presence of a large number of 'clear' cells (weakly stained by conventional histological reagents) to which he attributed a local endocrine (paracrine) role.[16] These cells were described in most organ systems, including the gut, pancreas, skin, adrenals, lung, pituitary and urogenital system. The modern concept of neuroendocrinology was introduced more than 30 years ago (*see* Scharrer's review[20]). The recent finding of identical active peptides in neurones and endocrine cells has unified these two systems into a single and versatile diffuse neuroendocrine system.[21] Support for this concept has been mostly derived from morphological (principally immunocytochemical) observations. Physiological proof of its existence as a uniform and co-ordinated system must await the further development of modern technology.

This massive system of regulatory peptides was first found in the mammalian gut and brain, similar peptides being present in the skin of amphibia, and it is now known to be present in almost every peripheral system. Our knowledge of the chemistry, actions and other properties of these regulatory peptides has grown considerably in the past few years (*see* several review articles and books dealing with the subject.[22–25] The main details of these regulatory peptides in mammals are summarized in *Table* 11.1. However, it must be taken into consideration that what the table format gains in terms of clarity and simplicity, it loses in terms of flexibility, and the subject is considerably more complex than it appears here.

2. IMMUNOCYTOCHEMICAL MEANS OF VISUALIZING THE DIFFUSE NEUROENDOCRINE SYSTEM IN ITS ENTIRETY

Several histological stains have been used for the separate demonstration of endocrine (APUD) cells and nerves. Some of these (silver impregnation,[26] lead haematoxylin)[27] have become quite popular because of their simplicity and satisfying results. However, not *all* endocrine or neural cell types can be readily demonstrated and the simultaneous staining of both components of the diffuse neuroendocrine system has not been achieved. Immunocytochemistry of regulatory peptides has played a key role in the demonstration of individual components of this system, lacking, in consequence, the ability to demonstrate it in its entirety.

In the 1960s a group of brain proteins were isolated and characterized.[28] Some of them turned out to be isomers of the glycolytic enzyme, enolase. One, in particular, was found to be present exclusively in neurones and termed neurone-specific enolase (NSE).[29] Interestingly, NSE was later found to be present in addition in endocrine (APUD) cells of the pancreas, thyroid, adrenal and

Table 11.1 **Mammalian regulatory peptides***

Peptide	*Main distribution*	*Established actions*	*Proposed mode of action*	*Sequence similarities with other peptides*
Gastrin	Gut	Stimulates gastric acid secretion	Endocrine	Cholecystokinin
Cholecystokinin	Gut Brain	Stimulates contraction of gall-bladder and secretion of pancreatic enzymes	Endocrine and neurotransmitter	Gastrin
Secretin	Gut	Stimulates secretion of pancreatic bicarbonate	Endocrine	Glucagon, VIP, GIP, PHI†
Glucagon and enteroglucagon	Pancreas Gut Brain	Regulates carbohydrate metabolism (pancreatic glucagon) Trophic to gut (enteroglucagon)	Endocrine and neurotransmitter	Secretin, VIP, GIP, PHI
Gastric inhibitory polypeptide (GIP)	Small intestine	Stimulates secretion of insulin Inhibits secretion of gastric acid	Endocrine	Secretin, Glucagon, VIP, PHI
Vasoactive intestinal polypeptide (VIP)	Central and peripheral nervous system	Stimulates muscle relaxation, vasodilation, secretion	Neurotransmitter	Secretin, Glucagon, GIP, PHI
Motilin	Small intestine	Stimulates gut motility	Endocrine	Not known
Pancreatic polypeptide	Pancreas	Inhibits pancreatic enzyme secretion and gall-bladder contraction	Endocrine	PYY‡ NPY§

Somatostatin	Gut Pancreas Thyroid Brain	Inhibits release and action of many peptides (? antagonistic to bombesin)	Endocrine Paracrine Neurotransmitter	Fish urotensin II
Substance P	Central and peripheral nervous system	Sensory Vasodilatory Stimulates muscle contraction	Neurotransmitter	Bombesin
Bombesin	Gut Brain Lung	Stimulates release of peptides (? antagonistic to somatostatin)	? Paracrine Neurotransmitter	Substance P
Neurotensin	Gut Adrenal Brain	Vasodilatory Inhibits gastric acid secretion	Endocrine Neurotransmitter	Amphibian xenopsin
Enkephalin	Gut Brain Sympathetic nervous system Carotid body Adrenal	Opiate	? Endocrine Neurotransmitter	Belongs to a large family of opioid peptides

*Only well-established peptides originally extracted from brain and gut are shown in this Table. †PHI *p*eptide *h*istidine *i*soleucine. ‡PYY *p*eptide *t*yrosine *t*yrosine. §NPY *n*eural *p*eptide *t*yrosine.

pituitary.[30] It has now been shown that antibodies to NSE are capable of demonstrating all components, both neural and endocrine, of the diffuse neuroendocrine system of most peripheral tissues including the gut,[31] pancreas (*Fig.* 11.1),[31] lung,[32] adrenal[30] and skin.[33] Thus, the use of NSE antibodies is the best available method of demonstrating the diffuse neuroendocrine system and its derivative neoplasms.

3. DISTRIBUTION OF REGULATORY PEPTIDES IN THE CENTRAL NERVOUS SYSTEM AND IN TWELVE PERIPHERAL ORGANS

We have recently been able to obtain information concerning the tissue distribution and precise quantities of a number of regulatory peptides both in the central nervous system and in most peripheral tissues, using a combination of immunocytochemistry and radioimmunoassay. Technological advances have greatly improved the reliability and consistency of these two methods. The recognition that peptide hormones are water soluble led to the development of a wide variety of fixatives suitable for peptide immunocytochemistry.[34] These are principally based on the use of cross-linking agents (e.g. diethylpyrocarbonate or *p*-benzoquinone) capable of rendering the peptide insoluble and yet preserving its antigenic sites free for subsequent antigen–antibody interactions. In addition, the development of improved immunization procedures has led to the production of many suitable antibodies (hetero- or monoclonal) of high affinity, avidity and titre for immunocytochemistry (for details *see* Chapters 2–4). Lastly, the availability of pure (synthetic) peptides or their fragments has not only ensured the specificity of the immune reaction (for details *see* Chapters 2 and 3), but has also permitted the accurate characterization, in terms of the specific antibody, to a determined segment of the peptide's amino acid sequence. The correct combination of a particular fixation procedure (e.g. cross-linking agents used in solution or as vapour fixatives), a variety of tissue preparations (e.g. cryostat sections, freeze-drying followed by vapour fixation and paraffin embedding) and the use of a panel of region-specific antibodies has greatly aided the accurate localization of a regulatory peptide within a particular tissue structure (for further details *see* Chapter 2 and the Appendix, p. 206).

In this section we shall describe the distribution of the peptides in the central nervous system and in twelve peripheral organ systems.

3.1 Brain

More than fifteen regulatory peptides have been found to be present in the brain, frequently in separate systems of cell bodies, many of which project to other brain areas in newly recognized peptidergic pathways.[18] Far from being a local curiosity in certain areas of the brain, peptide-containing neurones are as widely distributed as those containing the classical neurotransmitters, acetylcholine and noradrenaline. The hypothalamus is particularly rich in peptides and every regulatory peptide so far discovered has been localized here.

The limbic system (hippocampus and amygdala) is the area of the brain involved in the setting of emotional tone, in memory and sexual drive and in modulation of

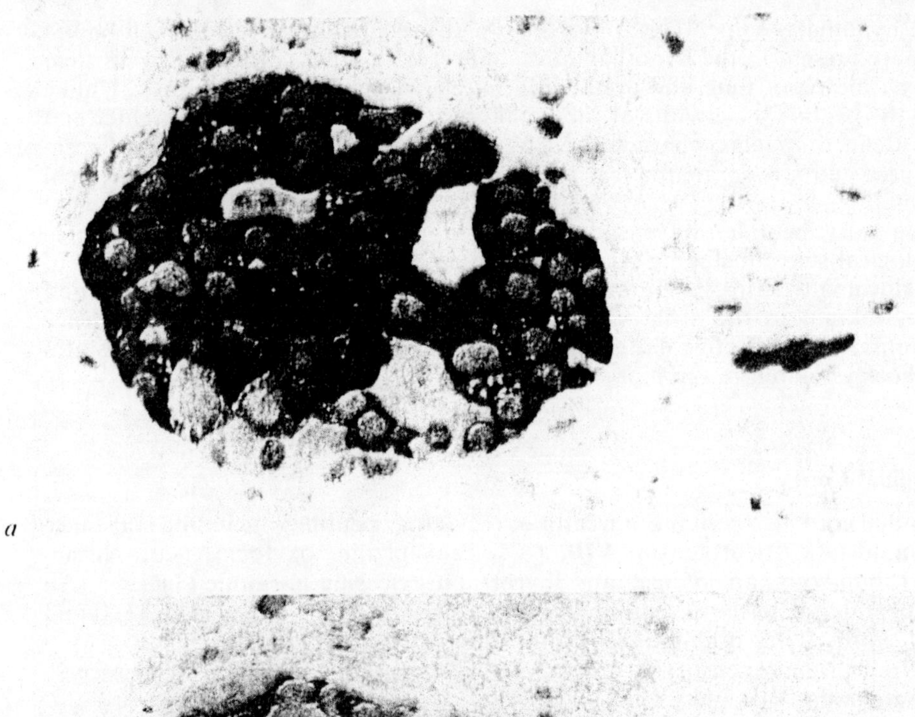

a

b

Fig. 11.1. Human pancreas, freeze-dried, fixed in formaldehyde vapour and resin-embedded (*see* the Appendix, Section II). Serial (1 μm) sections immunostained for: (*a*) neurone-specific enolase and (*b*) insulin. Note the delineation of pancreatic islets by the antibodies to neurone-specific enolase which immunostain all endocrine cell types of the islets.

autonomic and visceral responses to external stimuli.[35] This area possesses dense and complex networks of nerves containing cholecystokinin (CCK), neurotensin, enkephalin, substance P, somatostatin and vasoactive intestinal polypeptide (VIP). The cortex also contains large amounts of VIP, CCK and somatostatin.

The hypothalamic peptides follow highly divergent trajectories; the cell bodies which are present in the hypothalamus connect with other cortical or brain stem areas by means of long and apparently unrelated peptidergic pathways. Unlike these, the peptidergic circuits of the limbic system are organized into tighter and better defined systems, characterized by their sharp delimitation from adjacent structures and strong anatomical relationships with particular brain functional systems.[18]

Regulatory peptides have shown to be abnormal in a number of neuropathological states including Alzheimer's disease or senile dementia (decrease of cortical somatostatin),[36] Huntington's chorea (reduction of substance P and cholecystokinin in the striatum[37] and schizophrenia associated with negative symptoms,[38] e.g. affective flattening, poverty of speech (decrease of somatostatin and cholecystokinin in the hippocampus).

3.2. Spinal Cord

The spinal cord also contains a wealth of regulatory peptides, including substance P, somatostatin, neurotensin, VIP, CCK, enkephalin, oxytocin, neurophysin, ACTH, bombesin, angiotensin and thyrotropin-releasing hormone (TRH).[39] All these peptides, with the exception of TRH, are found in the dorsal (*Fig*. 11.2) and ventral horns. They orginate from three different sources:

a. from primary sensory neurones of the dorsal root ganglion (substance P, somatostatin, VIP and CCK);

b. from intrinsic cell bodies of the spinal cord (enkephalin, neurotensin, and possibly oxytocin, ACTH and bombesin);

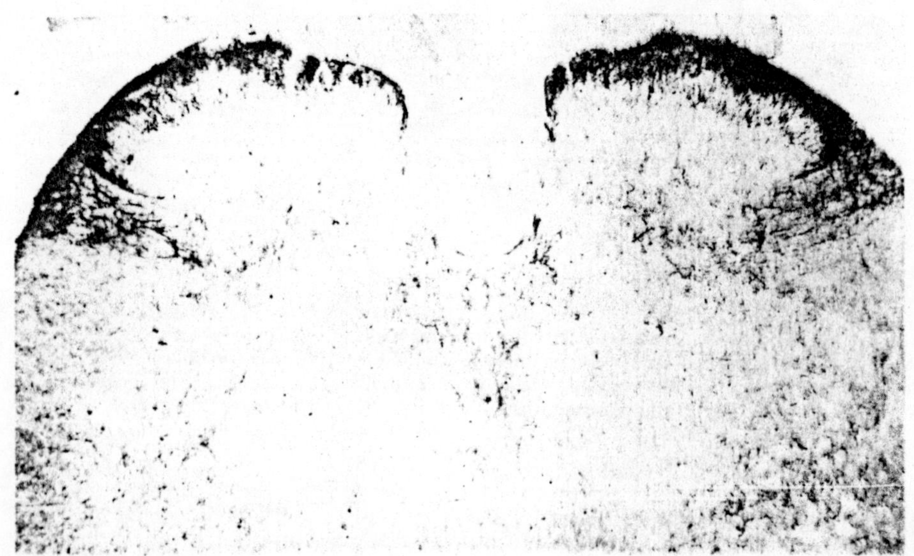

Fig. 11.2. Rat spinal cord (lumbar segment) fixed by perfusion in a solution of *p*-benzoquinone (*see* the Appendix, p. 207) Substance P-immunoreactive material in the dorsal horn (laminae I and II).

c. from cell bodies localized in the upper regions of the spinal cord or the basal ganglia of the brain, forming descending peptidergic pathways. Microsurgical manipulation of the spinal cord has demonstrated, in addition, ascending pathways.

3.3. Gastrointestinal Tract and Pancreas

3.3.1. *Gut*

ENDOCRINE CELLS. The gut and pancreas are known to contain the largest number of regulatory peptides; in fact, a significant proportion of them were first discovered in the gut. Their localization in mucosal endocrine cells or enteric nerves has been fully elucidated by immunocytochemistry and the use of specific antibodies. Those regulatory peptides localized in mucosal endocrine cells have a limited and well-defined anatomical distribution in the bowel, whereas those localized in enteric nerves are found in almost every area and layer of the gut wall. Mucosal endocrine cells intermingled with non-endocrine epithelial cells show a

Fig. 11.3. Human colonic mucosa fixed by immersion in a solution of *p*-benzoquinone (*see* Appendix, p. 207) and immunostained using antibodies to enteroglucagon. Note the classical, irregular shape of the enteroglucagon cell, with its luminal connection (thin arrow) and basal prolongation lying on the basement membrane (thick arrow).

characteristic triangular shape with the narrow end connected to the intestinal lumen and the base resting on the basal lamina of the epithelium (*Fig*. 11.3). A characteristic dendritic process or tail elongation (*Fig*. 11.3) has been described in a number of endocrine cell types.[41] Electron immunocytochemistry, principally by the serial semithin/thin sectioning procedure,[42] has played a key role in correlating the peptide product of a given cell type with the particular appearance of its electron-dense secretory granules, a characteristic feature of peptide-producing endocrine cells. This light/electron microscopical peptide/granule correlation has led to the reclassification of the gut endocrine cell types, with emphasis on peptide production as a primary criterion.[43] The distribution of the different endocrine cell types in the various anatomical regions of the gut matches well with the information obtained by radioimmunoassay of tissue extracts from similar areas.[44]

Mucosal endocrine cells mostly produce peptide hormones acting via the circulation. Basal and postprandial gut hormone levels have been shown to be abnormal in a number of gastrointestinal diseases.[45] In spite of this, the information available concerning the state of the endocrine cell is fragmentary or non-existent. This is partly due to the fact that quantitative immunocytochemistry of the various endocrine cell types has been carried out so far on tissue sections.[46–48] Thus only parts of the endocrine cells can be assessed, necessitating subsequent tedious and possibly inaccurate extrapolation to whole gut. In order to overcome this, we have been able to develop a staining procedure employing whole mounts of microdissected crypts and villi.[49] The specifically immunostained endocrine cells are thus visualized and counted *in toto*.

NERVES. Peptide-containing enteric nerves are very well demonstrated by immunocytochemistry (*Fig*. 11.4).[50] Nerves containing substance P and VIP are the

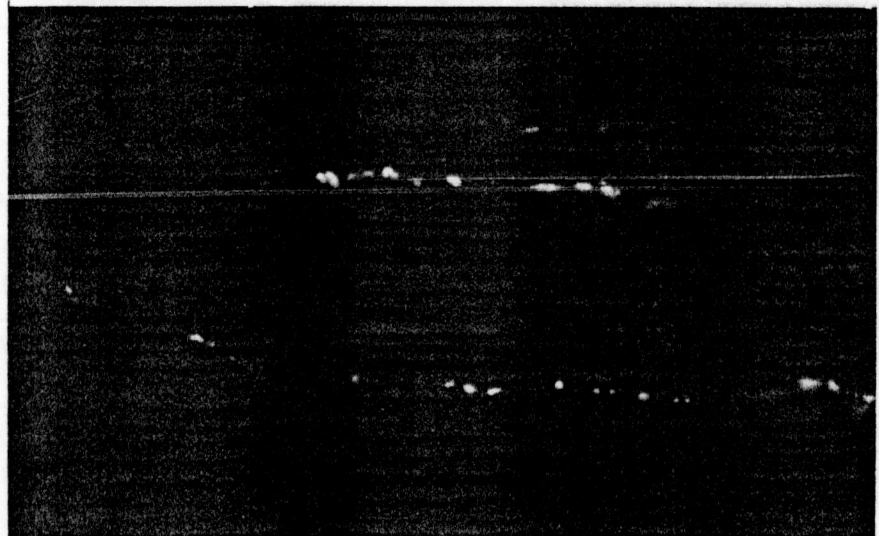

Fig. 11.4. Human colonic wall (muscle), fixed in a solution of *p*-benzoquinone, containing enteric nerves with classical varicosities immunostained in this case using antibodies to VIP.

most abundant, while nerves containing somatostatin and enkephalins and possibly CCK are found in lesser numbers. Peptidergic nerves form an intricate and complex network in every layer of the bowel wall, each nerve showing the classical beaded or varicose appearance. Numerous and extensive animal studies have been reported in the literature concerning this rich peptide-containing network.[51–53] Delicate surgical procedures, e.g. myotomy,[54] extrinsic denervation and separate cultures of enteric plexus,[53–55] have permitted the accurate mapping of individual peptide pathways, their intrinsic origin from neurones located in the myenteric or submucous plexus and their putative extrinsic pathways,[51] connecting the brain and the gut via relay systems (i.e. sympathetic chain).

It has been known for some time that regulatory peptides in enteric nerves are stored in large p-type electron-dense granules of a non-adrenergic non-cholinergic type.[56,57] The application of immunocytochemistry at ultrastructural level has further supported the reputed heterogeneity of these p-type neurosecretory granules[58] and allowed the subclassification of this large peptidergic component of the enteric nervous system.[59] Using the immuno-gold staining procedure (*see* Chapter 6) and specific antibodies to substance P and VIP, we have recently demonstrated their localization in two separate subclasses of large dense p-type neurosecretory granules (*Fig.* 11.5*a, b*), recognizable by their significant ($P < 0{\cdot}01$) differences in size (substance P 85 ± 15 nm, VIP 98 ± 19 nm) and electron density. Many other subclasses of peptidergic granules remain unlabelled by the above procedure, suggesting that a number of other regulatory peptides (e.g. enkephalin, somatostatin) remain to be ultrastructurally localized.

Until recently, the non-neuronal component (glial and Schwann cells) of the enteric nervous system was little investigated at light microscopal level, due to the lack of specific markers. Both glial and Schwann cells have, however, characteristic features at the ultrastructural level.[60] In 1980 Jessen et al.[61] (*see also* Chapter 10) using antibodies to gliofibrillary acidic protein (GFAP) were able to visualize glial cells at the light microscopical level and recently we have been able to immunostain S-100 (*Fig.* 11.6), a newly discovered brain protein, both at light and electron microscopical level (in the latter using the PAP procedure on vibratome sections; (for details *see* Chapter 2). S-100 is now regarded as a useful marker for both non-neuronal components, glial and Schwann cells.

PATHOLOGY OF THE ENTERIC NERVOUS SYSTEM. Peptidergic nerves, in particular those containing VIP and substance P, have been shown to be significantly decreased in diseases of the bowel, e.g. Chagas' disease,[62] Hirschsprung's disease[63] in man, and grass sickness disease of the horse, all associated with intractable chronic constipation and absence or degeneration of intrinsic neuronal cell bodies (*Fig.* 11.7). In contrast, peptidergic nerves appear normal in a generalized autonomic neuropathy like that found in the Shy–Drager syndrome which shows no involvement of the gut neuronal cell bodies.[62] Unlike Chagas' or Hirschsprung's disease, Crohn's disease is characterized by strikingly increased and highly abnormal VIPergic innervation of the bowel.[64] Interestingly, these changes are particularly marked in the mucosa and submucosa of both the granulomatous and non-granulomatous areas, thus making the examination of endoscopic biopsies a potentially useful diagnostic tool.

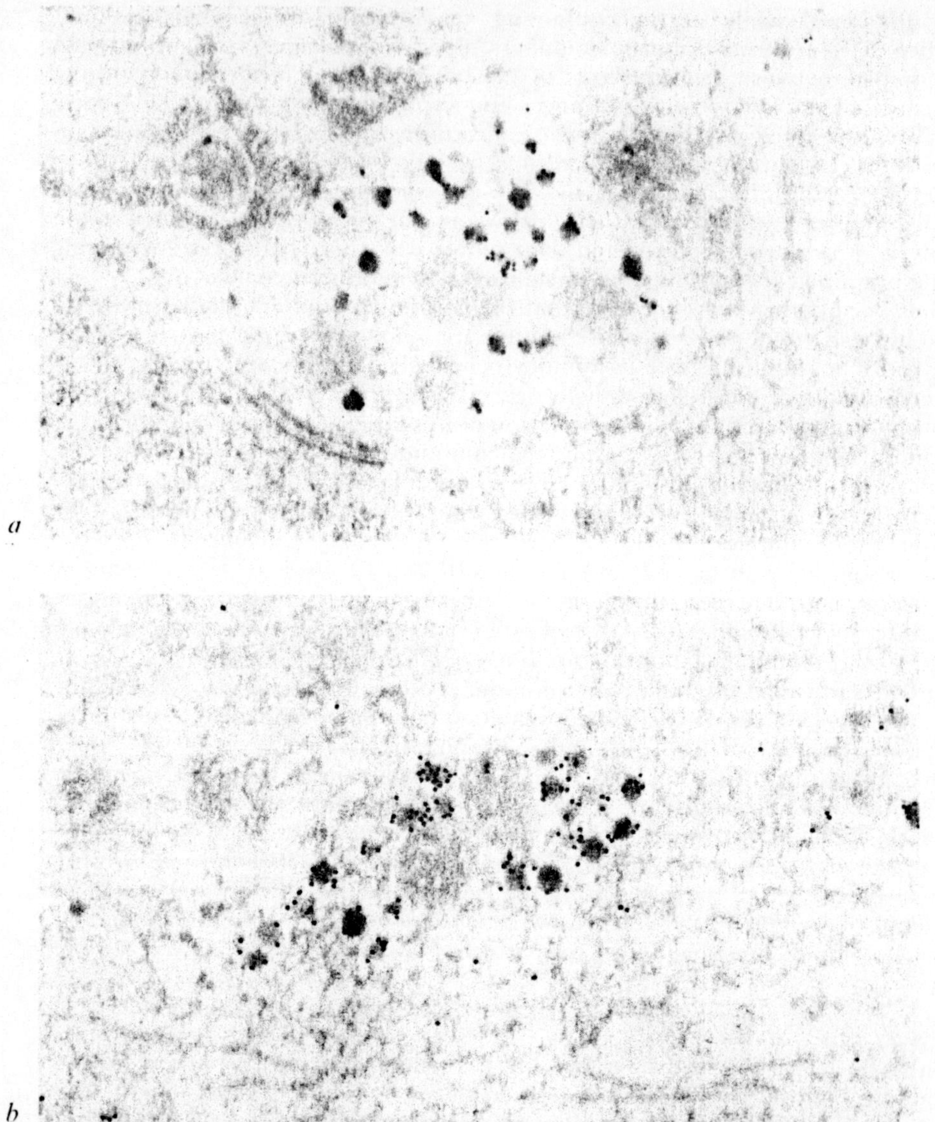

Fig. 11.5. Guinea-pig colon fixed in 3 per cent purified glutaraldehyde for 2 h. Immuno-gold staining was carried out on the thin sections using antibodies to substance P (A) and to VIP (B). Note the different sizes of the secretory granules (*see text*). Magnification × 50 000.

3.3.2. *Pancreas*

The immunolocalization of four regulatory peptides (insulin, glucagon, pancreatic polypeptide[65] and somatostatin[9] (*Fig.* 11.8) in four different endocrine cell types of the pancreatic islets is beyond dispute.[43] A fifth endocrine cell type, recognized at the ultrastructural level by the presence of small (mean diameter 100–110 nm)

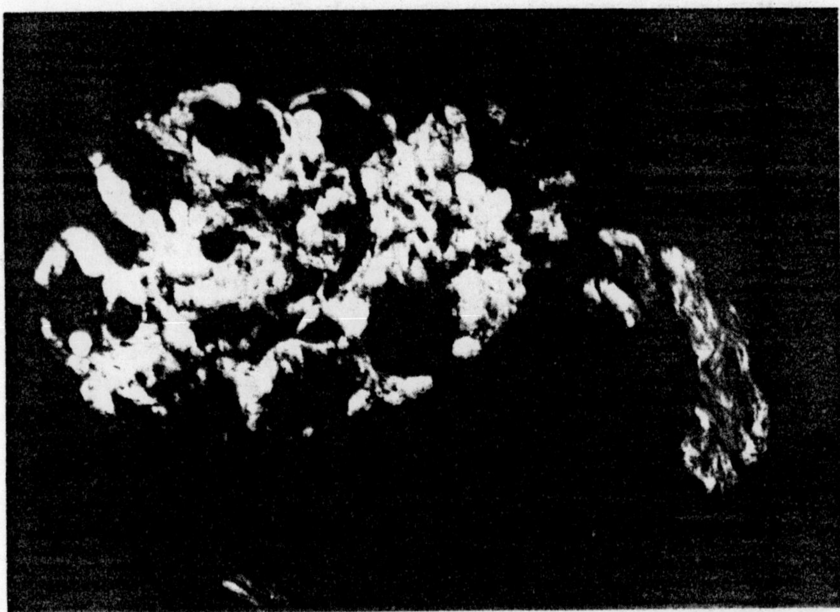

Fig. 11.6. Human colon fixed in *p*-benzoquinone and immunostained using antibodies to the brain protein S-100. Immunoreactive glial cells are seen delineating unreactive cell bodies in the myenteric plexus.

round, dense (D_1 type) secretory granules, still awaits the elucidation of its peptide product. At least four regulatory peptides (VIP,[66] substance P,[67] enkephalin[67] and tetrin/gastrin/CCK[68]) have been localized by immunocytochemistry to autonomic nerves. Of these, the most convincingly demonstrated is VIP. VIP nerves are present in the exocrine pancreas and in and around the pancreatic islets (*Fig.* 11.9). In this latter localization, VIP-containing nerves are particularly concentrated in dense networks. This is further validated by comparing the results of radioimmunoassay of VIP in total pancreas and in isolated pancreatic islets. The islets contain at least 20 times more extractable VIP than the same weight of pancreatic tissue treated *in toto*. The immunolocalization of VIP in the pancreas[66] is of interest in view of the reported actions of VIP in the exocrine pancreas, as a stimulant of pancreatic bicarbonate secretion and as a potent insulinotropic peptide.

TUMOURS. Although pancreatic endocrine tumours (APUDomas) can be regarded as rare pancreatic neoplasms, the knowledge of their precise nature has greatly advanced after the advent of immunocytochemistry and electron microscopy. At least four characteristic clinical syndromes are nowadays well recognized as due to markedly increased circulating levels of the corresponding hormone.[69] They are the insulinoma syndrome, the glucagonoma syndrome, the VIPoma [Verner–Morrison or *W*atery *D*iarrhoea *H*ypokalaemia and *A*chlorhydria (WDHA)] syndrome and the gastrinoma or Zollinger–Ellison syndrome. Although many of these tumours produce more than one hormone (these tumours are

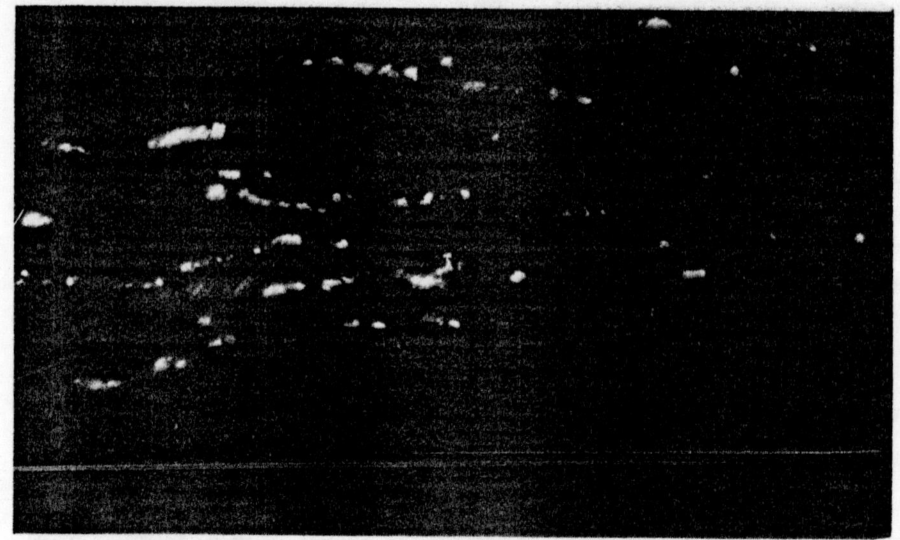

a

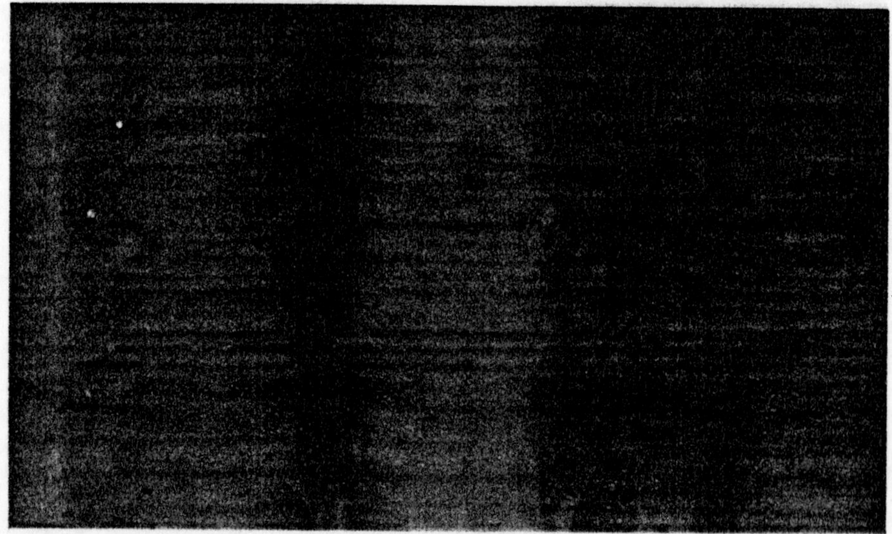

b

Fig. 11.7. Horse ileum fixed in *p*-benzoquinone and immunostained for VIP. Note the normal number and distribution of VIP-containing nerves in (*a*) 'control' animal and the significant decrease of VIP nerves in (*b*) a horse affected by 'grass sickness'.

nowadays called islet cell tumours), one regulatory peptide is produced predominantly as shown by immunocytochemistry (*Fig.* 11.10). These features are further validated by electron microscopy, which can also distinguish several endocrine cell types (*Fig.* 11.10*b*) one of which, usually corresponding to the hormone responsible for the clinical features, is the majority cell type. In addition,

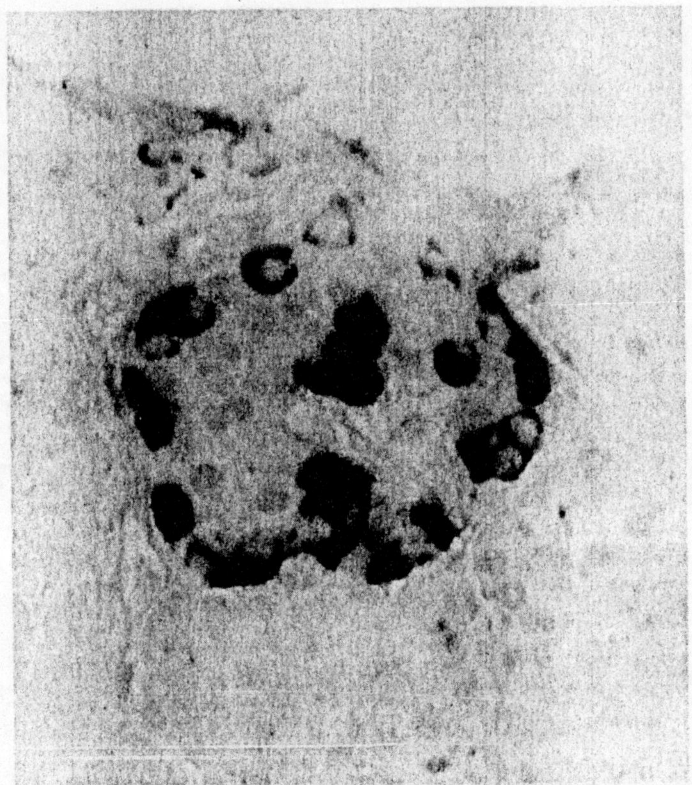

Fig. 11.8. Human pancreas freeze dried and fixed in formaldehyde vapour. Somatostatin immunoreactive cells are seen at the periphery of the islet.

immunocytochemistry for neurone-specific enolase (NSE) (*see above*; *Fig.* 11.11) can be of excellent diagnostic value[70] (*see* the Appendix, p. 206 and Chapter 20).

3.4. Respiratory Tract

Five regulatory peptides, VIP, substance P, bombesin, somatostatin and CCK, have been shown to be present in the respiratory tract of man and other mammals.[71] VIP, substance P and bombesin are easily localized by immunocytochemistry, whereas somatostatin and CCK, possibly in view of their low tissue concentration, can only be demonstrated with confidence by radioimmunoassay of tissue extracts. VIP is found in autonomic nerves, most of which originate locally from neuronal cell bodies found in the tracheal wall. VIP-containing nerves are found prinicpally in the upper respiratory tract, including the nasal mucosa, in close contact with seromucous glands and blood vessels (*Fig.* 11.12). This anatomical distribution fits well with the proposed vasodilatory and secretomotor actions of VIP.[72] Substance P is also localized to autonomic nerves found throughout the bronchial tree, in particularly close contact with the bronchial

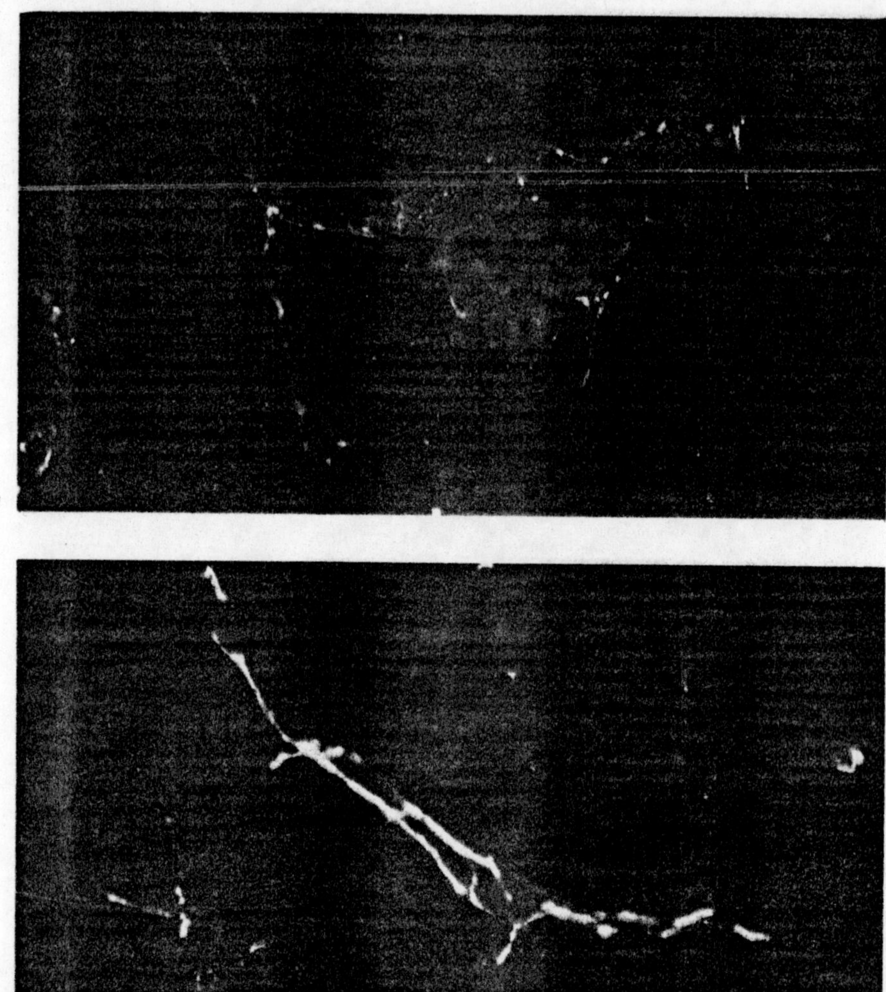

Fig. 11.9. Human pancreas fixed in a solution of *p*-benzoquinone and immunostained with antibodies to VIP. Numerous VIP-containing nerves are seen around the pancreatic islets (*a*) and in the exocrine tissue (*b*).

epithelium and smooth muscle as well as around blood vessels in the same area. Substance P nerve fibres originate outside the lung, from cell bodies (primary sensory neurones) present in the nodose ganglion. Substance P is a putative sensory neurotransmitter, as well as displaying potent broncho- and vasoconstrictor actions.[73]

Unlike VIP and substance P, bombesin[74] is predominantly localized to a subclass of mucosal endocrine cells. The presence of significant concentrations of bombesin in lung neuroendocrine tumours, has also recently been described.[75]

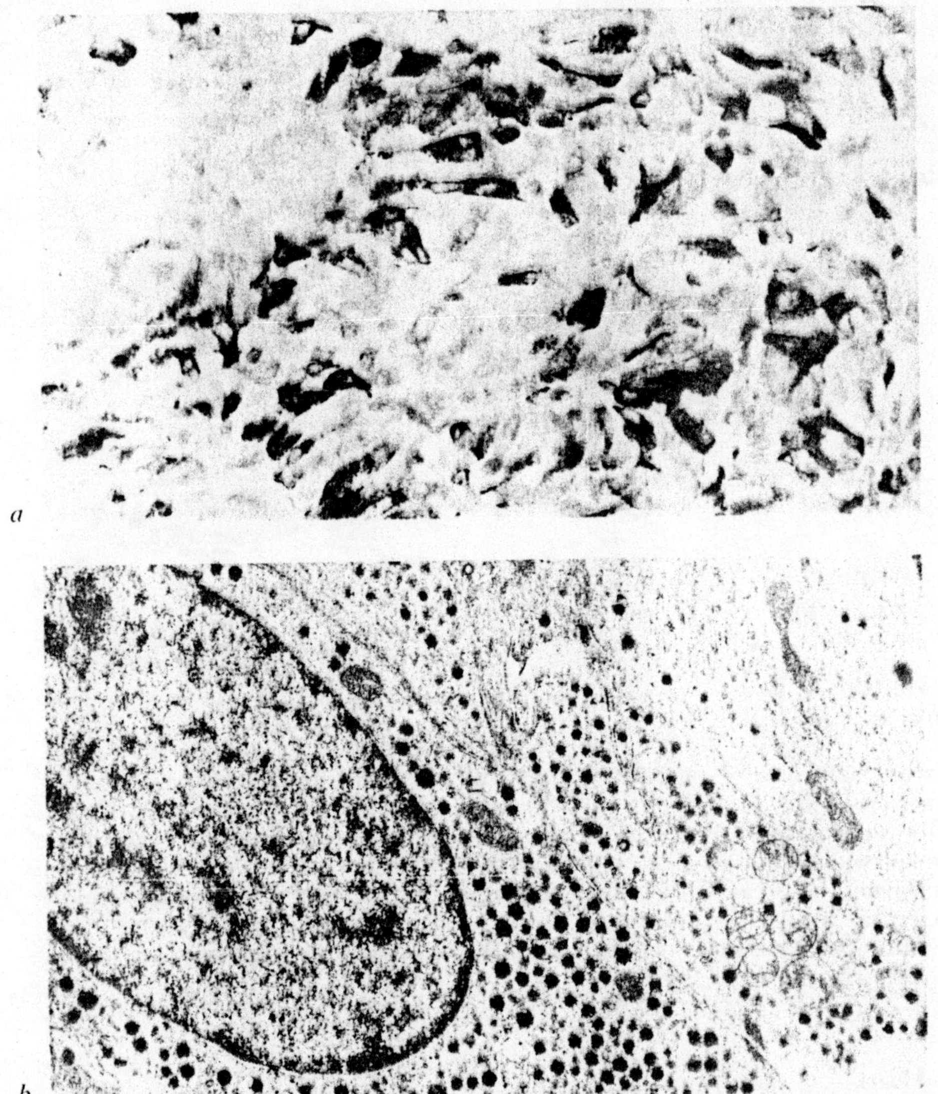

Fig. 11.10. Section of an islet cell tumour immunostained using antibodies to pancreatic polypeptide (PP). Pancreatic endocrine tumours frequently produce more than one regulatory peptide and PP is present in over 50 per cent of islet cell tumours. The 'mixed' endocrine nature of these neoplasm is also seen at the electron microscopical level where more than one type of secretory granule can be found. In (b) two separate granule types measuring 150 ± 23 nm (PP) and 105 ± 11 nm can be distinguished. Magnification: (*b*) × 5000.

3.5. Genital Tract

Both the female and male genital tract are richly innervated by peptidergic nerves, as demonstrated by immunocytochemistry and radioimmunoassay.[76,77] VIP and substance P are the most abundant peptides, although lesser concentrations of somatostatin can also be detected. Both VIP and substance P are present in

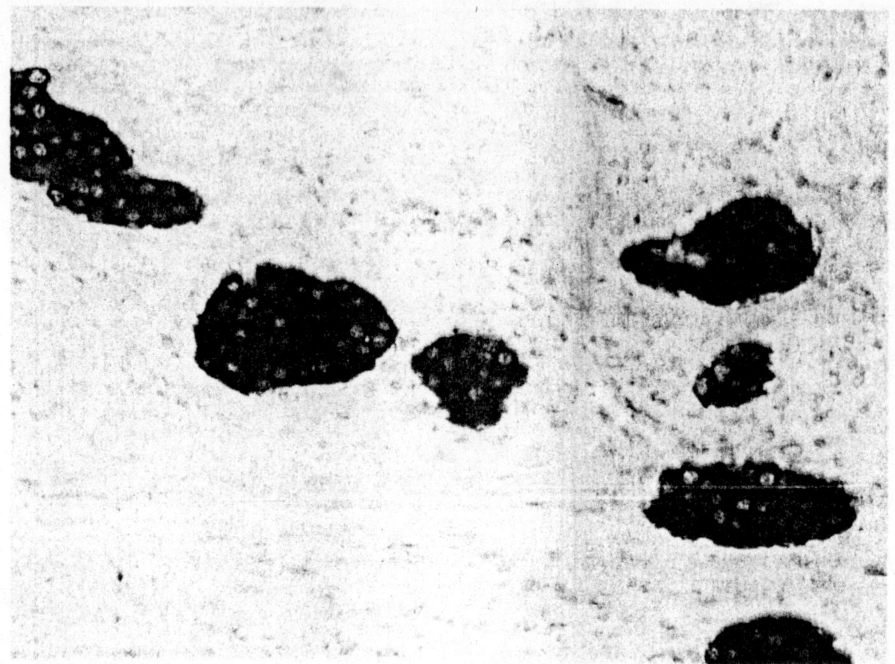

Fig. 11.11. Ovarian carcinoid fixed in 10 per cent formalin. Note the strong immunoreactivity for neurone-specific enolase.

separate networks of peptidergic nerve fibres. Their distribution, as in other tissues, fits well with their proposed sets of actions. Thus, the sensory neurotransmitter, substance P, is found in particularly high concentrations beneath the vaginal epithelium and, most importantly, around the sensory corpuscles of the glans penis, whereas VIP, known for its powerful vasodilatory and secretory properties, is found in the erectile tissue of the penis (*Fig.* 11.13*a*) and in the cervical canal and cervicovaginal region, in close proximity to mucous glands and blood vessels (*Fig.* 11.13*b*).

3.6. Heart

The most abundant regulatory peptide found in the heart is substance P,[78] although lesser concentrations of VIP are also detected. Immunolocalization of substance P reveals its presence in nerve fibres very near the large cardiac vessels and coronary arteries as well as in the conducting system and the myocardium itself.

3.7. Thyroid

The localization of calcitonin in the thyroid C-cells and in medullary carcinomas is beyond dispute.[79] In addition, somatostatin has recently been shown to be

Fig. 11.12. Upper trachea from a cat fixed in *p*-benzoquinone and immunostained for VIP. VIP immunoreactive nerves show their classical appearance and common distribution around seromucous glands.

colocalized in some of the same thyroid C-cells[12,43] and the presence of somatostatin in medullary thyroid carcinomas has also been reported.[80] In addition, VIPergic nerves innervate the thyroid follicles, as has recently been reported by one group of workers in an interesting parallel immunocytochemical and neurophysiological study.[81]

3.8. Adrenal Medulla and Carotid Body

Large concentrations of an immunoreactive material resembling enkephalin, a peptide of the opiate family, have been found in both the adrenal medulla[14,15] and the carotid body.[82] Immunocytochemistry, at light and electron microscopical levels (the latter by the immuno-gold staining procedure; *see* Chapter 6), with antibodies to the enkephalins and to dopamine β-hydroxylase (DβH) for the detection of catecholamines reveals the costorage of amine and peptide in the same secretory granule.[15] Another regulatory peptide, neurotensin, known to exert potent hypotensive and vasodilatory actions, has also lately been localized to the adrenal medulla,[83] to a subgroup of noradrenaline-containing adrenal medullary cells, separate from the large subclass of enkephalin-immunoreactive phaeochromocytes.[15] Phaeochromocytomas frequently contain enkephalin-immunoreactive material[84] (*Fig.* 11.14) and the presence of neurotensin-like material has lately been shown in a phaeochromocytoma cell line maintained in culture.[85]

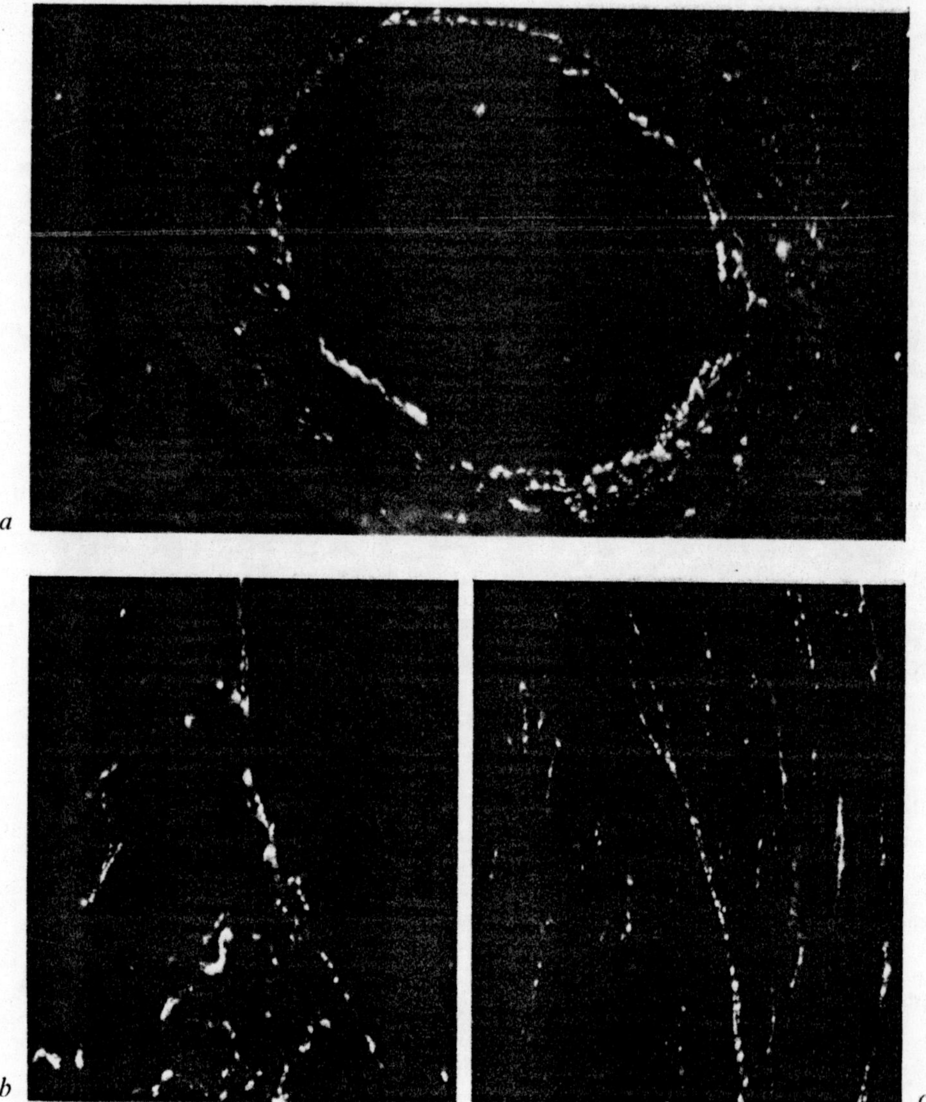

Fig. 11.13. Male (human) (*a*) and female (cat) (*b* and *c*) genital tract fixed in *p*-benzoquinone and immunostained for VIP. Note the typical distribution of VIP-containing nerves around arteries of the corpus cavernosum (*a*) in the uterine muscle (*b*) and cervical glands (*c*).

3.9. Eye

A number of regulatory peptides (including VIP, substance P, somatostatin, enkephalin, neurotensin and CCK) have been found in the eye, mostly in the retina[86–88] (in different types of amacrine cell) and the uvea.[89] The distribution of substance P and VIP in the uvea is interesting as this tissue is known to be involved

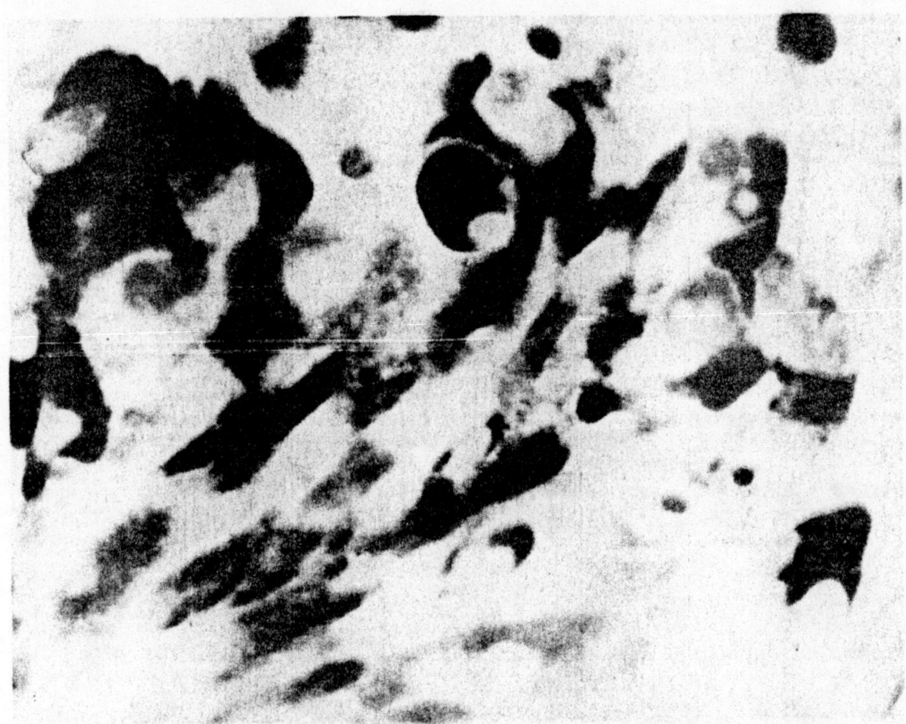

Fig. 11.14. Enkephalin-like immunoreactive material in noradrenaline-containing cells of a human phaeochromocytoma freeze dried and fixed in formaldehyde vapour.

in responses to noxious stimuli. VIP and substance P are found in entirely separate peptidergic systems.[90] VIP is present in an intrincate plexus around the choroid vessels (*Fig.* 11.15*a*) and substance P in a delicate network of fibres in the iris (*Fig.* 11.15*b*). The abrupt interruption of substance P-containing nerve fibres at the level of the ciliary body provides an excellent anatomical demarcation for this structure at the level of the annular formation. Interruption of the trigeminal pathway produces a remarkable decrease in the number of substance P-containing nerve fibres in the anterior portion of the uvea,[91,92] but no changes of the VIP-containing nerves of the choroid. VIP apparently originates locally, from neuronal cell bodies which are best demonstrated after treatment of animals with colchicine, a neurotransmitter transport blocker.

3.10. Skin

Substance P, VIP, bombesin and somatostatin are present in the skin.[93] Regional variations in the distribution of some of these regulatory peptides can be detected. Substance P shows the greatest variations with larger concentrations [mean ± s.e. (*mean*)] in the more sensitive areas like the anus (8 ± 1·6 pmol/g) and snout (13 ± 3·6 pmol/g) than in other less specialized areas, e.g. back

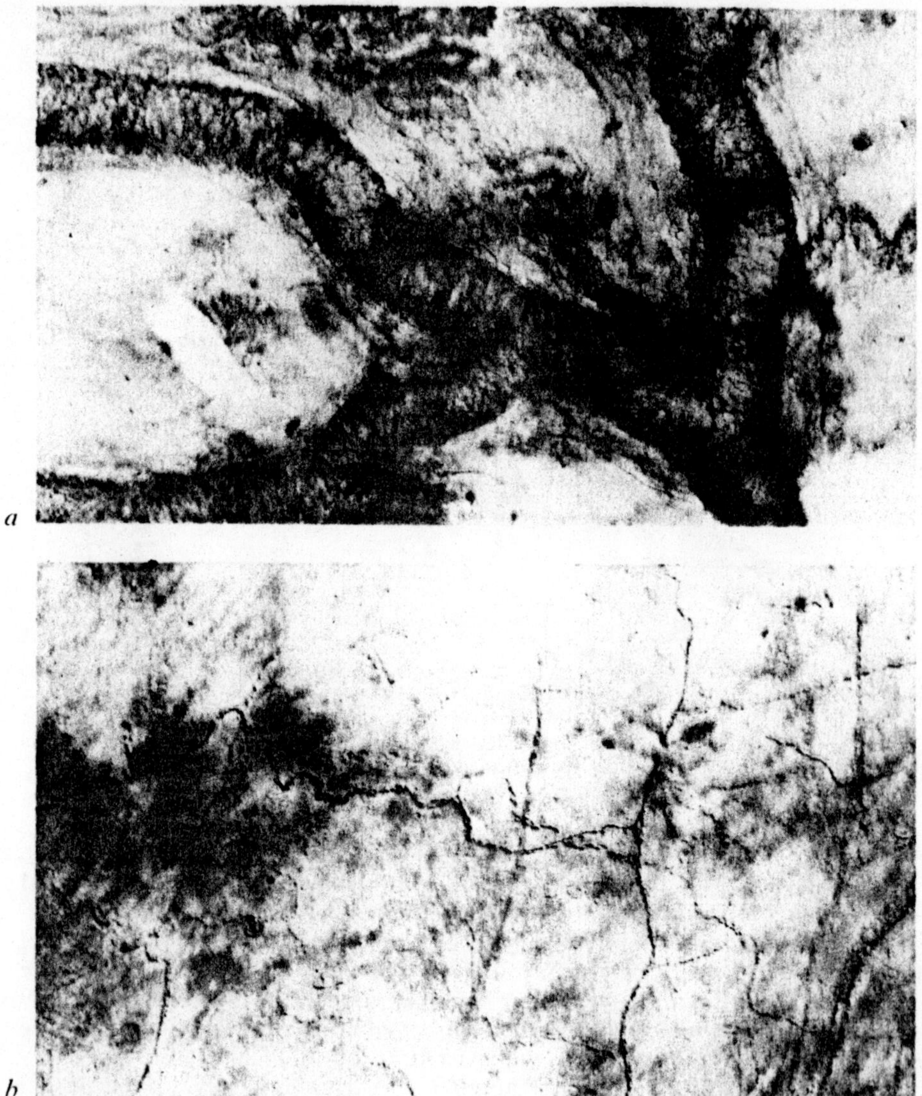

Fig. 11.15. A stretch preparation of guinea-pig uvea fixed in a solution of *p*-benzoquinone and immunostained for VIP (*a*) and substance P (*b*). VIP-containing nerves are particularly concentrated around choroid vessels (*a*) whereas substance P nerves are especially numerous in the iris (*b*).

(1·4 ± 0·7 pmol/g). While immunocytochemistry localizes substance P and VIP with confidence, the lower concentrations of somatostatin and bombesin (the latter ranging from 0·4 ± 0·1 to 3 ± 0·8 pmol/g) makes their localization a more difficult task. VIP-containing nerve fibres are detected in close association with blood vessels and sebaceous glands (*Fig.* 11.16*a*), especially in areas where radioimmunoassay detects the highest concentrations (e.g. the snout 12 ± 5 pmol/g). Substance P nerve fibres, in agreement with the data obtained from tissue extracts,

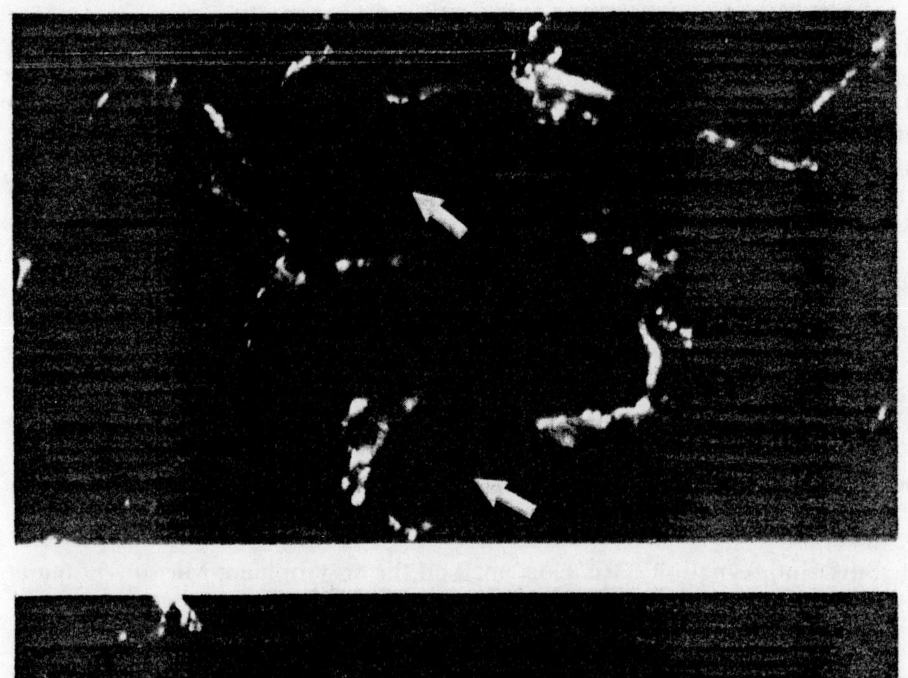

a

b

Fig. 11.16. Pig skin fixed in a solution of *p*-benzoquinone and immunostained for VIP (*a*) and for substance P (*b*). Note the close associations of VIP nerves with sebaceous glands (arrows) and blood vessels (*a*) and of substance P nerves with the epidermis (*b*).

are mostly found in more sensitive areas, distributed beneath the epithelium and often penetrating the epidermal layers (*Fig.* 11.16*b*).

4. CONCLUSION

The finding of identical active peptides in both the endocrine and the neural parts of the diffuse neuroendocrine system has added a new dimension to endocrinology.

The leading role played by immunocytochemistry in most of these new and exciting discoveries is beyond dispute. The increasing availability of monoclonal antibodies and pure peptides or their derivative fragments for ensuring specificity and the possibility of correlating immunocytochemical findings with precise quantitative analysis of tissue peptides makes the investigation of regulatory peptide distribution both rewarding and enjoyable. The demonstration of abnormal peptide distribution in a considerable number of common human disorders adds respectability to our efforts and there is no doubt that immunocytochemistry of regulatory peptides will not only be with us for many years to come, but will even exceed our expectations as the essential tool for these investigations.

Appendix

I. USEFUL DIAGNOSTIC METHODS FOR APUDOMAS

Conventional haematoxylin and eosin stain (features of an islet cell tumour or of a carcinoid)

Special stains

- *A.* Silver impregnation (Grimelius' method for argyrophilia; Masson–Fontana method for argentaffinity)
- *B.* Lead haematoxylin
  Masked metachromasia
- *C.* Formaldehyde-induced fluorescence (biogenic amines)[94]

Immunocytochemistry

- Peptide (wide variety)
- Amine (5-hydroxytryptamine)
- Enzyme (neurone-specific enolase, dopamine β-hydroxylase)

Electron microscopy (secretory granules)

- *A.* Conventional
- *B.* Immunocytochemistry

II. PREPARATION OF EPOXY RESIN-EMBEDDED TISSUE FOR LIGHT MICROSCOPICAL IMMUNOSTAINING

Tissues are treated in the routine manner for liquid or vapour fixation and then one of the following schedules is adopted:

Freeze-dried, vapour-fixed tissue

1. After removal from vapour, cut into pieces (max. in any direction 2–3 mm)
2. Immerse in Araldite I (recipe below)
3. Put into evacuated oven for 60 min at 60 °C (*without lid on*)
4. Remove from oven and replace cap

Liquid-fixed tissue

1. After washing in buffer, dehydrate in alcohol (50 per cent, 70 per cent, 100 per cent
2. Clear in xylene (two changes)
3. Cut into small pieces (max. in any direction 2–3 mm)
4. Wash in two changes of propylene oxide (2 × 5 min)
5. Discard propylene oxide and replace with Araldite I (recipe below)

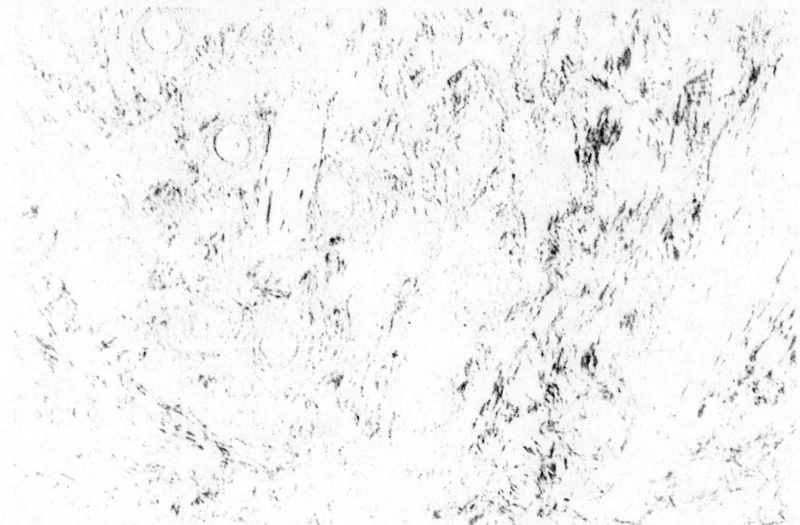

Plate 1. Paraffin section, 7 µm, through Bouin-fixed rat nucleus ruber stained brown with the anti-neurofibrillar group monoclonal antibody 06-17 and blue with the anti-perikaryal-neurofibrillar group monoclonal antibody 02-135. Dual antigen PAP method.* Although both antibodies react on electroblot transfer with a 200 kilodalton band, staining does not overlap. 02-135 stains selected perikarya and their projections. 06-17 stains many neurofibrils, but no perikarya and none of the projections stained by 02-135. (*Chapter 1*)

* Sternberger L. A. and Joseph E. A. The unlabeled antibody method. Contrasting color staining of paired pituitary hormones without antibody removal. *J. Histochem. Cytochem.* 1979 **27**, 1424–1429.

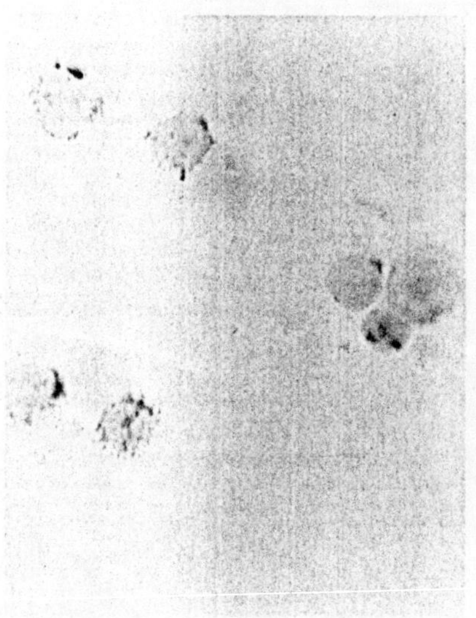

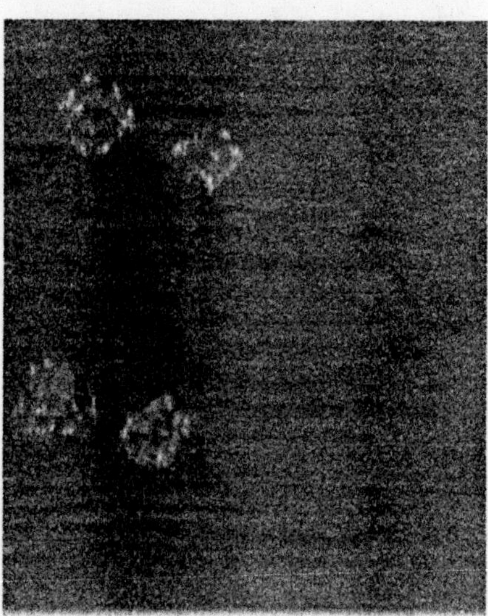

Plate 2. Mononuclear cell suspension labelled with OKT 3 (Orthomune) and GAM G40 (Janssen Life Sciences Products). The preparation is stained for acid alpha-naphthyl acetate esterase and counter-stained with methyl green. In the lymphocytes, three different staining patterns can be distinguished: a dot-like activity, a diffuse granular pattern or no reaction at all[36].
Left: viewed with transmitted light. Right: viewed with polarized light epi-illumination combined with weak transmitted light. The light, back-scattered by the gold particles, is easily detectable and helps to make quantification more accurate and convenient. From joint work with Dr M. De Waele (Brussels, Belgium).

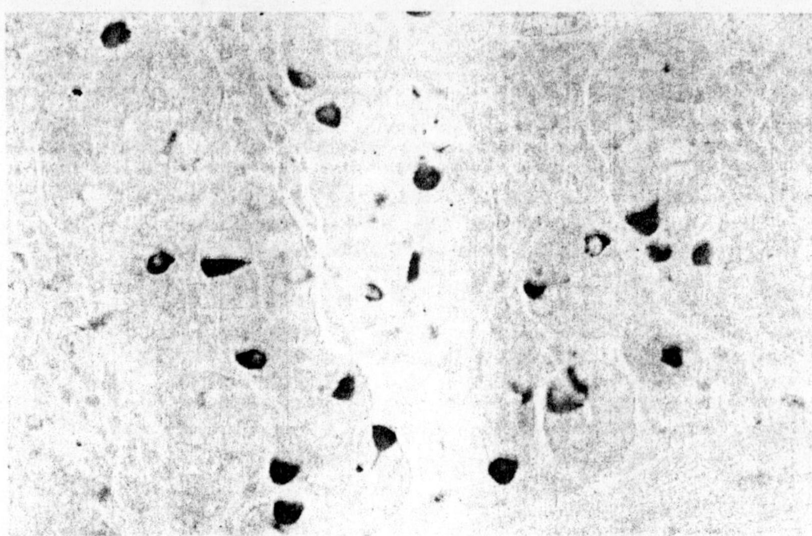

Plate 3. Pig duodenum, freeze dried, and fixed in formaldehyde vapour and embedded in paraffin; 5-μm section double immunostained for gastrin by the PAP method, using 4-chloro-1-naphthol as coupler and for serotonin by the indirect method using gold–goat anti-rabbit IgG as the second layer. Gastrin cells are stained blue and serotonin cells are stained red. From joint work with Drs J. M. Polak and J. Gu (London, UK). (*Chapter 2*)

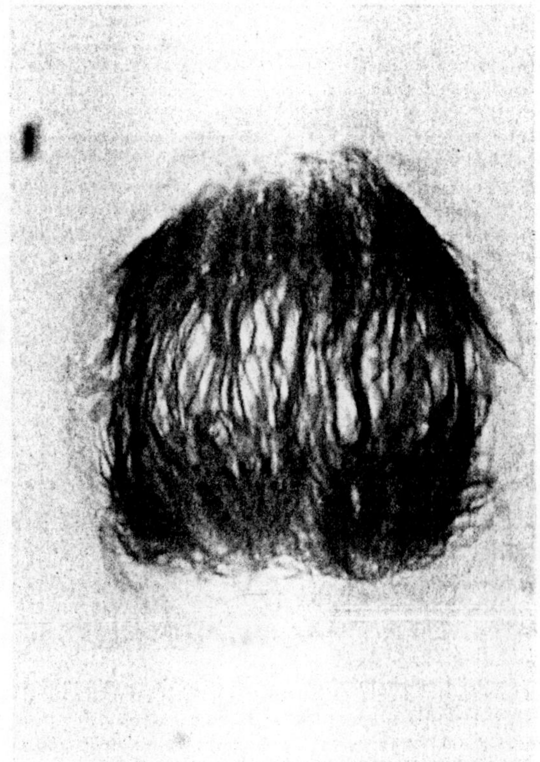

Plate 4. Light microscopical view of a dividing plant endosperm cell (*see* De Mey et al.[60]). Microtubules are stained red by rabbit anti-tubulin and GAR G20 (Janssen Life Sciences Products). Chromosomes are stained with toluidine blue. From joint work with Dr M. De Brabander (Beerse, Belgium), Dr A.-M. Lambert (Strasbourg, France) and A. Bajer (Eugene, Oregon, USA). (*Chapter 2*)

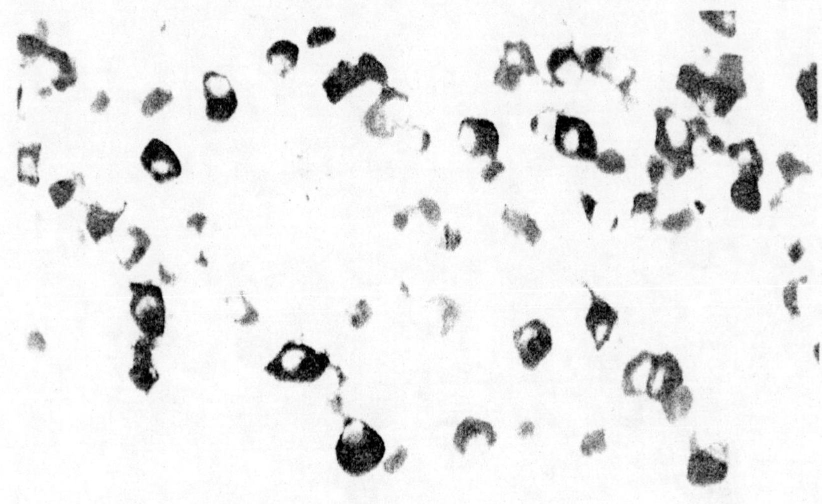

Plate 5. Double immunoenzymatic staining of plasma cells in paraffin-embedded human tonsil. Kappa chains have been revealed in blue, using alkaline phosphatase as antibody label. Lambda chains are revealed using peroxidase to give a brown reaction product. *(Chapter 7)*

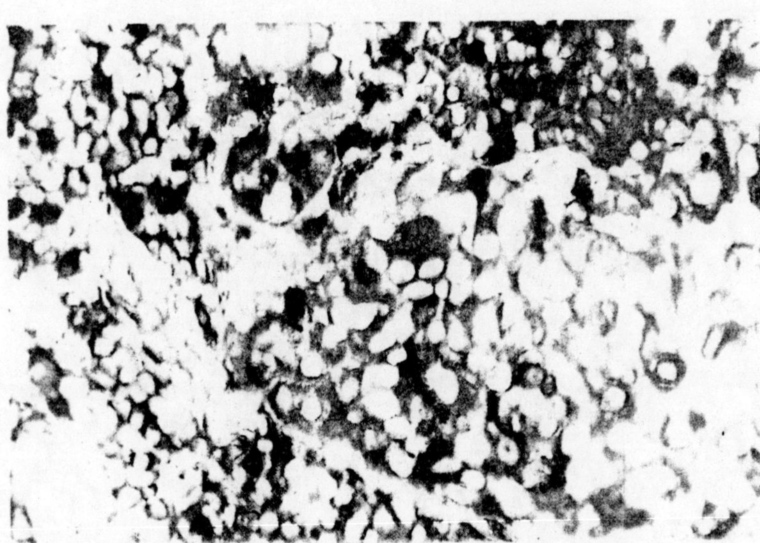

Plate 6. Double immunoenzymatic staining of a deposit of metastatic carcinoma in a lymph node. The carcinoma cells have been stained with an immuno-alkaline phosphatase unlabelled antibody sandwich based upon a rabbit antiserum reactive with a human epithelial membrane antigen (kindly provided by Dr M. Ormerod). The surrounding lymphoid tissue has been stained by an immuno-peroxidase technique using a monoclonal anti-HLA-DR antibody. Note that some of the carcinoma cells appear to show mixed staining for both antigens. *(Chapter 7)*

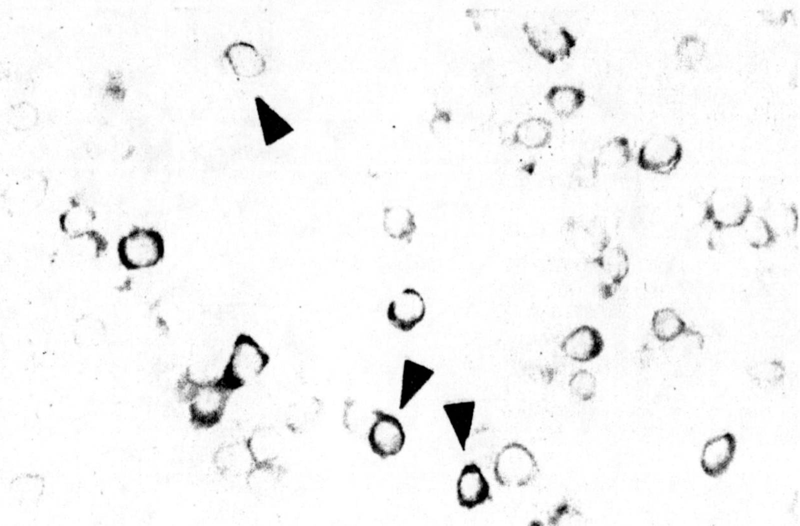

Plate 7. Double immunoenzymatic stain of a paraffin-embedded lymph node for J chain (blue) and for Ig (brown). Note the presence of cells (arrowed) which stain for J chain alone. This pattern of labelling is rare in reactive lymphoid tissue but may be found in non-Hodgkin's lymphomas.[32] (*Chapter 7*)

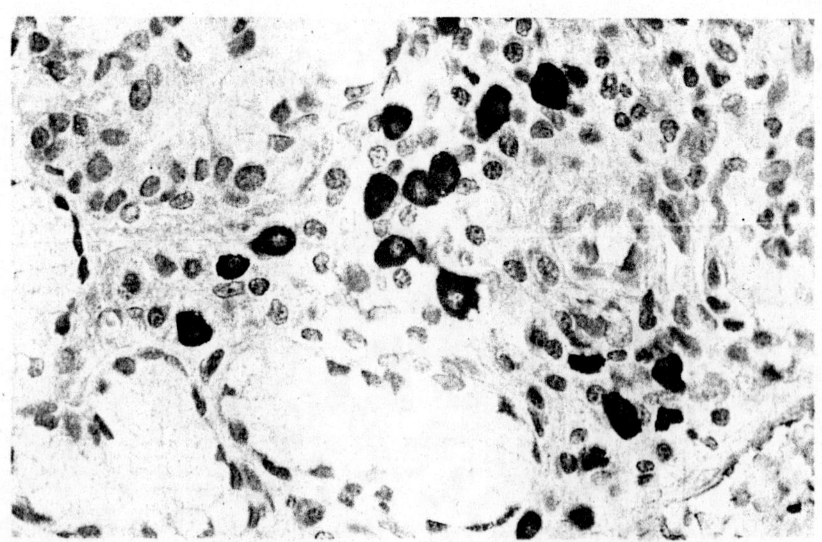

Plate 8. Immuno-alkaline phosphatase staining of lambda light chains in paraffin-embedded human tonsil using mononclonal anti-lambda chains, followed by sheep anti-mouse Ig and complexes of monoclonal anti-alkaline phosphatase and alkaline phosphatase (APAAP complexes). Note the intensity and sharp localization of the label, which compares well with that achieved using immunoperoxidase techniques. The enzyme label has been revealed with naphthol AS–MX plus Fast Red, thus providing a good contrast with the haematoxylin counterstain. (*Chapter 7*)

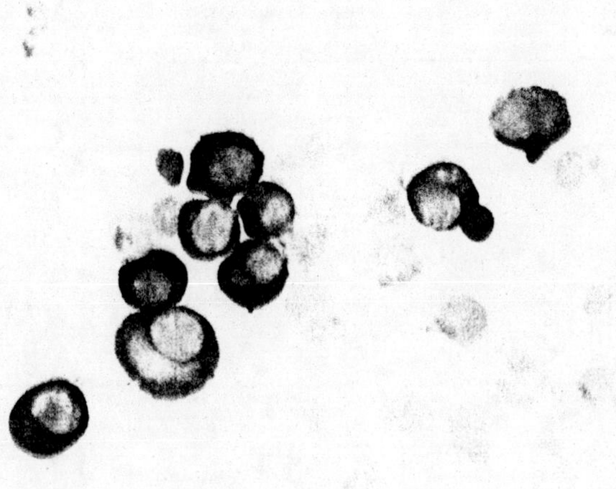

Plate 9. Staining of myeloma cells in a marrow aspirate for J chain using a monoclonal antibody against this constituent followed by an immuno-alkaline phosphatase sandwich (as in *Plate 8*). (*Chapter 7*)

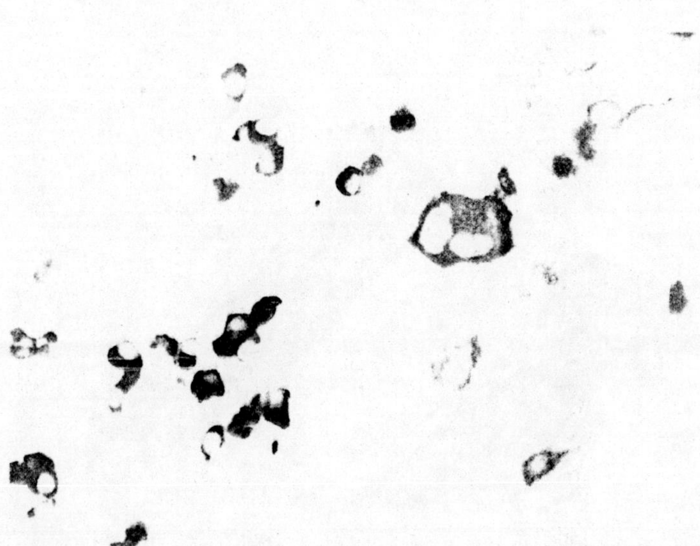

Plate 10. Double immunoenzymatic staining of Hodgkin's tissue (paraffin section) for kappa and lambda light chains showing that Reed–Sternberg cells give a mixed labelling reaction which contrasts clearly with the single-staining pattern of plasma cells. This reaction is accounted for by the presence of polyclonal serum Ig within Reed–Sternberg cells. (*Chapter 7*)

Plate 11. Double immunoenzymatic staining of tonsil tissue (paraffin embedded) for kappa and lambda light chains showing the presence of numerous scattered cells which have passively absorbed serum Ig (probably during tissue processing). Note the absence of background staining (presumably because fixation has denatured extracellular immunoglobulin) and the clarity with which the few single stained Ig-containing cells (which have synthesized this material rather than absorbed it from their environment) stand out in contrast to the double stained cells. (*Chapter 7*)

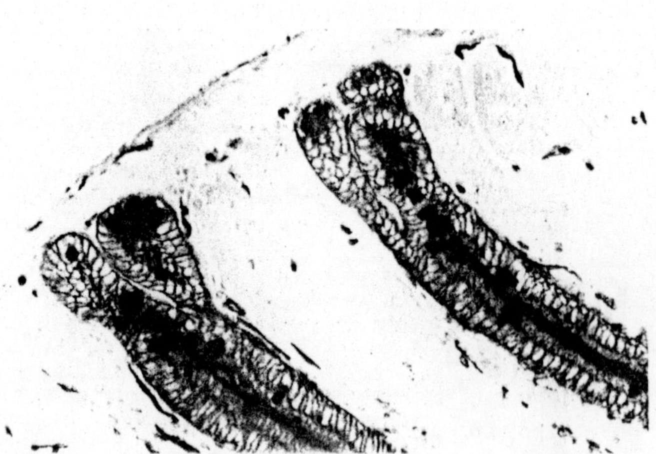

Plate 12. Mosaic cell populations in small intestine of a DDK ↔ C57BL/6 mouse chimera. Frozen section. 5-μm, counterstained with haemalum.

DBA–peroxidase (brown) binds to epithelial cells and mucin of C57BL/6 origin, and to endothelium of DDK origin. Note that the epithelial cells in any individual crypt appear to be all the same type. There is also a polymorphism between the strains for the binding of DBA–peroxidase to Paneth cell granules (at the bottom of the crypts). (*Chapter 8*)

Plate 13. RAN-2 immunoreactivity on the surface of a cultured astrocyte. Rat cerebellum, 10 days in culture. *(Chapter 10)*

Plate 14. Cryostat section of rheumatoid synovial membrane stained with a combination of Mc Abs OKT 4 and OKT 8. The green (fluorescein) stained cells are OKT 4 +ve and the orange (rhodamine) stained cells are OKT8 +ve. Such combined staining clearly shows the high ratio of OKT4 +ve to OKT 8 +ve cells in this perivascular infiltrate; to produce this picture, the same area is photographed once using barrier filters to reveal only the fluorescein staining and then again using filters to reveal the rhodamine +ve cells (normally red) that appear orange. *(Chapter 13)*

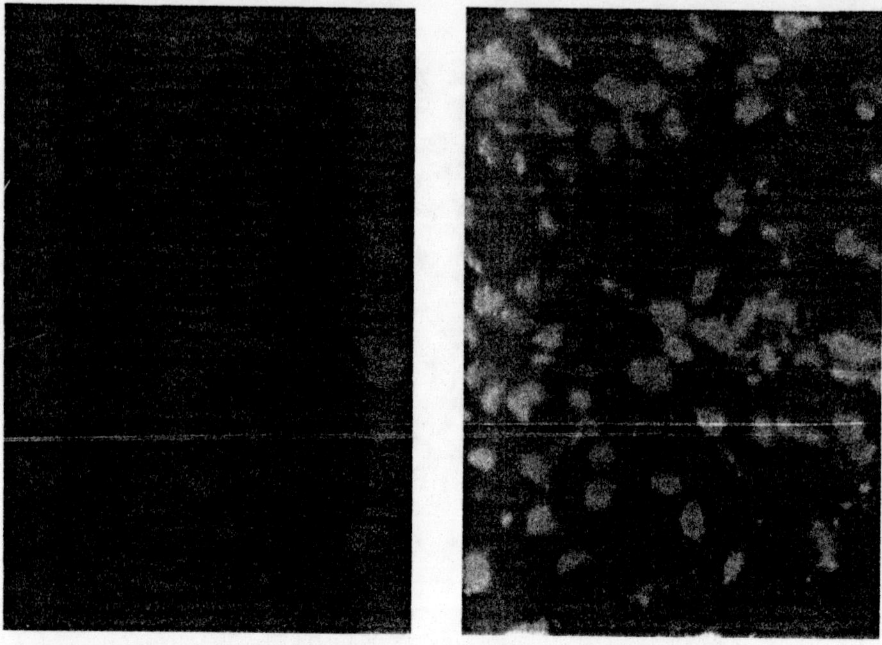

Plate 15. Combined immunofluorescence staining for the analysis of macrophage subsets. (*a*) Cells stained with an anti-macrophage monoclonal (FMC-17) and (*b*) simultaneous anti-HLA-DR staining on the same cells.

(*Chapter 13*)

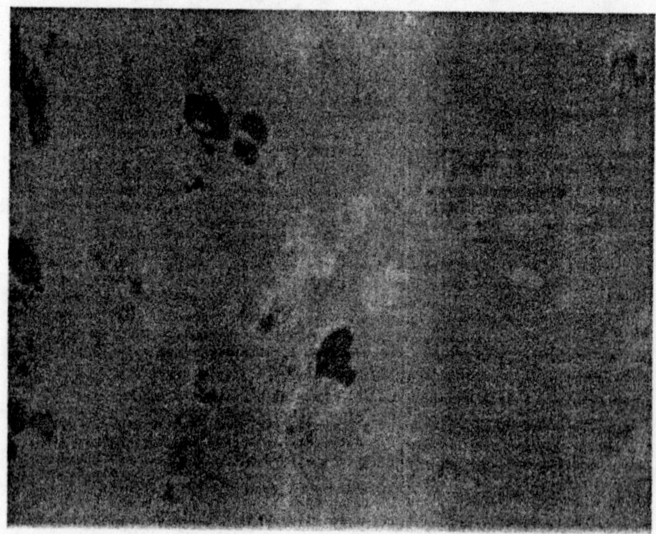

Plate 16. Combined staining with FMC-17 to demonstrate cells of the macrophage lineage and a cytochemical reaction for acid phosphatase activity. The section is photographed using epifluorescence to demonstrate FMC-17 staining and then with Kohler illumination to demonstrate the acid phosphatase activity. This double exposure results in the FMC-17 fluorescence appearing 'pink' and the acid phosphatase activity appearing red. It can be seen that only a proportion of the FMC-17 +ve cells are ACP +ve.

(*Chapter 13*)

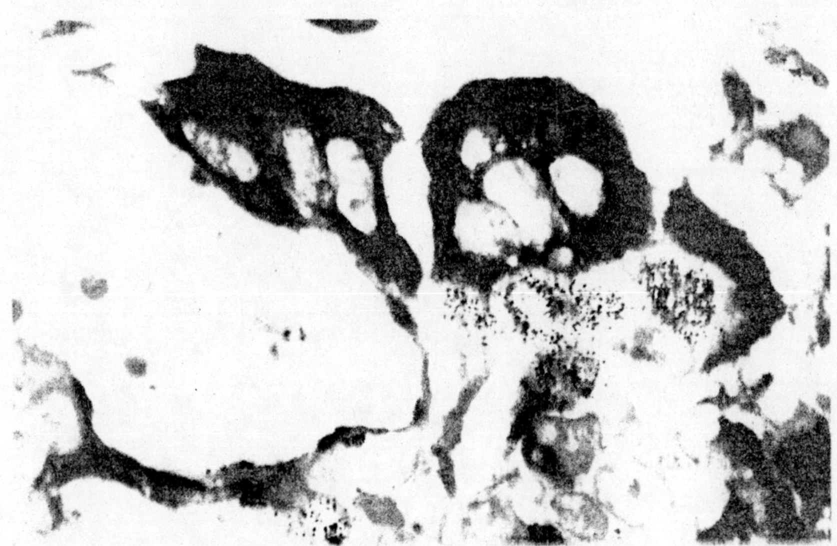

Plate 17. Metastatic deposit of testicular choriocarcinoma injected with tritium-labelled thymidine 20 min before biopsy. Fixed section stained with indirect immunoperoxidase technique for the beta subunit of human chorionic gonadotropin (HCG) followed by autoradiography. The black grains represent thymidine labelling in the malignant mononuclear cytotrophoblast (the dividing layer) while HCG is in the syncytiotrophoblast. (*Chapter 15*)

Plate 18. Resin-embedded section of moderately differentiated carcinoma of the colon stained for carcinoembryonic antigen (CEA). The staining is mainly localized on the luminal membrane. (*Chapter 15*)

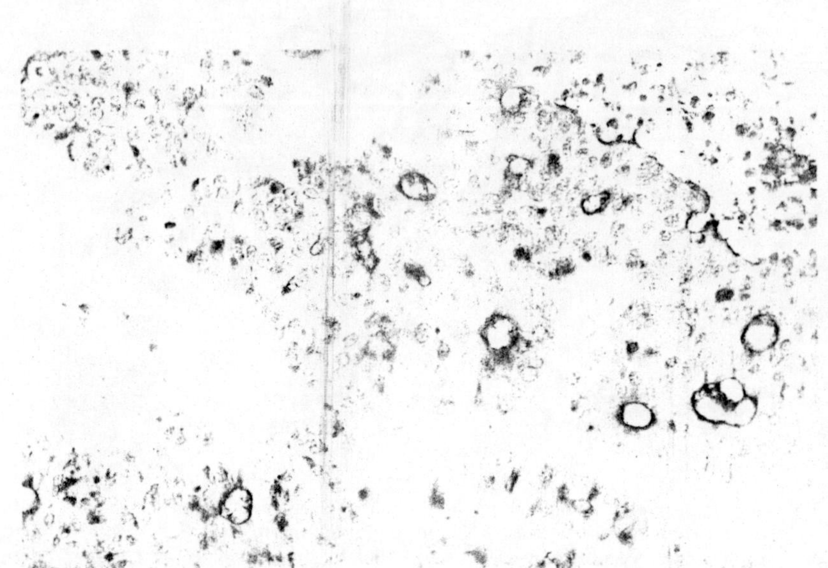

Plate 19. Moderately differentiated carcinoma of the lung stained with antiserum raised against milk fat globule membrane and demonstrating epithelial membrane antigen (EMA). *(Chapter 15)*

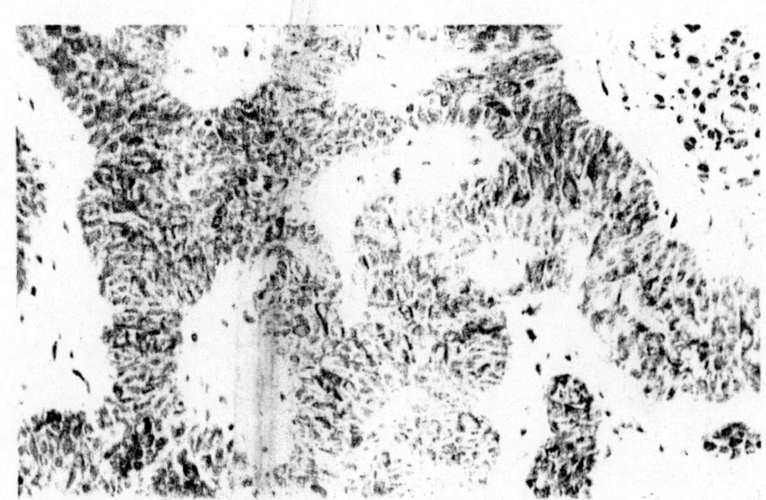

Plate 20. Staining for keratin in basal cell papilloma. Immunoperoxidase. *(Chapter 16)*

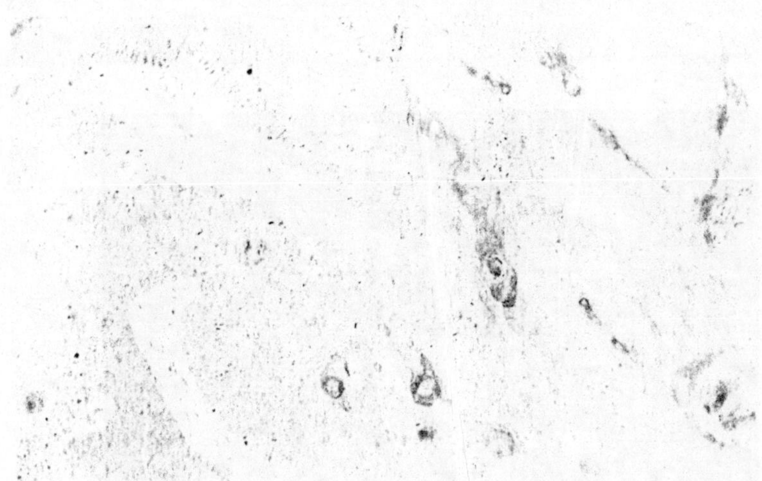

Plate 21. Patchy presence of keratin in a carcinoma of the skin. Immunoperoxidase. *(Chapter 16)*

Plate 22. Positive staining for keratin of an epithelial thymoma. Immunoperoxidase. *(Chapter 16)*

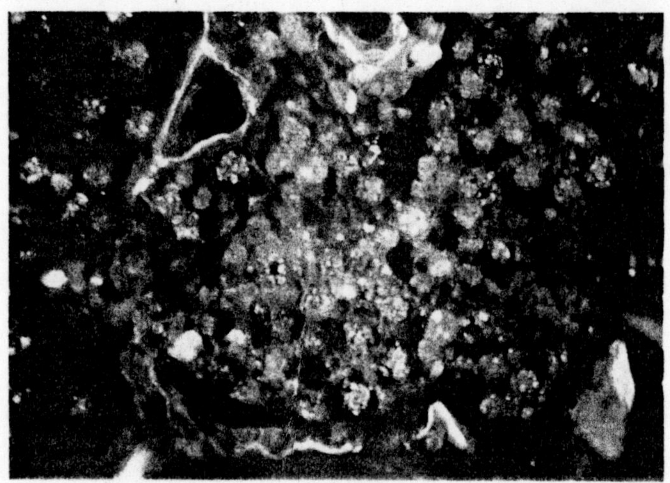

Plate 23. Fluorescent stain of lung infected with *Legionella*. The organism is visible as intra- and extracellular fluorescent granules. (*Chapter 18*)

Plate 24. *Treponema pallidum* stained by fluorescent method in formalin-fixed paraffin section. (*Chapter 18*)

Plate 25. Macrophages of Whipple's disease in rectal biopsy stained with an anti-streptococcal serum. (*Chapter 18*)

Plate 26a. Cryostat section of human antrum treated with patient's serum and FITC anti-human IgG showing bright-green cells scattered in the antral mucosa. *(Chapter 19)*

Plate 26b. Same section stained with rabbit anti-gastrin serum and counterstained with rhodaminated goat anti-rabbit antiserum. The red cells have the same distribution as in (*a*). In the double exposure photograph the green and red cells appeared yellow thus confirming the identity of the cells stained by the patient's serum and the rabbit anti-gastrin serum. *(Chapter 19)*

Plates 26, 27. Double immunofluorescence test used to identify gastrin cell antibodies in sera of patients with 'autoimmune' antral (Type B) gastritis. *(Chapter 19)*

Plate 27. Double exposure photography[1] of a section of human antrum first treated with the same patient's serum as in *Plate* 26*a* and subsequently stained with rabbit anti-somatostatin antiserum. The red somatostatin cells are clearly separate from the green cells (gastrin cells) which reacted with the patient's serum. (*Chapter 19*)

Plate 28. Corticotropin-producing cells in a rat pituitary. Post-embedding staining, unlabelled antibody-enzyme method, fixation in glutaraldehyde and osmium tetroxide, embedding in epon; semithin section (1 μm). Counterstained with thionin. (*Chapter 20*)

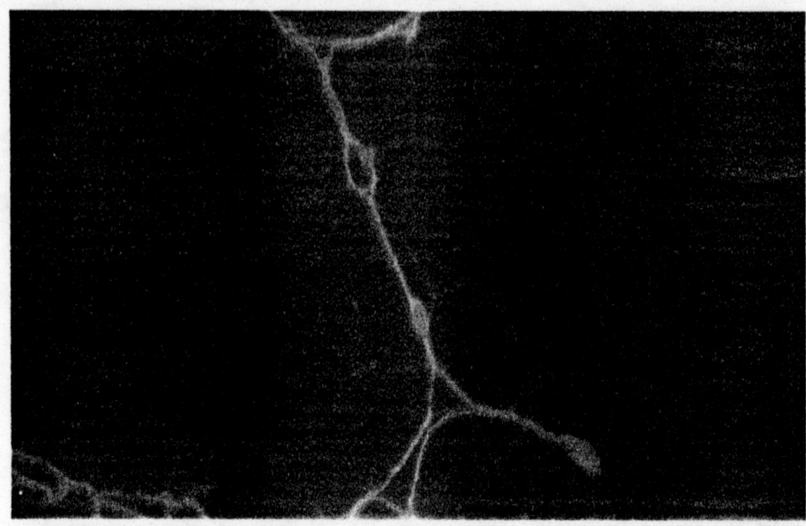

Plate 29a. A semithin frozen section of the mammary gland stained for mouse serum albumin. Note the circular arrangement of the spots of fluorescence. *(Chapter 5)*

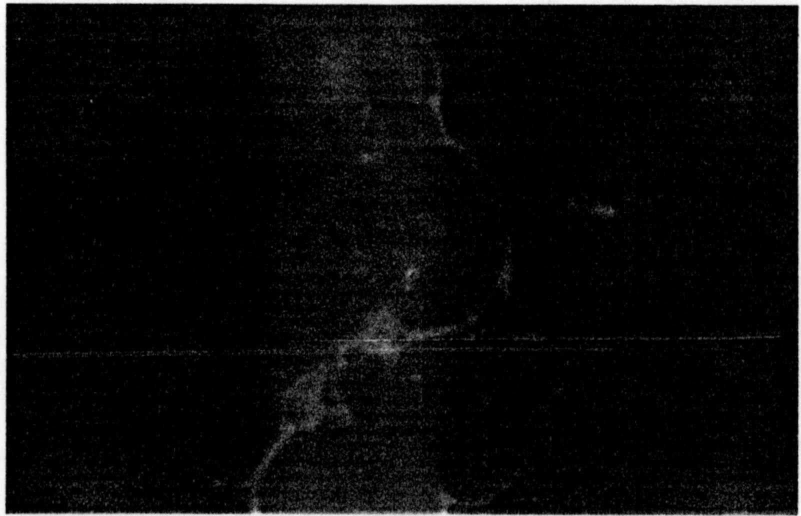

Plate 29b. A semithin frozen section of the mammary epithelium stained for mouse serum albumin. The punctate spots of fluorescence are arranged around a very large lipid droplet in the apex of a cell. *(Chapter 5)*

Common procedure

1. Rollamix at room temperature overnight
2. Warm to 45 °C
3. Discard Araldite I
4. Immerse in Araldite II (recipe below)
5. Rollamix for 3–4 h at room temperature
6. Warm to 45 °C for 10–20 min
7. Arrange pieces of tissue in prelabelled 'BEEM' capsules and top up with Araldite II
8. Cure at 60 °C for two days
9. Microtomy (0·5–2 μm sections)
10. Resin removal (15–30 min in sodium ethoxide)
11. Routine immunostaining

Araldite resin

Araldite I
1 part of CY212 (epoxy resin)
1 part of dodecenyl succinic anhydride
0·025 parts of dibutylphthalate
Mix for 20 min before use

Araldite II
To the above mixture (Araldite I) add 0·04 parts of benzyldimethylamine
Mix for 20 min before use

III. BENZOQUINONE SOLUTION FIXATION (BQS)

1. Prepare a 0·4 per cent solution of recrystallized benzoquinone in 0·01 M phosphate-buffered saline (PBS) pH 7·1–7·4. Allow a tissue: solution volume ratio of approximately 1 : 10, using specimens no larger than 2 × 2 cm.

2. Add PBS to benzoquinone crystals and agitate until all crystals are in solution.

NB. If solution is brown, discard it. The solution must be bright yellow when it is used. Prepare the solution a maximum of 20 min before it is to be used. Store the crystals in a cool dry dark place, otherwise they will turn brown and no longer be of use.

3. Fix the tissue by immersing it in the solution for the appropriate time period (*see below*).

4. Following fixation, transfer the tissue into PBS containing 7 per cent sucrose and 0·01 per cent sodium azide. The solution need not be prepared freshly each time. It can be prepared in bulk and stored at 4 °C.

5. Rinse in PBS–sucrose at 4 °C for several hours

6. Freeze a block of the tissue for cryostat sectioning

7. Spare fixed tissue may be stored in PBS–sucrose solution at 4 °C. Check the tissue frequently for microbial growth.

Fixation times (ordinary histological block size)

Small endoscopic biopsies	30 min
Thin-walled gut (child or animal)	1 h
Thick-walled gut	2 h
Pancreas	2 h
Tumour	2 h

REFERENCES

1. Bayliss W. M. and Starling E. H. The mechanism of pancreatic secretion. *J. Physiol.* (*Lond.*) 1902, **28**, 325–353.
2. Van Noorden S. and Polak J. M. Advances in immunocytochemistry. In: Bloom S. R. and Polak J. M., ed., *Gut Hormones*, 2nd ed. Edinburgh, Churchill Livingstone, 1981; 80–89.
3. Pearse A. G. E. and Polak J. M. Immunocytochemical localisation of substance P in mammalian intestine. *Histochemistry* 1975, **41**, 373–375.
4. Nilsson G., Larsson L-I., Brodin E., Pernow P. and Sundler F. Localisation of substance P-like immunoreactivity in mouse gut. *Histochemistry* 1975, **43**, 97–99.
5. Bryant M. G., Bloom S. R., Polak J. M., Albuquerque R. H., Modlin I. and Pearse A. G. E. Possible dual role for VIP as gastrointestinal hormone and neurotransmitter substance. *Lancet 1976*, **i**, 991–993.
6. Larsson L-I., Fahrenkrug J., Schaffalitzky de Muckadell O., Sundler F., Håkanson R. and Rehfeld J. F. Localisation of vasoactive intestinal polypeptide (VIP) to central and peripheral neurons. *Proc. Natl. Acad. Sci. USA* 1976, **73**, 3197–3200.
7. Hökfelt T., Johansson O., Efendic S., Luft R. and Arimura A. Are there somatostatin-containing nerves in the rat gut? Immunohistochemical evidence for a new type of peripheral nerves. *Experientia* 1975, **31**, 852–854.
8. Elde, R., Hökfelt, T., Johansson, O. and Terenius L. Immunohistochemical studies using antibodies to leucine-enkephalin. Initial observations on the nervous system of the rat. *Neuroscience* 1976, **1**, 349–351.
9. Polak J. M., Pearse A. G. E., Grimelius L., Bloom S. R. and Arimura A. Growth-hormone release-inhibiting hormone in gastrointestinal and pancreatic D cells. *Lancet* 1975, **i**, 1220–1225.
10. Polak J. M., Sullivan S. N., Bloom S. R., Facer P. and Pearse A. G. E. Enkephalin-like immunoreactivity in the human gastrointestinal tract. *Lancet* 1977, **i**, 972–974.
11. Alumets J., Håkanson R., Sundler F. and Chang K. J. Leu-enkephalin-like material in nerves and enterochromaffin cells in the gut. *Histochemistry* 1978, **56**, 187–196.
12. Hökfelt T., Efendic S., Hellerström C., Johansson O., Luft, R. and Arimura A. Cellular localisation of somatostatin in endocrine cells and neurons of the rat with special reference to the A_1 cells of the pancreatic islet and to the hypothalamus. *Acta Endocrinol. Suppl. 200* 1975, **80**, 5–41.
13. Van Noorden S., Polak J. M. and Pearse A. G. E. Single cellular origin of somatostatin and calcitonin in the rat thyroid gland. *Histochemistry* 1977, **53**, 243–247.
14. Schultzberg M., Hökfelt T., Lundberg J. M., Terenius L., Elfvin L. G. and Elde R. Enkephalin-like immunoreactivity in nerve terminals in sympathetic ganglia and adrenal medulla and in adrenal medullary gland cells. *Acta Physiol. Scand.* 1978, **103**, 475–478.
15. Varndell I. M., Tapia F. J., De Mey J., Rush R. A., Bloom S. R. and Polak J. M. Electronimmunocytochemical localisation of enkephalin-like material in catecholamine-containing cells of the carotid body, the adrenal medulla, and in phaeochromocytomas of man and other mammals. *J. Histochem. Cytochem.* 1982, **30**, 682–690.
16. Feyrter F. *Über diffuse endokrine epitheliale Organe*. Leipzig, J. A. Barth, 1938: 6–17.
17. Pearse A. G. E. The cytochemistry and ultrastructure of polypeptide hormone-producing cells of the APUD series, and the embryonic, physiologic and pathologic implications of the concept. *J. Histochem. Cytochem.* 1969, **17**, 303–313.
18. Roberts G. W., Crow T. J. and Polak J. M. Neuropeptides in the brain. In: Bloom S. R. and Polak J. M., ed., *Gut Hormones*, 2nd ed. Edinburgh, Churchill Livingstone, 1981: 457–463.
19. Polak J. M. and Bloom S. R. Peripheral localisation of regulatory peptides as a clue to their function. *J. Histochem. Cytochem.* 1980. **28**, 918–924.
20. Scharrer E. Principles of neuroendocrine integration. *Res. Publ. Assoc. Nerv. Ment. Dis.* 1966, **43**, 1–35.
21. Polak J. M. and Bloom S. R. The diffuse neuroendocrine system. *J. Histochem. Cytochem.* 1979, **27** 1398–1400.
22. Solcia E., Capella C., Buffa R., Usellini L., Fiocca R. and Sessa F. Endocrine cells of the digestive system. In: Johnson L. R., ed., *Physiology of the Gastrointestinal Tract*. New York, Raven Press, 1981: 39–58.
23. Polak J. M. and Bloom S. R. Regulatory peptides: new aspects. In: Williams E. D., ed., *Recent Advances in Endocrinology*. London, W. B. Saunders Ltd, 1982: – .
24. Walsh J. H. Gastrointestinal hormones and peptides. In: Johnson L. R., ed., *Physiology of the Gastrointestinal Tract*. New York, Raven Press, 1981: 59–144.

25. Bloom S. R. and Polak J. M. (ed.) *Gut Hormones*, 2nd ed. Edinburgh, Churchill Livingstone, 1981.
26. Grimelius L. and Wilander E. Silver stains in the study of endocrine cells of the gut and pancreas. *Invest. Cell Pathol.* 1980, **3**, 3–12.
27. Solcia E., Vasallo G. and Capella C. Lead–haematoxylin as a stain for endocrine cells. *Histochemie* 1969, **20**, 116–126.
28. Moore B. W. Chemistry and biology of two proteins, S-100 and 14-3-2, specific to the nervous system. In: Pfeiffer C. C. and Smythies J. R., ed., *International Review of Neurobiology*, Vol. 15. New York, Academic Press, 1972: 215–225.
29. Marangos P. J. and Zomzely-Neurath C. Determination and characterisation of neuron-specific protein (NSP) associated enolase activity. *Biochem. Biophys. Res. Commun.* 1976, **68**, 1309–1316.
30. Schmechel D. E., Brightman M. W. and Marangos P. J. Neurone-specific enolase is a molecular marker for peripheral and central neuroendocrine cells. *Nature* 1979, **276**, 834–836.
31. Bishop A. E., Polak J. M., Facer P., Ferri G-L., Marangos P. J. and Pearse A. G. E. Neuron-specific enolase: a common marker for the endocrine cells and innervation of the gut and pancreas. *Gastroenterology*, 1982, in press.
32. Wharton J., Polak J. M., Cole G. A., Marangos P. J. and Pearse A. G. E. Neuron-specific enolase as an immunocytochemical marker for the diffuse neuroendocrine system in human foetal lung. *J. Histochem. Cytochem.* 1981, **29**, 1359–1364.
33. Gu J., Polak J. M., Tapia F. R., Marangos P. J. and Pearse A. G. E. Neuron-specific enolase in the Merkel cells of mammalian skin: the use of specific antibody as a simple and reliable histologic marker. *Am. J. Pathol.* 1981, **104**, 63–68.
34. Pearse A. G. E. and Polak J. M. Bifunctional reagents as vapour and liquid-phase fixatives for immunocytochemistry. *Histochem. J.* 1975, **7**, 179–186.
35. Roberts G. W., Woodhams P. L., Polak N. M. and Crow T. J. Distribution of neuropeptides in the limbic system of the rat: the amygdaloid complex. *Neuroscience* 1982, 7, 99–131.
36. Davies P., Katzman R. and Terry R. D. Reduced somatostatin-like immunoreactivity in cerebral cortex from cases of Alzheimer disease and Alzheimer senile dementia. *Nature* 1980, **288**, 279–280.
37. Emson P. C., Arregui A., Clement V., Sandberg B. E. B. and Rossor M. Regional distribution of methionine enkephalin and substance P-like immunoreactivity in normal human brain and in Huntington's disease. *Brain Res.* 1980, **199**, 147–160.
38. Roberts G. W., Ferrier I. N., Lee Y. C., Adrian T. E., O'Shaughnessey D. J., Crow T. J., Polak J. M. and Bloom S. R. Schizophrenia—peptidergic deficits in the limbic system. *Reg. Pept.* 1982, **3**, 81.
39. Gibson S. J., Polak J. M., Bloom S. R. and Wall P. D. The distribution of nine peptides in rat spinal cord with special emphasis on the substantia gelatinosa and on the area around the central canal (lamina *X*). *J. Comp. Neurol.* 1981, **201**, 65–79.
40. Polak J. M. and Bloom S. R. The neuroendocrine design of the gut. In: Buchanan K. (ed.) *Clinics in Endocrinology and Metabolism, Vol. 8. No. 2.* London, Saunders, 1979: 313–330.
41. Larsson, L-I. Somatostatin cells. In Bloom S. R. and Polak J. M. (ed.) *Gut Hormones*. Edinburgh, Churchill Livingstone, 1981: 350–353.
42. Polak J. M., Pearse A. G. E. and Heath C. M. Complete identification of endocrine cells in the gastrointestinal tract using semithin–thin sections to identify motilin cells in human and animal intestine. *Gut* 1975, **16**, 225–229.
43. Solcia E., Polak J. M., Larsson L.-I., Buchan A. M. J. and Capella C. Update on Lausanne classification of endocrine cells. In: Bloom S. R. and Polak J. M. (ed.) *Gut Hormones*, 2nd ed. Edinburgh, Churchill Livingstone, 1981: 96–100.
44. Bloom S. R. and Polak J. M. Gut hormone overview. In: Bloom S. R. and Polak J. M. (ed.) *Gut Hormones*, Edinburgh, Churchill Livingstone, 1978: 3–19.
45. Bloom S. R. and Polak J. M. Plasma hormone concentrations in gastrointestinal disease. In: Creutzfeldt W. (ed.) *Gastrointestinal Hormones*, Vol. 9. London, Saunders, 1980: 785–798.
46. Creutzfeldt W., Arnold R., Creutzfeldt C. and Track N. S. Mucosal gastrin concentration, molecular forms of gastrin, number and ultrastructure of G-cells in patients with duodenal ulcer. *Gut* 1976, **16**, 745–754.
47. Sjolund, K., Alumets J., Berg N-O., Håkanson R. and Sundler F. Duodenal endocrine cells in adult coeliac disease. *Gut* 1979, **20**, 547–552.
48. Overgaard Nielsen H., Lauritsen K. and Christiansen L. A. The antral gastrin-producing cells in duodenal ulcer patients. *Acta Pathol. Microbiol. Scand.* (*Sect. A*) 1981, **89**, 293–296.
49. Ferri G-L., Harris A., Wright N. A. and Bloom S. R. Quantification of endocrine cells in whole intestinal crypts and villi. *Histochem. J.* 1982, **14**, 692–695.

50. Polak J. M. and Bloom S. R. Organisation of the gut peptidergic innervation. In: Bloom S. R. and Polak J. M. (ed.) *Gut Hormones*. Edinburgh, Churchill Livingstone, 1981: 487–494.
51. Schultzberg M., Hökfelt T., Nilsson G, Terenius L., Rehfeld J. F., Brown M., Elde R., Goldstein M. and Said S. Distribution of peptide and catecholamine-containing neurons in the gastrointestinal tract of rat and guinea pig. Immunohistochemical studies with antisera to substance P, vasoactive intestinal polypeptide, enkephalins, somatostatin, gastrin/cholecystokinin, neurotensin and dopamine β-hydroxylase. *Neuroscience* 1980, **3**, 689–744.
52. Furness J. B., Costa M., Franco R. and Llewellyn-Smith I. J. Neuronal peptides in the intestine: distribution and possible functions. In: Costa E. and Trabucchi M. (ed.) *Neural Peptides and Neuronal Communications*. New York, Raven Press, 1980: 601–617.
53. Jessen K. R., Saffrey M. J., Van Noorden S., Bloom S. R., Polak J. M. and Burnstock G. Immunohistochemical studies of the enteric nervous system in tissue culture and in situ: localization of vasoactive intestinal polypeptide (VIP), substance P and enkephalin-immunoreactive nerves in the guinea-pig gut. *Neuroscience* 1980, **5**, 1717–1735.
54. Furness J. B. and Costa M. Types of nerves in the enteric nervous system. *Neuroscience* 1980, **5**, 1–20.
55. Jessen K. R., Polak J. M., Van Noorden S., Bloom S. R. and Burnstock G. Peptide-containing neurons connect the two ganglionated plexuses of the enteric nervous system. *Nature* 1980, **293**, 391–393.
56. Baumgarten H. G., Holstein A. F. and Owman C. H. Auerbach's plexus of mammals and man: electron microscopical identification of three different types of neuronal processes in myenteric ganglia of the large intestine from rhesus monkeys, guinea pigs and man. *Z. Zellforsch. Mikrosk. Anat.* 1970, **106**, 376–397.
57. Larsson L.-I. Ultrastructural localisation of a new neuronal peptide (VIP). *Histochemistry* 1977, **54**, 173–176.
58. Cook R. D. and Burnstock G. The ultrastructure of Auerbach's plexus in the guinea-pig. 1. Neuronal elements. *J. Neurocytol.* 1976, **5**, 171–194.
59. Probert L., De Mey J. and Polak J. M. Distinct subpopulations of enteric p-type neurones contain substance P and vasoactive intestinal polypeptide. *Nature* 1981, **294**, 470–471.
60. Gabella G. Structure of muscles and nerves in the gastrointestinal tract. In: Johnson L. R. (ed.) *Physiology of the Gastrointestinal Tract*. New York, Raven Press, 1981: 197–241.
61. Jessen K. R. and Mirsky R. Glial cells in the enteric nervous system contain glial fibrillary protein. *Nature* 1980, **286**, 736–737.
62. Long R. G., Bishop A. E., Barnes A. J., Albuquerque R. H., O'Shaughnessey D. J., McGregor G. P., Bannister R., Polak J. M. and Bloom S. R. Neural and hormonal peptides in rectal biopsy specimens from patients with Chagas' disease and chronic autonomic failure. *Lancet* 1980, **i**, 559–562.
63. Bishop A. E., Polak J. M., Lake B. D., Bryant M. G. and Bloom S. R. Abnormalities of the colonic regulatory peptides in Hirschsprung's disease. *Histopathology* 1981, **5**, 679–688.
64. Bishop A. E., Polak J. M., Bryant M. G., Bloom S. R. and Hamilton S. Abnormalities of vasoactive intestinal polypeptide-containing nerves in Crohn's disease. *Gastroenterology* 1980, **79**, 853–860.
65. Heitz Ph., Polak J. M. and Pearse A. G. E. Identification of the D_1 cell as the source of human pancreatic polypeptide (HPP). *Gut* 1976, **17**, 755–758.
66. Bishop A. E., Polak J. M., Green I. C., Bryant M. G. and Bloom S. R. The location of VIP in the pancreas of man and rat. *Diabetologia* 1980, **18**, 73–78.
67. Larsson L-I. Innervation of the pancreas by substance P, enkephalin, vasoactive intestinal polypeptide and gastrin/CCK immunoreactive nerves. *J. Histochem. Cytochem.* 1979, **27**, 1283–1284.
68. Larsson L-I. and Rehfeld J. F. A peptide resembling COOH-terminal tetrapeptide amide of gastrin from a new gastrointestinal endocrine cell type. *Nature* 1979, **277**, 575–577.
69. Welbourn R. B., Wood S. M., Polak J. M. and Bloom S. R. Pancreatic endocrine tumours. In: Bloom S. R. and Polak J. M. (ed.) *Gut Hormones*. Edinburgh, Churchill Livingstone, 1981: 547–554.
70. Tapia F. J., Polak J. M., Barbosa A. J. A., Boom S. R. Marangos P. J., Dermody C. and Pearse A. G. E. Neuron-specific enolase is produced by endocrine tumours. *Lancet* 1981, **i**, 808–811.
71. Polak J. M. and Bloom S. R. Regulatory peptides in the respiratory tract of man and other mammals. *Exp. Lung Res.* 1982, in press.

72. Said S. I. Vasoactive intestinal peptide (VIP): isolation, distribution, biological actions,
structure–function relationships and possible functions. In: Glass J. (ed.) *Gastrointestinal*
Hormones. New York, Raven Press, 1980: 273–346.
73. Skrabanek P. and Powell D. (ed.) *Substance P*. Vol. 2. Edinburgh, Churchill Livingstone, 1980.
74. McDonald T. J. Non-amphibian bombesin-like peptides. In: Bloom S. R. and Polak J. M. (ed.) *Gut*
Hormones, 2nd ed. Edinburgh, Churchill Livingstone, 1981: 407–412.
75. Moody T. W., Pert C. B., Gazdar A. F., Carney D. N. and Minna J. D. High levels of intracellular
bombesin characterize human small-cell lung carcinoma. *Science* 1981, **214**, 1246–1248.
76. Lynch E. M., Wharton J., Bryant M. G., Bloom S. R., Polak J. M. and Elder M. G. The differential
distribution of vasoactive intestinal polypeptide in the normal human female genital tract.
Histochemistry 1980, **67**, 169–177.
77. Polak J. M., Gu J., Mina S. and Bloom S. R. VIPergic nerves in the penis. *Lancet* 1981, **ii**, 217–219.
78. Wharton J., Polak J. M., McGregor G. P., Bishop A. E. and Bloom S. R. The distribution of
substance P-like immunoreactive nerves in the guinea pig heart. *Neuroscience* 1981, **6**, 2193–2204.
79. Austin L. A. and Heath H. Calcitonin: Physiology and pathophysiology. *N. Engl. J. Med.* 1981,
304, 269–278.
80. Sundler F., Alumets J., Håkanson R., Bjorklund L. and Ljundberg O. Somatostatin-
immunoreactive cells in medullary carcinoma of the thyroid. *Am. J. Pathol.* 1977, **88**, 381–386.
81. Ahren B., Alumets J., Ericsson K., Fahrenkrug J., Fahrenkrug L., Håkanson R., Hedner P., Loben
I., McLander A., Rerys C. and Sundler F. VIP occurs in intrathyroidal nerves and stimulates
thyroid hormone secretion. *Nature* 1980, **287**, 343.
82. Wharton J., Polak J. M., Pearse A. G. E., McGregor G. P., Bryant M. G., Bloom S. R., Emson
P. C., Bisgard G. E. and Will J. H. Enkephalin, vasoactive intestinal polypeptide (VIP) and
substance P-like immunoreactivity in the carotid body. *Nature* 1980, **284**, 269–271.
83. Terenghi G., Varndell I. M., Wharton J., Lee Y. C., Bloom S. R. and Polak J. M. Neurotensin-like
immunoreactivity in adrenal medulla. *J. Pathol.* 1982, in press.
84. Sullivan S. N., Bloom S. R. and Polak J. M. Enkephalin in peripheral neuroendocrine tumours.
Lancet 1978, **i**, 1155.
85. Tischler A. S., Lee Y. C., Slayton V. W. and Bloom S. R. Content and release of neurotensin in
PC12 phaeochromocytoma cell cultures: modulation by dexamethasone and nerve growth
factor. *Reg. Pept.* 1982, in press.
86. Uddman R., Alumets J., Ehringer B., Håkanson R., Loren I. and Sundler F. Vasoactive intestinal
polypeptide nerves in ocular and orbital structures of the cat. *Invest. Ophthalmol. Vis. Sci.* 1980,
19, 878–885.
87. Brecha N. and Karten J. H. Localisation of enkephalin, substance P, neurotensin and somatostatin
immunoreactivity within the amacrine cells of the retina. *Anat. Rec.* 1980, **196**, 225–228.
88. Loren I., Tornquist K. and Alumets J. VIP (vasoactive intestinal polypeptide)-immunoreactive
neurons in the retina of the rat. *Cell. Tissue Res.* 1980, **210**, 167–170.
89. Unger W. G., Butler J. M., Cole D. R., McGregor G. P. and Bloom S. R. Substance P, vasoactive
intestinal polypeptide (VIP) and somatostatin levels in ocular tissue of normal and sensorily
denervated rabbit eye. *Exp. Eye Res.* 1981, **32**, 797–801.
90. Butler J. M., Terenghi G., Polak J. M., Bloom S. R. and Cole D. F. Distribution of substance P and
VIP-containing neurons in the uvea. *Ophthalmic Res.* 1982, in press.
91. Polak J. M. and Bloom S. R. The peripheral substance P-ergic system. *Peptides* 1981, **2**, Suppl 2,
133–148.
92. Butler J. M., Powell D. and Unger W. G. Substance P levels in normal and sensorily denervated
rabbit eyes. *Exp. Eye Res.* 1980, **30**, 311–313.
93. O'Shaughnessy D. J., McGregor G. P., Mina S., Ghatei M. A., Gu J., Polak J. M. and Bloom S. R.
Regulatory peptides in the skin. Soc. for Endocrinology. 1981, 164th Meeting. p. 26.
94. Falk B. and Owman C. A detailed methodological description of the fluorescence method for the
cellular demonstration of biogenic monoamines. *Acta Univ. Lund* 1965, Section II, no. 7, 1–23.
95. Tatemoto K. New peptides, their structure and function. In: Bloom S. R., Polak, J. M. and
Lindenlaub, E. (ed.) *Systemic Role of Regulatory Peptides*. Stuttgart, Schattauer, 1982, in press.

12

Immunocytochemistry of Pituitary Hormones

P. Petrusz and
P. Ordronneau

The primary objective of pituitary immunocytochemistry is the identification of the individual cells or cell types responsible for the production and release of pituitary hormones. Implicit in this goal is the assumption that each hormone is synthesized, stored and secreted by one distinct cell type and, conversely, that one cell contains only one hormone. Although it is clear today that none of these assumptions is completely true (cf. the literature[1–6]), it is also clear that one of the greatest achievements of immunocytochemistry during the past decade has been the nearly total elimination of the past confusion concerning the functional classification and nomenclature of the hormone-producing cells of the adenohypophysis. Thus, there is general agreement today that immunocytochemical staining with suitable antisera will identify distinct populations of cells, with well-defined light and electron microscopical characteristics, as containing and, therefore, presumably producing each particular hormone. This consensus, and the simplicity and clarity of the functional classification (thyrotropes, somatotropes etc.) are entirely due to information obtained with immunocytochemical techniques. The correspondence of the classical tinctorial cell types with the immunocytochemically identified functional cell types is textbook information[7] and has been the subject of many excellent reviews during the past few years.[8–19]

1. THE TECHNOLOGY OF PITUITARY IMMUNOCYTOCHEMISTRY

1.1. Tissue Preparation

The general principles of preparation of pituitary tissue for immunocytochemical staining are not different from those worked out for use with any other animal tissue. Pituitary tissue can be fixed either by immersion or by in situ perfusion (the former is preferred for electron microscopy). Good results have been demonstrated with a wide range of fixatives but objective and systematic studies on this subject are very few.[20] Detailed guidelines are given by Sternberger[21] and Pearse.[22] As suggested by the latter author, the best strategy today is to try out a suitable series of fixatives when a new problem is being considered. Such a series may include:

1. Buffered formaldehyde (2–5 per cent, freshly prepared from paraformaldehyde), with low amounts (not more than 0·5 per cent) of glutaraldehyde added for electron microscopy. In our experience, such mixtures are the most satisfactory general fixatives for all pituitary hormones.
2. Picric acid–formaldehyde.[11]
3. Bouin's fluid or one of its variants.[23]
4. Periodate–lysine–paraformaldehyde.[24]
5. Osmium tetroxide alone, in combination with, or following aldehyde fixation.[25,26]

Fixation must be followed by extensive washing in buffered saline (48 h at 4 °C in several changes) in order to remove excess fixative. At this stage, it is possible to prepare frozen or Vibratome (Oxford Instruments) sections and stain them floating in the immunocytochemical reagents. It is generally agreed that such *pre-embedding immunocytochemistry* is one of the most powerful approaches available today,[27–30] especially since it may be followed by embedding in plastic resins and sectioning for electron microscopy. Alternatively, fixation and washing may be followed by routine paraffin or plastic embedding and processing for *post-embedding immunocytochemistry*.

Most of the electron microscopical immunocytochemistry of the pituitary has been done using post-embedding techniques. Recently, pre-embedding methods became more popular.[5,31–33] Theoretically, these methods could result in increased sensitivity since the antigen is localized before exposure to the organic solvents required for embedding. When comparing pre- and post-embedding staining in the pituitary, the cell types which stain with a given antiserum are the same with both techniques.[31–33] However, subcellular localization of hormones may be improved, e.g. immunoreactive hormone has been found in the endoplasmic reticulum with the pre-embedding methods only.[31,32] It might also be reasonably expected that the contents of the Golgi cisternae would be immunoreactive. However, this is never the case.[5,31–33] It may be that a protein is so modified in the Golgi that an antibody does not recognize it. Further studies should clarify this situation. There are two other advantages of the pre-embedding methods: (1) the immunocytochemically stained material can be viewed before embedding and stained areas selected for electron microscopy and (2) these techniques usually result in superior ultrastructural detail.

Penetration of the antibodies through the entire thickness of the section is essential for optimal results. The organic solvents used during paraffin processing render such sections permeable; however, when fixed frozen or Vibratome sections or cell monolayers are used, a short treatment with detergents[34,35] or ethanol (P. Petrusz, 1981, unpublished results) is advisable before application of primary antibody.

1.2. Unmasking of Tissue Antigens with Proteolytic Enzymes

In addition to permeabilization with detergents or alcohol, tissue antigens can be effectively unmasked, or rendered more accessible to the antibodies, by carefully controlled pretreatment of the sections with proteolytic enzymes. In fact, this procedure was first used for the demonstration of human pituitary hormones[36] and its usefulness has been confirmed in our laboratory. The treatment results both in intensification of the subsequent specific staining and in reduction of the non-specific background. Trypsin,[37] pepsin,[38] and pronase[39] have all been recommended and used with success. The required conditions seem to vary with the nature of the antigen localized, the method of tissue preparation (e.g. degree of cross-linking during fixation), and with the enzyme used. Details of these procedures have been recently reviewed.[40] Since it has now been demonstrated that pretreatment with proteolytic enzymes improves the immunocytochemical localization of a great variety of tissue antigens, it can be recommended as one of the routine procedures that should be available in every immunocytochemistry laboratory.

1.3. Immunocytochemical Staining Methods

Practically all available immunocytochemical techniques have been used successfully on pituitary tissue. A review of the literature indicates that the great majority of investigators preferred either the indirect (two-layer sandwich) technique with fluorescent marker as first described by Weller and Coons[41] and more recently with horseradish peroxidase, introduced by Nakane and Pierce,[42] or the peroxidase–anti-peroxidase (PAP) method of Sternberger.[21,43] Since these techniques are generally known and are extensively covered in other chapters in this volume, detailed protocols will not be given here. Our own laboratory has obtained very satisfactory results with the original variant of the unlabelled antibody–enzyme techniques, the 'bridge' method of Avrameas,[44] Mason et al.[45] and Sternberger and Cuculis.[46] Our modification[47] appears to be as practical and efficient as the PAP method, provided that a high-titre anti-peroxidase serum is used in the third layer.[48] Step-by-step protocol for its use for both light and electron microscopical (post-embedding) staining is given in the Appendix.

Additional intensification can be achieved with the formation of extended bridges (larger complexes) by simply repeating the incubations with the second and third-layer antibodies.[49,50] The 'double bridge' and 'double PAP' variants have been used successfully for pituitary hormones in our laboratory.[50,51] Especially for light microscopy on paraffin sections, the double bridge is now our routine method unless special considerations dictate otherwise. Detailed protocols for both light and electron microscopy are given in the Appendix.

Protein A may be used instead of the anti-IgG as the second-layer component[52,53] either coupled to a suitable marker or as a bridge to bind the third-layer anti-peroxidase antibodies. In recent years, the protein A–colloidal gold technique has emerged as one of the most promising innovations, especially for electron microscopical immunocytochemistry (for review, *see* Roth[54]). Additional variants of the bridge principle include the avidin–biotin system to replace immunological binding.[55] The so-called ABC modification of this technique[56] has been satisfactory in our hands for localization of pituitary and other hormones (P. Petrusz et al. (1982), unpublished observations), although we could not substantiate the claims that it is much more 'sensitive' than the PAP technique.

1.4. Interpretation of Immunocytochemical Results

Besides a reliable technique, nothing is more important in immunocytochemistry than the design of experiments to incorporate appropriate and informative controls and the correct interpretation of the results. As to criteria of validity, the reader is referred to several publications which provide detailed discussions and guidelines.[21,22,47,48,57,58] In summary, the minimum requirements for valid immunocytochemical staining should include *all* of the following:

1. Careful preliminary standardization of all steps of the method.
2. Running increasing dilutions (titration) of the primary antibody up to a dilution where all immunostaining disappears (the latter also serves as a control equivalent to omission of the primary antibody)

(1) and (2) together will ensure what we term 'method specificity.' They will also help to avoid disturbing background staining, and establish the optimal dilution of the primary antibody.

3. Parallel tests of 'antiserum specificity' to demonstrate inhibition of the staining with homologous antigen and, if feasible, lack of such inhibition with heterologous (but relevant) antigens.
4. Use of all available other (biochemical, physiological etc.) information to support the validity of the localization (e.g. known presence or absence of the antigen in certain tissues, physiological changes etc.)

In interpreting the results of the above validity tests, the following considerations must be kept in mind: first of all, that antibodies recognize only antigenic *sites* or *regions*, not larger biochemical or biological entities. It follows that specificity tests also reflect only such site or regional specificity, and not molecular specificity. If the recognized site or a site very similar to it occurs in several different molecules (true cross-reaction), it is not reasonable to expect that immunocytochemistry with antibodies directed to that *site* will be able to discriminate between such *molecules*. In such a case, one either has to find antibodies directed against dissimilar portions of the antigenic molecules in question (*see* Chapters 2 and 4), or has to resort to techniques other than immunocytochemistry. These considerations apply equally to both conventional and monoclonal antibodies, and are especially important with regard to the pituitary hormones, several of which are known to share common antigenic determinants (e.g. the common α-subunit of glycoprotein hormones; extensive homology between prolactin and growth hormone; repeated occurrence of the α-melanocyte-stimulating hormone (α-MSH) sequence in pro-opiomelanocortin). Secondly, radioimmunoassay characterization of the primary antiserum does not permit any conclusions as to the specificity of that antiserum in immunocytochemistry. This fact is amply demonstrated in the literature and is probably due to the different conditions prevailing in the two techniques. Finally, it is important to keep in mind that the value of the absorption test crucially depends on the purity of the antigen preparation used.

It is clear that in some cases, due to unavailability of materials (tissue, antisera, purified antigen), even the above minimal criteria cannot be met in practice. In such cases, more emphasis may be placed on indirect evidence, such as in criterion (4), and it will become more important than ever to interpret the results with caution.

2. IMMUNOCYTOLOGY OF THE PITUITARY

Among all species, the rat's pituitary has been studied most extensively with immunocytochemical techniques. Hormone-producing cell types have been best characterized in this species.[8–12,47] For non-mammalian species, the reader is referred to the excellent recent review of Doerr-Schott.[18] It is quite clear from these and other studies that, in spite of the similarities, information obtained in one species is usually not directly applicable to another. The following brief description of the recognized functional cell types is based mainly on studies with the rat pituitary, with only a few references to other species. A separate brief section will be devoted to the immunocytology of the human pituitary, due to its great importance in pathological diagnosis.

2.1. The Gonadotropic Cells

The gonadotropic cells of the rat pituitary have been identified with immunocytochemistry.[2,5,12,32,41,59–68] These cells are well known to represent a subclass of the basophils. Under the light microscope, they appear large, ovoid or somewhat angular. Rarely, they send out slender processes towards capillaries. A population of follicle-stimulating hormone (FSH) cells (*see below*) has been described as stellate.[8,11] The nucleus is, as a rule, eccentric and a prominent Golgi complex is visible even with the light microscope at the centre of the cell, adjacent to the nucleus. Although gonadotropes are found throughout the pars distalis, they are somewhat more numerous in a medial–dorsal–anterior region, the so-called sex zone, near the pars intermedia; in this region the cells are arranged in clusters around capillaries. Occasionally a few 'aberrant' cells are found within the pars intermedia.

Early electron microscopical immunocytochemistry indicated that both FSH and luteinizing hormone (LH) were present in one cell type and only FSH was present in another.[9] In later studies it was shown that there are many forms of gonadotropes and that some can contain more than one hormone.

One gonadotropic cell type is large and oval with two populations of secretory granules: one is about 200 nm in diameter and the other ranges from 400–600 nm.[9,11,62] This cell type is found in both male and female pituitaries,[62] and always contains both FSH and LH.[62,65,69] In serial sections from the same cell both hormones can be found in both classes of secretory granules and even within the same secretory granule.[32,65]

Another gonadotrope is smaller and angular with a single population of secretory granules that range from 200 to 250 nm in diameter.[62,67] In the male these cells are rare, but they can contain both hormones.[65] In the female this cell type is more frequent and may also contain both FSH and LH.[62,69]

The third gonadotropic cell type is morphologically identical to the corticotrope (*see below*).[69] Though this cell type rarely contained LH, it was demonstrated to contain both FSH and adrenocorticotropin (ACTH), with both hormones in the same granule.[2] A brief description of the human gonadotropes can be found in Pelletier et al.[13]

In conclusion, there is an almost unanimous agreement among investigators today that the majority of the gonadotropes (70 per cent or more) contain both FSH and LH, while the remainder contain only one of these hormones. Variations in cell morphology related to age, sex, and stage of sexual cycle as well as variations unrelated to any identifiable cause have been described.[65,66,70] The question of how the differential release of the two hormones from a single cell is controlled remains one of the most intriguing problems of pituitary research.

2.2. The Thyrotropic cells

Thyrotropic cells have been difficult to distinguish from gonadotropes with immunocytochemistry because of the nearly identical structure of the α-subunit of thyroid-stimulating hormone (TSH) and the α-subunit of the gonadotropins FSH and LH. However, availability of antisera to the hormone-specific β-subunit and recognition of certain morphological characteristics of these cells now make such distinction relatively easy.[12,32,59,63,71,72]

Individual thyrotropes are scattered throughout the pars distalis but they are more frequent and occur in clusters in the centro-medial regions of the gland. They are rather variable in size and shape (polygonal to stellate), with a central nucleus and a Golgi complex which, unlike that in the gonadotropes, is not obvious with light microscopy.

Electron microscopical immunocytochemistry confirms that the thyrotrope is stellate or polygonal. The ultrastructure of the thyrotrope can be confused with that of the corticotrope; however, the thyrotrope has more cytoplasm and smaller secretory granules (100–150 nm in diameter) than the corticotrope. The secretory granules of the thyrotrope are found either in rows along the plasma membrane[8,9,11] or filling the cytoplasm.[73] Descriptions of the human thyrotrope can be found by Pelletier et al.[13] and Moriarty and Tobin.[73]

2.3. The Corticotropic Cells

Prior to the availability of reliable immunocytochemical methods and recent information on the chemistry of the hormone(s) produced by these cells, this was the most controversial cell type of the adenohypophysis. Even today, its correspondence to any of the old tinctorial cell types cannot be determined with certainty, except that it belongs to the basophils. The immunocytochemistry of the corticotropes can now be understood in the light of the recent discovery that the peptide hormones α-MSH, ACTH, β-lipotropin (β-LPH), and β-endorphin are all fragments of a single large polypeptide precursor (molecular weight 31 000 daltons), pro-opiomelanocortin.[74] In the cells of the pars intermedia, ACTH is preferentially cleaved to α-MSH and corticotropin-like intermediate-lobe peptide (CLIP).[75] β-LPH and its fragment, β-endorphin, also contain the five amino acid sequence of the endogenous opiate peptide methionine-enkephalin.[76]

In practical terms, this means that region-specific antibodies directed to any of the characteristic amino acid sequences occurring in α-MSH, CLIP, ACTH, β-LPH, β-endorphin and Met–enkephalin should be expected to label all cells of the pars intermedia as well as the corticotropes of the pars distalis. Indeed this appears to be the case in the great majority of the numerous studies carried out to date.[4,47,63,72,77–84] In addition, Bloom et al.[85] and Pelletier et al.[86] also localized a more recently recognized portion of the same 31-kilodalton precursor, γ-MSH, to the same cells and, in fact, to the same secretory granules, as those containing ACTH.

However, some unresolved problems and inconsistencies still remain. The enzymatic processing of this family of peptides, perhaps not surprisingly, seems to be under differential control in anatomically distinct regions of the gland. Biologically active and inactive (acetylated) forms are stored in variable proportions, and the latter are especially abundant in the pars intermedia.[84,87,88] Earlier studies localized β-LPH, in addition to corticotropes, to somatotropes and lactotropes[81] or to somatotropes and gonadotropes, as well as to TSH-producing tumour cells.[89] Dissociation, i.e. less than complete overlap, was repeatedly reported between cells specifically stained for ACTH and those stained for endorphin in the normal pituitary[90–93] as well as in pituitary adenomas.[94] In addition, several groups reported the detection of calcitonin-like immunoreactivity in the rat and human pituitary.[95–98] Most investigators localized this antigen in the pars intermedia and in the corticotropes of the pars distalis; however, Watkins and Moore[99] found it to be localized in the thyrotropes.

The corticotropes of the adenohypophysis are considered part of the APUD (amine or amine precursor uptake and decarboxylation) system.[100–102] They are rather evenly distributed in the pars distalis. The cell body appears polyhedral or stellate with the light microscope and sends out long slender processes between other cells towards capillaries. Nakane[8] noted the overlapping distribution and preferential juxtaposition of ACTH and growth hormone (GH) cells.

Ultrastructural immunocytochemistry has identified the corticotrope as a small angular or stellate cell with sparse cytoplasm and secretory granules which measure 220 nm in diameter. These granules are located mainly at the periphery of the cell.[8,9,72,79] The mitochondria are slender, twisted and have a dense matrix.[11] Information on the ultrastructural characteristics of the human corticotropes are given by Moriarty and Garner[1] and Moriarty.[11]

2.4. The Lactotropic Cells

The lactotropes, also called prolactin (PRL) cells, mammotropes, or luteotropes, have been identified as a cell type distinct from the somatotropes in the rat[12,25,26,47,103–105] as well as in primates including man.[14,106–109]

Lactotropes are distributed uniformly in the pars distalis and appear distinctly more numerous in the normal female than in the male. Their morphology with the light microscope appears extremely variable. They are frequently cup shaped and surround large oval gonadotropic cells,[8] the nucleus is centrally located. One characteristic feature of these cells is the uneven distribution of the PRL-containing secretory granules in the cytoplasm, resulting in similarly uneven, spotty, fragmented immunostaining. For these reasons, the true shape of these cells is difficult to recognize with the light microscope. Their unique association with the gonadotropes is especially striking in the pituitaries of castrated rats[110] and may have interesting functional implications. From ultrastructural studies the lactotrope usually appears cup shaped, since it extends cytoplasmic processes between and around other cells, particularly the gonadotropes.[8] The ends of these processes reach to the endothelium of the sinusoids. Lactotropes are only moderately granulated. However, the polymorphic secretory granules are the largest of all the cell types, often exceeding 800 nm.[8,9,11] Recently, four morphologically distinct cell types have been found to contain prolactin.[6] These may represent different functional states of the same cell. Descriptions of the human lactotrope are given by Pelletier et al.[13] and Duello and Halmi.[111]

2.5. The Somatotropic Cells

Somatotropes have been described in the hypophyses of all vertebrates studied. Their identity as the major class of acidophils had been clear before the application of immunocytochemical techniques. Somatotropes have been identified with immunocytochemistry by several groups.[12,14,25,26,47,103,106,108,112] The somatotropes of the pars distalis are numerous and evenly dispersed. They are characteristically fairly uniform in size and shape, ovoid or cuboidal, with a central nucleus and closely packed and evenly distributed cytoplasmic granules. They are often arranged in rows alongside capillaries, one GH cell immediately next to the other—a rare occurrence with other cell types. As mentioned earlier, they are often associated with ACTH cells.[8] From electron microscopical immunocytochemistry the somatotrope appears as an oval or pyramidal cell with a slightly eccentric nucleus. The abundant secretory granules fill the cytoplasm and measure about 300–350 nm in diameter.[8,9,11] Information on the human somatotrope is given by Pelletier et al.[13] and Duello and Halmi.[111]

3. SPECIAL APPLICATIONS OF PITUITARY IMMUNOCYTOCHEMISTRY

Immunocytochemistry is ideally suited for the identification of a few individual, widely scattered cells containing pituitary hormones in a larger mass of tissue where their detection by other (e.g. biochemical) techniques would be impossible. This is well illustrated by immunocytochemical studies carried out on the pars tuberalis of the adenohypophysis and on the pharyngeal pituitary. The pars tuberalis (PT) surrounds the pituitary stalk and covers the basal surface of the median eminence. Early immunocytochemical investigations of the PT of the rat revealed a sparse distribution of gonadotropes only. The rest of the PT was negative for all of the other adenohypophyseal hormones.[113] Similar results were found in the monkey[114,115] and the human.[116] Later it was shown that the gonadotropes of the PT were able to respond to castration[117,118] and steroid hormone replacement.[118] Due to its size and location, the PT remains intact following hypophysectomy. With immunocytochemistry on hypophysectomized rats, the PT has been shown to contain not only numerous gonadotropes but active thyrotropes, somatotropes, lactotropes, and corticotropes as well.[50] This response to hypophysectomy occurs within 2 days (P. Ordronneau, 1980, unpublished results). Therefore, the PT might provide a mechanism to maintain minimal amounts of adenohypophyseal hormones in the absence of a functioning pars distalis. At any rate, the hypophysectomized animal can no longer be thought of as a model free of pituitary hormones.

Similarly, several pituitary hormones (GH, PRL, TSH and ACTH) were identified in cells of the human pharyngeal pituitary.[119-122] The implications of these findings are similar to those regarding the PT.

Immunocytochemistry has also been very useful in investigating the embryonic development and differentiation of the various hormone-producing cells of the pituitary. Again, such studies could not have been carried out with any of the other techniques available today. As a result of this work, it is now abundantly clear that both the rat and the human fetal adenohypophysis are capable of synthesizing all the classical pituitary hormones.[123-128] Corticotropes are the first functional cell type to appear, with immunocytochemically recognizable form and quantity of ACTH in their cytoplasm (day 15 of pregnancy in the rat, eighth week in the human). As to the subsequent events, there is some disagreement (probably due to individual and technical variations) with regard to the precise time and order of the appearance of the other hormones in the fetal pituitary; however, the following composite picture probably reflects the events rather faithfully: in the rat, ACTH is followed by the appearance of TSH, α-MSH, FSH/LH, GH and PRL, in that order;[125-129] in the human, the order is GH, PRL, α-subunits of the glycoprotein hormones, β-subunit of TSH and β-subunit of FSH/LH.[107,123,124,130,131] Based on studies with anencephalic human fetuses, it has been shown that at least the early stages of this well-ordered differentiation (i.e. the appearance of ACTH, LPH and endorphin) can occur normally even in the absence of the hypothalamus.[132,133] Further studies are needed to resolve the remaining controversial issues and to clarify the role of the brain and possible other factors in the entire process of pituitary cytodifferentiation.

In recent years, much interest has been focused on the presence and possible functions of various pituitary hormones in the brain. The largest number of papers deal with the peptides of the ACTH/LPH family (α-MSH;[83,134] β-LPH;[135] β-endorphin[136-139] γ-MSH[85]). The limited neuronal system identified in these studies, with cell bodies restricted mainly to the hypothalamic arcuate nucleus, is clearly different from the much more extensive system demonstrated with antisera to enkephalins.[140] In addition to the ACTH-like and opioid peptides, other pituitary hormones such as gonadotropins[141,142] and prolactin[143-145] have been localized in the brain with immunocytochemistry or by bioassay or radioimmunoassay (for review *see* Krieger and Liotta[146]). These and many other similar recent discoveries have far-reaching implications for neurobiology and endocrinology. They indicate that the nervous system and the endocrine system are related to each other much more closely than previously thought, and form a unified regulatory and information processing system relying on many of the same chemical messengers.

4. IMMUNOCYTOCHEMISTRY OF THE HUMAN PITUITARY

Although in the preceding sections references have been made to the general immunocytological characteristics of the functional cell types of the human pituitary gland, the substantial species differences and the great practical importance of the topic justify a separate brief summary.

The histopathological examination of the human pituitary has been revolutionized by the introduction of sensitive and reliable routine immunocytochemical techniques. We fully share the conviction of Kovacs et al.[19] that immunocytochemical techniques must be introduced 'in every laboratory that deals with problems related to pathology of the pituitary gland.'

The applications of immunocytochemistry of the human pituitary fall into three broad categories.

a. Characterization of the hormone-producing cells of the normal pituitary. This is, of course, a prerequisite for the recognition and definition of physiological, pathological or experimental changes. Sources of fundamental information on this subject, including both light and electron microscopy, are summarized in *Table* 12.1.

Table 12.1 **Immunocytochemical characterization of the hormone-producing cell types of the normal human adenohypophysis**

Cell type	*Light microscopy*	*Electron microscopy*
Gonadotropes	Pelletier et al.[147]	Pelletier et al.[13]
Thyrotropes	Phifer and Spicer[148]	Moriarty and Tobin[73]
Corticotropes	Phifer et al.[78]	Moriarty[11]
Lactotropes	Halmi[14]	Duello and Halmi[111]
Somatotropes	Halmi[14]	Duello and Halmi[111]

b. Characterization of pituitary tumours. Many different tumours may arise in or around the region of the sella turcica and some of these may present difficult diagnostic problems. Most tumours of the adenohypophysial endocrine cells (adenomas) cannot be properly classified without immunocytochemistry, since the old tinctorial classification (acidophilic, basophilic, chromophobic) proved to be unsatisfactory and of little value in indicating possible hormonal activity.[19] With the application of immunocytology, one caution should be kept in mind: here as well as in any other application of immunocytochemistry, negative results are of no decisive value. Otherwise, the value of immunocytochemistry in pituitary tumour diagnosis cannot be overestimated. According to Halmi:[14] '. . . in the appraisal of the functional status of pituitary tumours, immunocytochemistry is a more valuable tool than any conventional histological technique available.' Further details on this subject are provided in Chapter 20 in this volume and in several authoritative publications.[14,94,149–151]

c. Localization of pituitary hormones as markers of non-pituitary tumours. This extremely important and interesting application of pituitary immunocytochemistry is discussed by Heyderman in Chapter 15.

5. FUTURE PROBLEMS

Pituitary immunocytochemistry will remain a valuable tool in detecting and monitoring functionally defined cell types and in studying secretory activity of the pituitary in many routine and research laboratories. The few problems which still exist with regard to physiological variations in cell morphology and hormone content will certainly be resolved in the near future.

There is another broad frontier of pituitary research where significant progress can be expected. A great number of substances *other than classical pituitary hormones* have been localized in the pituitary, although their occurrence there cannot always be explained with current concepts of endocrine cell biology. First of all, these include several *neuropeptides* such as thyrotropin-releasing hormone[152,153] gonadotropin-releasing hormone,[154–156] enkephalins,[158,157] substance P[159] and neurotensin.[160] Most of these localizations have been confirmed in our laboratory, but much more work is needed before their true significance can be understood.

Another class of substances which have been identified in the pituitary is that of the various *growth factors* (GF). Several of these are available in highly purified form, but no cellular localization studies have been reported to date as far as we could ascertain. The known pituitary growth factors include fibroblast GF,[161] chondrocyte GF,[162,163] ovarian GF,[164,165] endothelial GF,[166] glial GF[167] and several others.

Even more striking is the list of the great variety of other substances which have been demonstrated to be of pituitary origin, but about which only fragmentary information is available. In this mixed group one may include the insulin stimulating hormones or β-cell tropins,[168,169] a new lipolytic–melanotropic peptide,[170] still another lipolytic peptide, peptide B, chemically related to neurophysins,[171] a factor regulating steroid metabolism in liver,[172] a hypercalcaemic factor,[173] a ubiquitin-related peptide,[174] a transcortin-like protein,[175] renotropic factor,[176] glial-specific S-100 protein, which actually has been localized to stellate cells in the pars distalis,[177,178] placental lactogen,[179,180] a low-molecular-weight sperm-releasing (LH-like) peptide,[77,181] a chorionic gonadotropin-like hormone,[182] renin,[183] aldosterone-stimulating factor[184] and a 'thymotropic' factor.[185] Most of these substances have been extracted and purified from the pituitary, and for many of them evidence has been presented that they are not identical to any of the known pituitary hormones chemically or immunologically, nor can their biological activity be reproduced with any of the known pituitary hormones alone or in combination. Immunocytochemistry is one of the techniques which can be expected to clarify the cellular origin of these and perhaps many other substances yet to be discovered in the pituitary gland.

Acknowledgements

Research in the authors' laboratory was supported by Research Grants NS14904 and HD13781 from the National Institutes of Health, and by the Neurobiology Program, University of North Carolina, Chapel Hill.

Appendix

I. SINGLE BRIDGE TECHNIQUE FOR LIGHT MICROSCOPY[47]

1. Rinse deparaffinized sections in PBS, then place in 2 per cent NSS for 10 min.
2. The primary antiserum for 46–50 h at 4 °C
3. Rinse in PBS for 3 min at room temperature (the rest of the method is done at room temperature)
4. 2 per cent NSS for 3 min
5. SA-RGG (1 : 50–1 : 100) for 10 min
6. Rinse in PBS for 3 min
7. A-HP (1 : 50–1 : 100) for 10 min
8. Rinse in PBS for 3 min
9. HP (0·5 mg/100 ml PBS with 0·1 per cent egg albumin) for 10 min
10. Rinse in 3 changes PBS, 1 min each
11. Filtered DAB (75 mg/100 ml 0·05 M Tris buffer with 0·002 per cent H_2O_2) for 10 min with constant stirring
12. Rinse in Tris followed by 2 changes of PBS
13. OsO_2 vapours for 10 min
14. Rinse in PBS for 5 min
15. Counterstain and mount

II. DOUBLE BRIDGE TECHNIQUE FOR LIGHT MICROSCOPY[50]

Steps 1–8 as Single Bridge technique then—

9. 2 per cent NSS for 3 min
10. SA-RGG for 10 min
11. Rinse in PBS for 3 min
12. A-HP for 10 min
13. Rinse in PBS for 3 min
14. HP (0·5 mg/100 ml PBS with 0·1 per cent egg albumin) for 10 min

Then as Single Bridge technique, steps 10–15

III. DOUBLE PAP TECHNIQUE FOR LIGHT MICROSCOPY[51]

Steps 1–6 as Single Bridge technique then—

7. PAP (1 : 50–1 : 100) for 10 min
8. Rinse in PBS for 3 min

Abbreviations used in the Appendix

PBS	Phosphate-buffered saline
NSS	Normal sheep serum
SA-RGG	Sheep anti-rabbit gamma globulin
A-HP	Anti-horseradish peroxidase
HP	Horseradish peroxidase
DAB	Diaminobenzidine
GMA	Glycol methacrylate
PAP	Peroxidase–anti-peroxidase

9. 2 per cent NSS for 3 min
10. SA-RGG for 10 min
11. Rinse in PBS for 3 min
12. PAP for 10 min

Then as Single Bridge technique, steps 10–15.

IV. POST-EMBEDDING SINGLE BRIDGE TECHNIQUE FOR ELECTRON MICROSCOPY[186]

1. Epoxy-embedded sections on Ni grids are placed on drops of 10 per cent H_2O_2 for 8–10 min (for GMA-embedded material go directly to Step 3)
2. Dip rinse in 3 beakers PBS
3. Float grids on drops of 2 per cent NSS for 5 min
4. Primary antiserum for 46–50 h at 4 °C
5. Rinse in 3 beakers PBS at room temperature (the rest of the method is done at room temperature)
6. 2 per cent NSS for 5 min
7. SA-RGG (1 : 50–1 : 100) for 10 min
8. Rinse in 3 beakers PBS
9. 2 per cent NSS for 5 min
10. A-HP (1 : 50–1 : 100) for 10 min
11. Rinse in 3 beakers PBS
12. HP (0·5 mg/100 ml PBS with 0·1 per cent egg albumin) for 10 min
13. Rinse in 3 beakers PBS
14. Substrate solutions
    a. DAB (12·5 mg/100 ml of 0·05 M Tris buffer with 0·0025 per cent H_2O_2) for 3–4 min
    Or
    b. 4-Chloro-1-naphthol (16 mg/100 ml 0·05 M Tris buffer with 0·02 per cent H_2O_2) for 60–70 s
    The solution should be freshly prepared and filtered (0·2 μm pore size) before use and the incubation should be done with constant stirring
15. Rinse in Tris followed by 2 beakers of PBS
16. Float on drops of 2 per cent OsO_4 in distilled H_2O for 10 min
17. Rinse in distilled-deionized H_2O and allow to dry

V. POST-EMBEDDING DOUBLE BRIDGE TECHNIQUE FOR ELECTRON MICROSCOPY[186]

Steps 1–11 as Post embedding Single Bridge technique
Then—

12. 2 per cent NSS for 5 min
13. SA-RGG for 10 min
14. Rinse in 3 beakers PBS
15. 2 per cent NSS for 5 min
16. A-HP for 10 min
17. Rinse in 3 beakers PBS

Then as Post-embedding Single Bridge technique, steps 12–17

VI. POST-EMBEDDING DOUBLE PAP TECHNIQUE FOR ELECTRON MICROSCOPY[51]

Steps 1–9 as Post-embedding Single Bridge technique
Then—

10. PAP (1 : 50–1 : 100) for 2–3 min
11. Rinse in 3 beakers PBS
12. 2 per cent NSS for 5 min
13. SA-RGG for 10 min
14. Rinse in 3 beakers PBS
15. 2 per cent NSS for 5 min
16. PAP for 2–3 min
17. Rinse in 3 beakers PBS

Then as Post-embedding Single Bridge technique, steps 14–17

VII. WEAK PAP STAINING INTENSIFICATION[51]

Should a PAP preparation be encountered which results in weak staining, an incubation in HP (0·5 mg/100 ml PBS with 0·1 per cent egg albumin) can be performed between the PAP step and the incubation in the substrate solution. This HP incubation is valuable for either single or double PAP stainings at both the light and electron microscopical levels.

REFERENCES

1. Moriarty G. C. and Garner L. L. Immunoelectron microscopical localization of ACTH/MSH peptides in rat and human pituitaries. *Front. Hormone Res.* 1977, **4**, 26–41.
2. Moriarty G. C. and Garner L. L. Immunocytochemical studies of cells in the rat adenohypophysis containing both ACTH and FSH. *Nature* 1977, **265**, 356–358.
3. Pelletier G., Leclerc R., Labrie F., Cote J., Chrétien M. and Lis M. Immunohistochemical localization of β-lipotropic hormone in the pituitary gland. *Endocrinology* 1977, **100**, 770–776.
4. Weber E., Voigt K. H. and Martin R. Concomitant storage of ACTH and endorphin-like immunoreactivity in the secretory granules of anterior pituitary corticotrophs. *Brain Res.* 1978, **157**, 385–390.
5. Beauvillain J. C., Tramu G. and Croix D. Electron microscopic localization of enkephalin in the median eminence and the adenohypophysis of the gunica-pig. *Neuroscience* 1980, **5**, 1705–1716.
6. Nogami H. and Yoshimura F. Fine structural criteria of prolactin cells identified immunohistochemically in the male rat. *Anat. Rec.* 1982, **202**, 261–274.
7. Halmi N. S. and Moriarty G. C. The hypophysis. In: Weiss L. and Greep R. O. (ed.) *Histology*, 4th ed. New York, McGraw-Hill, 1977: 1039–1066.
8. Nakane P. K. Classifications of anterior pituitary cell types with immunoenzyme histochemistry. *J. Histochem. Cytochem.* 1970, **18**, 9–20.
9. Nakane, P. K. Identification of anterior pituitary cells by immunoelectron microscopy. In: Tixier-Vidal A. and Farquhar M. G. (ed.) *Ultrastructure in Biological Systems*, Vol. 7. *The Anterior Pituitary*. New York, Academic Press, 1975: 45–61.
10. Leleux P. and Robyn C. Immunohistochemistry of individual adenohypophysial cells. *Acta Endocrinol. [Suppl.] (kbh.)* 1971, **153**, 168–189.
11. Moriarty G. C. Adenohypophysis: ultrastructural cytochemistry. A review. *J. Histochem. Cytochem.* 1973, **21**, 855–894.
12. Baker B. L. Functional cytology of the hypophysial pars distalis and pars intermedia. In: Knobil E. and Sawyer W. H. (ed.) *Handbook of Physiology* Section 7, *Endocrinology*, Vol. IV Part 1. Washington DC, Am. Physiol. Soc. 1974: 45–80.
13. Pelletier G., Robert F. and Hardy J. Identification of human anterior pituitary cells by immunoelectron microscopy. *J. Clin. Endocrinol. Metab.* 1978, **46**, 534–542.

14. Halmi N. S. Immunostaining of growth hormone and prolactin in paraffin-embedded and stored or previously stained materials. *J. Histochem. Cytochem.* 1978, **26**, 486–495.
15. Halmi N. S. Immunoperoxidase staining of primate pituitaries with antibodies against the β subunits of human pituitary glycoprotein hormones. *J. Histochem. Cytochem.* 1981, **29**, 837–843.
16. Baker B. L. and Gross D. S. Cytology and distribution of secretory cell types in the mouse hypophysis as demonstrated with immunocytochemistry. *Am. J. Anat.* 1978, **153**, 193–215.
17. Dubois M. P. Immunocytochemistry of polypeptide hormones: a review. *Acta Histochem.* [*Suppl.*] 1980, **XXII**, 141–177.
18. Doerr-Schott J. Immunohistochemistry of the adenohypophysis of non-mammalian vertebrates. *Acta Histochem.* [*Suppl.*] 1980, **XXII**, 185–223.
19. Kovacs K., Horvath E. and Ryan N. Immunocytology of the human pituitary. In: Sternberg S. S. and DeLellis R. A. (ed.) *Diagnostic Immunocytochemistry*. New York, Masson, 1981: 17–35.
20. Brandtzaeg P. Tissue preparation methods for immunohistochemistry. In: Bullock G. R. and Petrusz P. (ed.) *Techniques in Immunocytochemistry*, Vol. 1. London, Academic Press, 1982: 1–75.
21. Sternberger L. A. *Immunocytochemistry*, 2nd ed. 1979. New York, John Wiley & Sons, 1979: 32.
22. Pearse A. G. E. *Histochemistry: Theoretical and Applied*. Vol. 1, 4th ed. New York, Churchill Livingstone, 1980: 202.
23. Humason G. L. *Animal Tissue Techniques*, 3rd ed. San Francisco, Freeman & Co., 1972.
24. McLean I. W. and Nakane P. K. Periodate–lysine–paraformaldehyde fixative. A new fixative for immunoelectron microscopy. *J. Histochem. Cytochem.* 1974, **22**, 1077–1083.
25. Baskin D. G., Erlandsen S. L. and Parsons J. A. Immunocytochemistry with osmium-fixed tissue. I. Light microscopic localization of growth hormone and prolactin with the unlabeled antibody–enzyme method. *J. Histochem. Cytochem.* 1979, **27**, 867–872.
26. Baskin D. G., Erlandsen S. L. and Parsons J. A. Influence of hydrogen peroxide and alcoholic sodium hydroxide on the immunocytochemical detection of growth hormone and prolactin after osmium fixation. *J. Histochem. Cytochem.* 1979, **27**, 1290–1292.
27. Smith R. E. Comparative evaluation of two instruments and procedures to cut non-frozen sections. *J. Histochem. Cytochem.* 1970, **18**, 590–591.
28. Hökfelt T., Fuxe K., Goldstein M. and Joh T. H. Immunohistochemical studies of three catecholamine synthesizing enzymes. Aspects on methodology. *Histochemie* 1973, **33**, 231–254.
29. Tougard C., Kerdelhuè B., Tixier-Vidal A. and Jutisz M. Light and electron microscope localization of binding sites of antibodies against ovine luteinizing hormone and its two subunits in rat adenohypophysis using peroxidase-labeled antibody technique. *J. Cell Biol.* 1973, **58**, 503–521.
30. Mazzuca M. and Dubois M. P. Detection of luteininzing hormone releasing hormone in the guinea pig median eminence with an immunohistoenzymatic technique. *J. Histochem. Cytochem.* 1974, **22**, 993–996.
31. Dacheux F. Ultrastructural localization of thyrotropin (TSH) in the porcine anterior pituitary. A comparison between pre-embedding and post-embedding methods. *Cell Tissue Res.* 1982, **222**, 299–311.
32. Tougard C., Picart R. and Tixier-Vidal, A. Immunocytochemical localization of glycoprotein hormones in the rat anterior pituitary. A light and electron microscope study using antisera against rat β subunits: a comparison between pre-embedding and postembedding methods. *J. Histochem. Cytochem.* 1980, **28**, 101–114.
33. Weber E., Voigt K. H. and Martin R. Granules and Golgi vesicles with differential reactivity to ACTH antiserum in the corticotroph of the rat anterior pituitary. *Endocrinology* 1978, **102**, 1466–1474.
34. Bohn W. A fixation method for improved antibody penetration in electron microscopical immunoperoxidase studies. *J. Histochem. Cytochem.* 1978, **26**, 293–297.
35. Hartman B. K., Zide D. and Udenfriend S. The use of dopamine β-hydroxylase as a marker for the central noradrenergic nervous system in the rat. *Proc. Natl Acad. Sci. USA* 1972, **69**, 2722–2726.
36. Brozman M. and Brozmanová E. Immunohistochemical methods to localize hormones in the human adenohypophysis. *Commun. Czechoslov. Soc. Histochem. Cytochem.* 1966, **1**, 25–30.
37. Huang S., Minassian H. and More J. D. Application of immunofluorescent staining on paraffin sections improved by trypsin digestion. *Lab. Invest.* 1976, **35**, 383–391.
38. Reading M. A digestion technique for the reduction of background staining in the immunoperoxidase method. *J. Clin. Pathol.* 1977, **30**, 88–90.

39. Denk H., Radaszkiewicz T. and Weirich E. Pronase pretreatment of tissue sections enhances sensitivity of the unlabelled antibody–enzyme (PAP) technique. *J. Immunol. Methods* 1977, **15**, 163–168.
40. Finley J. C. W. and Petrusz P. The use of proteolytic enzymes for improved localization of tissue antigens with immunocytochemistry. In: Bullock G. R. and Petrusz (ed.) *Techniques in Immunocytochemistry*. Vol. 1. London, Academic Press, 1982: 239–249.
41. Weller T. H. and Coons A. H. Fluorescent antibody studies with agents of varicella and herpes zoster propagated in vitro. *Proc. Soc. Exp. Biol. Med.* 1954, **86**, 789–794.
42. Nakane P. K. and Pierce G. B. Jr. Enzyme-labeled antibodies: preparation and application for the localization of antigens. *J. Histochem. Cytochem.* 1966, **14**, 929–931.
43. Sternberger L. A., Hardy P. H. Jr., Cuculis J. J. and Meyer H. G. The unlabeled antibody enzyme method of immunohistochemistry. Preparation and properties of soluble antigen–antibody complex (horseradish peroxidase–antihorseradish peroxidase) and its use in identification of spirochetes. *J. Histochem. Cytochem.* 1970, **18**, 315–333.
44. Avrameas S. Indirect immunoenzyme techniques for the intracellular detection of antigens. *Immunochemistry* 1969, **6**, 825–831.
45. Mason T. C., Phifer R. F., Spicer S. S., Swallow R. A. and Dreskin R. B. An immunoglobulin-enzyme bridge method for localizing tissue antigens. *J. Histochem. Cytochem.* 1969, **17**, 563–569.
46. Sternberger L. A. and Cuculis J. J. Method for enzymatic intensification of the immunocytochemical reaction without use of labeled antibodies. *J. Histochem. Cytochem.* 1969, **17**, 190.
47. Petrusz P., DiMeo P., Ordronneau, P., Weaver C. and Keefer D. A. Improved immunoglobulin-enzyme bridge method for light microscopic demonstration of hormone-containing cells of the rat adenohypophysis. *Histochemistry* 1975, **46**, 9–26.
48. Petrusz P., Ordronneau P. and Finley J. C. W. Criteria of reliability for light microscopic immunocytochemical staining. *Histochem. J.* 1980, **12**, 333–348.
49. Vacca L. L., Rosario S. L., Zimmerman E. A., Tomashefsky P. and Ng P.-Y. Application of immunoperoxidase techniques to localize horseradish peroxidase tracer in the central nervous system. *J. Histochem. Cytochem.* 1975, **23**, 208–215.
50. Ordronneau P. and Petrusz P. Immunocytochemical demonstration of anterior pituitary hormones in the pars tuberalis of long-term hypophysectomized rats. *Am. J. Anat.* 1980, **158**, 491–506.
51. Ordronneau P., Lindström P. B.-M. and Petrusz P. Four unlabeled antibody bridge techniques: a comparison. *J. Histochem. Cytochem.* 1981, **29**, 1397–1404.
52. Notani G. W., Parsons J. A. and Erlandsen S. L. Versatility of *Staphylococcus aureus* protein A in immunocytochemistry. Use in unlabeled antibody enzyme system and fluorescent methods. *J. Histochem. Cytochem.* 1979, **27**, 1438–1444.
53. Celio M. R., Lutz H., Binz H. and Fey H. Protein A in immunoperoxidase techniques. *J. Histochem. Cytochem.* 1979, **27**, 691–692.
54. Roth J. The protein A-gold (pAg) technique. Qualitative and quantitative approach for antigen localization on thin sections. In: Bullock G. R. and Petrusz P. (ed.) *Techniques in Immunocytochemistry, Vol* 1. London, Academic Press, 1982: 107–133.
55. Guesdon J. L., Ternynck T. and Avrameas S. The use of avidin–biotin interaction in immunoenzymatic techniques. *J. Histochem. Cytochem.* 1979, **27**, 1131–1139.
56. Hsu S.-M., Raine L. and Fanger H. Use of avidin–biotin–peroxidase complex (ABC) in immunoperoxidase techniques: a comparison between ABC and unlabeled antibody (PAP) procedures. *J. Histochem. Cytochem.* 1981, **29**, 577–580.
57. Petrusz P., Sar M., Ordronneau P. and DiMeo P. Specificity in immunocytochemical staining. *J. Histochem. Cytochem.* 1976, **24**, 1110–1112.
58. Larsson L.-I. Peptide immunocytochemistry. *Prog. Histochem. Cytochem.* 1981, **13** no. 4, 1–83.
59. Bugnon C., Fellman D., Lenys D. and Bloch B. Etude cytoimmunologique des cellules gonadotropes et des cellules thyréotropes de l'adénohypophyse du Rat. *C. R. Soc. Biol. (Paris)* 1977, **171**, 907–913.
60. Monroe S. E. and Midgley A. R. Jr. Localization of luteinizing hormone in the rat pituitary gland by a cross-reaction with antibodies to human chorionic gonadotropin. *Fed. Proc.* 1966, **25**, 315.
61. Rennels E. G. Gonadotrophic cells of rat hypophysis. In: Benoit J. and DaLage C. (ed.) *Cytologie de l'Adénohypophyse*. Paris, Editions du Centre National de la Recherche Scientifique, 1963: 201–213.
62. Moriarty G. C. Electron microscopic immunocytochemical studies of rat pituitary gonadotrophs: a sex difference in morphology and cytochemistry of LH cells. *Endocrinology* 1975, **97**, 1215–1225.

63. Yoshimura F. and Nogami H. Immunohistochemical characterization of pituitary stellate cells in rats. *Endocrinol. Jap.* 1980, **27**, 43–52.

64. Yoshimura F., Nogami H., Shirasawa N. and Yashiro T. A whole range of fine structural criteria for immunohistochemically identified LH cells in rats. *Cell Tissue Res.* 1981, **217**, 1–10.

65. Childs (Moriarty) G. V., Ellison P. G. and Garner L. L. An immunocytochemist's view of gonadotropin storage in the adult male rat: cytochemical and morphological heterogeneity in serially sectioned gonadotropes. *Am. J. Anat.* 1980, **158**, 397–410.

66. Childs (Moriarty) G., Ellison D., Foster L. and Ramaley J. A. Postnatal maturation of gonadotropes in the male rat pituitary. *Endocrinology* 1981, **109**, 1683–1692.

67. Tixier-Vidal A., Tougard C., Kerdelhué B. and Jutisz M. Light and electron microscopic studies on immunocytochemical localization of gonadotropic hormones in the rat pituitary gland with antisera against ovine FSH, LH, LHα, and LHβ. *Ann. N.Y. Acad. Sci.* 1975, **254**, 433–461.

68. Herbert D. C. Localization of antisera to LHβ and FSHβ in the rat pituitary gland. *Am. J. Anat.* 1975, **144**, 379–385.

69. Moriarty G. C. Ultrastructural–immunocytochemical studies of rat pituitary gonadotrops in cycling female rats. *Gunma Symp. Endocrinol.* 1976, **13**, 207–219.

70. Matsumura H. and Daikoku S. Sexual difference in LH-cells of the neonatal rats as revealed by immunocytochemistry. *Cell Tissue Res.* 1977, **182**, 541–548.

71. Baker B. L. and Yu Y. Y. The thyrotropic cell of the rat hypophysis as studied with peroxidase-labelled antibody. *Am. J. Anat.* 1971, **131**, 55–72.

72. Bowie E. P., Williams G., Shiino M. and Rennels E. G. The corticotroph of the rat adenohypophysis: a comparative study. *Am. J. Anat.* 1973, **138**, 499–520.

73. Moriarty G. C. and Tobin R. B. Ultrastructural immunocytochemical characterization of the thyrotroph in rat and human pituitaries. *J. Histochem. Cytochem.* 1976, **24**, 1131–1139.

74. Eipper B. A. and Mains R. E. Analysis of the common precursor to corticotropin and endorphin. *J. Biol. Chem.* 1978, **253**, 5732–5744.

75. Scott A. P., Lowry P. J., Bennett H. P. J. McMartin C. and Ratcliff J. G. Purification and characterization of porcine corticotrophin-like intermediate lobe peptide. *J. Endocrinol.* 1974, **61**, 369–380.

76. Hughes J., Smith T. W., Kosterlitz H. W., Fothergill L. A., Morgan B. A. and Morris H. R. Identification of two related pentapaptides from the brain with potent opiate agonist activity. *Nature* 1975, **258**, 577–579.

77. Baker B. L., Pete S., Midgley A. R. Jr. and Gersten B. E. Identification of the corticotropin cell in rat hypophyses with peroxidase-labeled antibody. *Anat. Rec.* 1970, **166**, 557–568.

78. Phifer R. F., Spicer S. S. and Orth D. R. Specific demonstration of the human hypophyseal cells which produce adrenocorticotropic hormone. *J. Clin. Endocrinol. Metab.* 1970, **31**, 347–361.

79. Moriarty G. C. and Halmi N. S. Electron microscopic study of the adrenocorticotropin-producing cell with the use of unlabeled antibody and the soluble peroxidase–antiperoxidase complex. *J. Histochem. Cytochem.* 1972, **20**, 590–603.

80. Dubois M. P. and Graf L. Demonstration by immunofluorescence of the lipotropic hormone LPH in bovine, ovine and porcine adenohypophysis. *Horm. Metab. Res.* 1973, **5**, 229.

81. Moon H. D., Li C. H. and Jennings B. M. Immunohistochemical and histochemical studies of pituitary β-lipotrophs. *Anat. Rec.* 1973, **175**, 529–537.

82. Scott A. P., Lowry P. J., Ratcliffe J. G., Rees L. H. and Landon J. Corticotrophin-like peptides in the rat pituitary. *J. Endocrinol.* 1974, **61**, 355–367.

83. Dubé D., Lissitzky J. C., Leclerc R. and Pelletier G. Localization of α-melanocyte stimulating hormone in rat brain and pituitary. *Endocrinology* 1978, **102**, 1283–1291.

84. Weber E., Truscot R. J., Evans C., Sullivan S., Angwin P. and Barchas J. D. α-*N*-acetyl β-endorphins in the pituitary: immunohistochemical localization using antibodies raised against dynorphin (1–13). *J. Neurochem.* 1981, **36**, 1977–1985.

85. Bloom F. E., Battenberg E. L., Shibasaki T., Benoit R., Ling N. and Guillemin R. Localization of γ-melanocyte stimulating hormone (γ-MSH) immunoreactivity in rat brain and pituitary. *Reg. Pept.* 1980, **1**, 205–222.

86. Pelletier G., Leclerc R., Benjannet S., Seidah N. G. and Chrétien M. Immunohistochemical localization of γ-melanocyte stimulating hormone in the rat pituitary gland. *Reg. Pept.* 1981, **2**, 81–89.

87. Smyth D. G., Massey D. E., Zakarian S. and Finnie M. D. A. Endorphins are stored in biologically active and inactive forms: isolation of α-*N*-acetyl-peptides. *Nature* 1979, **279**, 252–253.

88. Zakarian S. and Smyth D. Distribution of active and inactive forms of endorphins in rat pituitary and brain. *Proc. Natl Acad. Sci. USA* 1979, **76**, 5972–5976.

89. Furth J., Chrétien M., Lis M., Bélanger A., Moy P. and Grauman J. Multipotent lipotropic hormones. In search of a pituitary cell producing multipotent LPH. *Arch. Pathol.* 1975, **99**, 572–581.
90. Deftos L. J. and Catherwood B. D. Dissociation between ACTH and β-endorphin immunoreactivity in cells of the rat pituitary gland. *Life Sci.* 1980, **27**, 223–228.
91. Osamura R. Y., Watanabe K., Nakai Y. and Imura H. Adrenocorticotropic hormone cells and immunoreactive β-endorphin cells in the human pituitary gland: normal and pathologic conditions studied by the peroxidase-labeled antibody method. *Am. J. Pathol.* 1980, **99**, 105–124.
92. Watkins W. B. Differential immunostaining of adenohypophysial cells with antisera to ACTH and β-endorphin. *Reg. Pept.* 1981, **1**, 375–385.
93. Watkins W. B., Yen S. S. and Moore R. Y. Presence of β-endorphin-like immunoreactivity in the anterior pituitary gland of rat and man and evidence for the differential localization with ACTH. *Cell Tissue Res.* 1981, **215**, 577–589.
94. Tramu G., Beauvillain J. C., Mazzuca M., Girard F., Laine E., Christiaens J. L., Wemeau J. L., Fossati P. and Linquette M. Adénomes hypophysaires avec cellules à α MSH et α et β-endorphines sans ACTH. *Ann. Endocrinol.* 1979, **40**, 561–562.
95. Deftos L. J., Burton D., Catherwood B. D., Bone H. G., Parthermore J. G., Guillemin R., Watkins W. B. and Moore R. Y. Demonstration by immunoperoxidase histochemistry of calcitonin in anterior lobe of rat pituitary. *J. Clin. Endocrinol. Metab.* 1978, **47**, 457–460.
96. Watkins W. B. and Moore R. Y. Immunoreactive calcitonin in the rat anterior pituitary and its localization in thyrotrophs. *Am. J. Anat.* 1980, **158**, 445–454.
97. Cooper C. W., Peng T.-C., Obie J. F. and Garner S. C. Calcitonin-like immunoreactivity in rat and human pituitary glands: histochemical, *in vitro*, and *in vivo* studies. *Endocrinology* 1980, **107**, 98–107.
98. Flynn J. J., Margules D. L. and Cooper C. W. Presence of immunoreactive calcitonin in the hypothalamus and pituitary lobes of rats. *Brain Res. Bull.* 1981, **6**, 547–549.
99. Watkins W. B. and Moore R. Y. Immunoreactive calcitonin in the rat anterior pituitary gland and its localization in thyrotrophs. *Am. J. Anat.* 1980, **158**, 445–454.
100. Takor-Takor T. and Pearse A. G. E. Cytochemical identification of human and murine pituitary corticotrophs and somatotrophs as APUD cells. *Histochemie* 1973, **37**, 207–214.
101. Ferrand R., LeDouarin N. M., Polak J. M. and Pearse A. G. E. APUD characteristics and immunocytochemistry of avian pituitary corticotrophs. *Experientia* 1975, **31**, 1096–1097.
102. Bugnon C., Gouget A. and Dessy C. Etude cyto-immunologique et histochimique des cellules rendues fluorescentes par la technique de Falck et Hillarp dans l'adénohypophyse du Chat. *C. R. Soc. Biol. (Paris)* 1975, **169**, 317–322.
103. Baker B. L., Midgley A. R. Jr., Gersten B. E. and Yu Y. Y. Differentiation of growth hormone and prolactin-containing acidophils with peroxidase-labeled antibody. *Anat. Rec.* 1969, **164**, 163–172.
104. Parsons J. A. and Erlandsen S. L. Ultrastructural immunocytochemical localization of prolactin in rat anterior pituitary by use of the unlabeled antibody enzyme method. *J. Histochem. Cytochem.* 1974, **22**, 340–351.
105. Sibley P. E., Joyce B., Groom G. V., Chandler J. A. and Griffith K. A technique for the quantitative determination of prolactin in pituitary and breast cells using immunocytochemistry, electron microscopy and X-ray microanalysis. *J. Endocrinol.* 1978, **77**, 58P–59P.
106. Pasteels J. L., Gausset P., Danguy A., Ectors P., Nicoll C. S. and Varavudhi P. Morphology of the lactotropes and somatotropes of man and rhesus monkeys. *J. Clin. Endocrinol. Metab.* 1972, **34**, 959–967.
107. Baker B. L. and Yu Y. Y. An immunocytochemical study of human pituitary mammotropes from fetal life to old age. *Am. J. Anat.* 1977, **148**, 217–239.
108. Herbert D. C. and Hayashida T. Histologic identification and immunocytochemical studies of prolactin and growth hormone in the primate pituitary gland. *Gen. Comp. Endocrinol.* 1974, **24**, 381–397.
109. Kovacs K., Ryan N., Horvath E., Penz G. and Ezrin C. Prolactin cells of the human pituitary gland in old age. *J. Gerontol.* 1977, **32**, 534–540.
110. Petrusz P. and Peng T.-C. Immunocytochemistry of pituitary prolactin (PRL) cells in gonadectomized rats. *Proc. V. Internatl Congr. Endocrinol.* 1976, **254**.
111. Duello T. M. and Halmi N. S. Ultrastructural–immunocytochemical localization of growth hormone and prolactin in human pituitaries. *J. Clin. Endocrinol. Metab.* 1979, **49**, 189–196.
112. Calderon L., Ryan N. and Kovacs K. Human pituitary growth hormone cells in old age. *Gerontology* 1978, **24**, 441–447.

113. Baker B. L. and Yu Y. Y. Immunocytochemical analysis of cells in the pars tuberalis of the rat hypophysis with antisera to hormones of the pars distalis. *Cell Tissue Res.* 1975, **156**, 443–449.
114. Baker B. L., Karsch F. J., Hoffman D. L. and Beckman W. C. Jr. The presence of gonadotropic and thyrotropic cells in the pituitary pars tuberalis of the monkey (*Macaca mulatta*). *Biol. Reprod.* 1977, **17**, 232–240.
115. Girod C., Dubois M. P. and Trouillas J. Immunohistochemical study of the pars tuberalis of the adenohypophysis in the monkey, *Macaca iris*. *Cell Tissue Res.* 1980, **210**, 191–203.
116. Baker B. L. Cellular composition of the human pituitary pars tuberalis as revealed by immunocytochemistry. *Cell Tissue Res.* 1977, **182**, 151–163.
117. Tramu G. and Dubois M. P. Identification par immunofluorescence des cellules à activité LH du cobaye mâle. *C. R. Acad. Sci.* 1972, **275**, 1159–1161.
118. Gross D. S. Effect of castration and steroid replacement on immunoreactive luteinizing hormone in the pars tuberalis of the rat. *Endocrinology* 1978, **103**, 583–588.
119. McPhie J. L. and Beck J. S. Growth hormone in the normal pharyngeal pituitary gland. *Nature* 1968, **219**, 625–626.
120. McPhie J. L. and Beck J. S. The histological features and human growth hormone content of the pharyngeal pituitary gland in normal and endocrinologically disturbed patients. *Clin. Endocrinol.* 1973, **2**, 157–173.
121. McGrath P. Prolactin activity and human growth hormone in pharyngeal hypophyses from embalmed cadavers. *J. Endocrinol.* 1968, **42**, 205–211.
122. McGrath P. Extrasellar adenohypophyseal tissue in the female. *Australas. Radiol.* 1970, **14**, 241–247.
123. Baker B. L. and Jaffe R. B. The genesis of cell types in the adenohypophysis of the human fetus as observed with immunocytochemistry. *Am. J. Anat.* 1975, **143**, 137–162.
124. Bugnon C., Lenys D., Bloch B. and Fellman, D. Exploration cyto-immunologique sur coupes semi-fines des phénomènes de différenciation précoce de diverse populations cellulaires dans l'adénohypophyse foetale humaine. *C. R. Soc. Biol. (Paris)* 1976, **170**, 975–982.
125. Watanabe Y. G. and Daikoku S. An immunohistochemical study on the cytogenesis of adenohypophysial cells in fetal rats. *Dev. Biol.* 1979, **68**, 557–567.
126. Chatelain, A., Dupouy J. P. and Dubois M. P. Ontogenesis of cells producing polypeptide hormones (ACTH, MSH, LPH, GH, prolactin) in the fetal hypophysis of the rat: influence of the hypothalamus. *Cell Tissue Res.* 1979, **196**, 409–427.
127. Swaab D. F., Visser M. and Tilders F. J. Stimulation of intra-uterine growth in rat by α-melanocyte-stimulating hormone. *J. Endocrinol.* 1976, **70**, 445–455.
128. Tougard C., Picart, R. and Tixier-Vidal A. Cytogenesis of immunoreactive gonadotropic cells in the fetal rat pituitary at light and electron microscope levels. *Dev. Biol.* 1977, **58**, 148–163.
129. Sétáló G. Vigh S. and Horváth J. Functional differentiation of the FSH-synthesizing cells in the pars distalis of the fetal pituitary gland of the rat. *Acta Biol. Acad. Sci. Hung.* 1976, **27**, 147–154.
130. Bugnon C., Bloch B. and Fellmann D. Cyto-immunological study of the ontogenesis of the gonadotropic hypothalamo-pituitary axis in the human fetus. *J. Steroid Biochem.* 1977, **8**, 565–575.
131. Li J. Y., Dubois ,M. P. and Dubois P. M. Somatotrophs in the human fetal anterior pituitary. An electron microscopic–immunocytochemical study. *Cell Tissue Res.* 1977, **181**, 545–552.
132. Chatelain A., Dubois M. P. and Dupouy J. P. Hypothalamus and cytodifferentiation of the foetal pituitary gland. Study *in vivo*. *Cell Tissue Res.* 1976, **169**, 335–344.
133. Begeot M., Dubois M. P. and Dubois P. M. Localisation par immunofluorescence de l'hormone β-lipotrope (β-LPH) et de la β-endorphine dans l'antéhypophyse de foetus humains normaux et anencéphales. *C. R. Acad. Sci. (Paris)* 1978, **286**, 213–215.
134. Jacobowitz D. M. and O'Donohue T. L. α-Melanocyte stimulating hormone: immunocytochemical identification and mapping in neurons of rat brain. *Proc. Natl Acad. Sci. USA* 1978, **75**, 6300–6304.
135. Watson S. J., Barchas J. D. and Li C. H. β-Lipotropin. Localization of cells and axons in rat brain by immunohistochemistry. *Proc. Natl Acad. Sci. USA* 1977, **74**, 5155–5158.
136. Bloom F., Battenberg E., Rossier J., Ling N. and Guillemin R. Neurons containing β-endorphin in rat brain exist separately from those containing enkephalin: immunocytochemical studies. *Proc. Natl Acad. Sci. USA* 1978, **75**, 1591–1595.
137. Finley J. C. W. Lindström, P. and Petrusz P. Immunocytochemical localization of β-endorphin-containing neurons in the rat brain. *Neuroendocrinology* 1981, **33**, 28–42.
138. Pelletier G. and Leclerc R. Immunocytochemical localization of adrenocorticotrophin in the rat brain. *Endocrinology* 1979, **104**, 1426–1433.

139. Watson S. J., Richard V. W. and Barchas J. D. Adrenocorticotrophin in rat brain. Immunocytochemical localization in cells and axons. *Science* 1978, **200**, 1180–1182.
140. Finley J. C. W., Maderdrut J. L. and Petrusz P. The immunocytochemical localization of enkephalin in the central nervous system of the rat. *J. Comp. Neurol.* 1981, **198**, 541–565.
141. Petrusz P. Localization and sites of action of gonadotropins in brain. In: Stumpf W. E. and Grant L. D. (ed.) *Anatomical Neuroendocrinology*. Basel, Karger, 1975: 176–184.
142. Hostetter G., Gallo R. V. and Brownfield M. S. Presence of immunoreactive luteinizing hormone in the rat forebrain. *Neuroendocrinology* 1981, **33**, 241–245.
143. Fuxe K., Hökfelt T., Eneroth P., Gustafsson J.-Å. and Skett P. Prolactin-like immunoreactivity: localization in nerve terminals of rat hypothalamus. *Science* 1977, **196**, 899–900.
144. Toubeau G., Desclin J., Parmentier M. and Pasteels J. L. Compared localizations of prolactin-like and adrenocorticotropin immunoreactivities within the brain of the rat. *Neuroendocrinology* 1979, **29**, 374–384.
145. Toubeau G., Desclin J., Parmentier M. and Pasteels J. L. Cellular localization of a prolactin-like antigen in the rat brain. *J. Endocrinol.* 1979, **83**, 261–266.
146. Krieger D. T. and Liotta A. S. Pituitary hormones in brain. Where, how, and why? *Science* 1979, **205**, 366–372.
147. Pelletier G., Leclerc R. and Labrie F. Identification of gonadotropic cells in the human pituitary by immunoperoxidase technique. *Mol. Cell. Endocrinol.* 1976, **6**, 123–128.
148. Phifer R. F. and Spicer S. S. Immunohistochemical and histologic demonstration of thyrotropic cells of the human adenohypophysis. *J. Clin. Endocrinol. Metab.* 1973, **36**, 1210–1221.
149. Kovacs K. Morphology of prolactin producing adenomas. *Clin. Endocrinol.* 1977, **6**(*Suppl.*), 71s–79s.
150. Duello T. M. and Halmi N. S. Pituitary adenoma producing thyrotropin and prolactin. An immunocytochemical and electron microscopic study. *Virchows Arch. Pathol. Anat. Physiol. Klin. Med.* 1977, **376**, 255–265.
151. Oliver L., Vila-Porcile E. and Dubois M. P. Cellular localization of peptides derived from proopiocortin in the normal human adenohypophysis and in tumors of Cushing's disease. *Horm. Res.* 1980, **13**, 211–229.
152. Childs (Moriarty) G. V., Cole D. E., Kubek M., Tobin R. B. and Wilber J. F. Endogenous thyrotropin-releasing hormone in the anterior pituitary: sites of activity as identified by immunocytochemical staining. *J. Histochem. Cytochem.* 1978, **26**, 901–908.
153. Childs G. V. and Ellison D. G. A critique of the contributions of immunoperoxidase cytochemistry to our understanding of pituitary cell function, as illustrated by our current studies of gonadotropes, corticotropes and endogenous pituitary GnRH and TRH. *Histochem. J.* 1980, **12**, 405–418.
154. Sternberger L. A. and Petrali J. P. Quantitative immunocytochemistry of pituitary receptors for luteinizing hormone-releasing hormone. *Cell Tissue Res.* 1975, **162**, 141–176.
155. Schreibman M. P., Halpern L. R., Goos H. J. and Margolis-Kazan H. Identification of luteinizing hormone-releasing hormone (LH-RH) in the brain and pituitary gland of a fish by immunocytochemistry. *J. Exp. Zool.* 1979, **210**, 153–159.
156. Bauer T. W., Moriarty C. M. and Childs G. V. Studies of immunoreactive gonadotropin releasing hormone (GnRH) in the rat anterior pituitary. *J. Histochem. Cytochem.* 1981, **29**, 1171–1178.
157. Weber E., Voigt K. H. and Martin R. Pituitary somatotrophs contain [Met]-enkephalin-like immunoreactivity. *Proc. Natl Acad. Sci. USA* 1978, **75**, 6134–6138.
158. Tramu G. and Leonardelli J. Immunohistochemical localization of enkephalins in median eminence and adenohypophysis. *Brain Res.* 1979, **168**, 457–471.
159. DePalatis L. R., Fiorindo R. P. and Ho R. H. Substance P immunoreactivity in the anterior pituitary gland of the guinea pig. *Endocrinology* 1982, **110**, 282–284.
160. Uhl G. R., Kuhar M. J. and Snyder S. H. Neurotensin: immunohistochemical localization in rat central nervous system. *Proc. Natl Acad. Sci. USA* 1977, **74**, 4059–4063.
161. Gospodarowicz D. Purification of a fibroblast growth factor from bovine pituitary. *J. Biol. Chem.* 1975, **250**, 2515–2520.
162. Corvol M. T., Malemud C. J. and Sokoloff, L. A pituitary growth-promoting factor for articular chondrocytes in monolayer culture. *Endocrinology* 1972, **90**, 262–271.
163. Malemud C. J. and Sokoloff L. Some biological characteristics of a pituitary growth factor (CGF) for cultured lapine articular chondrocytes. *J. Cell. Physiol.* 1974, **84**, 171–180.
164. Gospodarowicz D., Jones K. L. and Sato G. Purification of a growth factor for ovarian cells from bovine pituitary gland. *Proc. Natl Acad. Sci. USA* 1974, **71**, 2295–2299.
165. Nishikawa K., Armelin H. and Sato G. Control of ovarian cell growth in culture by serum and pituitary factors. *J. Cell Biol.* 1976, **70**, 231a.

166. Maciag T., Cerundolo J., Ilsley S., Kelley S. P. R. and Forand E. Identification and partial characterization of an endothelial growth factor from bovine hypothalamus. *Proc. Natl Acad. Sci. USA* 1979, **76**, 5674–5678.
167. Brockes J. P., Lemke G. E. and Balzer D. R. Jr. Purification and preliminary characterization of a glial growth factor from the bovine pituitary *J. Biol. Chem.* 1980, **255**, 8370–8373.
168. Chieri R. A., Bosabe J. C. and Farina J. M. S. Evidence for a hypophyseal factor that stimulates insulin secretion by pancreas (insulotrophine). *Horm. Metab. Res.* 1976, **8**, 329–332.
169. Beloff-Chain A., Dunmore S. and Morton J. β-cell tropin, a peptide of the pituitary pars intermedia which stimulates insulin release. *FEBS Lett.* 1980, **117**, 303–307.
170. Smith G., Gilardeau C., Lis M. and Chrétien M. Isolation of a new lipolytic-melanotropic peptide from human pituitary glands. *Rev. Can. Biol.* 1976, **35**, 51–61.
171. Lorén I., Schwandt P., Alumets J., Håkanson R. and Neureuther G. Evidence that lipolytic peptide B occurs in the ACTH/MSH cells of the pituitary and in the brain. An immunocytochemical study of its distribution in correlation to the distribution of neurophysin. *Cell Tissue Res.* 1980, **205**, 349–359.
172. Skett P., Eneroth P., Gustafsson J.-Å., Sonnenschein C., Stenberg A. and Ahlen A. Evidence for an unidentified factor from pituitary gland which affects steroid metabolism in isolated hepatocytes and hepatoma cells of rat. *Mol. Cell. Endocrinol.* 1978, **10**, 249–262.
173. Parsons J. A. Evidence for a hypercalcemic factor in the fish pituitary, immunologically related to mammalian parathyroid hormone. In: Copp D. H. and Talmage R. V. (ed.) Endocrinology of Calcium Metabolism. Amsterdam, Excerpta Medica, 1978: 111–114.
174. Scherrer H., Seidah N. G., Benjannet C., Crine P., Lis M. and Chrétien M. Biosynthesis of a ubiquitin-related peptide in rat brain and in human and mouse pituitary tumors. *Biochem. Biophys. Res. Commun.* 1978, **84**, 874–885.
175. Koch B., Lutzbucher B., Briaud B. and Mialhe C. Immunoidentification of transcortin-like material within adenohypophysis. *Horm. Metab. Res.* 1978, **10**, 174.
176. Nicholson W. E., Barton R. N., Puett D., Orth D. N. and Liddle G. W. Evidence for renotropic activity in ovine pituitaries. *Endocrinology* 1979, **105**, 16–20.
177. Cocchia D. and Miani N. Immunocytochemical localization of the brain-specific S-100 protein in the pituitary gland of adult rat. *J. Neurocytol.* 1980, **9**, 771–782.
178. Nakajima T., Yamaguchi H. and Takahashi K. S100 protein in folliculostellate cells of the rat pituitary anterior lobe. *Brain Res.* 1980, **191**, 523–531.
179. Campbell G. T., Wagoner J., Gregerson K. A. and Joyner W. L. Immunohistochemical visualization of prolactin, growth hormone, and a substance resembling placental lactogen in the in situ and ectopic pituitary in the hamster. *Endocrinology* 1979, **105**, 905–910.
180. Witorsch R. J. Evidence for human placental lactogen immunoreactivity in rat pars distalis. *J. Histochem. Cytochem.* 1980, **28**, 1–9.
181. Hall H. and Dubois M. P. Immunochemical comparison between a low molecular weight andenohypophyseal constituent and the gonadotrophins. *Acta Endocrinol.* 1977, **86**, 273–287.
182. Robertson D. M., Suginami H., Hernandez-Montes H., Puri C. P., Choi S. K. and Diczfalusy E. Studies on a human chorionic gonadotrophin-like material present in non-pregnant subjects. *Acta Endocrinol.* 1978, **89**, 492–505.
183. Naruse K., Takii Y. and Inagami T. Immunohistochemical localization of renin in luteinizing hormone-producing cells of rat pituitary. *Proc. Natl Acad. Sci. USA* 1981, **78**, 7579–7583.
184. Sen S., Valenzuela R., Smeby R., Bravo E. and Bumpus F. M. Localization, purification, and biological activity of a new aldosterone-stimulating factor. *Hypertension* 1981, **3** (*pt. 2*), 181–186.
185. Saxena R. K. and Talwar G. P. An anterior pituitary factor stimulates thymidine incorporation in isolated thymocytes. *Nature* 1977, **268**, 57–59.
186. Ordronneau, P. Post-embedding unlabeled antibody techniques for electron microscopy. In: Bullock G and Petrusz P. (ed.) *Techniques in Immunocytochemistry*, Vol. 1. New York, Academic Press, 1982: 269–281.

13

Immunofluorescence Membrane Staining and Cytochemistry, applied in Combination for Analysing Cell Interactions in situ

L. W. Poulter, M. Chilosi, G. J. Seymour, S. Hobbs and G. Janossy

1. INTRODUCTION

The identification and characterization of the various cell types involved in complex biological mechanisms should ideally be carried out in situ using histological preparations for it is only in their natural microenvironment that the inter-relationship of these cell populations can be properly observed. The dilemma is how to 'breathe life' (i.e. allocate function) into a literally frozen 'landscape' of interacting cells. This consideration applies most notably to immunological mechanisms. Previously, the immunohistologist had to rely primarily on morphology as the major discriminating factor between cell types. However, within a tissue section it is frequently difficult, and in some cases impossible, to distinguish differing cell populations on morphological grounds alone. Because of this it was necessary to analyse cells circulating in the peripheral blood and/or to prepare cell suspensions from the tissues under analysis. Distinctive membrane markers and functional tests were developed for such analysis, which, when viewed together, contributed to the image of a delicate spider's web of immunoregulation (reviewed in the literature[1-3]). These techniques have been found invaluable in revealing information concerning functional properties and cellular interaction. Whether such 'artificial' techniques *in vitro* fully reflect relationships in the body has not, however, been satisfactorily determined.

There is clearly a need to relate these *in vitro* observations, as a matter of policy, to the corresponding situation *in vivo*. This kind of systematic comparative approach is likely to modify both the immunologists' and the histologists' views. It suffices to mention only two obvious examples of how these changes of opinion may take place. First, the observation that interdigitating cells seed from the bone marrow to the thymic medulla[4–6] and may therefore influence the education of thymocytes has already changed the interpretation of previous experiments about thymic education.[7] These previous studies ignored the histologically verifiable fact of cell migration.

Second, the role of HLA-DR antigens (*h*uman *l*eucocyte *a*ntigens) in the control of immune responses, an assertion of immunologists,[8,9] has a histological connotation. This is apparently manifested in the fact that Langerhans cells of the skin and their close relatives, the interdigitating cells of lymph nodes (which are acknowledged to be mobile cells with antigen-presenting capacity), both express very high quantities of HLA-DR antigens.[10–12] The important point to realize from this example, is that the comparative analysis of cultures *in vitro* versus microenvironments *in vivo* might provide mutually confirmative and elucidating results. This hope is clearly supported by the great similarity of cellular interactions observed in normal bone marrow and in long-term bone marrow cultures.[13]

Over the last few years two important advances in our knowledge of the immune system have led many investigators to realize the importance of a more critical analysis of the immunological response systems in situ. Firstly, the concept of the immune response simply involving interactions of T- and B-lymphocytes with macrophages has been refined. It is now established that there are subpopulations of T-lymphocytes, with different functional capacities[12,14,15] as well as different populations of B-lymphocytes. Furthermore, the concept of the macrophage as the non-lymphoid accessory cell of the immunological response is far too simplistic. It has now been revealed that the 'macrophage family' is an extremely heterogeneous population with widely differing functional capacities.[10,11,17] Secondly, it is clear that the spatial relationships of these many different cell populations both within normal tissues and in pathological situations is highly organized.[11,18,19] The analysis of these microenvironments is, therefore, of considerable importance.

The remaining dilemmas we address are many-fold. For example, subpopulations of T-lymphocytes, which show different functions *in vitro* and have been characterized by reliable *immunological* reagents (e.g. monoclonal antibodies; Mc Abs) must be identified within tissues. Helpful in this respect is the fact that these Mc Abs can now be used in informative combinations. This review briefly summarizes the methods of using Mc Abs produced in mice, tagged with different haptens and labelled with a non-cross-reactive anti-hapten additional antibody layer, using double immunofluorescence combinations in tissue sections. These methods give sufficiently strong staining and are useful in establishing the relationships of different lymphocyte subsets in their natural microenvironments.[18,20,21] Secondly, there are as yet no reliable antisera to analyse the whole spectrum of cells which belong to the macrophage family. It is therefore important also to establish the morphological and cytochemical characteristics of the cells, both being parameters which can contribute to cell identification. An essential point, however, is that the abundant expression of HLA-DR antigens most probably delineates the cells which play immunoregulatory roles[10,16,18] and for this reason this chapter describes how the cytochemical methods can be used together (*in combination*) with immunofluorescence techniques. The purpose of this recently standardized cytochemical–immunofluorescence technology is, as will be shown below, two-fold. It delineates, with great precision, the macrophage subsets present in normal tissues and inflammatory reactions, and it is the best available approach to find and characterize new Mc Abs reactive with these subsets.

Thus the study below falls into four parts: (*a*) use of Mc Abs in double combinations; (*b*) standardization of cytochemical staining together with immunofluorescence; (*c*) the demonstration of the search for new macrophage-reactive Mc Abs; finally (*d*) we give a short example of the applications of these techniques in analysing pathological tissues.

In formalin-fixed and paraffin-embedded tissues the tissue antigens and the activity of certain enzymes is frequently partially or totally destroyed.[22, 23] For this reason, studies are performed on cryostat sections of frozen tissue samples in order to optimally preserve antigenicity and enzyme activity.[23, 25] The techniques for tissue preparation and sectioning are given in the Appendix, as are the detailed accounts of the methods described below.

2. USE OF MONOCLONAL ANTIBODIES IN DOUBLE COMBINATIONS

The need for immunohistological analysis of tissue sections has resulted in the development of both immunofluorescence (IF) and immunoperoxidase (IP) methods to detect specific cell populations. These methods are strictly complementary and are frequently used in our laboratory. We tend however to label more slides with IF than with IP. The reason for this is that we are interested not only in the characterization of individual cells with different Mc Abs but also in the interactions in situ between different cell types (*Fig*. 13.1). [6, 18, 21] To this end, it is convenient to use combinations of reagents labelled with fluorescein–isothiocyanate (FITC, green) and tetramethylrhodamine–isothiocyanate (TRITC, red). IP is used, as a permanent record, to establish the reactivity of Mc Abs in haematoxylin–eosin-stained histological preparations and is applied more frequently than IF in histopathology laboratories. [26, 27]

The ultimate sophistication is to use the combination of IP and immuno-alkaline–phosphatase double labelling.[27] These techniques are described in depth elsewhere in this book (Chapter 7). We concentrate here on the IF methods of immunohistological analysis.

2.1. Handling of Antibodies

Mc Abs stored frozen in ascites or in diluted form (without carrier protein) for longer than 1–2-month periods deteriorate rapidly: the specific staining decreases and the non-specific staining increases. This is a rather peculiar phenomenon; the stored Mc Abs which perform so poorly in sections are still excellent when used for labelling cells in suspension. In contrast, there are three types of storage where Mc Abs keep their power for longer than 9 months: (i) frozen tissue culture supernatants of Mc Ab-producing clones (which contain 10–20 per cent fetal calf serum); (ii) ascites fluid (1 part) plus glycerol (2 parts) kept cool (and *not* frozen) at −20 °C; (iii) lyophilized Mc Abs (which, when reconstituted, are stored with glycerol as in (ii)).

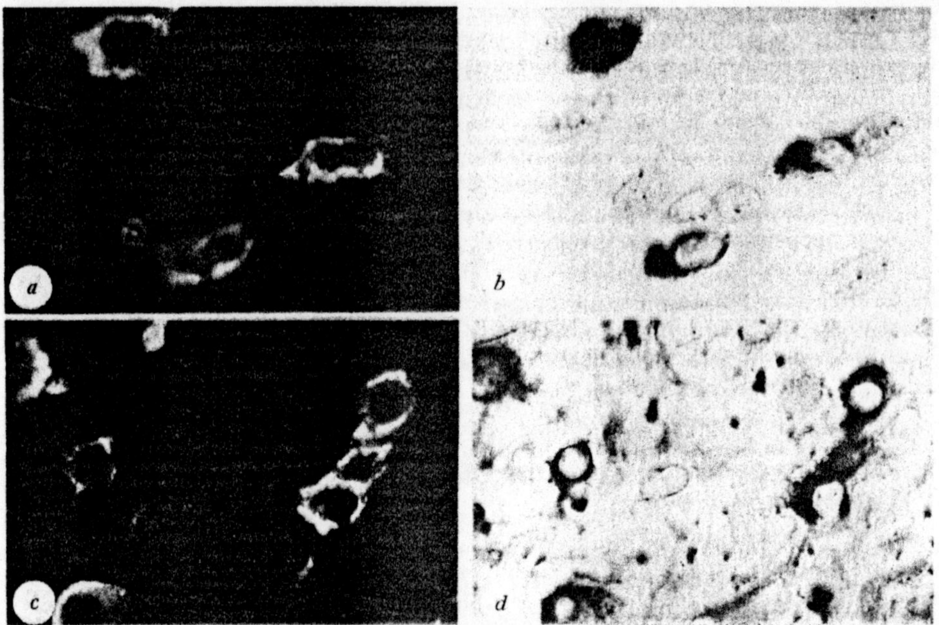

Fig. 13.1. Areas of two cryostat sections of rheumatoid synovial membrane. The first (*a, b*) shows HLA-DR +ve cells identified by indirect immunofluorescence (*a*) and acid phosphatase activity detected histochemically in the same cells (*b*).

The second (*c, d*) shows a combination of HLA-DR staining and adenosine triphosphatase activity on the same cells (*see text* for methods).

2.2. Use of Combined Antibodies in Tissue Sections

The immunofluorescence techniques for staining specific cells within tissue sections are given in the Appendix. *Combinations* of antisera can be used providing they are of different species origin, or are of different subclass (i.e. IgG_1 and IgG_2). In such circumstances, the fluorochrome-labelled second layers can be either directed against the different species of immunoglobulin used, or against the different subclasses used in the second combination (*Table* 13.1). This latter system has worked well when cell suspensions have been studied, but does not always give good results with tissue sections. As an alternative, the first layer antibodies can be conjugated to haptens (arsanilic acid, Ars) and glutamic acid (Glu),[28] thus allowing the use of second-layer antibodies directed against these haptens and not against the species immunoglobulin of the primary layer antibodies (*Table* 13.2).

An important separate question is how to use immunochemical manipulations in order to facilitate strong histological staining. Firstly, Mc Abs which, on their own, give weak staining (e.g. OKT 4) can be mixed with other Mc Abs which react with different epitopes on the *same* antigen (e.g. OKT 4A, OKT 4B, and OKT 4D).[29] This increases staining intensity. Second, these antibodies can be coupled to biotin[30] (unfortunately, not suitable for analysing liver sections owing to the presence of endogeneous biotin) and labelled avidin, which has a strong affinity for biotin, can be used as a second layer. Finally the Ars + Glu combination (described

Table 13.1. ***Examples of combination staining of surface markers***

Targets	*1st layers*	*2nd layers*	*3rd layers*	*Result*
1. T inducers	OKT 4*[46]—Biotinated	Avidin + FITC	—	T inducers green
+				
T suppressors	OKT 8[47]—Arsanilated	Rabbit anti-Ars	Sheep anti-rabbit immunoglobulin-TRITC	T suppressors red
2. T inducers	OKT 4* (Mouse anti-human)	Goat anti-mouse immunoglobulin-FITC	—	T inducers green
+				
HLA-DR +ve cells	Chicken anti-human[48] HLA-DR	Sheep anti-chicken immunoglobulin-TRITC	—	HLA-DR +ve cells red
3. B-lymphocytes	Sheep anti-human immunoglobulin-TRITC	—	—	B-cells red
+				
T-lymphocytes	UCHT1[49]	Goat anti-mouse immunoglobulin-FITC		T-cells green

* For use on tissue sections a mixture of OKT 4 clones is necessary (*see text*).

*Table 13.2. **Triple marker systems to amplify antigen labelling with Mc Abs in tissue sections****

First layer	*Coupled to*	*Second layer*	*Third layer*
OKT 8	Arsanilic acid (Ars)[28]	Rabbit anti-Ars	Swine anti-rabbit—TRITC
OKT 4A, OKT 4B, OKT 4D } mixture	Glutamic acid (Glu)[28]	Goat anti-Glu	Swine anti-goat—FITC
OKT 3, UCHT1	Glutamic acid (Glu)[28]	Goat anti-Glu	Swine anti-goat—FITC
OKT 11, OKT 11A } mixture	Arsanilic acid (Ars)[28]	Rabbit anti-Ars	Swine anti-rabbit—TRITC

* These triple layers are used only when Mc Abs are used in combinations with each other. All these reagents are absorbed with human liver powder and human, rabbit and goat immunoglobulins when necessary. The second layers are either Rivanol (IgG) preparations of hyperimmune antisera or affinity purified antibodies eluted from immunoadsorbant columns. These triple layer systems are not required when Mc Abs are used in combinations with *conventional* antisera such as chicken anti-human HLA-DR (Ia-like) antiserum or goat anti-human Ig against the various isotypes. In these cases two layer systems (Mc Ab + good, selected goat anti-mouse Ig–TRITC or FITC) yield adequately strong staining (*see text*).

above) can be applied with triple layer amplification (as shown in *Table* 13.2) thus giving a most impressive, but still specific, staining (*Plate* 14).

3. STANDARDIZATION OF CYTOCHEMICAL STAINING TOGETHER WITH IMMUNOFLUORESCENCE

It has been pointed out above that the various cells of the macrophage family are heterogeneous by cytochemical criteria (summarized in *Table* 13.3) as well as by the expression of HLA-DR antigens (*Table* 13.4). The question therefore arises as to whether the combination of *cytochemistry* and *immunofluorescence* together would yield a meaningful classification of the macrophage series. On the basis of the observations shown in *Tables* 13.3 and 13.4 it is likely that immunoregulatory cells (including interdigitating reticular cells and Langerhans cells) will be strongly HLA-DR positive and acid phosphatase negative (or weakly positive in the Golgi

*Table 13.3. **Heterogeneity of cytochemical activity in members of the 'macrophage family'***

	Acid phosphatase	*Adenosine triphosphatase*	*Non-specific esterase*
Monocytes	± [52]	– [51]	++ [52]
Tissue histiocytes	++ [35]	– [35]	++ [35]
Inflammatory macrophages	+++ [41,43]	± [41,43]	+++ [40]
Interdigitating cells	± [35,41,53]	+++ [41,53]	– [35]
Follicular dendritic cells	– [35]	± [41]	± [35]
Langerhans cells	± [45]	++ [45,54]	± [45]
Giant cells	++ [40]	++ [40]	++ [40]

Table 13.4. **Expression of HLA-DR antigens by cells of the 'macrophage family'**

1. *Positive populations*	*Reference*
Small proportion of circulating monocytes	50, 51
Variable proportion of tissue histiocytes	38, 43
Inflammatory macrophages	43
Giant cells	40
Interdigitating cells	6, 12, 41
Dendritic cells	36
Langerhans cells	37

2. *Negative or very weakly positive populations*	*Reference*
Most circulating monocytes	50, 51
Some tissue histiocytes	38, 43
Follicular dendritic reticulum cells	16, 37

area), while another distinct macrophage population would show strong acid phosphatase positivity but express HLA-DR antigens relatively weakly. Those populations with similar characteristics (e.g. giant cells and inflammatory macrophages) can readily be distinguished on morphological grounds.

Although it is quite feasible to stain serial sections for HLA-DR and ACP it is obvious that the direct demonstration of specific surface antigen and enzyme activity on the same cells (*i.e. in the same section*) would yield a more precise representation of the coincidence (or lack of coincidence) of these two characteristics. Therefore techniques were established that incorporated the use of immunofluorescence and cytochemistry on individual tissue sections.

Preliminary experiments have revealed that staining for acid phosphatase (ACP) with the pararosaniline method gave strong autofluorescence on both the FITC and TRITC channels and could not be combined with IF. For this reason the Gomori lead method[31] was used to detect ACP activity (*see* the Appendix).

Using such methods it was shown that tissue macrophages are, as predicted, heterogeneous (*Table* 13.5). Two major cell types characterized are HLA-DR strongly positive, acid-phosphatase weakly positive (Type I), and acid-phosphatase strongly positive, HLA-DR weakly positive (Type II). In normal tissues such as thymic medulla, spleen and normal synovium, Type I cells appear to be the population of interdigitating cells which also express high adenosine triphosphatase (ATPase) activity[35] (*Table* 13.3). In the normal skin, Langerhans cells also show the same Type I pattern of HLA-DR and enzyme reactivity. Some

Table 13.5. **Heterogeneity of tissue macrophages demonstrated using combined techniques for cytochemical and immunofluorescence analysis***

Cells	*Acid phosphatase*	*HLA-DR fluorescence staining*
Type I	±	+ + +
Type II	+ + +	±
Type III	+ + +	+ + +

* Preliminary observations using new methods (*see text*).

diffusely distributed dendritic cells in the mesangial areas of kidney and perivascular cells in the pancreas also correspond to Type I cells. Furthermore, in the lamina propria of the normal jejunum the majority ($>$80 per cent) and in the colon a large proportion (50 per cent) of macrophage-like cells have Type I characteristics. In contrast, Type II cells correspond to the diffusely distributed macrophages in the thymic cortex and medulla, scattered cells in the paracortical area and a large proportion of sinus histiocytes in lymph nodes, a few tissue macrophages in synovium, 20 per cent of macrophages in jejunum and 50 per cent of macrophages in the large bowel.

An interesting third type of cell (Type III in *Table* 13.5) was also observed which strongly expressed both HLA-DR antigens and acid-phosphatase activity. This cell type was most prominent in the synovial lining of the rheumatoid synovium (*see also below*), in the lung, probably representing alveolar macrophages, and during delayed-type hypersensitivity reactions in the skin. It is possible that this population represents immunologically activated macrophages and/or cell types involved in the defence of exposed surfaces. The reason we call attention to this cell type here is a technical observation. It is shown in *Fig.* 13.1*a, b* that there is only minimal, if any, interference between the HLA-DR staining and the lead deposits identifying ACP activity. The HLA-DR staining is visible in the cytoplasmic areas of individual cells where ACP activity can also be observed, indicating that the ACP does not 'cover' the HLA-DR activity. The observations above therefore clearly show that this interesting combination staining is feasible and further work is now necessary to map out the exact details of macrophage heterogeneity with this new technology.

Table 13.6. ***Heterogeneity of macrophage-like cells in the synovial lining of patients with rheumatoid arthritis***

Cells	*FMC 17*	*HLA-DR*	*ACP*	*ATPase*	*Percentage of total**
Round	+	+	+†	+	60–70
Giant cells	+	+	+	+	2–5
Stellate	+	+	+	—	20–30
Stellate	+	+	—	+	2–5
Elongated	+	+	+	—	1–5

* Percentages given are an overall impression. There is frequently considerable variability from one area to another within the same specimen.
† No attempt has been made to quantitate enzyme activity. Designations represent positive or negative reactions using standardized histochemical techniques.
ACP, acid phosphatase

Additional combinations such as combined staining for acid phosphatase and ATPase have also been performed with moderate success. These studies have confirmed the heterogeneity of gut macrophages (80 per cent Type I + 20 per cent Type II cells in the jejunum; 50 per cent Type II cells in the ileum and colon.[32, 35]) This combination method is, however, capricious and sometimes it is difficult to obtain sufficiently strong ATPase staining in the interdigitating cell types when the section is counterstained for ACP. It is also interesting that ATPase, unlike the lysosomal ACP, is a membrane-associated enzyme and the precipitated products obtained during its detection apparently interfere with the intensity of the similarly membrane-associated HLA-DR staining. Thus the interference between HLA-DR and ATPase staining is stronger than the interference between HLA-DR and ACP staining (*Fig.* 13.1*c, d*).

4. THE USE OF COMBINATION STAINING FOR STANDARDIZING NEW MONOCLONAL ANTIBODIES

It is of extraordinary advantage to stain tissue sections with a known Mc Ab (using, for example, a FITC-labelled second layer) and also stain with a Mc Ab of unknown specificity (using TRITC; *see* Table 13.2). With this technique it can be established whether the two Mc Abs detect the same, overlapping or different populations. The reliability of the method results from the fact that in tissue sections of easily available tissues, such as tonsil, many different cell types are present, (subsets of T- and B-lymphocytes, different types of macrophages, epithelium and endothelium etc). The differential reactivity of the two reagents can immediately be assessed on all those cell types. We routinely use the double staining methods to analyse the reactivity pattern of new Mc Abs in comparison with already standardized known reagents.

In this part of the chapter we would like to describe a further step in characterizing the reactivity of Mc Abs with macrophage subpopulations. An interesting observation, made in a number of laboratories using both conventional antisera and Mc Abs, is that reagents against *blood monocytes* frequently fail to react with tissue macrophages (although they may stain a few cells present in the tissue sections inside the blood vessels).[34] Thus antibodies which recognize well-characterized populations of tissue macrophages are rare. Further reasons for the scarcity of anti-macrophage reagents are two-fold. It is difficult to obtain cell suspensions from the sessile tissue macrophage populations (in order to immunize against human tissue macrophages) and when the new reagents are studied for activity only in cell suspensions, the Mc Abs selectively reacting with tissue macrophages are discarded. We suggest here that the most convenient method of finding new reagents for macrophages is to test recently produced Mc Abs in tissue sections in combination staining with acid phosphatase on one hand and with anti-HLA-DR staining on the other. We use the example of FMC-17, a Mc Ab made against peripheral monocytes by Dr D. Brookes and Dr H. Zola at the Flinders Medical Centre, Australia, in order to demonstrate the principle of Mc Ab standardization with these new methods.

In suspensions of peripheral blood leucocytes, FMC-17 stained all morphologically identifiable monocytes but was negative on lymphoid cells. When used on sections of normal tissues it stained many macrophage-like cells in the marginal zone, interdigitating cells in the T-lymphocyte areas of the tonsil and at the corticomedullary junction of the normal thymus and a proportion of synovial lining cells. In inflamed tissues, such as the synovium in rheumatoid arthritis (RA), many cells in the lining layers and diffusely distributed cells in the deeper stroma showed positive FMC-17 staining. With this distribution of staining how could the specificity of FMC-17 be determined? The answer to this question (as stated above) lay in combined staining with FMC-17 and other antibodies of known specificity, as well as cytochemical reactions to determine the individual identity of the FMC-17 +ve cells.

These combined techniques revealed that the FMC-17 +ve cells of the tonsil paracortex were HLA-DR +ve and ATPase +ve, yet negative for ACP. Such characteristics are consistent with those of the antigen present in interdigitating cells.[16,35,37] The same phenotype was expressed by most of the FMC-17 +ve cells of the thymic medulla and also a large proportion of the FMC-17 +ve cells in the RA synovium (*Plate* 15).

However, in this latter tissue and in sections of biopsy material from cutaneous delayed-type hypersensitivity reactions a proportion of FMC-17 +ve HLA-DR +ve cells were ACP +ve (*Plate* 16). Furthermore, in these tissues some of the FMC-17 +ve ACP +ve cells expressed no HLA-DR staining. One can conclude therefore that this new Mc Ab stains the highly differentiated interdigitating cells (both in normal tissues, e.g. tonsil paracortex, and in pathological tissue, e.g. RA synovium) and is also positive on tissue histiocytes and inflammatory macrophages. With the cellular specificity of this new Mc Ab established using these staining combinations, its value in the analysis of macrophage heterogeneity and in studies of pathological microenvironments has been increased.[38] Further investigations are now necessary to find Mc Abs which react with defined histiocyte–macrophage subpopulations.

5. COMBINED ANALYSIS OF PATHOLOGICAL TISSUES

Mononuclear cell infiltrates are characteristic of many immune-associated inflammatory reactions. These infiltrates will include large numbers of macrophage-like cells ranging from the classical sessile tissue histiocyte to the interdigitating dendritic cells shown to be involved with antigen presentation.[37,39] The diverse functional capacities of these heterogeneous populations make it of importance to be able to identify specific cell types and thus determine their relationship with the lymphoid populations present.

As mentioned above, a typical example of this situation is found in the inflamed synovial membrane of patients with rheumatoid arthritis. Conventional histology of such tissue reveals a hyperplastic synovial lining with extensive perivascular infiltration composed of lymphocytes and macrophage-like cells. Such specimens have been subjected to analysis with combined immunofluorescence and cytochemical methods. The cells of the RA synovial lining were almost all the HLA-DR +ve cells which are also FMC-17 +ve. Morphological and cytochemical observations however, revealed that up to five cell types could be distinguished (*Table* 13.4).[40] The majority were ACP +ve, ATPase +ve and round/oval in shape. Smaller numbers were stellate in appearance and were ACP +ve, ATPase −ve or conversely ACP −ve, ATPase +ve. The latter population thus expressed all the characteristics of the interdigitating cells of the tonsil paracortex and thymic cortico-medullary junction.[6] Although few in number within the lining layers, cells with identical characteristics were present in large numbers interspaced among the lymphocytes of the perivascular infiltrates (*Plate* 15).[41] Such an observation might have significant value in determining the functional relationships between different cells. These observations raise the possibility that the close association of large numbers of antigen-presenting cells and infiltrating lymphocytes contributes to the continuous immunostimulation within the rheumatoid synovium.[42] Inflammatory (activated) macrophages (HLA-DR +ve FMC-17

+ve, ACP +ve, ATPase −ve) were also seen, but these cells tended to be diffusely dispersed throughout the stroma and not directly related to lymphoid aggregates. Another cell type identified (FMC-17 +ve, HLA-DR −ve, ACP +ve) was taken to be the classical tissue histiocyte.

None of these cells expressed strong staining with a Mc Ab directed against the C_3b receptor (kindly supplied by Dr D. Y. Mason). In rare specimens where germinal centres had developed, C_3b positive cells were found in association with these structures. Such cells however were HLA-DR −ve, ACP +ve/−ve, ATPase +ve/−ve thus expressing the phenotype of the follicular dendritic cells found in normal lymph node germinal centres.

Similar analysis of delayed-type hypersensitivity (DTH) reactions in the skin has shown that the majority of HLA-DR +ve, FMC-17 +ve cells accumulating in the dermis are ACP +ve (thus inflammatory macrophages), and not the ATPase +ve interdigitating cells.[43] As the DTH reaction develops (12–24 h after antigenic challenge), small numbers of HLA-DR +ve, FMC-17 +ve, ATPase +ve, ACP −ve interdigitating cells are present in association with the perivascular lymphocytic infiltrates, but unlike the RA synovium these cells never dominate the DTH reaction; indicating that different control processes from those in RA may be operating in DTH.

The Langerhans cells in the epidermis of normal skin were HLA-DR +ve, FMC-17 +ve, C_3b +ve, ATPase +ve, ACP −ve. Other Mc Abs are already available for analysing further subpopulations of macrophages/histiocytes. It has been a fortuitous observation that both Mc Abs to cortical thymocyte antigen (mol. wt 48 000), OKT 6 and NA1/34, react with Langerhans cells in skin.[44] In addition, the Langerhans cells also react with the Mc Ab to C_3b receptor (Mason[55]) confirming previous observations which utilized the cumbersome isolation of these cells and a technique for C_3b rosetting. Combination staining with Mc Abs to these markers revealed that the number of Langerhans cells in the epidermis increased as the DTH reaction developed. Furthermore, cells with the characteristics OKT 6 +ve, NA1/34 +ve, HLA-DR +ve, C_3b +ve, ATPase +ve, ACP −ve could be found in the dermis of DTH skin after 24–48 h. As these characteristics are not associated with interdigitating cells (OKT 6 −ve, C_3b −ve) or inflammatory macrophages (ATPase −ve, ACP +ve) one would appear to be identifying Langerhans cells that were no longer localized in the epidermis.[37]

More recently, such techniques have been applied to studies of the skin lesions from patients with histiocytosis X.[45] Results demonstrated that the major cell type in these lesions expressed the same phenotype as normal epidermal Langerhans cells. Using these combination techniques this proliferating population could readily be distinguished from tissue macrophages and giant cells which were HLA-DR −ve, ACP +ve and are also present in the lesions.[45] This phenotypic analysis helps to establish that in the skin and lymph nodes of patients with histiocytosis X the same cell type can be seen. This study proves previous suggestions made on the basis of ultrastructural studies, that the aberrant cell populations in this disease do indeed derive from Langerhans cells.

Finally, it is important to emphasize that, although in many respects the characteristics of Langerhans and interdigitating cells are similar, they also show differences. The interdigitating cells as opposed to Langerhans cells are OKT 6 −ve, NA1/34 −ve, C_3b −ve. With the availability of more Mc Abs, finer divisions of the macrophage/histiocyte system will be possible.

PROSPECTS

We have limited our discussion in this chapter to those combinations of immunofluorescence and cytochemical techniques that have already proved of value in advancing our understanding of immunological mechanisms in situ.

With an ever widening panel of Mc Abs and the multiplicity of cytochemical methods not as yet tested in these techniques of *combination analysis*, it goes without saying that we have as yet only scratched the surface of the potential of such analysis. There is no doubt that such methods will contribute greatly to our understanding of the immune response in situ. Thus, improvement of diagnostic procedures and a more rational basis for therapeutic regimes may well emerge.

Appendix

METHODS

I. Tissue Preparation

a. Cut specimens of fresh material into small blocks ($<1\,cm^3$)
b. Orientate on a cork disc and cover in O.C.T. (Ames Corp.)
c. Freeze in isopentane cooled in a bath of liquid nitrogen
d. Store in a precooled tube at −70 °C or in a liquid nitrogen tissue storage container

II. Section Preparation

a. Cut 6-μm sections on a cryostat maintained at −25 °C
b. Air-dry the sections for 1 h at room temperature
c. Fix in chloroform/acetone (1 : 1) for 5 min at room temperature
d. Allow sections to dry, and freeze at −70 °C overnight
e. Freeze-dry for 2 h

III. Immunofluorescence Staining Basic Technique

a. Wash sections for 10 min in phosphate-buffered saline (PBS) at room temperature
b. Place the slides horizontally in a humid chamber and apply 5–10 μl of appropriate antisera onto section
c. Incubate at room temperature for 30 min
d. Wash in PBS for 30 min
e. Apply second layer antibody that has been conjugated to either fluorescein–isothiocyanate (FITC, green fluorescence) or tetraethylrhodamine–isothiocyanate (TRITC, red fluorescence), and incubate for 30 min
f. Wash in PBS for 30 min and mount in buffered glycerol

IV. Combined Immunofluorescence (Double Staining)

With the availability of the green and red fluorochromes (FITC and TRITC), it is possible to stain two different antigens on one section. The technique is the same as that described above with the following two alterations.

1. At stage (*b*) two antisera, mixed together can be applied to the sections. They must of course be directed against different surface markers and themselves be of a different origin or hapten-conjugated.

2. At stage (*e*) a mixture of two second-layer antibodies are applied, each directed against the species of origin of one of the primary antibodies.

As one of these second layer antibodies can be conjugated to FITC and the other to TRITC it becomes possible to 'stain' two different surface markers within the same section (*see text*).

V. Amplification

Weak positive staining can be enhanced by using triple layer systems (*see text*). Here, the second antibody applied is not conjugated to a fluorochrome but is then followed by a third antibody, directed against the species immunoglobulin of this second layer; this third antibody is conjugated to a fluorochrome.

VI. Examination

All fluorescent preparations are examined under an ultraviolet microscope equipped with an epifluorescence condenser containing selective filters for FITC and TRITC. The filters used are set 10 (450–490 SB) with conversion filter OB 12 (FITC) and set 15 (H546) (TRITC).

VII. Acid Phosphatase Activity (Lead Method)[31]

a. Fix sections in calcium formalin pH 7 for 5 min at 4 °C
b. Wash in distilled water for 5–10 min and air dry
c. Incubate at 37 °C in medium (given below) for 1–2 h (time must be standard for comparative studies)
d. Wash *thoroughly* in distilled water
e. Immerse in 0·3 per cent ammonium sulphide solution for 30 s
f. Wash in distilled water and mount in glycerol–formalin

Incubation Medium

Dissolve 0·125 g of lead nitrate in 40 ml of acetate buffer pH 5·0; to this add 10 ml of 3 per cent sodium β-glycerophosphate. Incubate this mixture for 16–24 h. After this time a precipitate will have formed. (If no precipitate is present, solution must be discarded.) Filter before incubation of tissue sections.

Active cells have black/brown dots or diffuse staining.

Control: preincubate slides for 10 min in acetate buffer containing 0·042 per cent sodium fluoride and include the same concentration in the incubation medium.

VIII. Adenosine Triphosphatase Method

a. Fix cryostat sections in calcium–formalin at 4 °C for 5–10 min
b. Wash in distilled water for 5 min and air dry

c. Incubate at 37 °C in medium (given below) for 60–120 min (time must be standard for comparative studies)
d. Wash for 5–10 min in distilled water
e. Immerse in 0·3 per cent ammonium sulphide solution for 30–60 s
f. Wash in distilled water for 5–10 min
g. Mount in glycerol–formalin or other aqueous medium

Incubation Medium

Dissolve 20 mg adenosine triphosphate in 20 ml of distilled water. Add 20 ml of Tris–maleate buffer pH 7·2. Slowly add 5 ml of 2·5 per cent magnesium sulphate and 2 ml of 2 per cent lead nitrate. Make up to 50 ml with the buffer. Filter before incubation.

Active cells appear with black/brown membrane staining.

Control: omit substrate from incubation medium. Use heat-inactivated sections as a control for non-specific sulphide binding.

IX. Combination Method for Immunofluorescence and Cytochemistry

a. Wash sections (*see* II *above*) in PBS for 10 min at room temperature
b. Incubate in primary antisera in a humid chamber for 30 min at room temperature
c. Wash in PBS for 10 min
d. Incubate in second (conjugated) antibody in a humid chamber for 30 min at room temperature (triple-layer immunofluorescence can also be used)
e. Wash in PBS for 10 min
One can prepare a 'wet mount' and examine it under the microscope to ensure that immunohistological staining is satisfactory
f. Wash slides for 5–15 min in distilled water
g. Refix sections in calcium–formalin pH 7·0 for 2 min at +4 °C
h. Wash in distilled water for 5 min
i. Proceed with selected cytochemical technique as described above
j. After final washing, sections must be mounted in 90 per cent glycerol in PBS containing 3 per cent formaldehyde (glycerol–formalin)

The fluorescent staining on the section is observed as described in (VI) above. The cytochemical reaction is examined with normal Kohler illumination.

REFERENCES

1. Cantor H. and Boyse E. A. Lymphocytes as models for the study of mammalian cellular differentiation. *Immunol. Rev.* 1977, 7, 47–69.
2. Reinherz E. L. and Schlossman S. F. The differentiation and function of human T lymphocytes. *Cell* 1980, **19**, 821–827.
3. Wigzell H. and Binz H. Lymphocyte receptors. *Prog. Immunol.* 1980, **4**, 94–104.
4. Rouse R. V. and Weissman I. L. Microanatomy of the thymus: its relationship to T cell differentiation. *Ciba Found. Symp.* 1981, **84**, 161–177.
5. Barclay A. N. and Mayrhofer G. Bone marrow origin of Ba-positive cells in the medulla of rat thymus. *J. Exp. Med.* 1981, **153**, 1666–1671.

6. Janossy G., Thomas J. A., Bollum F. J., Grainger S., Pizzolo G., Bradstock K. F., Wong L., Ganeshaguru J. and Hoffbrand A. V. The human thymic microenvironment: an immunohistological study. *J. Immunol.* 1980, **125**, 202–212.

7. Zinkernagel R. M. Callahan G. N., Althage A., Cooper S., Klein P. A. and Klein J. On the thymus in the differentiation of 'H-2 self-recognition' by T cells: evidence for dual recognition? *J. Exp. Med.* 1978, **147**, 882–896.

8. McDevitt H. O. The role of H-2 I region genes in regulation of the immune response. *Prog. Immunol.* 1980, **4**, 503–512.

9. Dausset J. and Contu L. The MHC and the immune system. *Prog. Immunol.* 1980, **4**, 573–579.

10. Special issue on dendritic and lymphocytic cells in the epidermis. *J. Invest. Dermatol.* 1980, **75**, 1–127. Proceedings of 29th Symposium on the Biology of the Skin.

11. Hoefsmit Ch. M., Kamperdijkm E. W. A., Hendrikes H. R., Beeten R. H. J. and Balfour B. M. Lymph node macrophages. In: Carr I. and Daems W. T. (ed.) *The Reticuloendothelial System.* Vol. I Morphology. New York, Plenum Press, 1980: 417–469.

12. Lampert I. A., Pizzolo G., Thomas J. A. and Janossy G. Immunohistochemical characterization of cells involved in dermatopathic lymphadenopathy. *J. Pathol.* 1980, **131**, 145.

13. Allen T. D. Haemopoietic microenvironments in vitro: ultrastructural aspects. *Ciba Found. Symp.* 1981, **84**, 38–67.

14. Reinherz E. L., Kung P. C., Goldstein G., Levey R. H. and Schlossman S. F. Discrete stages of human intrathymic differentiation: analysis of normal thymocytes and leukemic lymphoblasts of T-Cell lineage. *Proc. Natl Acad. Sci. USA* 1980, **77**, 1588–1592.

15. Ledbetter J. A., Evans R. L., Lipinksi M., Cunningham C. R., Good R. A. and Herzenberg L. A. Evolutionary conservation of surface molecules that distinguish T lymphocyte helper–inducer and cytotoxic/suppressor subpopulations in mouse and man. *J. Exp. Med.* 1981, **153**, 310–323.

16. Humphrey J. H. Differentiation of function among macrophages. *Ciba Found. Symp.* 1981, **84**, 302–321.

17. van Furth R., Langevoort H. L. and Schaberg A. Mononuclear phagocytes in human pathology—proposal for an approach to improved classification. In: van Furth R. (ed.) *Mononuclear Phagocytes in Immunity, Infection and Pathology.* Oxford, Blackwell Scientific Publications, 1975: 1–15.

18. Janossy G., Thomas J. A. and Pizzolo G. The analysis of lymphoid subpopulations in normal and malignant tissues by immunofluorescence technique. *J. Cancer Res. Clin. Oncol.* 1981, **161**, 1–11.

19. Lennert K. *Malignant Lymphomas other than Hodgkin's Disease.* Berlin, Springer-Verlag, 1978.

20. Tidman N., Janossy G., Bodger M., Granger S., Kung P. C. and Goldstein G. Delination of human thymocyte differentiation pathways utilizing double staining techniques with monoclonal antibodies. *Clin. Exp. Immunol.* 1981, **45**, 457–467.

21. Janossy G., Tidman N., Selby W. S., Thomas J. A., Granger S., Kung P. C. and Goldstein G. Human T lymphocytes of inducer and suppressor type occupy different microenvironments. *Nature* 1980, **288**, 81–84.

22. Warnke R., Pederson M., Williams C. and Levy R. A study of lymphoproliferative diseases comparing immunofluorescence with immunohistochemistry. *Am. J. Clin. Pathol.* 1978, **70**, 867–875.

23. Curran R. C. and Gregory J. Demonstration of immunoglobulin in cryostat and paraffin sections of human tonsil by immunofluorescence and immunoperoxidase techniques. *J. Clin. Pathol.* 1978, **31**, 974–983.

24. Chilosi M., Pizzolo G., Menestrina F., Iannucci A. M., Bonetti F. and Fiore-Donati L. Enzyme histochemistry on normal and pathologic paraffin-embedded lymphoid tissues. *Am. J. Clin. Pathol.* 1981, **76**, 729–736.

25. Nachlas M. M., Prinn W. and Seligman A. M. Quantitative estimation of lyo- and desmoenzyme in tissue sections with and without fixation. *J. Biophys. Biochem. Cytol.* 1956, **2**, 487–502.

26. Stein H., Bonk A., Tolksdorf G., Lennert K., Rodt H. and Gerdes J. Immunohistologic analysis of the organization of normal lymphoid tissue and non-Hodgkin's lymphomas. *J. Histochem. Cytochem.* 1980, **28**, 746–760.

27. Mason D. Y., Stein H., Naiem M. and Abdulaziz A. Immunohistological analysis of human lymphoid tissue by double immunoenzymatic labelling. *J. Cancer Res. Clin. Oncol.* 1981, **101**, 13–22.

28. Simmonds R. G., Smith W. and Marsden H. 3-Phenylazo-4-hydroxyphenylothiocyanate: versatile reagent for the efficient haptenation of Ig and other carrier molecules. *J. Immunol. Methods* 1982, in press.

29. Granger S. (1982) PhD Thesis, University of London.

30. Warnke R. and Levy R. Detection of T and B cell antigens with hybridoma monoclonal antibodies: a biotin–avidin–horseradish peroxidase method. *J. Histochem. Cytochem.* 1980, **28**, 771–776.
31. Gomori G. *Microscopic Histochemistry*. Chicago, University Press, 1952.
32. Selby W. S., Janossy G., Goldstein G. and Jewell D. P. T lymphocyte subsets in human intestinal mucosa the distribution and relationship to MHC-derived antigens. *Clin. Exp. Immunol.* 1981, **44**, 453–458.
33. Selby W., Poulter L. W., Janossy G., Bofill M. and Hobbs S. Heterogeneity of HLA-DR +ve macrophage-like cells in the human gut. 1982, *J. Clin. Pathol.* in press.
34. Greaves M. F., Falk J. A. and Falk R. E. A surface antigen marker for human monocytes. *Scand. J. Immunol.* 1975, **4**, 555–561.
35. Müller-Hermelink H. K., Heusermann U. and Strutte H. J. Enzyme histochemical observations on the localisation and structure of the T cell and B cell regions in the human spleen. *Cell. Tissue Res.* 1974, **154**, 167–179.
36. Steinman R. M., Kaplan G., Witmer M. D. and Cohn Z. A. Identification of a novel cell type in peripheral lymphoid organs of mice. V. Purification of spleen dendritic cells, new surface markers and maintenance *in vitro*. *J. Exp. Med.* 1979, **149**, 1–16.
37. Balfour B. M., Drexhage H. A., Kamperdijk E. W. A. and Hoefsmitt E. M. Antigen-presenting cells, including Langerhans cells, veiled cells and interdigitating cells. *Ciba Found. Symp.* 1981, **84**, 281–301.
38. Seymour G., Poulter L. W., Janossy G., Bofill M., Zola H. and Bradley J. Characterization of a new monoclonal antibody (FMC 17) directed against human monocytes. 1982. Submitted.
39. Steinman R. H. and Witmer M. D. Lymphoid dendritic cells are potent stimulators of the primary mixed lymphocyte reaction in mice. *Proc. Natl Acad. Sci. USA* 1978, **75**, 5132–5137.
40. Poulter L. W., Duke O., Hobbs S., Janossy G. and Panayi G. Histochemical discrimination of HLA-DR positive cell populations in the normal and arthritic synovial lining. *Clin. Exp. Immunol.* 1982, **48**, 381–388.
41. Poulter L. W., Duke O., Hobbs S., Janossy G. and Panayi G. The involvement of interdigitating cells in the pathogenesis of rheumatoid arthritis, 1982, *J. Clin. Pathol.* in press
42. Janossy G., Panayi G., Duke O., Bofill M., Poulter L. W. and Goldstein G. Rheumatoid arthritis: a disease of T lymphocyte/macrophage immunoregulation. *Lancet* 1981, **ii**, 839–842.
43. Poulter L. W., Seymour G. J., Duke O., Janossy G. and Panayi G. Immunohistological analysis of delayed-type hypersensitivity reactions in man. 1982, *Cell. Immunol.* In press.
44. Murphy G. F., Bhan A. K., Sato S., Mihm M. C. and Harriet T. J. A new immunological marker for human Langerhans cells. *N. Engl. J. Med.* 1981, **304**, 791–792.
45. Thomas J. A., Janossy G., Chilosi M., Pritchard J. and Pincott J. R. Combined immunological and histochemical analysis of skin and lymph node lesions in histiocytosis X. *J. Clin. Pathol.* 1982, **35**, 327–337.
46. Kung P. C., Goldstein G., Reinherz F. L. and Schlossman S. F. Monoclonal antibodies defining distinctive human T cell surface antigens. *Science* 1979, **206**, 347–349.
47. Reinherz E. L., Kung P. C., Goldstein G. and Schlossman S. F. A monoclonal antibody reactive with the human cytotoxic/suppressor T cell subset previously defined by a heteroantiserum dermal TH2. *J. Immunol.* 1979, **124**, 1301–1307.
48. Janossy G., Thomas J. A., Pizzolo G., Granger S. M., McLaughlin J., Habeshaw J. A., Stansfield A. G. and Sloan J. Immuno-histological diagnosis of a lymphoproliferative disease by selected combinations of antisera and monoclonal antibodies. *Br. J. Cancer* 1980, **42**, 224–242.
49. Callard R. E., Smith C. M., Worman C., Linch D., Cawley J. C. and Beverley P. C. L. Unusual phenotype and function of an expanded subpopulation of T cells in patients with haemopoietic disorders. *Clin. Exp. Immunol.* 1981, **43**, 497–505.
50. Janossy G. Membrane markers in leukaemia. In: Catovsky D. (ed.) *Methods in Haematology. The Leukaemic Cell* Edinburgh, Churchill-Livingstone, 1981: 129–183.
51. Duke O., Panayi G., Poulter L. W. and Janossy G. Significance of the expression of HLA-DR antigen on the circulating monocytes of patients with rheumatoid arthritis. 1982. Submitted.
52. Beckstead J. H., Halverson P. S., Ries C. A. and Baunton D. F. Enzyme histochemistry and immunohistochemistry on biopsy specimens of pathologic human bone marrow. *Blood* 1981, **57**, 1088–1098.
53. Heusermann U., Strutte H. J. and Müller-Hermelink H. K. Interdigitating cells in the white pulp of the human spleen. *Cell Tissue Res.* 1974, **153**, 415–417.
54. MacKenzie I. C. and Squier C. A. Cytochemical identification of ATPase positive Langerhans cells in EDTA-separated sheets of mouse epidermis. *Br. J. Dermatol.* 1975, **92**, 523–533.
55. Gerdes J., Naiem M., Mason D. Y. and Stein H. Human complement (C_3b) receptors defined by a mouse monoclonal antibody. *Immunology* 1982, **45**, 645–653.

14

Immunocytochemistry of Lymphoreticular Tumours

P. Isaacson and
D. H. Wright

1. INTRODUCTION

Histopathological analysis of lymphoreticular neoplasms is greatly hampered by the essentially similar morphology of the different lymphoreticular cells as they appear in histological sections. It is to the credit of histopathologists that, despite this, great advances have been made in the understanding and classification of malignant lymphomas on morphological grounds alone. Patterns of cell distribution and subtle variations in nuclear size and shape have been exploited to the full yielding useful sets of diagnostic criteria and several soundly based classifications of this group of tumours.[1,2] Nevertheless, observer error is still significant and differences in the quality and nature of fixation and staining can radically change the histological appearances of a malignant lymphoma. Recognizing this, pathologists have long sought some other means of accurately identifying the small blue cells that confront them in sections of malignant lymphoma in order both to substantiate their diagnoses and to provide a scientific basis for their classifications. Enzyme histochemistry, mainly carried out on frozen sections, was the first step in this direction. This was followed by immunological studies of dispersed cells from lymphoma tissue. The limitations of both these procedures lay in the uncertainty as to which of the numerous cell types identified were the malignant cells since in most lymphoreticular tumours there is a large component of benign reactive and supportive cells. Immunocytochemistry, nevertheless, was clearly of great value in defining subtypes of lymphoproliferative diseases and it soon became the aim of histopathologists to apply these techniques to histological sections in such a way that the morphological and immunological properties of lymphoreticular tumours could be studied simultaneously. The impact of the first immunohistochemical studies of lymphoma[3,4] was dramatic and has been followed by an almost exponential growth of the subject. Using immunoenzyme (principally immunoperoxidase) techniques many cellular antigens can be accurately labelled and viewed with the light microscope in either frozen or paraffin sections offering the histopathologist the setting with which he is most familiar. Variations in methods abound and each group engaged in immunohistochemistry has its favoured techniques which it will often champion as the most suitable. It is not the purpose of this chapter to compare the various techniques available. We will, however, briefly describe the techniques used in our laboratory and give our reasons for using them.

2. METHODS

The tissue antigens useful in the study of lymphoreticular tumours can be divided into two broad groups. The first consists of cytoplasmic antigens which can be demonstrated in routinely fixed paraffin-embedded tissue and the second of membrane antigens which are best demonstrated in frozen sections. Following the application of primary antiserum one or other variety of the immunoenzyme techniques is used to demonstrate the binding sites of the primary antibody. Again, each technique has its adherents and in giving our preferred methods for the immunoperoxidase techniques (*see* the Appendix, p. 270) we are in no way assuming that other methods are of less value.

The cytoplasmic antigens which can assist in the diagnosis of lymphoreticular tumours include the different heavy and light chains of immunoglobulin (Ig), J chain, alpha-1-antitrypsin (α1-antiT), α1-antichymotrypsin (α1-antiCT), lysozyme and albumin. The method used to demonstrate these is critically dependent on the fixative used. The routine fixative in our laboratory has for many years been neutral buffered formol–saline (NBFS). Our initial interest in studying a large volume of retrospective material, the wide use of NBFS throughout the world and the suitability of NBFS for electron microscopy has led us to concentrate our efforts on demonstrating antigens in NBFS-fixed tissue. Initial results both from our own laboratory and others were disappointing but following the discovery that antigens 'masked' by NBFS fixation could be 'unmasked' by prior treatment of sections with proteolytic enzymes,[5,6] the excellent preservation of antigens by NBFS soon became apparent. Crisp reproduceable results, with very little 'background' staining can be obtained with the method given in *Table* 14.1. All steps in the method require great care and must be controlled. It is particularly important to realize that the degree of proteolytic digestion necessary is directly related to the time that the tissue has been exposed to the fixative. The use of other fixatives such as Bouin's, formol mercury, B5 and formol–acetic[7] obviates the need for the use of proteolytic enzymes but, in our opinion, each of these fixatives has one or more disadvantages when compared with NBFS, not the least being that the cytoplasmic disruption that they cause obscures the precise cytoplasmic location of antigens being demonstrated and makes them unsuitable fixatives for electron microscopy.

Fixation and paraffin embedding destroys or obscures many membrane antigens. As cell suspension studies have shown these antigens are, nevertheless, of great importance in characterizing lymphoreticular cells and have become even more important since the advent of monoclonal antibodies to cell membrane constituents. A wide variety of membrane antigens can be demonstrated in frozen sections and in our laboratory these at present include membrane Ig, and a range of unknown determinants defined by monoclonal antibodies. These include antibodies to a common white cell antigen, to T-cells and their helper and suppressor subsets, to HLA-DR antigens (*h*uman *l*eucocyte *a*ntigens) and to an antigen which appears to define dendritic reticulum cells. All these monoclonal antibodies have been kindly made available to us by Dr Peter Beverley of the ICRF Human Tumour Immunology Unit, University College, London. Monoclonal antibodies specific for other antigens are rapidly being developed and will be of increasing use in the characterization of malignant lymphomas. There are a number of different approaches to the immunostaining of frozen sections[8] but the method devised by Stein et al.[9] (*see* the Appendix, p. 270) appears to yield the optimal combination of immunoreactivity and preservation of morphological detail.

3. IMMUNOHISTOCHEMISTRY OF NORMAL LYMPHOID TISSUE

The histogenesis of lymphoreticular tumours can be closely correlated with the cells comprising normal lymphoid tissue. Thus, to appreciate the significance of immunohistochemical results obtained in cases of malignant lymphoma it is first necessary to be familiar with the staining reactions of these normal cells. Normal lymphoid tissue (*Fig*. 14.1) is comprised of B-cells, which include cells of the follicle centre, the mantle zone and plasma cells, T-cells, which are found principally in the

Fig. 14.1. Section of reactive lymph node showing B-cell areas consisting of follicle centres (FC) and mantle zones (M). Most T-lymphocytes are found in the paracortex (PC). (Haematoxylin and eosin.)

Table 14.1. ***Immunohistochemistry of normal lymphoid tissue***

	FCC	*Mantle cells*	*Plasma cells*	*T-cells*	*Macrophages*	*Reticulum* cells*
SIg	±	+	−	−	−	−
HLA-DR	+	+	−	−	+	+
Other membrane antigens	+†	+†	−	+†‡	+†	+†§
CIg	±	−	+	−	(±)	−
J chain	±	−	±	−	−	−
Lysozyme	−	−	−	−	+	−
α1-antiT (α1-antiCT)	−	−	−	−	+	−

* Interdigitating reticulum cells (IDRC), dendritic reticulum cells (DRC).
† Common white cell antigen.
‡ T-cell antigens as defined by monoclonal antibodies.
§ Antigens to DRC defined by monoclonal antibodies.

paracortical areas, and cells of the monocyte–macrophage system and related 'reticulum' cells. The immunohistochemical characteristics of these various components are listed in *Table* 14.1 and will be briefly summarized here.

3.1. B-cells

B-cells are identified by their production of either surface membrane Ig (SIg) or cytoplasmic Ig (CIg). With the exception of plasma cells, B-cells carry HLA-DR antigens on their surfaces. The small B-lymphocytes of the mantle zone of the follicle synthesize SIgM and SIgD and show polytypic membrane staining with respect to their light chain class (*Fig.* 14.2). The centrocytes and centroblasts of the follicle centre synthesize SIg but this is largely obscured in frozen sections by the large amount of extracellular Ig bound to the processes of dendritic reticulum cells (*Fig.* 14.3). In paraffin sections of hyper-reactive follicles, CIg can be demonstrated in a population of cells which in thin (1 μm) sections can be clearly seen to be centrocytes and centroblasts (*Fig.* 14.4). This appears to represent a pathway of CIg synthesis distinct from that shown by plasma cells.[10,11] These CIg-synthesizing follicle centre cells also synthesize J chain[12] (*Fig.* 14.5). J or joining chain, is the molecule which links the IgA dimer or IgM pentamer and is synthesized both by cells producing these classes of CIg, and by immature CIg-synthesizing cells which may subsequently switch from the class of heavy chain they are synthesizing to another.[13] The heavy and light chains of CIg in immunoblasts and plasma cells, which are found outside the follicles, can be well demonstrated in paraffin sections (*Fig.* 14.6) and J chain can be shown in immunoblasts and immature plasma cells as well as plasma cells synthesizing dimeric IgA and IgM.

3.2. T-cells

T-cells cannot be identified immunohistochemically in paraffin sections but the use of monoclonal antibodies to membrane antigens of T-cells permits their identifica-

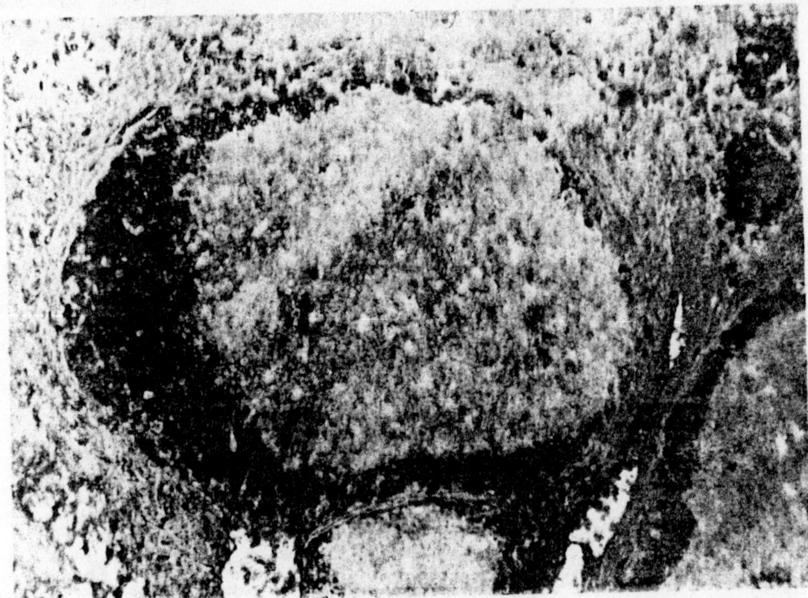

Fig. 14.2. Frozen section of reactive lymph node stained for IgD. Note strong surface staining of mantle cells.

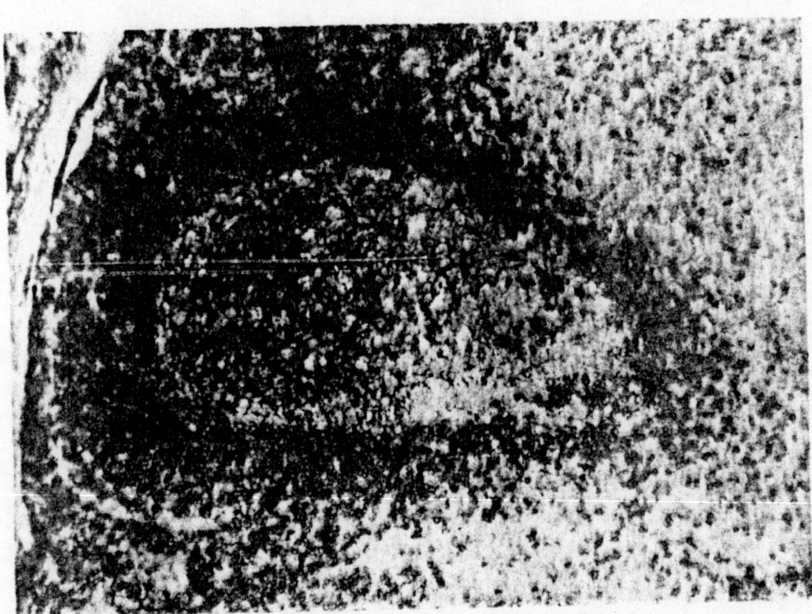

Fig. 14.3. Frozen section of reactive lymph node stained for IgM. There is staining of mantle cells and extracellular 'network' in the follicle centre.

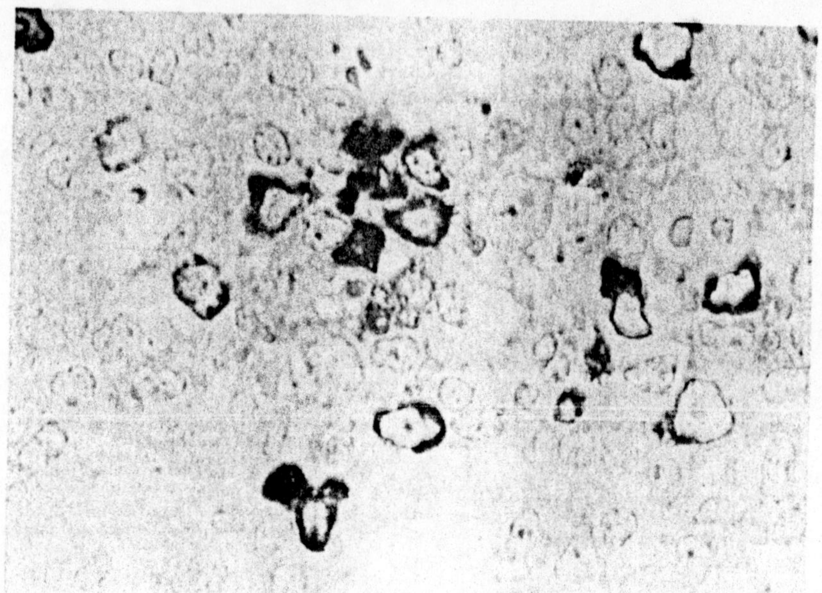

Fig. 14.4. Follicle centre of human tonsil stained for κ chain. In this 1-µm section of plastic-embedded tissue the FCC nature of the positively staining cells is clearly evident.

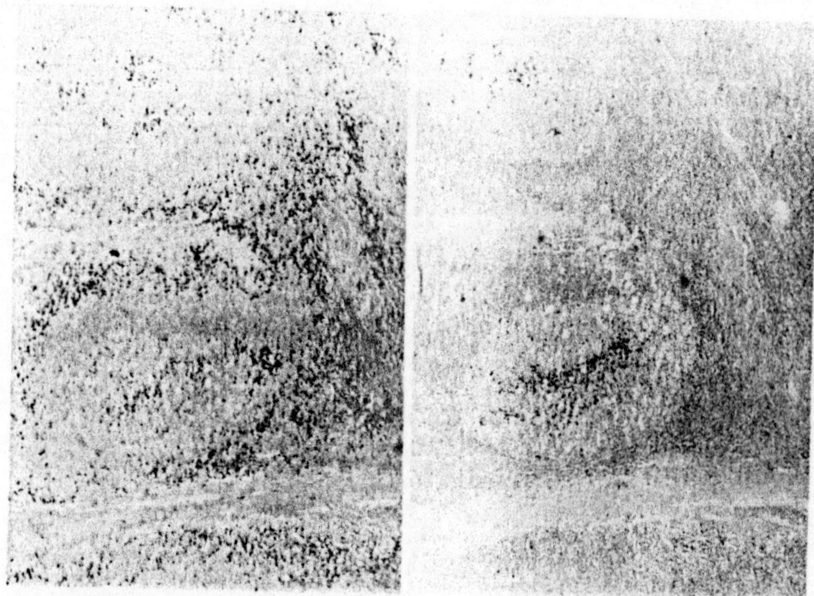

Fig. 14.5 Paraffin section of human tonsil stained for IgG (left) and J chain (right). A population of FCCs stains for both IgG and J chain while surrounding plasma cells are positive for IgG alone.

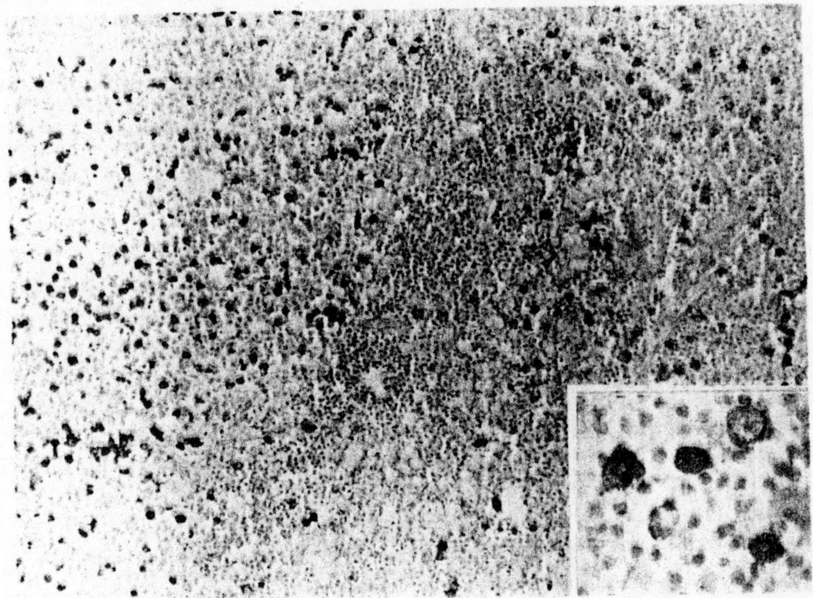

Fig. 14.6. Paraffin section of reactive lymph node stained for IgG. Many FCCs in the follicle (left) contain IgG. Positively staining cells outside the follicle (right) are a mixture of plasma cells and immunoblasts. Inset shows darkly staining plasma cells in contrast to larger immunoblasts showing a perinuclear rim of cytoplasmic staining.

Fig. 14.7. Frozen section of reactive lymph node stained with a monoclonal antibody to T-cells. The majority of T-cells are confined to the paracortex, some are seen in the follicle centres. The characteristic zonal distribution of T-cells is seen in the follicle centre at lower right.

tion in frozen sections (*Fig.* 14.7). They are found predominantly in the paracortical zone although some T-cells are also found within the follicle centre where they are distributed in a characteristic zonal pattern. Monoclonal antibodies can distinguish T-helper and T-suppressor populations.

3.3. Macrophages

Macrophages are found within the sinuses of lymphoid tissue and also in follicle centres as the so-called tingible body macrophages. Lysozyme is synthesized by macrophages and can be demonstrated within them in paraffin sections. These cells can also be stained with antisera to α1-antiT and α1-antiCT (*Fig.* 14.8). Both these antiproteases are probably synthesized by macrophages although this has only been conclusively shown for α1-antiT.[14] Both dendritic reticulum cells (DRC) which are found in follicle centres, and interdigitating reticulum cells (IDRC) which are found in paracortical tissue, are thought to be of monocyte/macrophage derivation. Neither of these cells can be demonstrated in paraffin sections but monoclonal antibodies to DRC have been developed and these cells can be specifically stained in frozen sections (*Fig.* 14.9).

3.4. Malignant Lymphoma

Having regard to the immunohistochemistry of normal lymphoid tissue intelligent choices can be made in studying cases of lymphoma with respect to the nature of the section (frozen or paraffin) and the most useful antisera to apply. Under ideal circumstances, when fresh tissue is received, a sample can be snap frozen in liquid nitrogen while paraffin sections are prepared from optimally fixed tissue. The histological appearances of the paraffin section will usually determine which immunohistochemical studies, if any, are most appropriate. To take extreme examples, it is pointless to study frozen sections of a plasmacytoma to determine whether or not the CIg in the plasma cells is monotypic and equally pointless, to stain paraffin sections of Burkitt's lymphoma to determine the nature of the SIg. Whichever method is chosen (and in some instances both frozen and paraffin sections should be studied), it is our practice to apply a panel of antisera and not simply those for which a positive result is expected. The reasons for this will become clearer later in this chapter but include the not infrequent finding of unexpected results, the need for continuous control of staining reactions of the different antisera and the exclusion of spurious staining reactions which are especially common in the case of immunoglobulin.

3.5. B-cell Lymphomas

The aim of the immunohistochemical study of B-cell lymphomas is to demonstrate monotypic immunoglobulin on the surface or within the cytoplasm of tumour cells. While more than one heavy Ig chain is occasionally synthesized by malignant B-cells, the finding of a single (i.e. monotypic) light chain is central to the diagnosis of a monoclonal and hence malignant proliferation of B-cells.[15] Biclonal tumours

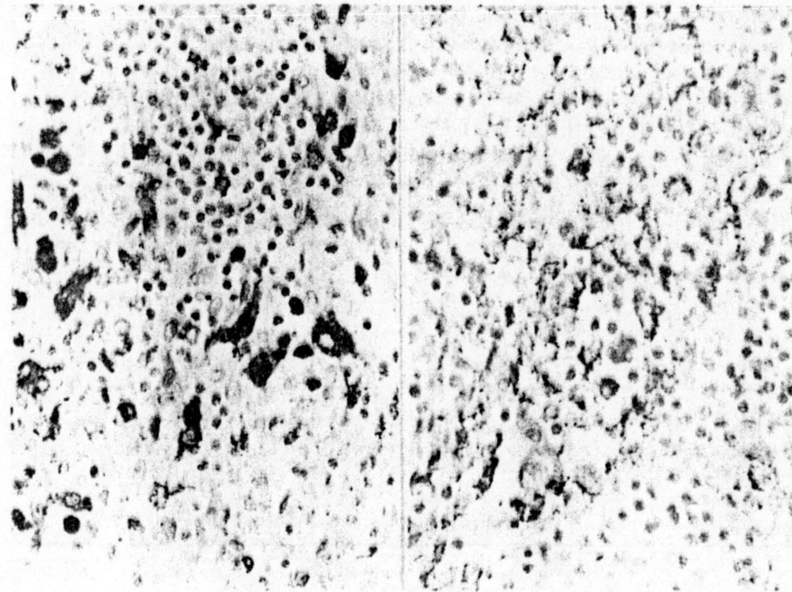

Fig. 14.8. Paraffin section of lymph node showing intrasinusoidal macrophages stained for lysozyme (left) and α1-antiT (right). There is great variation in intensity of lysozyme staining in contrast to the uniform content of α1-antiT positive granules.

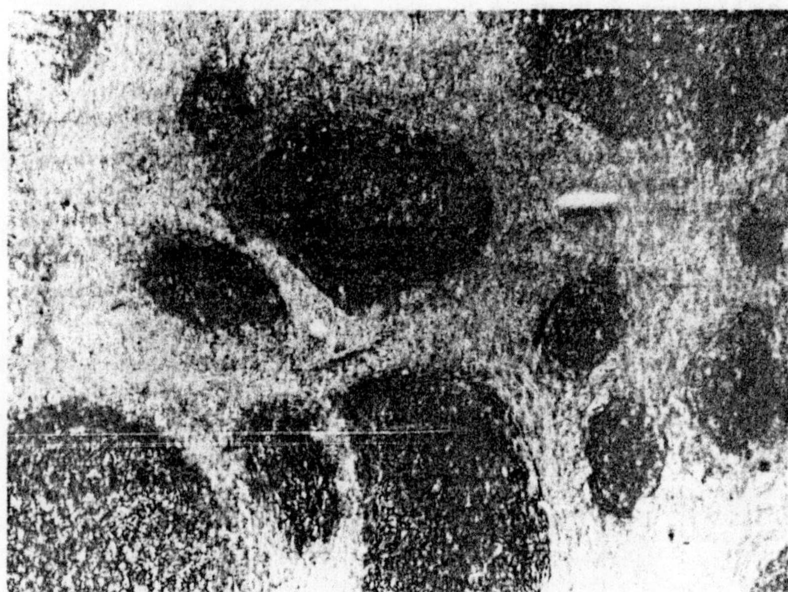

Fig. 14.9. Frozen section of reactive lymph node stained with a monoclonal antibody to DRCs. A strong network staining pattern defines the follicles.

synthesizing both κ and λ light chains have been reported,[16] but are very rare indeed. Despite this, numerous reports describing staining for both κ and λ light chains in paraffin sections of B-cell lymphomas have appeared in the literature.[17,18] It cannot be sufficiently stressed that if immunohistochemical studies show both κ and λ light chains in the tumour cells then either the B-cell proliferation is not neoplastic or the staining results are spurious. This problem of anomalous CIg staining in paraffin sections[10,19] deserves detailed discussion.

When a polytypic CIg pattern of staining (i.e. cells staining for both κ and λ light chains) is detected in a malignant lymphoma the following causes for non-specific staining should be sought (*Table* 14.2).

Table 14.2. ***Causes of anomalous (polytypic) CIg staining***

A. Reactive plasma cell population

B. Uptake of Ig from tissue fluid
- i. By dead or damaged cells
- ii. By macrophages
- iii. By Reed–Sternberg cells or other uncharacterized multinucleated giant cells

1. There may be a large reactive B-cell population in the tumour. Usually the number of reactive cells containing CIg is far less than the tumour cells but this is not always the case. The reactive cells are usually plasma cells (not follicle centre cells, FCCs) and stain much more intensely and uniformly than the tumour cells.

2. Polytypic immunoglobulin may be taken up from the tissue fluids either non-specifically by dead or damaged cells or possibly specifically by macrophages, including neoplastic macrophages, Reed–Sternberg cells and multinucleated cells of uncertain nature that are often found in FCC lymphomas.

While Ig in the reactive cell population does not usually lead to problems in interpretation it can be very difficult to distinguish between synthesized Ig and Ig that has been taken up from tissue fluids. Careful attention to the pattern of Ig staining will, however, permit this distinction to be made. The main features distinguishing these two types of staining are summarized in *Table* 14.3.

Table 14.3. ***Immunoglobulin-containing cells in malignant lymphoma***

	Cells synthesizing Ig	*Cells taking up Ig*
Pattern of staining	Granular	Smooth
	Concentration in Golgi region	Concentration at periphery of cytoplasm
	Staining of perinuclear space	Diffuse, cytoplasmic staining
Immunoglobulin	Monotypic	Polytypic
J chain	+	−
Other plasma proteins	−	±

The earliest site of CIg synthesis is in the perinuclear space[20] and the resulting perinuclear pattern of staining is characteristically seen in many large B-cell lymphomas. As cells mature, Ig accumulates in the region of the Golgi apparatus. Accumulation of Ig in dilated profiles of rough endoplasmic reticulum gives rise to a characteristic granular staining pattern with the immunoperoxidase stain (*Fig.* 14.10). These features are highlighted in thin (1 μm) sections of either paraffin[21] or plastic-embedded[22] tissue (*Fig.* 14.11). Positive staining for J chains is a reliable indicator of Ig synthesis and the majority of CIg positive lymphomas synthesize J chain.[12]

Cells taking up Ig from tissue fluids sometimes, but by no means always, show morphological evidence of death or degeneration. The positively staining cells tend to occur in broad swathes across the section with concentration around blood vessels (*Fig.* 14.12). Other plasma proteins such as albumin or α1-antitrypsin can also be demonstrated within these cells (*Fig.* 14.13). The quality of staining of α1-antiT taken up non-specifically in this way is quite different from synthesized α1-antiT in macrophages (*see below*). Since there is very little J chain in the plasma, cells taking up Ig do not stain for J chain. Cells of the monocyte/macrophage system, whether reactive or neoplastic, sometimes take up polytypic IgG, probably specifically. The identity of these cells will be apparent if antibodies to macrophage markers are included in the panel of antisera used in the study of lymphomas (*see below*). Reed–Sternberg cells and other multinucleate cells[10] also appear to take up IgG specifically and so stain polytypically for Ig in the same way as macrophages. When a cell has taken up Ig from its environment the quality of the Ig staining is quite different from an Ig-synthesizing cell (*Fig.* 14.14). The cytoplasm is diffusely stained without the characteristic granularity of synthesized Ig and there is frequently an increased intensity of staining at the periphery of the cell.

In the description of the characteristic immunohistochemical staining reactions of B-cell lymphomas that follows, the need for critical interpretation of the results must always be borne in mind. Anomalous staining patterns due to uptake of Ig are especially common in paraffin sections of large cell tumours of FCC origin and immunoblastic lymphomas and, if misinterpreted, can lead to basic conceptual errors in the understanding of B-cell neoplasia.

3.6. Lymphocytic Lymphoma

Malignant lymphoma of lymphocytic type consists of a uniform population of small lymphocytes with occasional foci of slightly larger cells (so-called proliferation centres). Most of the patients whose lymph nodes show this picture have underlying chronic lymphocytic leukaemia. The malignant cells in this condition do not synthesize CIg and thus are best characterized in frozen sections. They almost always synthesize IgM of single light chain type.

3.7. Lymphoplasmacytic and Lymphoplasmacytoid Lymphoma

Small lymphocytic lymphoma may show plasmacytic or plasmacytoid differentiation. In this event staining of paraffin sections will show monotypic CIg in the plasma cells or plasmacytoid cells. This Ig will be of the same class as the SIg on the small lymphocytes. The positively staining cells in lymphoplasmacytic lymphoma

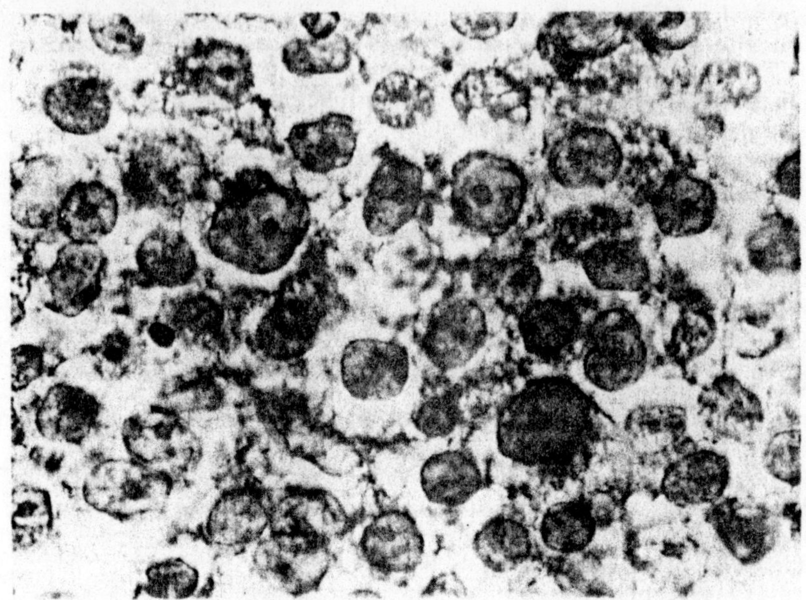

Fig. 14.10. Routine (5 μm) paraffin section of a FCC (centroblastic) lymphoma stained for IgM. Perinuclear staining is present in most of the cells with concentration in the Golgi region evident in a few cells.

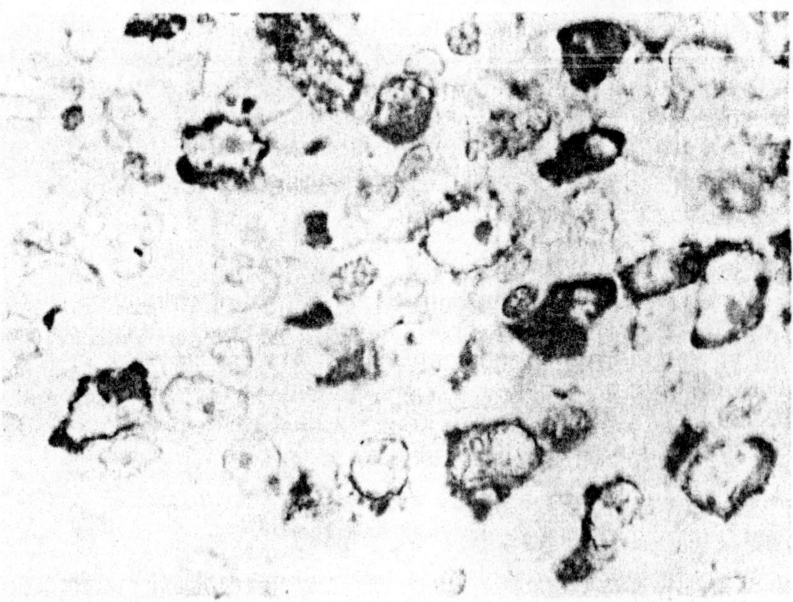

Fig. 14.11. Thin (1 μm) paraffin section of a FCC (centroblastic/centrocytic) lymphoma stained for κ chain. Staining of the perinuclear space and the granular consistency of cytoplasmic staining are well seen.

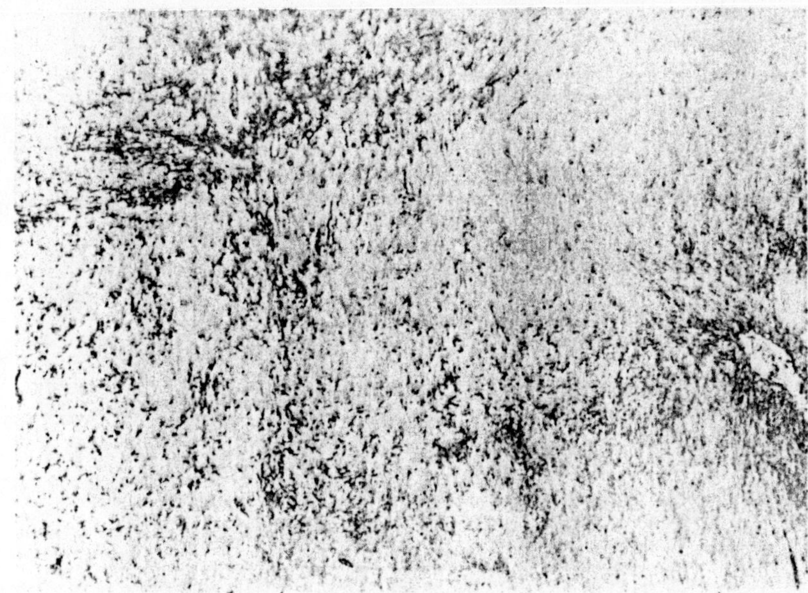

Fig. 14.12. Paraffin section of a FCC (centroblastic/centrocytic) lymphoma stained for κ chain. Swathes of positively staining cells are present with concentration around blood vessels. Stain for λ chain produced an identical picture.

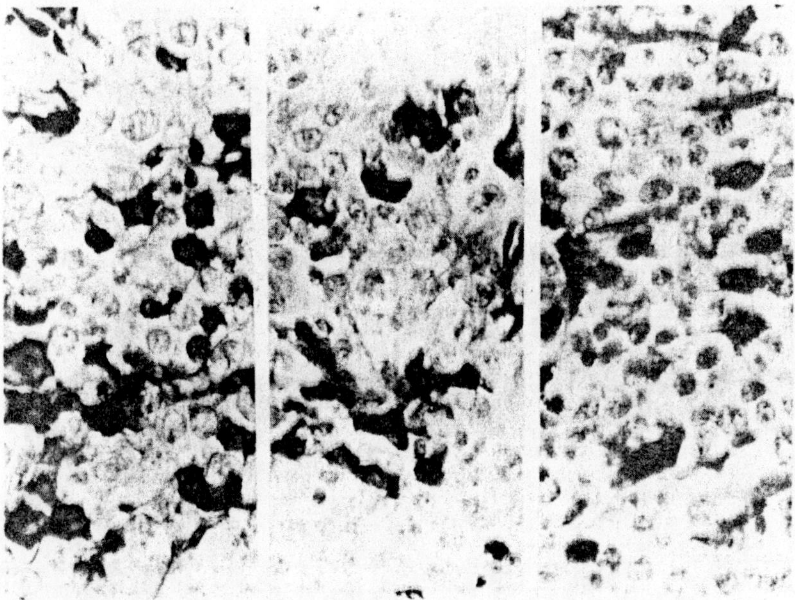

Fig. 14.13. Paraffin sections of the same tumour illustrated in Fig. 14.12 stained for κ chain (left), λ chain (centre) and albumin (right). Note diffuse non-granular polytypic staining of tumour cells which also stain for albumin.

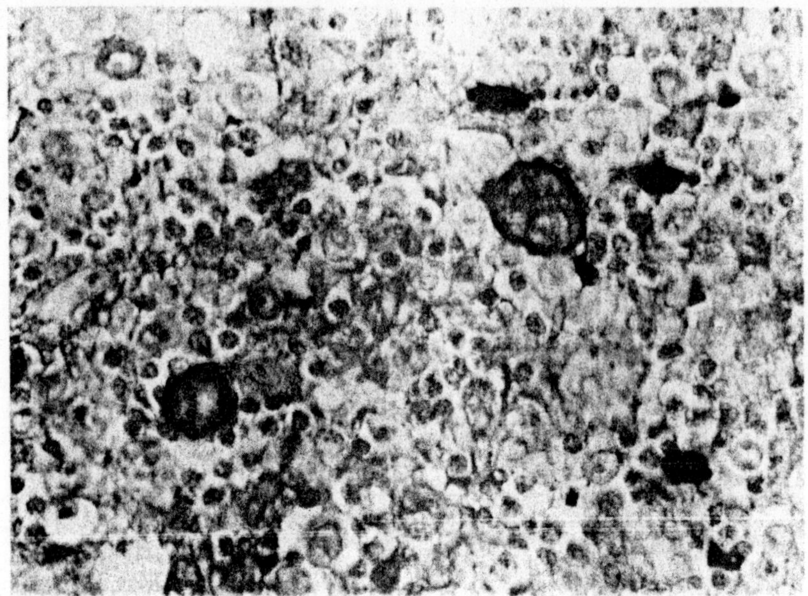

Fig. 14.14. Large multinucleated cells in a paraffin section of a FCC lymphoma (IgM κ) stained for IgG. Tumour cells are negative but there is diffuse staining of the large multinucleated cells with intensification at the periphery. Both light chains were present in these cells. Occasional reactive IgG plasma cells are also seen.

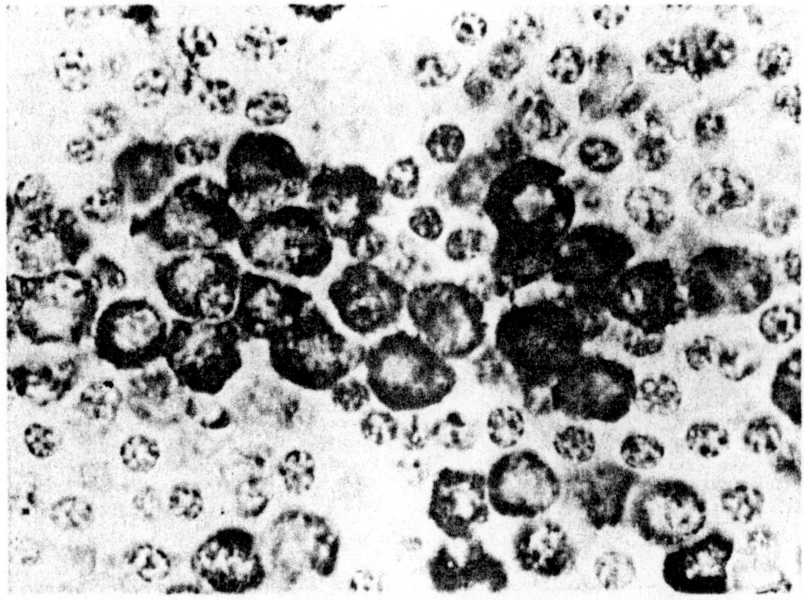

Fig. 14.15. Paraffin section of a lymphoplasmacytic lymphoma showing monotypic (κ) cytoplasmic staining of plasma cells.

resemble mature plasma cells (*Fig.* 14.15), while in lymphoplasmacytoid tumours positive staining is often in the form of cytoplasmic and nuclear inclusions.

3.8. Plasmacytoma

Plasma cell tumours, whether medullary or extramedullary, characteristically show uniform cytoplasmic staining of all the cells for a single heavy and light Ig chain. In common with most other tumours synthesizing CIg, plasma cell tumours stain positively for J chain regardless of the class of heavy chain being synthesized.

3.9. Follicle Centre Cell Tumours

Over 50 per cent of non-Hodgkin's malignant lymphomas are of follicle centre cell (FCC) origin. They may be follicular, follicular and diffuse, or diffuse in their pattern of growth. These tumours can be usefully studied for both membrane and cytoplasmic antigens and the characteristic staining patterns for each will be described separately.

Staining of frozen sections of FCC lymphomas of follicular pattern will show monotypic SIg on the tumour cells (*Fig.* 14.16). In contrast to reactive follicles, the follicles of a FCC lymphoma show poorly developed or absent mantle zones which, when present, consist of small lymphocytes bearing surface IgM and IgD of both light chain classes. While the characteristic network pattern of extracellular Ig is not seen in neoplastic follicles, numerous dendritic reticulum cells are present and can be demonstrated using an appropriate monoclonal antibody (*Fig.* 14.17). T-cells which are zonally distributed in reactive benign follicles tend to be evenly distributed in neoplastic follicles and can be shown to be predominantly of the helper variety (*Fig.* 14.18). Vast numbers of T-cells are sometimes present between the follicles together with varying numbers of tumour cells. The FCC lymphomas of diffuse type exhibit uniform monotypic Ig surface staining of the neoplastic cells (*Fig.* 14.19). Dendritic reticulum cells can be shown in these tumours either as single cells or in small groups. T-cells, predominantly of the helper type, are distributed throughout the tumour, usually in far fewer numbers than in the follicular tumours. The cells of both follicular and diffuse FCC lymphomas stain positively for HLA-DR antigens. Monotypic CIg together with J chain can be demonstrated in paraffin sections of up to two-thirds of cases of FCC lymphoma. The Ig is found predominantly in large centrocytes and centroblasts but only occasionally in small centrocytes and thus, depending on the cytology of the tumour, may be found in very few cells. It should be stressed that the monotypic CIg is found in FCCs and not in plasma cells except uncommonly when the tumour shows plasmacytic differentiation. The characteristic features of cells staining positively for CIg and the pitfalls in the interpretation of Ig staining in this group of lymphomas have already been discussed and the reader is referred to *Figs.* 14.10 and 14.11.

3.10. Immunoblastic Lymphoma

Large cells with central nucleoli and pyroninophilic cytoplasm, so-called immunoblasts, may be present in considerable numbers in FCC lymphomas. When

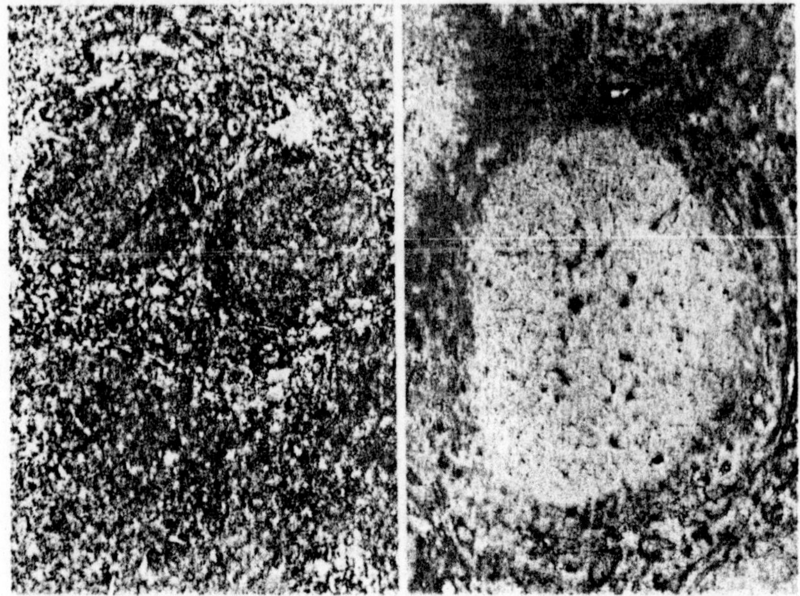

Fig. 14.16. Frozen section of a FCC lymphoma, follicular, showing positive surface staining for κ chain (left) and absence of staining in section stained for λ chain (right).

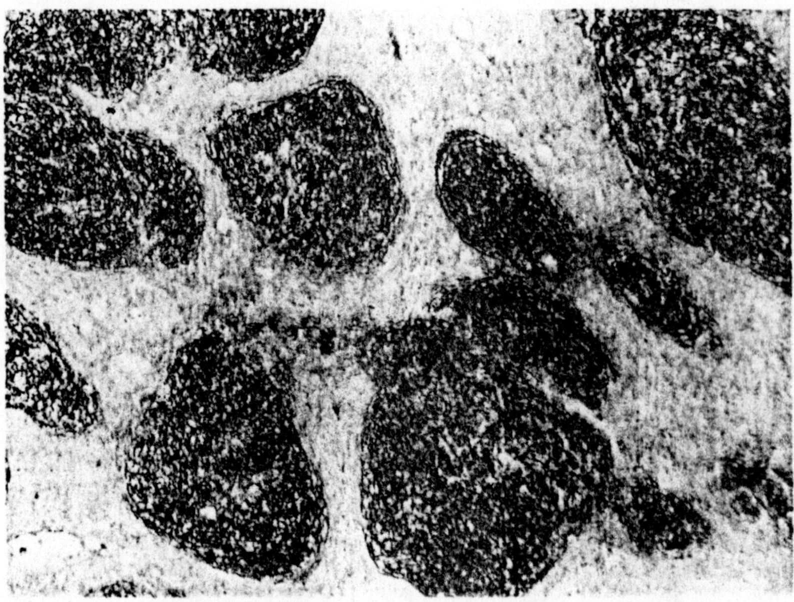

Fig. 14.17. Frozen section of a FCC lymphoma, follicular, stained with a monoclonal antibody to DRCs. There is intense 'network' staining of the malignant follicles.

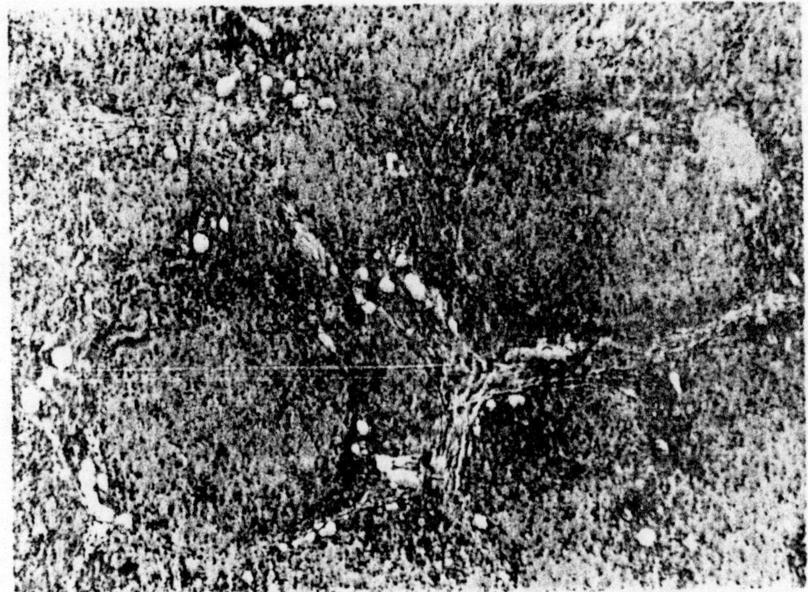

Fig. 14.18. Frozen section of a FCC lymphoma, follicular, stained with a monoclonal antibody to T-cells. Large numbers of T-cells are present, mostly between the follicles.

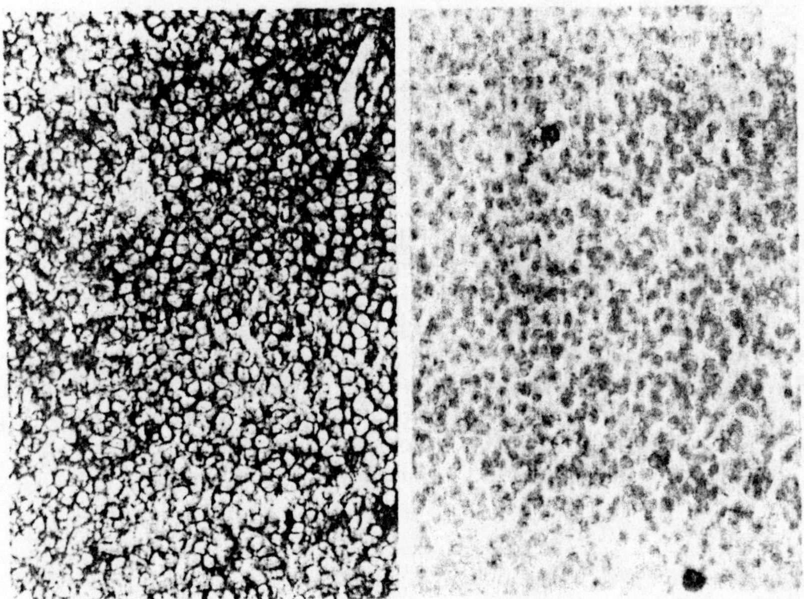

Fig. 14.19. Frozen section of a FCC lymphoma, diffuse, stained for κ chain (left) and λ chain (right). Positive surface staining for κ chain is present on all cells in contrast to only occasional cells positive for λ chain.

they are the predominant cell, and especially when clear plasmacytic differentiation is present, the term immunoblastic sarcoma (IBS) is justified. Immunoblasts will synthesize SIg which can be demonstrated in frozen sections and CIg synthesis is also a feature permitting the characterization of these tumours in paraffin sections (*Fig*. 14.20).

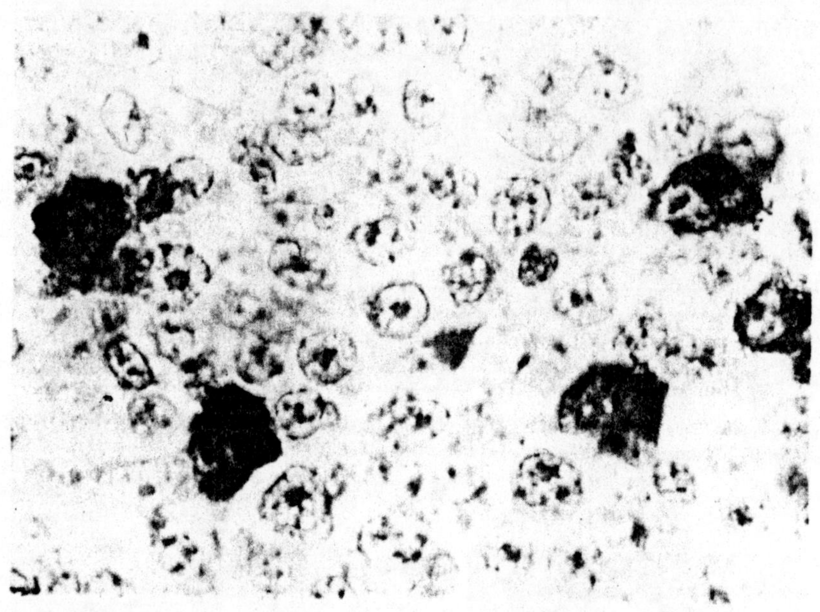

Fig. 14.20 Paraffin section of an immunoblastic lymphoma showing monotypic (κ chain) cytoplasmic staining of a minority of tumour cells.

3.11. T-cell Lymphomas

With the development of monoclonal antibodies to T-cell subsets these can now be positively identified in frozen sections. Despite the availibility of antibodies to T-helper and T-suppressor cells, the problem of determining monoclonality in T-cell proliferations remains. Certain cutaneous T-cell lymphomas show a helper phenotype,[23] but this is not always the case in the nodal tumours where both helper and suppressor phenotypes may be expressed. In contrast to B-cell tumours most, but not all, T-cell lymphomas do not express HLA-DR antigen. Large numbers of T-cells can be present in lymphomas of other types necessitating caution in the interpretation of sections stained for T-cell subsets.

3.12. Histiocytic (Monocyte/Macrophage) Lymphoma

The diagnosis of malignant lymphoma of true histiocytic (i.e. monocyte/ macrophage) origin can be very difficult on morphological grounds alone since the histological spectrum of these tumours, which include histiocytic lymphoma,[24] malignant histiocytosis,[25] and malignant histiocytosis of the intestine,[26] is very broad. Initial enthusiasm for lysozyme as the diagnostic immunohistochemical marker of this group of tumours[27] has ebbed somewhat[28,29] and in our experience, although sometimes positive, it is an unreliable marker of malignant histiocytes. Intracytoplasmic α1-antiT, which is synthesized by macrophages, is, at present, the best marker of these tumours.[14] We have found that α1-antiT is best demonstrated in sections of NBFS-fixed paraffin-embedded tissue following trypsin digestion. Other fixatives require stronger concentration of antibody and do not result in such consistent staining. As with CIg the nature of the positive-staining reaction is important since, along with other proteins, α1-antiT can be taken up from tissue fluids non-specifically by dead or damaged cells. The staining pattern of synthesized α1-antiT is granular often with concentration of staining in the Golgi zone. Alpha-1-antiT which has been absorbed stains the cytoplasm diffusely and often stains the nucleus as well. The amount of α1-antiT varies considerably within the cells of any individual histiocytic tumour and from case to case. Thus cells may be stuffed with positively staining granules (*Fig*. 14.21) or only isolated granules may be present which can best be identified with an oil immersion objective (*Fig*. 14.22). Malignant histiocytes also contain α1-antiCT which appears to mirror α1-antiT in its distribution, although the intensity of staining of the two antigens is not always the same within the same tumour. While α1-antiT is also present in neutrophils, these cells do not contain α1-antiCT and eosinophils contain neither. Malignant histiocytes sometimes stain polytypically for Ig and appear to take up IgG specifically, possibly via their Fc receptors. The quality of the staining, its polytypic nature and the absence of J chain distinguish these cells from malignant B-cells.

3.13. Hodgkin's Disease

Immunohistochemistry has, until now, proved disappointing with respect to Hodgkin's disease. Reed–Sternberg cells often contain polytypic IgG and a proportion stain positively for α1-antiT.[30] The development of cell lines apparently composed of Reed–Sternberg cells[31] should permit the development of monoclonal antibodies to these cells.

SUMMARY

The advent of immunohistochemical techniques has greatly increased our knowledge of the histogenesis of lymphoproliferative diseases and aided in the development of rational classifications. The increasing number of antibodies available and differences in the nature of the tissue (fixed or fresh) which can be investigated mean that careful choices must be made when studying any individual case. In *Table* 14.4 the staining reactions in frozen and paraffin sections of the principal categories of malignant lymphoma are summarized. It cannot be overemphasized that interpretation of the results is as important as the immunohistochemical technique itself. By applying a range of antisera to any given case an appreciation of the histogenesis of most cases of malignant lymphoma is now possible.

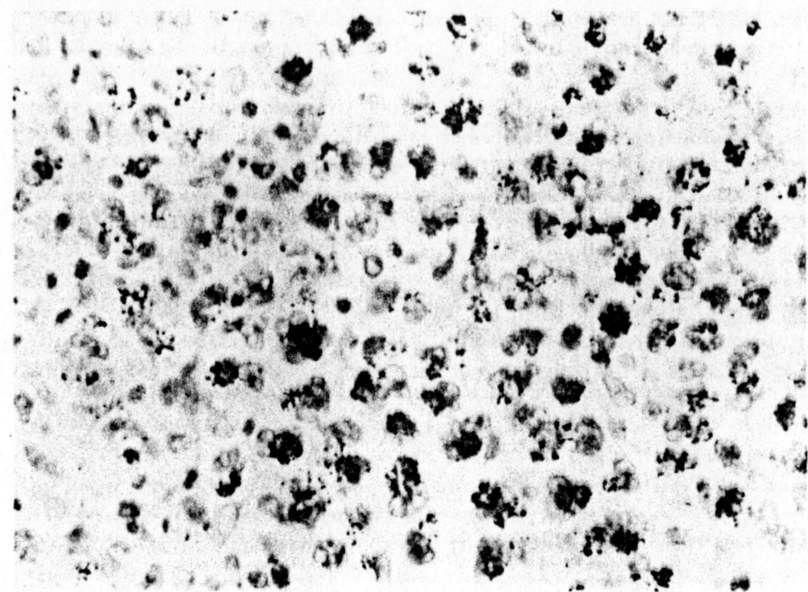

Fig. 14.21. Paraffin section from a case of malignant histiocytosis of the intestine stained for α1-antiT. Abundant positively staining granules are present in the tumour cells.

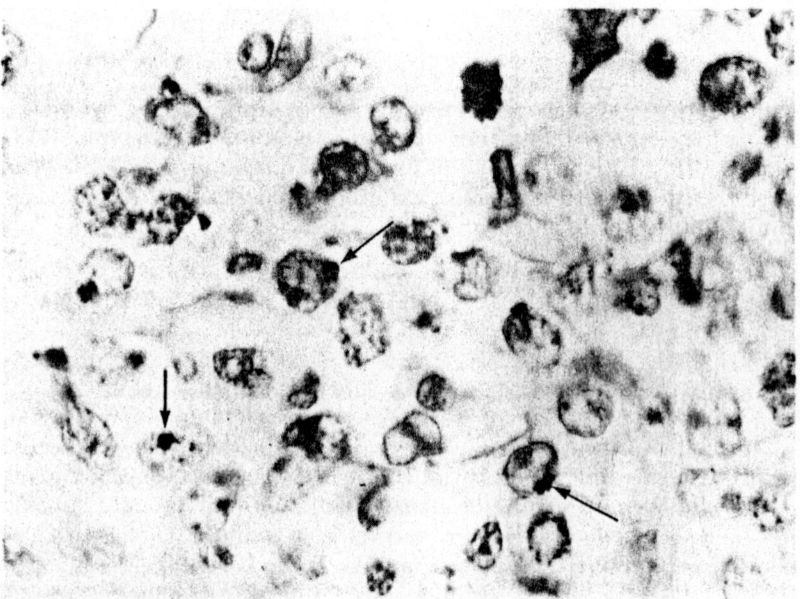

Fig. 14.22. Paraffin section of a histiocytic lymphoma stained for α1-antiT. Positive staining in the tumour cells is confined to a single area close to the nucleus (arrows).

Table 14.4. **Immunohistochemistry of malignant lymphomas**

		Lymphocytic	*Lymphoplasmacytic/ lymphoplasmacytoid*	*Plasma-cytic*	*Follicle centre cell*	*Immuno-blastic*	*T-cell*	*Histiocytic*	*Hodgkin's disease (RS cells)*
Frozen sections	SIg	Monotypic	Monotypic on lymphoid cells	–	Monotypic	Monotypic	–	–	–
	Membrane antigens	+*	+*	–	+* †	+*	+‡	+*	–
Paraffin sections	CIg	–	Monotypic in plasma cells and plasmacytoid cells	Monotypic	Monotypic in up to 2/3 of cases	Monotypic	–	– or polytypic (IgG)	Polytypic (IgG)
	J chain	–	+ in plasma cells and plasmacytoid cells	+	+ in CIg +ve cells	–	–	–	–
	Lysozyme	–	–	–	–	–	–	Occasionally +ve	–
	α1-antiT (α1-antiCT)	–	–	–	–	–	–	+	±

* HLA-DR antigens.
† DRC, dendritic reticulum cell antigen as defined by monoclonal antibodies.
‡ T-cell antigens as defined by monoclonal antibodies.

Acknowledgements

The immunoperoxidase stains on frozen sections illustrated in this chapter were performed by Miss Najat Al-Saffar.

Appendix

I. IMMUNOPEROXIDASE TECHNIQUE FOR DEMONSTRATION OF ANTIGENS IN PARAFFIN SECTIONS

1. Deparaffinize for 10 min each in two changes of xylol and take to alcohol
2. Inhibit endogenous peroxidase by treating with freshly prepared 0·5 per cent H_2O_2 in methanol for 10 min
3. Wash well in tapwater
4. Place in distilled water at 37 °C for 10 min
5. Treat with 0·1 per cent trypsin in 0·1 per cent $CaCl_2$ (adjust to pH 7·8 with 0·1 N NaOH) for 15–30 min at 37 °C*
6. Rinse in cold distilled water 2–3 min with agitation
7. Wash in Tris-buffered saline (TBS) two changes, 10 min each (see Formula I, p. 271)
8. Normal swine serum diluted 1 : 5 with TBS, 10 min. Drain off (use to reduce excessive background staining—an optional step)
9. Rabbit anti-human serum (at current dilution in TBS†) for 30 min
10. TBS wash three times for 10 min each
11. Swine anti-rabbit IgG (at current dilution in TBS†) 30 min
12. TBS wash 3 times for 10 min each
13. PAP (peroxidase–rabbit anti-peroxidase, at current dilution in TBS†) 30 min
14. TBS wash 3 times for 10 min each
15. Diaminobenzidine (DAB) (*see* Formula II, p. 271) 10 min or longer
16. Wash in TBS followed by a wash in running tap water 5 min
17. Counterstain with haematoxylin, wash well 2–3 min. Differentiate in 1 per cent acid alcohol, blue by washing in tap water for at least 10 min
18. Dehydrate, clear and mount

II. IMMUNOPEROXIDASE TECHNIQUE FOR DEMONSTRATION OF MEMBRANE ANTIGENS

1. Cut cryostat sections at 6 μm
2. Air dry 30 min at room temperature (18–20 °C)
3. Freeze dry at −63 °C and 10^{-2} torr for 18 h
4. Wrap slides in aluminium foil and seal
5. Store at −20 °C until required
6. Bring wrapped slides to room temperature
7. Fix in acetone for 20 min at room temperature
8. Transfer to Tris-buffered saline (TBS) without allowing sections to dry
9. Rinse in TBS

* Time may vary with batch of trypsin.
† Determined by titration.

A. Monoclonal Antibodies

10. Apply mouse anti-human serum (at current dilution in TBS*) 30 min
11. Wash in TBS 3 times for 2 min each
12. Apply rabbit anti-mouse Ig peroxide conjugate (at current dilution in TBS†) 30 min
13. Wash in TBS 3 times for 2 min each
14. Apply DAB 10 min or longer
15. Rinse in TBS followed by a wash in running tap water 5 min
16. Counterstain in Harris's haematoxylin 5 min
17. Wash in tap water, differentiate in 1 per cent acid alcohol, blue in running tap water
18. Dehydrate, clear and mount

B. Surface Immunoglobin

10. Apply rabbit anti-human serum (at current dilution in TBS†) 30 min
11. Wash in TBS 3 times for 2 min each
12. Apply goat anti-rabbit IgG (at current dilution in TBS†) 30 min
13. Wash in TBS 3 times for 2 min each
14. Apply rabbit peroxidase–anti-peroxidase (at current dilution in TBS*) 30 min
15. Wash in TBS 3 times for 2 min each
16. Apply DAB 10 min or longer
17. Rinse in TBS followed by a wash in running tap water 5 min
18. Counterstain in Harris's haematoxylin 5 min
19. Wash in tap water, differentiate in 1 per cent acid alcohol, blue in running tap water
20. Dehydrate, clear and mount

Formula I

Tris-HCl-buffered saline, pH 7·6 (TBS)

Sodium chloride	80 g
Tris [tris(hydroxymethyl)-methylamine]	6·05 g
1 N Hydrochloric acid	38 ml
Distilled water to	10 000 ml

Check pH and adjust to 7·6 if necessary.

Formula II

DAB (Graham and Karnovsky)

For use:
carefully weigh out 5 mg 3,3′-diaminobenzidine tetrahydrochloride (possibly carcinogenic) into a glass vial
Add 10 ml Tris-HCl buffer pH 7·6
Immediately before use add 0·1 ml of freshly prepared 1 per cent H_2O_2
(0·1 ml H_2O_2 (30 per cent, 100 volume) in 2·9 ml distilled water)

Buffer 0·2 M Tris (24·228 g/l)	12 ml
0·1 N HCl	19 ml
Distilled water	19 ml
	50 ml pH 7·6

† Determined by titration.

REFERENCES

1. Lennert K. Follicular lymphoma. A tumor of the germinal centres. In: Akazaki K., Rappaport H., Berard C. W., Bennet J. M. and Ishikawa E. (ed.) *Malignant Diseases of the Hematopoietic System, GANN Monograph on Cancer Research*, No. 15. Tokyo, University of Tokyo Press, 1973: 217–231.
2. Lukes R. J. and Collins R. D. New observations in follicular lymphoma. In: Akazaki K. et al. (ed.) *Malignant Diseases of the Hematopoietic System, GANN Monograph on Cancer Research*, No. 15. Tokyo, University of Tokyo Press, 1973: 209–215.
3. Taylor C. R. and Burns J. The demonstration of plasma cells and other immunoglobulin-containing cells in formalin fixed paraffin-embedded tissues using peroxidase-labelled antibody. *J. Clin. Pathol.* 1974, **27**, 14–20.
4. Garvin A. J., Spicer S. S. and McKeever P. E. The cytochemical demonstration of intracellular immunoglobulin. *Am. J. Pathol* 1976, **82**, 457–470.
5. Curran R. C. and Gregory J. The unmasking of antigens in paraffin sections of tissue by trypsin. *Experientia* 1977, **33**, 1400.
6. Mepham B. L., Frater W. and Mitchell B. S. The use of proteolytic enzymes to improve immunoglobulin staining by the PAP technique. *Histochem. J.* 1979, **11**, 345–357.
7. Curran R. C. and Gregory J. Effects of fixation and processing on immunohistochemical demonstration of immunoglobulin in paraffin sections of tonsil and bone marrow. *J. Clin. Pathol.* 1980, **33**, 1047–1057.
8. Judd M. A. and Britten K. J. M. Tissue preparation for the demonstration of surface antigens by the immunoperoxidase technique. *Histochem. J.* 1982 (in press).
9. Stein H., Bonk A., Tolksdorf G., Lennert K., Rodt H. and Gerdes J. Immunohistologic analyses of the organisation of normal lymphoid tissue and non-Hodgkin's lymphomas. *J. Histochem. Cytochem.* 1980, **28**, 746–760.
10. Isaacson P., Wright D. H., Jones D. B., Payne S. V. and Judd M. A. The nature of the immunoglobulin-containing cells in malignant lymphoma. An immunoperoxidase study. *J. Histochem. Cytochem.* 1980, **28**, 761–770.
11. Stein H. and Tolksdorf G. Development and differentiation of the T-cell and B-cell systems: a perspective. In: van den Tweel J. G. (ed.) *Malignant Lymphoproliferative Diseases*. Leiden, Leiden University Press, 1980: 13–29.
12. Isaacson P. Immunochemical demonstration of J chain: a marker of B-cell malignancy. *J. Clin. Pathol* 1979, **32**, 802–807.
13. Brandtzaeg P. Presence of J chain in human immunocytes containing various immunoglobulin classes. *Nature* 1974, **252**, 418–419.
14. Isaacson P., Jones D. B., Millward-Sadler G. H., Judd M. A. and Payne S. Alpha-1-antitrypsin in human macrophages. *J. Clin. Pathol.* 1981, **34**, 982–990.
15. Warnke R. and Levy R. Immunopathology of follicular lymphomas: a model of B-lymphocyte homing. *N. Engl. J. Med.* 1978, **298**, 481–486.
16. Sarasombath S., Mestecky J. and Skvaril F. Monoclonal IgG-κ and IgG-1-λ proteins with different idiotypic determinants present in a single patient. *Clin. Exp. Immunol.* 1977, **29**, 67–74.
17. Taylor C. R., Russell R. and Chandar S. An immunohistologic study of multiple myeloma and related conditions using an immunoperoxidase method. *Am. J. Clin. Pathol.* 1978, **70**, 612–622.
18. Diktor M. Immunoblastic sarcoma after thymus epithelial graft in an immunodeficient child. *Histopathology* 1980, **4**, 661–668.
19. Isaacson P. and Wright D. H. Anomolous staining patterns in immunohistologic studies of malignant lymphoma. *J. Histochem. Cytochem.* 1979, **27**, 1197–1199.
20. Avremeas S. and Leduc E. H. Detection of simultaneous antibody synthesis in plasma cells and specialised lymphocytes in rabbit lymph nodes. *J. Exp. Med.* 1970, **131**, 1137–1168.
21. Judd M. A. One micrometre paraffin sections: an aid to interpretation of immunoperoxidase staining of immunoglobulins. *J. Microsc.* 1980, **120**, 201–206.
22. Giddings J., Griffin R. L. and MacIver A. G. Demonstration of immunoproteins in araldite-embedded tissues. *J. Clin. Pathol.* 1982, **35**, 111–114.
23. Lawrence E. C., Broder S., Jaffe E. S., Braylan R. C., Dobbins W. O., Young R. C. and Waldmann T. A. Evaluation of a lymphoma with helper T-cell characteristics in Sezary's syndrome. *Blood* 1978, **52**, 481–492.
24. Byrne G. E. and Rappaport H. Malignant histiocytes. In: Akazaki K. et al. (ed.) *Malignant Diseases of the Hematopoietic System, GANN Monograph on Cancer Research*, No. 15 Tokyo, University of Tokyo Press, 1973: 145–162.

25. Isaacson P., Wright D. H. and Jones, D. B. Malignant lymphoma of 'true' histiocytic (monocyte/macrophage) origin. *Cancer* 1982, (in press).
26. Isaacson P. and Wright D. H. Malabsorption and intestinal lymphomas. In: Wright R. (ed.) *Recent Advances in Gastrointestinal Pathology*. London, W. B. Saunders, 1980: 193–212.
27. Taylor C. R. Immunoperoxidase techniques: theoretical and practical aspects. *Arch. Pathol. Lab. Med.* 1978, **102**, 113–121.
28. Risdall R. J., Sibley R. K., McKenna R. W., Brunning R. D. and Dehner L. P. Malignant histiocytosis: a light and electron microscopic and histochemical study. *Am. J. Surg. Pathol.* 1980, **4**, 439–450.
29. Mendelsohn G., Eggleston J. C. and Mann R. B. Relationship of lysozyme (muramidase) to histiocytic differentiation in malignant histiocytosis. *Cancer* 1980, **45**, 273–279.
30. Payne S. V., Wright D. H., Jones K. J. M. and Judd M. A. The macrophage origin of Reed–Sternberg cells. An immunohistochemical study. *J. Clin. Pathol.* 1982, **35**, 159–166.
31. Schaadt M., Diehl V., Stein H., Fonatsch C. and Kirchner H. H. Two neoplastic cell lines with unique features derived from Hodgkin's disease. *Int. J. Cancer* 1980, **26**, 723–731.

15 Tumour Markers

E. Heyderman

Tumour markers are tumour-derived or associated products which may be used either to diagnose or to monitor malignant disease. In immunocytochemistry, the term is confined to those substances which are secreted and stored by the tumour or its stroma, and which may be demonstrated in tissue sections, smears, monolayer cultures or cell suspensions by immunocytochemical techniques. Some, like most of the hormones, are present in the cytoplasm. Others, like carcinoembryonic antigen (CEA)[1] and blood group substances, are generally expressed on the cell membrane. Markers, like CEA, when found in moderately well differentiated adenocarcinomas tend to be mainly localized to the luminal surface of the malignant acini. In squamous cell carcinomas the distribution is pericellular[2] while in poorly differentiated tumours it may be cytoplasmic.[3] Markers which are not produced by the tumour itself, like C-reactive protein, but are of value as circulating markers of disease progression, are excluded from this definition.

The criteria for circulating tumour markers have recently been reviewed.[4] As far as tissue markers are concerned, since endocytosis of hormones and other substances into the Golgi[5] or rough endoplasmic reticulum (RER)[6] may occur, demonstration at ultrastructural level in the Golgi apparatus or in the rough endoplasmic reticulum (RER) is insufficient evidence of secretion. In tissue sections, therefore, it is only possible to demonstrate storage. To prove synthesis, other techniques need to be employed. These include incorporation of radioactive precursors into the tumour *in vivo* or *in vitro* or secretion of the marker into the medium by cultured cells.[7] It may also be possible to show continued secretion of a tumour marker in a human tumour xenograft in an immune-deprived mouse[8] or into the circulation of the mouse host.[9]

1. POTENTIAL VALUE OF IMMUNOPEROXIDASE TUMOUR MARKER PROFILES

The value of tumour marker localization falls into five main categories:

a. *Determination of the site of an occult primary tumour* which presents with metastatic disease.[10]

b. *Differential diagnosis* of similar lesions, as in deciding whether a single cell malignant infiltrate represents lymphomatous infiltration or secondary deposits of lobular carcinoma of the breast[10] or confirming that sarcoma-like renal lesions represent sarcomatoid change in hypernephroma.[11]

c. *Functional, as well as morphological, classification* of malignant disease. This is of value, for example, in the classification of testicular tumours where it may be difficult on morphological grounds to determine whether giant cells are of trophoblastic lineage, or are of undetermined tumour giant cell type or are foreign-body type giant cells in a granulomatous response to necrosis, and where the demonstration of intracellular human chorionic gonadotropin (HCG)[12] taken together with the morphology is conclusive.

d. *Prognosis.* While there have been conflicting reports on the prognostic significance of the presence of a variety of markers in breast carcinomas,[13–16] Bulman's study of 44 breast carcinomas has failed to confirm any prognostic significance.[17] Nor have larger studies on the presence of HCG in testicular tumours[18–20] confirmed the suggestion that the presence of HCG may impart a worse prognosis.[21] The best candidates for markers correlating with prognosis seem to be blood group substances at least in the bladder,[22] although a study of blood group substance A in breast cancer failed to show any adverse prognostic significance in its loss.[23]

e. *Marker prediction.* There is general agreement that only a proportion of breast carcinomas contain CEA.[24] If this indicates failure of some breast cancers to secrete CEA, routine measurement of circulating levels in all mammary cancer patients would result in failure to detect recurrence in those patients who have widespread metastatic disease, but whose tumours do not secrete CEA. Examination of a biopsy or the mastectomy specimen could help to define this subset.

There is evidence from tissue culture studies[19] that some tumour cells secrete CEA but have a failure of their transport mechanism and are unable to excrete it. The localization of CEA in all sections of colorectal carcinomas so far studied,[25] while only a proportion have raised levels on metastasis, may be partly explained on this basis. The same seems to be true for gastric cancer.[17]

In the case of those patients with malignant testicular teratomas who do not have raised postorchidectomy levels of human chorionic gonadotropin (HCG) or alpha-fetoprotein (AFP), demonstration of HCG or AFP in the original orchidectomy specimen would support a clinical view that there was no residual disease, since the primary tumour was capable of secretion. Where neither can be demonstrated, negative levels are less reliable and patients with such tumours would be at higher risk, since biochemical determination of elevated levels is often possible long before recurrence is clinically detectable. In marker-silent patients the clinician would have to depend on physical examination, X-rays, CAT scan or ultrasound. At present all of these are poor determinants of recurrence compared with biochemical assessment.

The demonstration of tissue markers by immunocytochemical technique in the biopsy or resection specimen would also help to decide on the most suitable antibody to use for the newer techniques of tumour localization using radiolabelled antibodies.[26–30]

Similarly, an indication of the most suitable marker would be of use in tumour-targeting using antibodies labelled with a variety of toxic agents,[31] such as chlorambucil,[32] diphtheria toxin[33] or monoclonal hybridoma antibodies.[34]

2. IMMUNOCYTOCHEMICAL METHODS

Localization of tumour products may be achieved by the use of antibodies labelled in a variety of ways. Fluorescein and rhodamine are used in the immunofluorescence technique.[35] The most widely used enzyme labels are horseradish peroxidase[36] and alkaline phosphatase.[37] The latter may be used either alone or together with a peroxidase-labelled antibody to demonstrate two antigens in the same tissue section.[38,39].

The immunoperoxidase technique has the advantage of an insoluble permanent reaction product so that by using suitable counterstains, good morphological detail is retained. With immunofluorescence, although the method is simpler, requiring no inhibition of endogenous enzyme or incubation in substrate, it is difficult to assess the surrounding morphology and with time and photographic exposure, the fluorescence fades. If the antigen under study survives fixation and processing, the immunoperoxidase technique, like other immunocytochemical techniques, may be used on formalin-fixed paraffin-embedded tissue many years old.

Excellent results using antibodies labelled with colloidal gold have been reported.[40] The colloidal gold technique is suitable for both light and electron microscopy and there are no problems with endogenous activity. It has not yet proved possible to label with gold particles of sufficient size to achieve as dense a reaction product as with peroxidase. This is no problem where the staining is intense and one is dealing with a well-established marker. Where the antigen under test is in small amount, particularly on the cell surface, and the distribution is unexpected, interpretation is more difficult.

Antibodies labelled with radioactive isotopes may be applied to tissue sections and the preparations subjected to autoradiography.[41] There are several problems with radiolabelled antibodies. Special safety facilities are necessary, non-specific scatter occurs and the incubation time for autoradiography may be unacceptably long. However, sensitivity is increased and the method may be combined with immunocytochemistry to demonstrate two products[42] in the same section. Uptake of tritium-labelled thymidine and HCG localization have been shown in a section of a skin deposit of metastatic testicular choriocarcinoma (*Plate* 17).[8]

Immunoperoxidase methods may employ directly labelled antibodies, double[43] or triple layers,[44] the peroxidase–anti-peroxidase (PAP)[45] or the labelled antigen method[46] (*Fig.* 15.1). Since it is the primary, specific antibody which is the most valuable and some activity is inevitably lost in conjugation, we prefer the indirect[47] to the direct method in which the first antibody is labelled with peroxidase. Repeated experiments comparing commercially available PAP reagents and indirect peroxidase conjugates prepared in our laboratory fail to confirm an

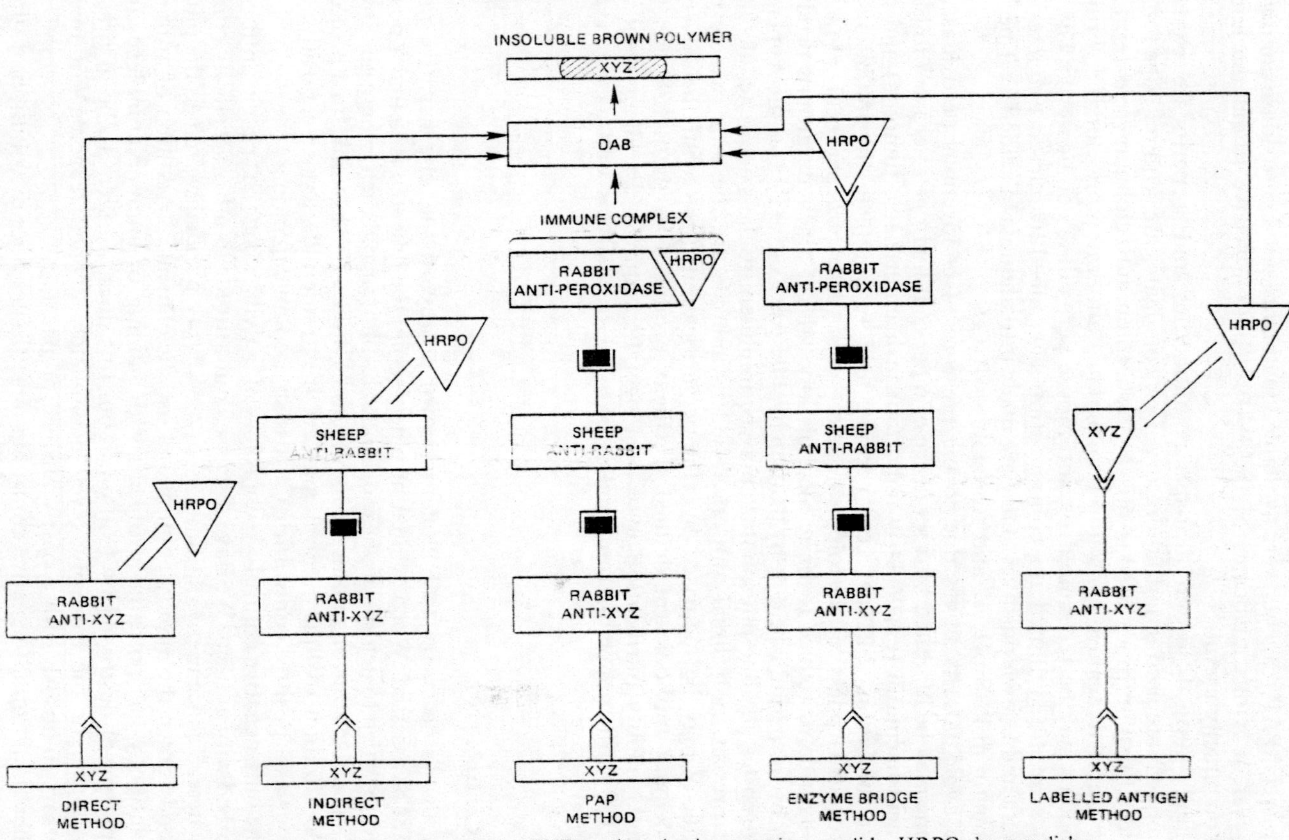

Fig. 15.1. Immunoperoxidase methods. XYZ, antigen in tissue section on slide; HRPO, horseradish peroxidase conjugate; DAB, diaminobenzidine. (Reprinted from E. Heyderman, *J. Clin. Pathol.* 1979, **32**, 971–978, with kind permission of the editor.)

increased sensitivity for the PAP method, and the indirect technique is shorter and less expensive. The PAP method is the most widely used, probably because good commercial PAP reagents were available long before good peroxidase conjugates. However, previous advocates of the PAP technique[48] have found the indirect method equally useful.[49]

Affinity-purified second anti-species antibodies and a periodate oxidation technique[50] are used in our laboratory. Few conjugates are required, one for each species of first antibody used. Antisera to sheep immunoglobulin cross-react with those directed against goat immunoglobulin, and rat monoclonal hybridoma antibodies may be recognized by anti-mouse peroxidase conjugates.[17] The conjugates are stored diluted at a concentration determined by reference to a series of antisera and known positives. They are made into aliquots in azide-free 1 per cent ovalbumin in PBS pH 7·2 and stored at −20 °C.

To avoid false positives due to the presence of endogenous peroxidase, the enzyme may be inhibited by a sequence of 6 per cent hydrogen peroxide to bleach any acid haematin (particularly important in haemorrhagic tumours) and commence inhibition, 2·25 per cent periodic acid to complete the inhibition and improve sharpness of histological detail and then 0·02 per cent potassium borohydride to reduce to basic alcohols any 'sticky' aldehyde groups produced either by the periodic acid or by fixation in aldehyde fixatives[47] (*see* the Appendix). Endogenous alkaline phosphatase may be inhibited by the same schedule.[51]

At present, polyclonal antisera raised in animals are the most widely used. Specificity may be increased by affinity purification[52] but with the advent of monoclonal antibodies of defined specificity produced by the hybridoma technique,[53] affinity purification may no longer be required. Excellent results are achieved with monoclonal antibodies on fixed tissues[54] as well as on cryostat material.

2.1. Controls

It is always necessary to include a known positive tissue, else negative results cannot be assessed. When evaluating a new serum the first step is to find a positive tissue on which to determine the most suitable dilution and then, where possible, to find a positive control among the tissues under study. For example, in a project on the localization of lung tumour markers it is preferable to use a positive lung tumour as a positive control for CEA than a colon carcinoma.

The only negative control we use is loss of activity when the specific antibody is absorbed with antigen. We have discussed our attitude to other specificity controls elsewhere.[39] Since many of our antisera are affinity-purified we repass the purified antibody down the affinity column and use the effluent.[55] Addition of antigen to the antibody has potential problems with binding of immune complexes to Fc receptors, not destroyed by fixation,[56] but this does not seem to be of practical importance. In the case of as yet incompletely defined reagents[57] it may be necessary to absorb with a partly purified preparation.[57]

There is an inbuilt negative control in most tissues since distribution of tumour markers is not uniform and if one is using several markers, they may well act as negative controls for each other. If the result is negative, the absorption control confirms the absence of endogenous peroxidase, of anti-human activity in the second anti-species antibody and of 'noise' in the system. Except in special circumstances, other controls such as using substrate alone, leaving out the first antibody or using so-called 'non-immune serum' are superfluous.[47]

3. TUMOUR MARKERS

Tumour markers have conventionally been divided into those which are appropriate to the tissue of origin (eutopic) such as human chorionic gonadotropin (HCG) in gestational choriocarcinomas and gut hormones in islet cell tumours of the pancreas, and ectopic for those products such as adrenocorticotropic hormone (ACTH) secreted by oat cell carcinomas of lung and medullary carcinomas of the thyroid, which are not thought to be a normal product of the tissue of origin. The concept of eutopic and ectopic tumour products is a useful one. However, it has become clear that our ideas of what is an appropriate or inappropriate product need to be revised. There is evidence of extraction of human placental lactogen (HPL) from the normal testis,[58] and of HCG by non-neoplastic extragonadal tissues.[59] The brain and gut are now considered a unified neuroendocrine system with many 'gut hormones' secreted by both.[60]

Substances secreted by tumours include hormones and other placental proteins, immunoglobulins, oncofetal antigens, membrane antigens, blood group substances and enzymes.

3.1. Hormones

3.1.1. *Human Chorionic Gonadotropin*

Human chorionic gonadotropin is a glycoprotein composed of two dissimilar non-covalently linked units—alpha and beta. The alpha unit is indistinguishable immunologically from the alpha subunits of the pituitary glycoprotein hormones, luteinizing hormone (LH), follicle-stimulating hormone (FSH) and thyroid stimulating hormone (TSH).[61] The beta subunits are different from each other and confer specificity, although neither the alpha nor the beta subunits are biologically active on their own.

HCG is secreted by the normal placenta and is measurable in the urine early in pregnancy. Its main function is the maintenance of the corpus luteum until the placenta takes over steroid synthesis. The circulating levels reach a maximum at around 8 weeks, declining at 16 weeks to a low level which is maintained until the end of pregnancy.

Most, if not all, gestational, ovarian and testicular choriocarcinomas secrete HCG. It is an excellent circulating tumour marker being secreted in a stoichometric relationship to the tumour mass. Together with adequate chemotherapy, the existence of a sensitive radioimmunoassay which detects small volume disease long before clinical detection would be possible has changed an appalling tumour into one of the most amenable to treatment.[62] It is often easier to demonstrate small foci of choriocarcinoma in testicular or ovarian tumours, or even to make the diagnosis in atypical metastatic skin deposits[10] with an immunostain for HCG, than with conventional haematoxylin and eosin preparations.[21]

One of the contributions made to tumour pathology by immunoperoxidase methods is in the demonstration that giant cells in malignant testicular tumours which have the morphology of syncytiotrophoblast cells contain HCG.[12,63] Previously this was disputed[64] and gynaecomastia or raised levels of placental proteins in testicular tumours not considered trophoblastic by conventional criteria were thought to be due to the occult presence of choriocarcinoma missed by sampling error.[65] Over 50 per cent of those testicular teratomas which would be classified as non-trophoblastic by conventional criteria and up to 17 per cent of seminomas have been found to contain HCG-positive cells.[66] Similar HCG-positive cells may be found in ovarian dysgerminomas and mixed germ-cell tumours.

Braunstein and his colleagues[67] have reported elevated levels of HCG in many kinds of extragonadal tumours. Muggia and co-workers[68] have collected a number of such tumours in which placental hormones may be demonstrated.[17] Preliminary results indicate a good correlation between elevated circulating levels and the presence of giant cells containing HCG in these tumours.

Studies on testicular tumours have employed antisera to the specific beta subunit of HCG, so excluding the demonstration of LH, FSH or TSH. Some tumours including islet cell tumours of the pancreas secrete the isolated alpha glycoprotein subunit.[69] In an examination of 44 breast carcinomas with antisera to HCG alpha previous reports of the presence of this substance in some of them[13,14] were confirmed, but it was not possible to demonstrate any adverse prognostic significance.[17]

3.1.2. *Human Placental Lactogen*

Human placental lactogen (HPL) is a polypeptide described by Josimovitch and Maclaren[70] and called by them human chorionic somatomammotropin (HCS), reflecting its growth hormone and prolactin-like properties. Original measurement was by pigeon-crop bioassay; now radioimmunoassay is available. HPL is the normal product of the syncytiotrophoblast and levels rise progressively during pregnancy. Monitoring circulating maternal levels during pregnancy is of value as an indication of fetal distress.[71]

There are reports of elevated levels in patients with a variety of tumours[72] but HPL has not proved of great value as a circulating tumour marker, even in the case of choriocarcinomas of testicular or gestational origin, although in a study of tissue sections of the orchidectomy specimen from 67 Stage I non-trophoblastic testicular teratomas, HPL was demonstrated in 80 per cent of the tumours which contained HCG,[21] as well as in gestational choriocarcinomas. In an ongoing study on the localization of placental proteins in extragonadal tumours associated with elevated circulating levels, HPL has been demonstrated in some.[17]

3.1.3. *Pregnancy-Specific Glycoprotein*

In a study on the wide variety of proteins which may be extracted from the human placenta, Bohn[73] described a polypeptide pregnancy-specific glycoprotein (SP1) which was shown to be a very early marker of pregnancy.[74] Horne and his colleagues[13] demonstrated the presence of SP1 in sections of breast cancers and claimed that its demonstration correlated with a worse prognosis. Others[17] have not been able to confirm this. SP1 may also be demonstrated in other tumours including colorectal carcinomas.[75]

3.1.4. *Placental Alkaline Phosphatase*

Placental alkaline phosphatase (Regan isoenzyme) is raised in some patients with trophoblastic and extragonadal disease and has been demonstrated in sections of a variety of tumours including lung and colon carcinoma.[76]

3.1.5. *Other Placental Proteins*

Bohn and his colleagues[77] have described a number of other placental proteins, including PP5, PP10, PP11 and PP12, which may be demonstrated in tissue sections of a variety of tumours.

3.1.6. *Calcitonin*

Calcitonin is the normal product of the C-cells of the thyroid and is secreted by thyroid medullary carcinomas. It is a polypeptide whose actions in the human are poorly understood.

The tissue localization of calcitonin is of value in the diagnosis of medullary carcinoma of the thyroid[78] and of C-cell hyperplasia.[79] In spite of reports of extraction from breast cancers[80] and elevation of circulating levels of calcitonin in patients with widespread breast cancer,[81] we have not found it possible to demonstrate calcitonin in 44 consecutively removed breast carcinomas.[17]

Calcitonin is secreted also by oat cell carcinomas[82] and may be demonstrated in some pancreatic tumours.[17]

3.1.7. *Growth Hormone*

Human growth hormone (HGH) is secreted by the eosinophils of the anterior pituitary. Levels are elevated in patients with eosinophil and some chromophobe adenomas, and the demonstration of HGH is of interest in confirming the presence of a tumour in hypophysectomy specimens.[83]

Elevation of growth hormone levels has been reported in association with a variety of lung tumours, mainly of adenocarcinomatous or of large cell type and has been extracted from 7 out of 18 lung cancers by Beck and Burger.[84] It has been suggested that growth hormone is involved in hypertrophic osteoarthropathy[85] though the weight of evidence is that hypertrophic osteoarthropathy is probably not the result of elevated HGH levels.[86] Sonksen and his colleagues[87] reported one case and referred to another of a bronchial carcinoid associated with acromegaly. In their case no growth hormone could be extracted from the tumour and it was suggested that the tumour might have been secreting HGH-releasing factor. No reports on immunocytochemical localization have been published.

3.1.8. *Adrenocorticotropic Hormone*

Since the report of Cushing's syndrome in association with the ectopic production of ACTH by a bronchial neoplasm[88] there have been numerous confirmatory reports of ectopic ACTH production by lung tumours.[89,90] Its production has mainly been associated with oat cell carcinomas, but it has also been found with other cell types.[91] Gewirtz and Yalow[92] found ACTH in 14 out of 15 assorted lung tumours, but Ratcliffe and his colleagues[93] have not confirmed this high prevalence.

ACTH may be localized by immunocytochemical techniques in bronchial carcinoids associated with severe Cushing's disease,[17] but not in oat cell carcinomas associated with raised circulating levels[94] although it has been demonstrated in other ectopic sites.[95] A possible explanation for the inability to localize ACTH in oat cell carcinomas is that though neurosecretory granules may readily be demonstrated in carcinoids of the lung they are much more sparse in oat cell carcinomas,[96] perhaps indicating a much lower level of storage. Immunocytochemical demonstration of ACTH in various skin carcinomas has been reported.[97]

3.1.9. *Antidiuretic Hormone*

Though elevated circulating levels of antidiuretic hormone (ADH) may be found in some cancers, especially oat cell carcinomas of the lung[98] and in cell lines derived from oat cell carcinomas,[99] localization studies in tissue sections of tumours have yet to be published.

3.1.10. *Thyroglobulin*

It is not uncommon for papillary or follicular thyroid carcinomas to present with metastatic disease. Thyroglobulin may be demonstrated in all but the most anaplastic[100,101] and is a useful adjunct to diagnosis.

3.1.11. *Steroid Hormones and Their Receptors*

Localization of steroid hormones in ovarian tumours has been reported,[102] and several workers have demonstrated oestrogen receptors or an oestrogen receptor-related material in tissue sections by immunofluorescence[103] or immunoperoxidase techniques,[104,105] though others have had little success in repeating the work.[106] While there is a basic biological interest in the determination of steroid receptors in breast and other oestrogen-dependent tumours, some receptor-poor tumours respond to endocrine manipulation and some receptor-rich tumours do not.[107] Anti-oestrogens and anti-adrenal agents have few undesirable side effects, so that in some centres a trial of therapy has come to replace receptor assay.

3.1.12. *Gut Hormones*

Localization of these is important in the precise classification of gut-related tumours and is discussed elsewhere in this book (*see* Chapters 11 and 20).

3.2. Oncofetal Antigens

These are substances expressed by the fetus, repressed during normal adult life and derepressed in malignancy. As with our concepts of what constitutes a truly ectopic product so our ideas on oncofetal antigens need to change, since it is apparent that CEA is expressed by many normal epithelia and AFP by the regenerating non-neoplastic liver and possibly other tissues.

3.2.1. *Carcinoembryonic Antigen*

Thought by the original workers, Gold and Freedman[1] to be a specific marker for colorectal cancer, elevated levels of carcinoembryonic antigen have been found in association with a wide variety of neoplastic and non-neoplastic lesions.[108,109] It has been localized in carcinomas of the large bowel (*Plate* 18),[3] in the small bowel,[110] and in tumours of the breast,[15,16,111] testis[21] and thyroid.[112] While of little value as a circulating marker in the initial diagnosis of cancer, there are many reports of its use as a marker of disease progression.[113]

CEA is an incompletely defined glycoprotein. One of the problems which has beset its localization is its cross-reaction with a similar but smaller glycoprotein, *n*ormal *c*ross-reacting *a*ntigen (NCA, CEX).[114,115] Efforts to absorb out the anti-NCA activity with spleen extract have had varied success.[17] NCA is present in cells of the monocytic and myeloid series,[116] so that its localization in polymorphs and macrophages using antiserum to CEA unabsorbed with NCA may give rise to difficulties in interpretation.

3.2.2. *Alpha-fetoprotein*

Alpha-fetoprotein (AFP) is a fetal equivalent of albumin with levels rising progressively during fetal development. Radioimmunoassay of alpha-fetoprotein is of value in the antenatal diagnosis of neural tube defects[117] and AFP is an excellent circulating marker of malignant testicular disease.[118] Levels are raised in two-thirds of patients with metastatic disease,[119,120] and it has been localized by immunofluorescence in tumours with yolk sac morphology.[121,122] Some tumours with no evidence of yolk sac differentiation secrete AFP and though it has been claimed that AFP may be demonstrated in undifferentiated and immature somatic elements,[63,123,124] many workers experience difficulty with localization in routinely fixed and processed material, so that there is still no unanimity on what constitutes yolk sac differentiation and which types of tissue are responsible for AFP production. Fixation in cold ethanol–acetic acid results in poor histological preservation. Good results have been achieved with Bouin fixation of a 'seminoma' grown as a xenograft in immune-deprived mice.[9]

AFP is much easier to demonstrate in hepatomas[125,126] and could be of value in discrimination from metastatic adenocarcinoma. Unfortunately some gastrointestinal tumours are associated with AFP secretion and some hepatomas secrete CEA.[127]

3.3. Immunoglobulins

The localization of immunoglobulins has been of value in the differential diagnosis of reactive and malignant lymphoproliferative conditions and is discussed elsewhere in this volume (*see* Chapter 14). The newer monoclonal antibodies to lymphocyte population subsets are of increasing interest for the precise classification of lymphomas. Unfortunately many of the most interesting antigens appear to be destroyed by fixation, so that cryostat material is necessary and retrospective studies are difficult.

3.4. Epithelial Membrane Antigen

Epithelial membrane antigen (EMA) is a large glycoprotein antigen or group of antigens present on most secretory epithelia and secreted by many carcinomas (*Plate* 19).[57] Its original demonstration was by antisera raised against a hepatic metastasis of a breast carcinoma, but antisera to milk fat globule membrane[128] appear to recognize the same determinant. Although it is present in so many sites, its demonstration is of value in diagnosis, since it has been found in all breast carcinomas so far studied.[57,129] Its absence in either a breast tumour or metastatic deposit where the differential diagnosis is an anaplastic carcinoma or a lymphoma excludes the diagnosis of mammary cancer. Similarly, its presence excludes the diagnosis of lymphoma since it has not been demonstrated in non-Hodgkin's lymphomas.[57] It has been used to demonstrate small foci of metastatic tumour in the bone marrow and lymph nodes since EMA is not normally present in these sites.

3.5. Other Milk Proteins

Casein or casein-like material has been demonstrated in a variety of mammary and other carcinomas.[130,131] It is possible that the immunogen contained EMA-like material since the distribution of staining with the antisera is very similar. We have demonstrated alpha-lactalbumin in the lactating breast[132] and in some breast cancers but there have been problems with negative controls.[17]

3.6. Fibronectin (LETS: Cold Insoluble Globulin)

When fibronectin[133] was first demonstrated in tissue cultures there was speculation that its demonstration would be of value in the assessment of malignant transformation. This has not proved to be the case.[134] It has been shown in a mouse xenograft of a human testicular teratoma and can readily be demonstrated in tissue sections of the tumour.[17,135] It has not, unfortunately, turned out to be any more reliable than AFP as a marker of yolk sac differentiation. Fibronectin has been demonstrated in the stroma of tumours and in the cytoplasm of soft-tissue neoplasms.[136]

3.7. Alpha-1-antitrypsin

Alpha-1-antitrypsin is stored in the livers of homozygotes suffering from alpha-1-antitrypsin deficiency and may be readily demonstrated by immunocytochemical techniques.[137] It may be demonstrated in hepatomas and testicular tumours[138] but while it has been claimed that this substance is a useful marker of yolk sac differentiation in testicular tumours,[123] it is found normally in macrophages, of which there are many in testicular tumours, and interpretation of the results may be difficult.

3.8. Other Yolk Sac Products

AFP and alpha-1-antitrypsin are only two of the many substances secreted by the yolk sac.[138] Haemoglobin A and F have been demonstrated in yolk sac tumours[139] as have a variety of other products including ferritin.[140] None have as yet proved to be clinically useful markers of yolk sac differentiation. Reports of the identification of atypical, possibly premalignant, intratubular germ cells by their ferritin content[141] await confirmation.

3.9. Prostatic Acid Phosphatase

The prostatic isoenzyme of acid phosphatase is an excellent marker of prostatic cancer[142] being elevated in many patients with metastatic disease. In 1978, Jöbsis and his colleagues[143] reported on the value of its localization in tumours as a marker of prostatic origin. It was not present in non-prostatic tumours but was demonstrated with ease in benign and malignant lesions of the prostate of varying degrees of differentiation. It is, however, present in some cells in pancreatic islets,[17] though origin from the latter can usually be excluded clinically when investigating a prostatic acid phosphatase-positive metastasis of unknown origin. Antisera to acid phosphatase may be used to distinguish rectal carcinomas infiltrating the prostate, bladder tumours infiltrating the prostate and prostatic carcinoma invading the rectum, since of these only prostatic carcinomas secrete prostatic acid phosphatase. Some prostatic tumours secrete CEA but all rectal carcinomas do so.[3] The absence of CEA excludes the diagnosis of colorectal carcinoma.

3.10. Prostate Antigen

Wang and colleagues[144] have described a new circulating and tissue marker of prostatic carcinoma extracted from benign and malignant prostatic tissue and from seminal plasma. The antiserum they have produced is of high titre and early results are very promising.[17]

3.11. Blood Group Substances

Since the demonstration of blood group substances A and B on normal human epithelia,[145] there have been many reports on the deletion of ABO antigens in neoplasia using their mixed red cell agglutination technique.[146] These include studies on carcinomas of the gastrointestinal tract,[147] cervix,[148,149] prostate,[150] breast,[151] and bladder.[22] In most hands the mixed red cell agglutination technique is rather fickle and recent reports have employed either ordinary human blood typing antisera[152] or affinity-purified rabbit antisera raised against the synthetic oligosaccharide determinants.[153] The association between loss of blood group antigens and invasiveness seems well established in the case of bladder cancer.[22,152] However, using rabbit antisera and an immunoperoxidase technique, we have failed to show any correlation between differentiation and prognosis in the breast[23] or cervix.[17] One of the problems we have found is that normal tissues show a patchy distribution of staining so that any assessment of deletion as a marker of malignant or premalignant change is difficult.

3.12. Ovarian Antigens

A variety of antigens have been extracted from human ovarian tumours.[154] Some like CEA[155] are found in many other tumours. Others like CX1[17,156] have more limited immunocytochemical specificity.

3.13. Melanoma Antigens

Melanomas may metastasize as pigmented or unpigmented (amelanotic) tumours. When they are amelanotic diagnosis may be very difficult, especially if there is either no history of a previous malignant melanoma or a dubious one. A specific marker would be of great value. Several have been described[157-159] but none are as yet of proven value in tissue diagnosis.

3.14. Glial Cell Markers

A variety of cell markers have been demonstrated in tumours of the central nervous system including glial fibrillary acidic protein (GFAP)[160,161] and carbonic anhydrase isoenzyme II (CA III)[162] and have proved of value in the classification of central nervous system tumours. Since neither CEA nor EMA have been demonstrated in intrinsic neural tumours their presence would suggest the existence of metastatic disease.

3.15. Leukaemia Antigens

The demonstration of differentiation antigens is part of the routine diagnosis of leukaemia, and cytofluorimetric analysis of peripheral blood or bone marrow is important for assessment and prognosis.[163]

3.16. 'Ubiquitins'

As yet there is no candidate for absolute tissue specificity. As mentioned above, the prostatic isoenzyme of acid phosphatase, which comes close, is found in some islet cells. The question is then whether tissue markers like EMA or CEA which are so widely distributed can be of value in tumour pathology.

There are several examples in the text. Staining with EMA can be used to demonstrate micrometastases in the bone marrow and lymph nodes.[57,129] A positive stain for CEA in one of those difficult pleural biopsies where the diagnosis lies between metastatic carcinoma and mesothelioma appears to preclude the diagnosis of the latter.[164] Skin infiltrates which are positive for EMA do not represent lymphomatous lesions.[10]

It seems clear that a battery of markers taken together with the morphology may help to elucidate the areas described: determination of the site of an occult primary tumour; differential diagnosis; functional classification of malignant disease; prognosis and marker prediction.

The advent of monoclonal hybridoma antibodies with their defined specificity holds great promise. However, they may be directed against commonly found determinants, and since they do not represent a mixed population of antibodies, like polyclonal antisera raised in animals, unwanted specificities cannot be diluted out. They need to be selected for immunocytochemistry and where there is an interesting cross-reaction as between neural and granulocytic cells making them unsuitable for immunocytochemical distinction between metastatic neuroblastoma and normal marrow cells,[165] alternative clones have to be sought. Once established, milligram quantities of pure immunoglobulin of defined specificity can be produced, particularly when the hybridomas are grown in mice as ascites tumours. Workers in widely separated centres will be able to obtain reasonable quantities of the same reagent and compare results in a more meaningful way. At present, some inconsistencies are easily explained by the fact that antisera are obtained from different sources. Even in single centres, antibody titre and specificity of antisera raised in animals vary from animal to animal of the same species and from bleed to bleed.

There is clearly a growing role for immunocytochemistry in the pathology of malignant disease, when taken together with conventional macroscopical and microscopical examination. Enzyme-labelled techniques are the most widely used, yielding a preparation which is also suitable for tissue recognition and diagnosis. Newer techniques, such as colloidal gold labelling, are of increasing interest and one looks forward to further developments in what has become an essential part of routine pathology.

Appendix

INDIRECT IMMUNOPEROXIDASE TECHNIQUE FOR PARAFFIN OR RESIN-EMBEDDED SECTIONS

1. Dewax through xylene and alcohols to water
For epoxy resin-embedded sections remove resin with saturated alcoholic sodium hydroxide (allowed to mature until dark brown—about 5 days). Wash off with absolute ethanol. Wash off with tap water — 5 min

2. Bleach acid haematin with 6 per cent hydrogen peroxide in distilled water. Wash off with tap water — 5 min

3. Inhibit endogenous peroxidase with 2·5 per cent periodic acid in distilled water. Wash off with tap water — 5 min

4. Block aldehyde groups with *fresh* 0·02 per cent potassium borohydride in distilled water. Wash off with tap water. Wash off with PBS-azide pH 7·2 containing 0·02 per cent sodium azide and 1 µl/ml of detergent (1 per cent BRIJ 96, Sigma). Blot dry	2 min
5. Apply 40 µl first antibody diluted in 1 per cent ovalbumin in PBS–azide, and cover with a 40 × 22 mm coverslip. Incubate in moist chamber to prevent coverslip adhering too strongly. Wash off with PBS–azide	1 h
6. Agitate in bath of PBS–azide containing 1µl/ml of detergent. Blot dry	15 min
7. Apply 40 µl of peroxidase conjugate diluted in 1 per cent ovalbumin in *azide-free* PBS. Cover with 40 × 22 mm coverslip and incubate in moist chamber. Wash off with PBS–azide	1 h
8. Agitate in PBS–azide bath containing detergent	15 min
9. Incubate in *fresh* diaminobenzidine (DAB) (100 mg in 200 ml 0·03 per cent hydrogen peroxide in PBS–azide). Wash in tap water	5 min
10. Counterstain in Mayer's haemalum; blue in lithium carbonate; dehydrate, clear and mount in Ralmount or other resinous mountant.	

REFERENCES

1. Gold P. and Freedman S. O. Demonstration of tumor-specific antigens in human colonic carcinomata by immunological tolerance and absorbtion techniques. *J. Exp. Med.* 1965, **121**, 439–462.
2. Goldenberg D. M., Sharkey R. M. and Primus F. J. Carcinoembryonic antigen in histopathology: immunoperoxidase staining of conventional tissue sections. *J. Natl Cancer Inst.* 1976, **57** 11–22.
3. Heyderman E. and Neville A. M. A shorter immunoperoxidase technique for the demonstration of carcinoembryonic antigen and other cell products. *J. Clin. Pathol.* 1977, **30**, 138–140.
4. Tate H. Assessing tumour markers. *Br. J. Cancer* 1981, **44**, 643–651.
5. King A. C. and Cuatrecasas P. Peptide hormone-induced receptor mobility, aggregation and internalization. *N. Engl. J. Med.* 1981, **305**, 77–88.
6. Mitra S. and Rao Ch. V. Gonadotropin and prostaglandin binding sites in rough endoplasmic reticulum and Golgi fractions of bovine corpora lutea. *Arch. Biochem. Biophys.* 1978, **191**, 331–340.
7. Ellison M. L., Lamb D., Rivett J. and Neville A. M. Quantitative aspects of carcinoembryonic antigen output by human lung carcinoma cell line. *J. Natl Cancer Inst.* 1977, **59**, 309–312.
8. Selby P. J., Heyderman, E., Gibbs J. and Peckham M. J. A human testicular teratoma serially transplanted in immune-deprived mice. *Br. J. Cancer* 1979, **39**, 578–583.
9. Raghavan D., Heyderman E., Monaghan P., Gibbs J., Ruoslahti E., Peckham M. J. and Neville A. M. Hypothesis: When is a seminoma not a seminoma? *J. Clin. Pathol.* 1981, **34**, 123–128.
10. McSween A. and Heyderman E. Immunocytochemical investigation of skin metastases. In: Filipe M. I. and Lake B. (ed.) *Histochemistry in Pathology*. Edinburgh, Churchill Livingstone, 1982, (In press).
11. Heyderman E. Immunocytochemistry in cancer diagnosis. In: Symington T., Williams A. E. and McVie J. G. (ed.) *The Tenth Pfizer International Symposium. Cancer Assessment and Monitoring*. Edinburgh, Churchill Livingstone, 1980: 147–171.
12. Heyderman E. and Neville A. M. Syncytiotrophoblasts in malignant testicular tumours. *Lancet* 1976, **ii**, 103.
13. Horne C. H. W., Reid I. N. and Milne G. D. Prognostic significance of inappropriate production of pregnancy proteins by breast cancers. *Lancet* 1976, **ii**, 279–282.
14. Walker R. A. Significance of α subunit HCG demonstrated in breast carcinomas by the immunoperoxidase technique. *J. Clin. Pathol.* 1978, **31** 245–249.
15. Walker R. A. Demonstration of carcinoembryonic antigen in human breast cancer by the immunoperoxidase technique. *J. Clin. Pathol.* 1980, **33**, 356–360.

16. Shousha S., Lyssiotis T., Godfrey V. M. and Scheuer P. J. Carcinoembryonic antigen in breast cancer tissue: a useful prognostic indicator. *Br. Med. J.* 1979, **1**, 777–779.

17. Bulman A. S. and Heyderman E; Heyderman E., Chapman D. V. and Richardson T. C. Unpublished data. 1982.

18. Parkinson C., Masters J. R. W., Krishnaswamy A. and Butcher D. N. HCG localisation of no prognostic value in testicular teratoma. *Lancet* 1981, **i**, 1059.

19. Heyderman E., Gibbons A. R. and Bulman A. S. Immunocytochemical identification of cells. In: Rosalki S. B. (ed.) *New Approaches to Laboratory Medicine*. VIIIth Merz and Dade International Symposium. Darmstadt, G.-I.-T. Verlag Ernst Giebeler, 1981: 135–148.

20. Raghavan D., Peckham M. J., Heyderman E., Tobias J. S. and Austin D. E. Prognostic factors in clinical Stage I non-seminomatous germ-cell tumours of the testis. *Br. J. Cancer* 1982, **45**, 167–173.

21. Heyderman E. Multiple tissue markers in human malignant testicular tumours. *Scand. J. Immunol. [Supp. 8.]* 1978, **8**, 119–126.

22. Lange P. H., Limas C. and Fraley E. E. Tissue blood-group antigens and prognosis in low stage transitional cell carcinoma of the bladder. *J. Urol.* 1978, **119**, 52–55.

23. Neville A. M. and Heyderman E. Tumour markers in human breast cancer. In: Boelsma E. and Rumke Ph. (ed.) *Tumour Markers: Impact and Prospects*. Amsterdam, Elsevier/North Holland Biomedical Press, 1979: 297–303.

24. Primus F. J., Clark C. A. and Goldenberg D. M. Immunohistochemical detection of carcinoembryonic antigen. In: DeLellis R. A. (ed.) *Diagnostic Immunohistochemistry*. New York, Masson Publishing USA Inc., 1981: 263–276.

25. Heyderman E. The role of immunocytochemistry in tumour pathology: a review. *J. R. Soc. Med.* 1980, **73**, 655–658.

26. Goldenberg D. M., DeLand F., Kim E., Bennet S., Primus F. J., van Nagell J. R., Estes N., DeSimone P. and Rayburn P. Use of radiolabeled antibodies to carcinoembryonic antigen for the detection and localization of diverse cancers by external photoscanning. *N. Engl. J. Med.* 1978, **298**, 1384–1388.

27. Mach J. P., Carrell S., Forni M., Donath A. and Alberto P. Tumor localization of radiolabeled antibodies against carcinoembryonic antigen in patients with carcinoma. A critical evaluation. *N. Engl. J. Med.* 1980, **303**, 5–10.

28. Begent R. H. J., Searle F., Stanway G., Jewkes R. F., Jones B. E., Vernon P. and Bagshawe K. D. Radioimmunolocalisation of tumours by external scintigraphy after administration of ^{131}I antibody to human chorionic gonadotrophin: preliminary communication. *J. R. Soc. Med.* 1980, **73**, 624–630.

29. Javadpour N., Kim E. E., DeLand F. H., Selyer J. R., Shah U. and Goldenberg D. M. The role of radio immunodetection in the management of testicular cancer. *J. Am. Med. Assoc.* 1981, **246**, 45–49.

30. Halsall A. K., Fairweather D. S., Bradwell A. R., Blackburn J. C., Dykes P. W., Howell A., Reeder A. and Hine K. R. Localisation of malignant germ-cell tumours by external scanning after injection of radiolabelled anti-alpha-fetoprotein. *Br. Med. J.* 1981, **283**, 942–944.

31. Olsnes S. Directing toxins to cancer cells. *Nature* 1981, **290**, 84.

32. Ghose T., Norvell S. T., Guclu A. and Macdonald A. S. Immunochemotherapy of human malignant melanoma with chlorambucil-carrying antibody. *Eur. J. Cancer* 1975, **11**, 321–326.

33. Thorpe P. E., Ross W. C. J., Cumber A. J., Hinson A. J., Hinson C. A., Edwards D. C. and Davies A. J. S. Toxicity of diphtheria toxin for lymphoblastoid cells is increased by conjugation to anti lymphocyte globulin. *Nature* 1978, **271**, 752–755.

34. Blythman H. E., Casellas P., Gros O., Jansen F. K., Paolucci F., Pau B. and Vidal H. Immunotoxins: hybrid molecules of monoclonal antibodies and a toxin subunit specifically kill tumour cells. *Nature* 1981, **290**, 145–146.

35. Coons A. H., Creech H. J. and Jones R. N. Immunological properties of an antibody containing a fluorescent group. *Proc. Soc. Exp. Biol. Med.* 1941, **47**, 200–202.

36. Graham R. C. and Karnovsky M. J. The early stages of absorption of injected horseradish peroxidase in the proximal tubules of mouse kidney: ultrastructural cytochemistry by a new technique. *J. Histochem. Cytochem.* 1966, **14**, 291–302.

37. Avrameas S. Coupling of enzymes to proteins with glutaraldehyde. Use of the conjugates for the detection of antigens and antibodies. *Immunochemistry* 1969, **6**, 43–52.

38. Mason D. Y. and Sammons R. E. Alkaline phosphatase and peroxidase for double immunoenzymatic labelling of cellular constituents. *J. Clin. Pathol.* 1978, **31**, 454–460.

39. Heyderman E., Bulman A. S. and Gibbons A. R. Immunoperoxidase techniques and controls. *J. Clin. Pathol.* 1981, **33**, 1219–1220.

40. De Mey J., Moeremans M., Geuens G., Nuydens R. and De Brabander M. High resolution light and electron microscopic localisation of tubulin with the IGS (immuno-gold staining) method. *Cell Biol. Int. Rep.* 1982, **5**, 889–899.
41. Meyer-Arendt J. R. Theory and application of autoradiography. *Acta Histochem.* 1962, **13**, 47–61.
42. Rooijen N. V. The application of autoradiography and other histochemical techniques to the same tissue section or smear. *J. Immunol. Methods* 1981, **40**, 247–252.
43. Coons A. H., Leduc E. H. and Connolly J. M. Studies on antibody production. I. A method for the histochemical demonstration of specific antibody and its application to a study of the hyperimmune rabbit. *J. Exp. Med.* 1955, **102**, 49–60.
44. Mason T., Pfifer R., Spicer S. and Dreskin R. An immunoglobulin–enzyme bridge method for localizing tissue antigens. *J. Histochem. Cytochem.* 1969, **17**, 563–569.
45. Sternberger L. A. *Immunocytochemistry*. 2nd ed. New York, John Wiley and Sons, 1979.
46. Mason D. Y. and Sammons R. E. The labeled antigen method of immunoenzymatic staining. *J. Histochem. Cytochem.* 1979, **27**, 832–840.
47. Heyderman E. Immunoperoxidase technique in histopathology: applications, methods and controls. *J. Clin. Pathol.* 1979, **32**, 971–978.
48. Burns J. Background staining and sensitivity of the unlabelled antibody–enzyme (PAP) method. Comparison with the peroxidase labelled antibody sandwich method using formalin fixed paraffin embedded material. *Histochemistry* 1975, **43**, 291–294.
49. Burns J. Personal communication. 1982.
50. Nakane P. K. and Kawaio A. Peroxidase-labeled antibody: a new method of conjugation. *J. Histochem. Cytochem.* 1974, **22**, 1084–1091.
51. Bulman A. S. and Heyderman E. Alkaline phosphatase for immunocytochemical labelling. Problems with endogenous activity. *J. Clin. Pathol.* 1981, **34**, 1349–1351.
52. Porath J., Axen R. and Ernback S. Chemical coupling of proteins to agarose. *Nature* 1967, **215**, 1491–1492.
53. Kohler G. and Milstein C. Derivation of specific antibody-producing tissue culture and tumour lines by cell fusion. *Eur. J. Immunol.* 1976, **6**, 292–295.
54. Cuello A. C., Wells C., Chaplin A. J. and Milstein C. Serotonin reactivity in carcinoid tumours demonstrated by a monoclonal antibody. *Lancet* 1982, **ii**, 771–773.
55. Heyderman E., Gibbons A. R. and Rosen S. W. Immunoperoxidase localisation of human placental lactogen: a marker for the placental origin of the giant cells in 'syncytial endometritis' of pregnancy. *J. Clin. Pathol.* 1981, **34**, 303–307.
56. McKeever P. E., Garvin A. J. and Spicer S. S. Immune complex receptors on cell surfaces. I. Ultrastructural demonstration on macrophages. *J. Histochem. Cytochem.* 1976, **24**, 948–955.
57. Heyderman E., Steele K. and Ormerod M. G. A new antigen on the epithelial membrane: its immunoperoxidase localisation in normal and neoplastic tissues. *J. Clin. Pathol.* 1979, **32**, 35–39.
58. Payne R. A. and Ryan R. J. Human placental lactogen in the male subject. *J. Urol.* 1972, **107**, 99–103.
59. Yoshimoto Y., Wolfsen A. R. and Odell W. D. Human chorionic gonadotropin-like substance in nonendocrine tissues of normal subjects. *Science* 1977, **197**, 575–577.
60. Polak J. M. and Bloom S. R. The neuroendocrine design of the gut. *Clin. Endocrinol. Metab.* 1979, **8**, 313–330.
61. Morgan F. J. and Canfield R. E. Nature of the subunits of human chorionic gonadotropin. *Endocrinology* 1971, **88**, 1045–1053.
62. Bagshawe K. D. and Begent R. H. J. Gestational trophoblastic tumours. In: Copplestone M. (ed.) *Gynecologic Oncology*. Edinburgh, Chuchill Livingstone, 1981: 757–772.
63. Kurman R. J., Scardino P. T., McIntire K. R., Waldmann T. A. and Javadpour N. Cellular localisation of alphafetoprotein and human chorionic gonadotropin in germ cell tumours of the testis using an indirect immunoperoxidase technique. A new approach to the classification utilising tumour markers. *Cancer* 1977, **40**, 2136–2151.
64. Pugh R. C. B. (ed.) *Pathology of the Testis*. Oxford, Blackwell Scientific Publications, 1976.
65. Hobson B. M. The excretion of chorionic gonadotrophin by men with testicular tumours. *Acta Endocrinol.* 1965, **49**, 337–348.
66. Heyderman E. Biological markers of human teratomas and related germ cell tumors. In: Damjanov I., Knowles B. B. and Solte D. (ed.). *Biology of Human Teratomas*. Clifton, N. Jersey, The Humana Press Inc. 1982 (In press).
67. Braunstein G., Vaitukaitis J. L., Carbone E. P. and Ross G. T. Ectopic production of human chorionic gonadotropin by neoplasms. *Ann. Intern. Med.* 1973, **78**, 39–45.

68. Muggia F. M., Rosen S. W., Weintraub B. D. and Hansen H. H. Ectopic placental proteins in nontrophoblastic tumors. Serial measurements following chemotherapy. *Cancer* 1975, **36**, 1327–1337.
69. Kahn C. R., Rosen S. W., Weintraub B. D., Fajans S. S. and Gorden P. Ectopic production of chorionic gonadotropin and its subunits by islet cell tumours. A specific marker for malignancy. *N. Engl. J. Med.* 1977, **297**, 565–569.
70. Josimovich J. B. and MacLaren J. A. Presence in the human placenta and term serum of a highly lactogenic substance immunologically related to pituitary growth hormone. *Endocrinology* 1962, **71**, 209–220.
71. Nielsen P. V., Egebo K., Find C. and Olsen C. E. Prognostic value of human placental lactogen (HPL) in an unselected obstetrical population. *Acta Obstet. Gynecol. Scand.* 1981, **60**, 469–474.
72. Rosen S. W. and Weintraub B. D. Ectopic placental lactogen In: Ruddon R. W. (ed.) *Biological Markers of Neoplasia: Basic and Applied Aspects*. Amsterdam, Elsevier/North Holland Inc., 1978.
73. Bohn H. Nachweis und Charakterisierung von Schwangerschafts-Proteinen in der menschlichen Placenta, sowie ihre quantitative immunologische Bestimmung im Serum schwangerer *Frauen. Arch. Gynak.* 1971, **210**, 440–457.
74. Grudzinskas J. G., Gordon Y. B., Jeffrey D. and Chard T. Specific and sensitive determination of pregnancy-specific β1-glycoprotein by radio-immunoassay. *Lancet* 1977, **i**, 333–335.
75. Heyderman E. and Monaghan P. Immunoperoxidase reactions in resin embedded sections. *Invest. Cell Pathol.* 1979, **2**, 119–122.
76. Miyayama H., Doellgast G. J., Memoli V., Gandbhir L. and Fishman W. H. Direct immunoperoxidase staining for Regan isoenzyme of alkaline phosphatase in human tumor tissues. *Cancer* 1976, **38**, 1237–1246.
77. Bohn H., Inaba N. and Luben G. New placental proteins and their potential diagnostic significance as tumour markers. *Oncodev. Biol Med.* 1981, **2**, 141–153.
78. Tashjian A. H., Wolfe H. J. and Voelkel E. F. Human calcitonin: immunologic assay, cytologic localization and studies on medullary thyroid carcinoma. *Am. J. Med.* 1974, **56**, 840–849.
79. Wolfe H. J., Melvin K. E. W., Cervi-Skinner S. J., Al Saadi A. A., Juliar J. F., Jackson C. E. and Tashjian A. H. C-cell hyperplasia preceding medullary thyroid carcinoma. *N. Engl. J. Med.* 1973, **289**, 437–441.
80. Hillyard C. J., Coombes R. C., Greenberg P. B., Galante L. S. and MacIntyre I. Calcitonin in breast and lung cancer. *Clin. Endocrinol.* 1976, **5**, 1–8.
81. Coombes R. C., Easty G. C., Detre S. I., Hillyard C. J., Stevens U., Girgis S. I., Galante L. S., Heywood L., MacIntyre I. and Neville A. M. Secretion of immunoreactive calcitonin by human breast carcinomas. *Br. Med. J.* 1975, **4**, 197–199.
82. Silva O. L., Becker K. L., Primack A., Doppman J. L. and Snider R. H. Increased serum calcitonin levels in bronchogenic cancer. *Chest* 1976, **69**, 495–499.
83. Ellis S. T., Beck J. S. and Currie A. R. The cellular localisation of growth hormone in the human foetal hypophysis. *J. Pathol. Bacteriol.* 1966, **92**, 179–183.
84. Beck C. and Burger H. G. Evidence for the presence of immunoreactive growth hormone in cancers of the lung and stomach. *Cancer* 1972, **30**, 75–79.
85. Steiner H., Dahlbach O. and Waldenstrom J. Ectopic growth hormone production and osteoarthropathy in carcinoma of the bronchus. *Lancet* 1968, **i**, 783–785.
86. Yesner R. Spectrum of lung cancer and ectopic hormones. *Pathol. Annual* 1978 Part 1, **13**, 217–240.
87. Sonksen P. H., Ayres A. B., Braimbridge M., Corrin B., Davies D. R., Jeremiah G. M., Oaten S. W., Lowy C. and West T. E. T. Acromegaly caused by pulmonary carcinoid tumours. *Clin. Endocrinol.* 1976, **5**, 503–513.
88. Brown H. W. A case of pluriglandular syndrome. 'Diabetes of bearded women.' *Lancet* 1928, **ii**, 1022–1023.
89. Liddle G. W., Givens J. R., Nicholson W. E. and Island D. P. The ectopic ACTH syndrome. *Cancer Res.* 1965, **25**, 1057–1061.
90. Rees L. H. and Ratcliffe J. G. Ectopic hormone production by non-endocrine tumours. *Clin. Endocrinol.* 1974, **3**, 263–299.
91. Imura H. Ectopic hormone syndromes. *Clin. Endocrinol. Metab.* 1980, **9**, 235–260.
92. Gewirtz G. and Yalow R. S. Ectopic ACTH production in carcinoma of the lung. *J. Clin. Invest.* 1974, **53**, 1022–1032.
93. Ratcliffe J. G., Podmore J., Stack B. H. R., Spilg W. G. S. and Gropp C. Circulating ACTH and related peptides in lung cancer. *Br. J. Cancer* 1982, **45**, 230–236.

94. Singer W., Kovacs K., Ryan N. and Horvath E. Ectopic ACTH syndrome: clinicopathological correlations. *J. Clin. Pathol.* 1978, **31**, 591–598.
95. Jarett L., Lacy P. E. and Kipwis D. M. Characterization by immunofluorescence of an ACTH-like substance in non-pituitary tumors from patients with hyperadrenocorticism. *J. Clin. Endocrinol.* 1964, **24**, 543–549.
96. Bensch K. G., Corrin B., Pariente R. and Spencer H. Oat-cell carcinoma of the lung. Its origin and relationship to bronchial carcinoid. *Cancer* 1968, **22**, 1163–1172.
97. Iwasaki H., Mitsui T., Kikuchi M., Imai T. and Fukushima K. Neuroendocrine carcinoma (trabecular carcinoma) of the skin with ectopic ACTH production. *Cancer* 1981, **48**, 753–756.
98. Schwartz W. B., Bennet W., Curelolop S. and Bartter F. C. Syndrome of renal sodium loss and hyponatremia probably resulting from inappropriate secretion of antidiuretic hormone. *Am. J. Med.* 1957, **23**, 529–542.
99. Pettengill O. S., Faulkner C. S., Wurster-Hill D. H., Maurer L. H., Sorenson G. D., Robinson A. G. and Zimmerman E. A. Isolation and characterization of a hormone-producing cell line from human small cell anaplastic carcinoma of the lung. *J. Natl Cancer Inst.* 1977, **58**, 511–518.
100. Lo Gerfo P., Li Volsi V., Colaccio D. and Feind C. Thyroglobulin production by thyroid cancers. *J. Surg. Res.* 1978, **24**, 1–6.
101. Burt A. and Goudie R. B. Diagnosis of primary thyroid carcinoma by immunohistological demonstration of thyroglobulin. *Histopathology* 1979, **3**, 279–286.
102. Kurman R. J., Andrade D., Goebelsmann U. and Taylor C. R. An immunohistological study of steroid localization in Sertoli–Leydig tumors of the ovary and testis. *Cancer* 1978, **42**, 1772–1783.
103. Pertschuk L. P., Tobin E. H., Brigati D. J., Kim D. S., Bloom N. D., Gaetjens E., Berman P. J., Carter A. C. and Degenshein G. A. Immunofluorescent detection of estrogen receptors in breast cancer: comparison with dextran-coated charcoal and sucrose gradient assays. *Cancer* 1978, **41**, 907–911.
104. Walker R. A., Cove D. H. and Howell A. Histological detection of oestrogen receptor in human breast carcinoma. *Lancet* 1980, **i**, 171–173.
105. Taylor C. R., Cooper C. L., Kurman R. J., Goebelsmann U. and Markland F. S. Detection of estrogen receptor in breast and endometrial carcinoma by the immunoperoxidase technique. *Cancer* 1981, **47**, 2634–2640.
106. Penney G. C. and Hawkins R. A. Histochemical detection of oestrogen receptors: a progress report. *Br. J. Cancer* 1982, **45**, 237–246.
107. Jensen E. V. Hormone dependence of breast cancer. *Cancer* 1981, **47**, 2319–2326.
108. Laurence D. J. R., Stevens U., Bettleheim R., Darcy D., Leese C., Turberville C., Alexander P., Johns E. W. and Neville A. M. Role of plasma carcinoembryonic antigen in the diagnosis of gastrointestinal, mammary and bronchial carcinoma. *Br. Med. J.* 1972, **iii**, 605–609.
109. Neville A. M. and Laurence D. J. R. Report of the workshop on the carcinoembryonic antigen (CEA): the present position and proposals for future investigation. *Int. J. Cancer* 1974, **14**, 1–18.
110. Isaacson P. and Judd M. A. Carcinoembryonic antigen (CEA) in the normal human small intestine: a light and electron microscopic study. *Gut* 1977, **18**, 786–791.
111. Wahren B., Lidbrink E., Wallgren A., Eneroth P. and Zajicek J. Carcinoembryonic antigen and other tumor markers in tissue and serum or plasma of patients with primary mammary carcinoma. *Cancer* 1978, **42**, 1870–1878.
112. Hamada S. and Hamada S. Localisation of carcinoembryonic antigen in medullary thyroid carcinoma by immunofluorescent techniques. *Br. J. Cancer* 1977, **36**, 572–576.
113. NIH Consensus Statement (Summary). Carcinoembryonic antigen: its role as a marker in the management of cancer. *Br. Med. J.* 1981, **282**, 373–375.
114. Mach J.-P. and Putztaszeri G. Carcinoembryonic antigen (CEA): demonstration of a partial identity between CEA and a normal glycoprotein. *Immunochemistry* 1972, **9**, 1031–1034.
115. Von Kleist S., Chavanel G. and Burtin B. Identification of a normal antigen that cross-reacts with the carcinoembryonic antigen. *Proc. Natl Acad. Sci. USA* 1972, **69**, 2492–2494.
116. Burtin P. and Fondaneche M.-C. Characterization of the non-specific cross-reacting antigen (NCA) in human monocytes. *Clin. Immunol. Immunopathol.* 1981, **20**, 146–156.
117. Brock D. J. H. and Sutcliffe R. G. Alpha-fetoprotein in the antenatal diagnosis of anencephaly and spina bifida. *Lancet* 1972, **ii**, 197–199.
118. Norgaard-Pedersen B. et al. Clinical use of AFP and HCG in testicular tumours of germ cell origin. *Lancet* 1978, **ii** 1042.
119. Grigor K. M., Detre S. I., Kohn J. and Neville A. M. Serum alpha 1-foetoprotein levels in 153 male patients with germ cell tumours. *Br. J. Cancer* 1977, **35**, 52–58.

120. Scardino P. T., Cox H. D., Waldmann T. A., McIntire R., Mittemeyer B. and Javadpour N. The value of serum tumor markers in the staging and prognosis of germ cell tumors of the testis. *J. Urol.* 1977, **118**, 994–999.

121. Teilum G., Albrechtsen R. and Norgaard-Pedersen B. Immunofluorescent localisation of alpha-fetoprotein synthesis in endodermal sinus tumour (yolk sac tumour). *Acta Pathol. Microbiol. Scand. Sect. A: Pathol.* 1974, **62**, 586–588.

122. Shirai T., Itoh T., Yoshiki T., Noro T., Tomino Y. and Hayasaka T. Immunofluorescent demonstration of alpha-fetoprotein and other plasma proteins in yolk sac tumor. *Cancer* 1976, **38**, 1661–1667.

123. Beilby J. O. W., Horne C. H. W., Milne G. D. and Parkinson C. Alpha-fetoprotein, alpha-1-antitrypsin and transferrin gonadal yolk-sac tumours. *J. Clin. Pathol.* 1979, **32**, 455–461.

124. Wagener C., Menzel B., Breuer H., Weissbach L., Tschubel K., Henkel K. and Gedigk P. Immunohistochemical localisation of alpha-fetoprotein (AFP) in germ cell tumours: evidence for AFP production by tissues different from endodermal sinus tumour. *Oncology* 1981, **38**, 236–239.

125. Engelhardt N. V., Goussev A. I., Shipova L. J. A. and Abelev G. I. Immunofluorescent study of alpha-fetoprotein (AFP) in liver and liver tumours. I. Technique of AFP localisation in tissue sections. *Int. J. Cancer* 1971, **7**, 198–206.

126. Palmer P. E. and Wolfe H. Immunocytochemical localization of oncodevelopmental proteins in human germ cell and hepatic tumors. *J. Histochem. Cytochem.* 1978, **26**, 523–531.

127. Thung S. N., Gerber M. A., Sarno E. and Popper H. Distribution of five antigens in hepatocellular carcinoma. *Lab. Invest.* 1979, **41**, 101–105.

128. Ceriani R. L., Thompson K., Peterson J. A. and Abraham S. Surface differentiation antigens of human mammary epithelial cells carried on the human milk fat globule. *Proc. Natl Acad. Sci. USA* 1977, **74**, 582–586.

129. Sloane J. P., Ormerod M. G. and Neville A. M. Potential pathological application of immunocytochemical methods to the detection of micrometastases. *Cancer Res.* 1980, **40**, 3079–3082.

130. Bussolati G., Pich A. and Alfani V. Immunofluorescence detection of casein in human mammary dysplastic and neoplastic tissues. *Virchows Arch. Abt. A. Pathol Anat.* 1975, **365**, 15–21.

131. Pich A., Bussolati G. and Carbonara A. Immunocytochemical detection of casein and casein-like proteins in human tissues. *J. Histochem. Cytochem.* 1976, **24**, 940–947.

132. Neville A. M., Grigor K. M. and Heyderman E. Biological markers and human neoplasia. In: Woolf N. and Anthony P. (ed.) *Recent Advances in Histopathology No. 10.* Edinburgh, Churchill Livingstone, 1978.

133. Yamada K. M. and Olden K. Fibronectins—adhesive glycoproteins of cell surface and blood. *Nature* 1978, **275**, 179–184.

134. McDonagh J. Fibronectin. A molecular glue. *Arch. Pathol. Lab. Med.* 1981, **105**, 393–396.

135. Ruoslahti E., Jalanko H., Comings D. E., Neville A. M. and Raghavan D. Fibronectin from human germ-cell tumors resembles amniotic fluid fibronectin. *Int. J. Cancer* 1981, **27**, 763–768.

136. Stenman S. and Vaheri A. Fibronectin in human solid tumours. *Int. J. Cancer* 1981, **27**, 427–435.

137. Palmer P. E., DeLellis R. A. and Wolfe H. J. Immunochemistry of liver in alpha-1-antitrypsin deficiency. *Am. J. Clin. Pathol.* 1974, **62**, 350–354.

138. Gitlin D. and Perricelli A. Synthesis of serum albumin, prealbumin, α-foetoprotein, α_1-antitrypsin and transferrin by the human yolk sac. *Nature* 1970, **288**, 995–997.

139. Albrechtsen R., Wewer, U. and Wimberley P. D. Immunohistochemical demonstration of a hitherto undescribed localization of hemoglobin A and F in endodermal cells of normal human yolk sac and endodermal sinus tumor. *Acta Pathol. Microbiol. Scand. Sect. A: Pathol.* 1980, **88**, 175–178.

140. Wahren B. Multiple fetal antigens in germ-cell tumors. *Scand. J. Immunol. [Suppl. 8.]* 1978, **8**, 131–136.

141. Jacobsen G. K., Jacobsen M. and Clausen P. P. Ferritin as a possible marker protein of carcinoma-in-situ of the testis. *Lancet* 1980, **ii**, 533–534.

142. Gutman A. B. The development of the acid phosphatase test for prostatic carcinoma. *Bull. N. Y. Acad. Med.* 1968, **44**, 63–76.

143. Jöbsis A. C., De Vries G. P., Anholt R. R. H. and Sanders G. T. B. Demonstration of the prostatic origin of metastases. An immunohistological method for formalin-fixed embedded tissue. *Cancer* 1978, **41**, 1788–1793.

144. Wang M. C., Papsidero L. D., Muriyama M., Valenzuela L. A., Murphy G. P. and Chu T. M. Prostate antigen: a new potential marker for prostatic cancer. *The Prostate* 1981, **2**, 89–96.

145. Coombs R. R. A., Bedford D. and Rouillard L. M. A and B blood-group antigens on human epidermal cells. *Lancet* 1956, **i**, 461–463.
146. Davidsohn I. Early immunologic diagnosis and prognosis of carcinoma. *Am. J. Clin. Pathol.* 1972, **57**, 715–730.
147. Davidsohn I., Kovarik S. and Ling Lee C. A, B, and O substances in gastro-intestinal carcinoma. *Arch. Pathol.* 1966, **81**, 381–390.
148. Davidsohn I., Kovarik S. and Ni L. Y. Isoantigens A, B and H in benign and malignant lesions of the cervix. *Arch. Pathol.* 1969, **87**, 306–314.
149. Stafl A. and Mattingley R. F. Isoantigens ABO in cervical neoplasia. *Gynecol. Oncol.* 1972, **1**, 26–35.
150. Gupta R. K., Schuster R. and Christian W. D. Loss of isoantigens A, B and H in prostate. *Am. J. Pathol.* 1973, **70**, 439–447.
151. Gupta R. K. and Schuster R. Isoantigens A, B and H in benign and malignant lesions of breast. *Am. J. Pathol.* 1973, **72**, 253–260.
152. Weinstein R. S., Coon J., Alroy J. and Davidsohn I. Tissue-associated blood group antigens in human tumors. In: DeLellis R. A. (ed.) *Diagnostic Immunohistochemistry*. New York, Masson Publishing USA Inc., 1981: 263–276.
153. Lemieux R. U. Human blood groups and carbohydrate chemistry. *Chem. Soc. Rev.* 1978, **7**, 423–452.
154. Raghavan D. and Neville A. M. The use of tumour markers in the management of ovarian malignancy. In: Boelsma E. and Rumke Ph. (ed.) *Tumour Markers: Impact and Prospects*. Amsterdam, Elsevier/North-Holland Biomedical Press, 1979: 289–296.
155. Van Nagell J. R., Donaldson E. S., Gay E. C., Sharkey P. R. and Goldenberg D. M. Carcinoembryonic antigen in ovarian epithelial cystadenocarcinomas. The prognostic value of tumor and serial plasma determinations. *Cancer* 1978, **41**, 2335–2340.
156. Wass M., Searle F. and Bagshawe K. D. Radioimmunoassay of the normal serum glycoprotien (CX 1) in monitoring ovarian malignancy. *Eur. J. Cancer Clin. Oncol.* 1981, **17**, 1267–1273.
157. Lewis M. G. and Ikonopisov R. L. Tumour-specific antibodies in human malignant melanoma. *Br. Med. J.* 1969, **3**, 547–552.
158. Werkmeister J., Edwards A., McCarthy W. and Hersey P. Prognostic significance of expression of antigens on melanoma cells. *Cancer Immunol. Immunother.* 1980, **9**, 233–240.
159. Nakajima T., Watanabe S., Sato Y., Kameya T. and Shimosato Y. Immunohistochemical demonstration of S100 protein in human malignant melanoma and pigmented nevi. *Gann* 1981, **72**, 335–336.
160. Eng L. F., Vanderhagen J. J., Bignami A. and Gerstl B. An acidic protein isolated from fibrous astrocytes. *Brain Res.* 1971, **28**, 351–354.
161. Bignami A. and Schoene W. C. Glial fibrillary acidic protein in human brain tumors. In: DeLellis R. A. (ed.) *Diagnostic Immunohistochemistry*. New York, Masson Publishing USA Inc. 1981: 213–227.
162. Ghandour M. S., Langley O. K., Vincendon G. and Gombos G. Double labeling immunohistochemical technique provides evidence of the specificity of glial cell markers. *J. Histochem. Cytochem.* 1979, **27**, 1624–1637.
163. Greaves M. F. and Janossy G. Patterns of gene expression and the cellular origins of human leukaemias. *Biochim. Biophys. Acta* 1978, **516**, 193–230.
164. Wang N., Huang S. and Gold P. Absence of carcinoembryonic antigen-like material in mesothelioma: an immunohistochemical differentiation from other lung cancers. *Cancer* 1978, **74**, 438–444.
165. Kemshead J. T., Bicknell D. and Greaves M. F. A monoclonal antibody detecting an antigen shared by neural and granulocytic cells. *Pediatr. Res.* 1981, **15**, 1282–1286.

16

Intermediate Filaments in Diagnostic Histopathology

D. J. Evans

1. INTRODUCTION

Filamentous components are ubiquitously present in the cytoplasm of cells and ultrastructural methods have succeeded in defining a number of subclasses in accordance with their size and morphology in sections. The smallest of these is actin, which has a fibre diameter of 6 nm. The larger components, myosin filaments and microtubules, have respective diameters of 15 and 22 nm. The class of fibre under discussion here has a diameter of 7–11 nm (hence the name 'intermediate') and was at first thought to be directly related to the other fibre classes. Subsequent work has shown, however, that the intermediate filaments are a distinct group, which can be further subclassified on the basis of chemical composition and immunological properties. Five major subclasses have been defined so far: the keratins, which are widely present in epithelial cells; desmin, which occurs in smooth muscle, cardiac muscle and skeletal muscle; vimentin, which is found in cells of mesenchymal origin; neurofilaments, which are seen in neurones, and glial fibrillary protein which is present in astrocytes.

The situation is, however, more complex than would appear from this list, since epithelial cells in tissue culture will produce both vimentin and prekeratin (which have a different distribution in the cytoplasm)[1] and in the course of development muscle cells produce vimentin, then both vimentin and desmin and then desmin alone.[2] There have also been reports recently that the major class of intermediate filament in vascular smooth muscle is vimentin.[3]

The identification of these components is of considerable theoretical interest. Their precise role in cellular morphology and metabolism remains to be determined, but should provide further insight into cell function. Here, however, we are concerned with the more mundane question of whether they will be of interest to the diagnostic pathologist, a question which has not as yet been fully answered though several interesting contributions have already appeared in the literature.[4–9]

2. NEUROFILAMENTS

With respect to the neurofilaments, no use has as yet been established in diagnostic practice for an antiserum. A possible area for exploration would be the recognition of axons in neurofibromas, where conventional silver staining methods are complicated by the widespread staining of connective tissue. However, this might well be regarded by the average surgical pathologist as a piece of academic nonsense since the diagnosis is quite straightforward on conventional sections. Neuroblastomas have been observed to retain their neurofilamentous component even when in culture so that staining for neurofilaments might provide a positive method of distinguishing these tumours from the other round cell neoplasms encountered in childhood.

3. GLIAL FIBRILLARY ACIDIC PROTEIN GFAP

With glial fibrillary acidic protein the predominant localization is in astrocytes.[10] Although some authors indicate its presence in oligodendroglia and microglia most workers are agreed that this is not so. It has recently been described in cells associated with the enteric ganglia.

It is difficult to envisage that an antiserum would have a widespread diagnostic use. Glial fibres are readily stained by conventional methods and the recognition of glial tissues is straightforward even in teratomas. It is true that when it occurs at unexpected sites, such as in a nasal biopsy (of a nasal glioma or encephelocele) the diagnosis may be missed by the unwary, but the availability of an antiserum would not guarantee its use on such a case.

4. THE KERATINS

We now come to the subject of the keratins, which is appreciably more difficult, since they represent a heterogeneous group of proteins solubilized from epithelial residues by the use of acid solvents, or urea and reducing agents. From epidermis it is possible to isolate six or seven major components.[11] As the filaments are composed of three subunits there are many possible combinations. The stratum corneum has a protein with a molecular weight of 63 000 daltons which is absent from the other layers and biochemical techniques demonstrate differences between keratins from different sites.[12]

Antibodies to keratins show strong species cross-reactivity, and their use has indicated that keratins are not confined to squamous epithelia but are present in other epithelial tissues. Within normal tissues, the distribution of keratin found by immunocytological methods depends on the nature of the antiserum used. A wide variety of antisera have been used ranging from guinea-pig antibodies to bovine hoof keratin and muzzle keratin, to sera raised against individual keratin polypeptides made from human stratum corneum. Löning et al.[13] reported that some of the antisera they made, in particular those directed at the 67 000 and 62 000-dalton components, did not stain the basal layer, whereas those directed at total keratin or 55 000-dalton component did so. The antibody used by Schlegel et al.[4,5] reacted with all the layers of the epidermis and in addition with salivary, sweat and pancreatic ducts, tongue, cervix and vaginal epithelium and urinary transitional epithelium. Breast ducts stained somewhat variably, mainly in the myoepithelium. Staining of basal and intermediate cells was observed in the lung and bronchus, and of basal cells in the prostate. In the kidney collecting tubules reacted and in the liver bile ducts were weakly positive.

The authors reported that the tumours examined by them[5] fell into three groups. As might be anticipated keratin was absent from sarcomas, lymphomas and neural tumours. It was also absent from adenocarcinomas of the prostate, kidney and colon. Strong staining was observed in skin carcinomas (both basal cell and squamous), and in squamous carcinomas from lung, cervix and mouth. It was also seen in transitional carcinomas of the bladder and rather surprisingly in malignant mesotheliomas of pleura and peritoneum. There was also a group of tumours which stained variably and this included breast carcinomas and non-squamous carcinomas of the lung.

Gabbiani and his colleagues[6] reported a series of tumours examined on cryostat sections for various intermediate filaments. They reported that with this technique and using antisera to bovine keratins, positive fluorescence was seen with all their carcinoma cases though it was less intense with adenocarcinomas of the colon, rectum and prostate.

Altmannsberger et al.[7] reported a similar range of specificity together with their finding that positive results could be obtained not only on cryostat sections but also on alcohol-fixed paraffin-embedded material. With formalin-fixed tissues, however, they obtained negative results.

However, Sieinski et al.[8] obtained staining on Bouin-fixed tissue and Gusterson et al.[9] on formalin-fixed tissue. Part of the difference may reside in the antibodies used: the affinity purification used by Gabbiani et al. and Altmannsberger may have resulted in loss of the high affintiy component in their sera.

Battifora[14] reported that using an antibody to human keratin, positive results were obtained with frozen sections of thymomas.

An antiserum which we prepared to keratin[15] by conventional methods[4] had a specificity similar to that described by Löning et al. for antisera directed against the 62 000 and 67 000-dalton components, in that it did not react with the basal layer of epidermis (*Fig*. 16.1). Despite the failure to stain normal basal cells it reacted weakly with the cells of basal cell papillomas (*Plate* 20) and basal cell carcinomas. Squamous carcinomas of the skin stained rather patchily, some of the cells giving strong positive staining and others being negative (*Plate* 21). In keratoacanthomas the staining was usually more orderly with negative basal layers. These results are in general similar to those reported by Löning et al.

The Jadassohn lesion or intra-epithelial epithelioma gave an interesting result, its cells being negative though the surrounding skin stained well. This argues against the view that this lesion is an irritated seborrhoeic keratosis, or a basal cell carcinoma as has sometimes been maintained.

We confirmed the findings of Löning et al. that in the lesions of lichen planus the

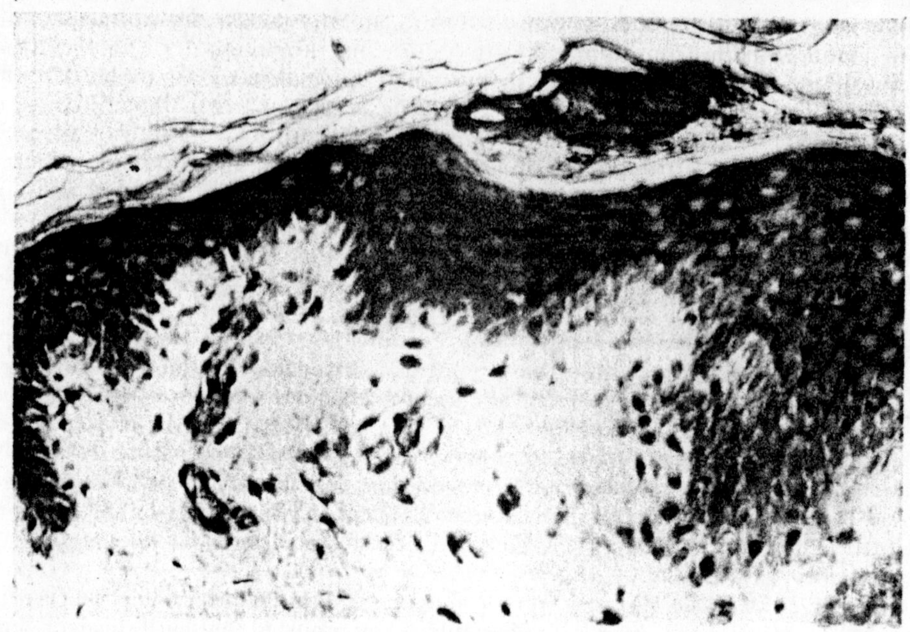

Fig. 16.1 Normal skin showing unstained basal layer and staining of other layers by keratin antiserum. Immunoperoxidase.

basal cells have acquired positivity. This antiserum gave a positive but patchy reaction with squamous carcinoma of the lung and was also positive with squamous carcinomas of oesophagus and uterine cervix and with epithelial thymomas (*Plate* 22). With prostatic carcinoma it was entirely negative but with most other carcinomas and with mesotheliomas it gave a rather weak but clearcut positive reaction. This serum appears somewhat less useful in its discrimination than that described by Schlegel[5] and has a narrower range of positivity than that of Gabbiani.

5. VIMENTIN AND DESMIN

With vimentin and desmin there are only a few studies available. Vimentin has been chemically isolated from an epithelioid sarcoma,[16] and using frozen sections Gabbiani et al.[6] demonstrated its presence in a wide variety of mesenchymal neoplasms ranging from uterine leiomyomas to malignant fibrous histiocytomas, and lymphomas. It has also been found in salivary tumours though its exact location in these is not clear.[17] Desmin has been shown to be present in a rhabdomyosarcoma and in a uterine myoma.[16] We have used antisera against desmin* and vimentin† to study a small series of smooth muscle tumours from the uterus and gastrointestinal tract.[18] The study was retrospective and the material available only as paraffin blocks but in a preliminary study it was found that both

* Kindly provided by Dr Badley.
† Kindly provided by Dr Virtanen.

antisera would react with formol and formol–mercury-fixed tissues up to 40 years old with the specificity indicated in *Tables* 16.1 and 16.2 which is similar to that previously described in the literature. The results are shown in *Table* 16.3. With uterine tumours all the benign examples studied contained desmin but this was absent from the leiomyosarcomas.

Table 16.1. **Distribution of immunostaining for desmin in normal tissues**

Desmin positive	*Desmin negative*
Gut smooth muscle	Fibroblasts
Uterine smooth muscle	Endothelium
Vascular smooth muscle	
Cardiac muscle	
Skeletal muscle	

Table 16.2. **Distribution of immunostaining for vimentin in normal tissues**

Vimentin positive	*Vimentin negative*
Fibroblasts	Cardiac muscle
Endothelial cells	Skeletal muscle
Vascular walls	Non-vascular smooth muscle

Table 16.3. **Distribution of immunostaining for desmin and vimentin in smooth muscle tumours**

Desmin positive		*Desmin negative*	
Uterine myomas	5/5	Uterine leiomyosarcomas	2/2
Gastrointestinal smooth muscle tumours	3/11	Gastrointestinal smooth muscle tumours	9/11
Vimentin positive		*Vimentin negative*	
Uterine myomas	5/5		
Gastrointestinal smooth muscle tumours*	9/11	Gastrointestinal smooth muscle tumours	2/11
Uterine leiomyosarcomas*	2/2		

* Sometimes weak.

The gastrointestinal tumours were, however, less straightforward. Although desmin was present in two oesophageal leiomyomas, it was not detected in any of the nine other smooth muscle tumours of the gastrointestinal tract even though four of these appeared benign on histological grounds. Interpretation of this is not easy as smooth muscle tumours of the gastrointestinal tract are notoriously unpredictable in their behaviour, and may metastasize even when they appear bland histologically. Probably the most logical deduction is that smooth muscle tumours of the gastrointestinal tract should be regarded with suspicion particularly when desmin is absent. The failure to demonstrate vimentin in two of the gastrointestinal smooth muscle tumours is probably attributable to the low levels of intermediate filaments demonstrable in a proportion of these tumours.

It should be noted in passing that the demonstration of desmin content in a tumour is by no means a general confirmation that it is benign, since desmin may be demonstrable in rhabdomyosarcomas.

To sum up the current position, it seems likely that antisera against intermediate filaments will establish a usage in histopathology. Possible areas are:

a. The recognition of neuroblastoma.
b. The distinction of carcinoma from malignant fibrous histiocytoma, atypical fibroxanthoma and malignant melanoma.
c. Discriminating between thymoma and lymphoma, or undifferentiated carcinomas and lymphomas.*
d. As an index of differentiation of smooth muscle tumours.
e. As an adjunct to other methods of recognizing the origin of an adenocarcinoma.

REFERENCES

1. Franke W. W., Schmid E., Winter S., Osborn M. and Weber K. Widespread occurrence of intermediate sized filaments of the vimentin type in cultured cells from diverse vertebrates. *Exp. Cell Res.* 1979, **123**, 25–40.
2. Gard D. L. and Lazarides E. The synthesis and distribution of desmin and vimentin *in vitro*. *Cell* 1980, **19**, 263–275.
3. Gabbiani G., Schmid E., Winter S., Chaponnier C., de Castonay D., Vandekerckhove J., Weber K. and Franke W. W. Vascular smooth muscle cells differ from other smooth muscle cells. Predominance of vimentin filaments and a specific α type actin. *Proc. Natl Acad. Sci. USA* 1981, **78**, 298–302.
4. Schlegel R., Banks Schlegel S. and Pinkus G. Immunohistochemical localisation of keratin in normal human tissues. *Lab. Invest.* 1981, **42**, 91–96.
5. Schlegel R., Banks Schlegel S., McLeod J. A. and Pinkus G. S. Immunoperoxidase localisation of keratin in human neoplasms. *Am. J. Pathol.* 1981, **101**, 41–50.
6. Gabbiani G., Kapanci Y., Barazzone P. and Franke W. W. Immunochemical identification of intermediate sized filaments in human neoplastic cells. *Am. J. Pathol.* 1981, **104**, 206–216.
7. Altmannsberger M., Osborn M., Schauer A. and Weber K. Antibodies to different intermediate filament proteins: cell type-specific markers on paraffin-embedded human tissues. *Lab. Invest.* 1981, **45**, 427–434.
8. Sieinski W., Dorsett B. and Ioachim H. L. Identification of prekeratin by immunofluorescence staining in the differential diagnosis of tumours. *Hum. Pathol.* 1981, **12**, 452–458.
9. Gusterson B., Mitchell D., Warburton M. and Sloane J. Immunohistochemical localisation of keratin in human lung tumours. *Virchows Arch. Abt. A Pathol. Anat.* 1982, **394**, 269–278.
10. Bignami A. and Dahl D. Specificity of the glial fibrillary acid protein for astroglia. *J. Histochem. Cytochem.* 1977, **25**, 466–469.
11. Sun T. T. and Green H. Keratin filaments of cultured human epidermal cells. Formation of intermolecular disulfide bridges during terminal differentiation. *J. Biol. Chem.* 1978, **253**, 2058–2060.
12. Lee L. D., Kubilus J. and Baden H. P. Intra-species heterogeneity of epidermal keratins isolated from bovine snout and hoof. *Biochem. J.* 1979, **177**, 187–196.
13. Löning T., Staquet M-J., Thivolet J. and Seifert G. Keratin polypeptides: distribution in normal and diseased human epidermis and oral mucosa. *Virchows Arch. Abt. A Pathol. Anat.* 1980, **388**, 263–288.
14. Battifora H., Sun T. T., Bahu R. M. and Rao S. The use of antikeratin antiserum as a diagnostic tool: thymoma versus lymphoma. *Hum. Pathol.* 1980, **11**, 635–641.

* Since this paper was prepared this use has been evaluated in nasopharyngeal tumours.[19]

15. Krausz T. and Evans D. J. Unpublished data, 1981.
16. Mills S. E., Fechner R. E., Bruns D. E., Bruns M. E. and O'Hara M. F. Intermediate filaments in eosinophilic cells of epithelioid sarcoma, a light microscopic ultrastructural and electrophoretic study. *Am. J. Surg. Pathol.*, 1981, **5**, 195–202.
17. Caselitz J., Osborn M., Seifert G. and Weber K. Intermediate-sized filament proteins (prekeratin, vimentin, desmin) in the normal parotid gland and parotid gland tumours. *Virchows Arch. Abt. A Pathol. Anat.* 1981, **393**, 273–281.
18. Evans D. J., Lampert I. A. and Jacobs M. Intermediate filaments in smooth muscle tumours. *J. Clin. Pathol.* 1982, in press.
19. Madri J. A. and Barwick K. W. An immunohistochemical study of nasopharyngeal neoplasms using keratin antibodies. *Am. J. Surg. Pathol.* 1982, **5**, 143–149.

17

Immunocytochemistry in Dermatology

A. C. Chu,
B. Bhogal and
M. M. Black

The last decade has seen major advances in our understanding of cutaneous diseases. The combined disciplines of clinical dermatology, histology and immunohistochemistry have enabled us to group disease processes together rationally and have given us some insight into the pathological mechanisms underlying a variety of dermatoses.

At St John's Hospital for Diseases of the Skin, the Institute of Dermatology, immunohistochemistry is routinely used to augment histopathology in differentiating the blistering skin disorders and to confirm the diagnosis of cutaneous lupus erythematosus. It is also proving valuable in the management of patients with the bullous disease, pemphigus, in which immunohistochemistry gives some indication of the activity of the disease. In the benign and malignant lymphocytic infiltrates of the skin, immunohistochemistry is also used to differentiate T- from B-cell neoplasms, and benign from malignant lymphocytic infiltrates.

In this chapter we will describe the techniques at present in use in our laboratories and will discuss the cutaneous diseases in which such tests are useful both in diagnosis and in the management of the patients.

1. SPECIMEN COLLECTION AND HANDLING

1.1. Skin Biopsy

The best site from which to take the skin biopsy will depend on the skin condition being investigated. In the dermatoses in which the cellular infiltrate is of primary interest, the most active and indurated lesions should be biopsied. In the majority of the bullous diseases, biopsies should be taken to incorporate part of the blister with perilesional skin (*Fig.* 17.1). The exception to this rule is dermatitis herpetiformis in which lesional skin may show no deposition of immunoglobulin. In these patients, only perilesional or clinically uninvolved skin should be examined as only these sites will show the characteristic immunohistochemical changes. Elliptical biopsies should be taken, where possible, for although punch biopsies are quicker and easier to perform, ellipses give a larger and more satisfactory cross sectional area for study.

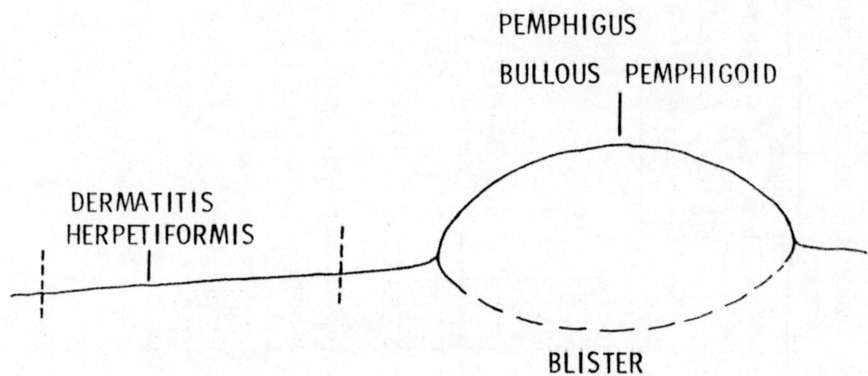

Fig. 17.1. Biopsy of blistering disorder.

Under optimal conditions, the skin biopsies should be immediately snap frozen and stored at −80 °C until used. Snap freezing by plunging the specimen wrapped in aluminium foil into liquid nitrogen is a widely used method of freezing tissue, but may result in freeze artefacts with distortion of the tissue architecture. One method of circumventing these freeze artefacts was advocated by Chayen et al.,[1] and is routinely used in our laboratory. This involves the use of *n*-hexane as a coolant which gives a slower but more even freezing of the tissue (*Fig.* 17.2). A drop of OCT compound (Lab Tek Products, Miles Labs Inc.) is mounted on a small cork board and the edges are frozen in *n*-hexane cooled in a bath of solid carbon dioxide in alcohol. The skin biopsy can then be orientated in the fluid central part of the OCT with the dermo-epidermal junction adjacent to the surface of the block. The whole block can then be plunged into the *n*-hexane leaving the skin biopsy frozen in a case of OCT. Using this method, the freeze artefacts are reduced, the tissue is properly orientated ready for cutting and the OCT protects the biopsy from the surface drying which results from prolonged storage.

If the skin biopsy is taken at a peripheral hospital, or needs to be transported and transportation in a frozen state is not possible, the biopsy can be maintained in a liquid fixative for a limited period. The fixative (*see* the Appendix, p. 330, Formula I) was introduced by Michel et al.[2] and was later evaluated by Skeete and Black.[3] In their study, Skeete and Black took biopsies from patients with various bullous diseases and cutaneous lupus erythematosus, simultaneously froze part of the biopsy and maintained part for a variable period in the liquid fixative. Their results indicated that the pH of the fixative was of primary importance. If the pH

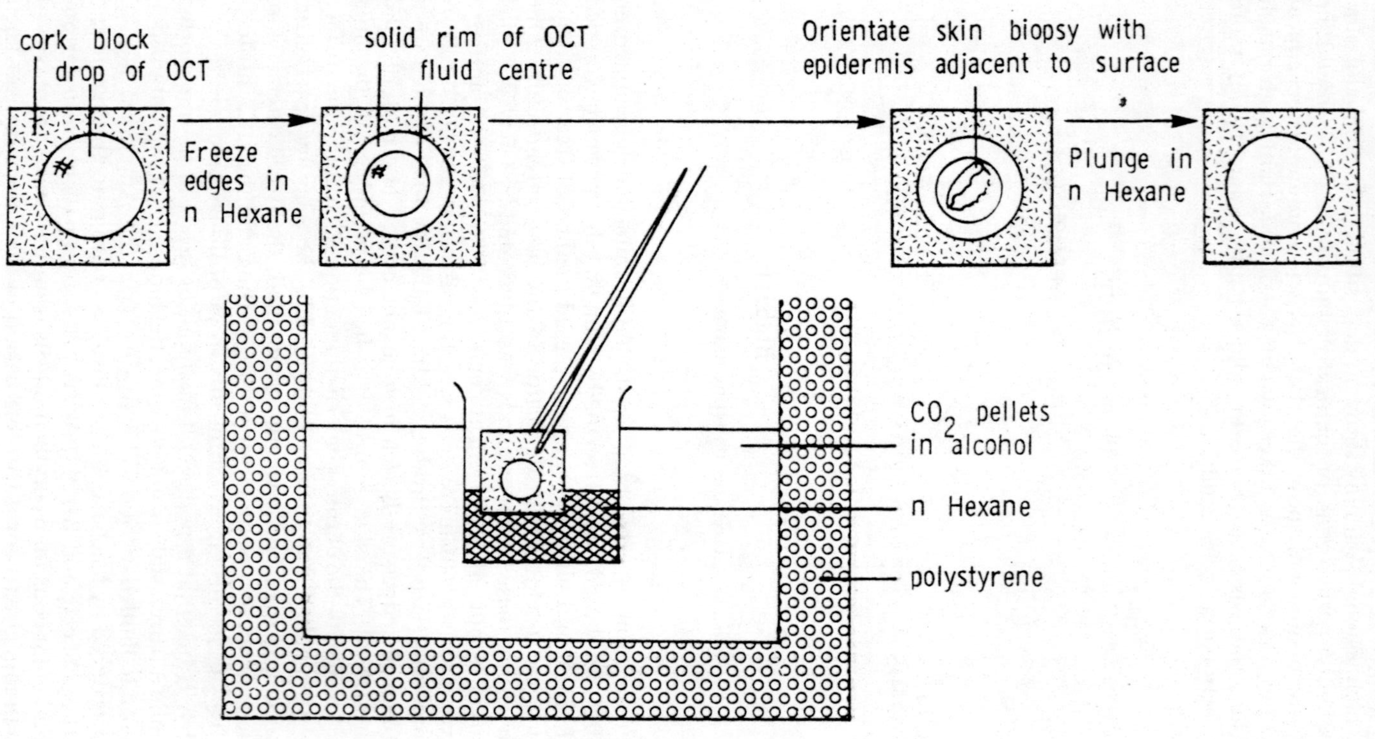

Fig. 17.2. Freezing skin biopsies.

was maintained above 7, the results obtained in direct immunofluorescence in tissue kept in liquid fixative for up to five days, were equivalent to those of the fresh frozen material. If the pH fell the results in direct immunofluorescence were consistently reproducible only in tissue kept in fixative for less than 24 hours. The pH is best maintained above 7 if the tissue is washed well in buffer before being put in the fixative and is then kept at 4 °C.

On arrival, the tissue sent in fixative should be washed three times in the buffer (Formula I) for 10 min each wash and then frozen and stored at −80 °C until used.

Liquid fixatives are not suitable for the transport of tissue in which the cellular infiltrates are to be examined using anti-T-cell antibodies. Biopsies for cellular evaluation should be snap frozen and transported on solid CO_2 or in liquid nitrogen.

1.2. Blood

Blood taken from patients for examination of circulating immunoglobulins is collected in glass tubes without anticoagulants. The tubes are incubated at 37 °C for 1 h and the serum pipetted from the blood clot. The serum is centrifuged at 1200 rev./min to pellet any cellular debris, and is then aliquotted and stored at −80 °C until used.

2. GENERAL TECHNIQUES

In our laboratory, immunohistochemical techniques are performed on cover slips rather than microscope slides. This has the advantage of being both space saving and requiring smaller volumes of wash buffer and fixatives. The cover slips are labelled on the reverse side and two cryostat sections are cut onto each. Drying of sections between washes is facilitated by the use of an ordinary electric fan.

Sections are washed by inserting the cover slips into a cover slip rack and placing this in a bath of buffer with continued stirring using a magnetic stirrer.

After the final wash, sections are fan dried and mounted on glass microscope slides in buffered glycerol (*see* the Appendix, p. 330, Formula II). The buffer used to dilute reagents and to wash sections in is Coons' phosphate-buffered saline pH 7·4 (*see* the Appendix, p. 330, Formula III).

3. BULLOUS DISORDERS

Three basic techniques are routinely used in the investigation of bullous disorders (*Fig.* 17.3):

a. Direct immunofluorescence—in which the patient's tissue is examined for the deposition of immunoglobulins, complement and fibrin.
b. Indirect immunofluorescence—in which normal epithelium is used as a substrate to detect circulating immunoglobulins in the patient's serum.
c. C_3 staining—in which complement-fixing antibodies are detected in the patient's serum.

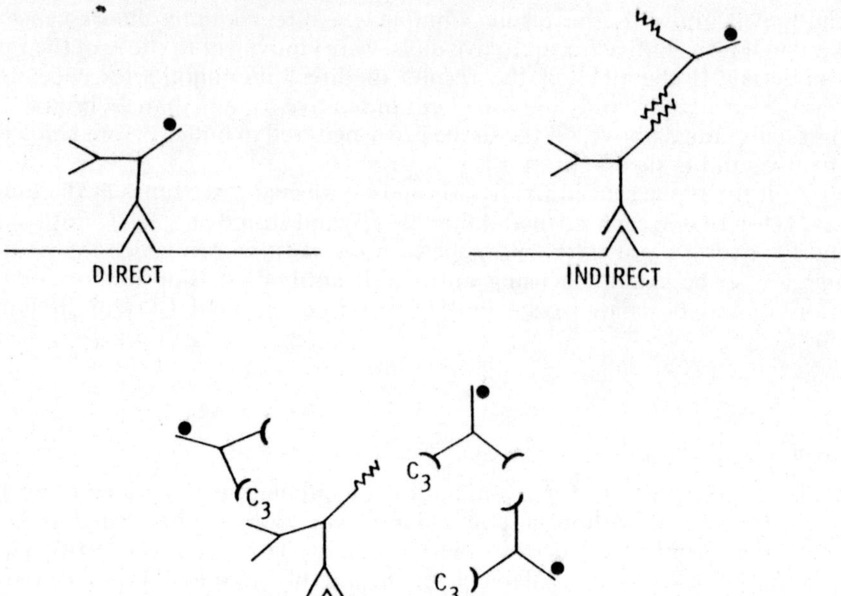

Fig. 17.3 Schematic representation of direct and indirect C_3 immunofluorescence techniques.

3.1. Direct Immunofluorescence

Sections are incubated with fluorescein-conjugated antisera to IgM, IgG, IgA, C_3 and fibrin. For details of the method, *see* the Appendix, p. 326.

3.2. Indirect Immunofluorescence

In this technique, the patient's serum is examined for the presence of circulating antibodies. A large number of tissue substrates have been used in this technique, including guinea-pig lip or oesophagus, normal human skin and monkey oesophagus. Guinea-pig tissue has been shown to give an unacceptably high rate of false negative results and is not recommended. Human skin is an obvious and satisfactory tissue but monkey oesophagus has been advocated as the best substrate to use in this test.

5-µm cryostat sections of the tissue substrate are incubated with the patient's serum followed by fluorescein-conjugated anti-human IgG. For details of the method, *see* the Appendix, p. 326.

3.3. Indirect C_3 Method

This method is useful in two situations:

a. The method amplifies the staining obtained by conventional immunofluorescence when complement-fixing antibodies are present. In conditions such as herpes gestationis where the circulating immunoglobulins are difficult to detect by indirect techniques, but fix complement, they can often be detected by the indirect C_3 method.

b. The method will demonstrate whether circulating antibodies are capable of fixing complement, which may give some clues as to the pathological mechanisms involved in the disease (*see* Section 4).

5-μm cryostat sections of normal human skin are incubated with the patient's serum followed by complement (normal human serum in complement-diluting buffer (*see* the Appendix, p. 331, Formula IV) and fluorescein-conjugated anti-C_3). For details of the method, *see* the Appendix, p. 326.

4. PEMPHIGUS VULGARIS

4.1. Clinical

Pemphigus vulgaris is a blistering skin disorder characterized by the formation of flaccid bullae which extend peripherally and rupture leading to areas of weeping erosions. Mucous membranes are commonly affected and may be the initial sites of involvement. The disease has a peak age incidence in the fourth and fifth decades and is more common in those of Jewish or Mediterranean ancestry. Before the advent of steroid therapy, pemphigus carried a high fatality and even now the mortality rate is about 10 per cent.

4.2. Histopathology

The characteristic histopathology of pemphigus is the presence of intra-epidermal bullae in which keratinocytes lose cohesion with each other and individual separated keratinocytes are present within the blister fluid (acantholysis). Acantholysis is not, however, pathopneumonic for pemphigus vulgaris as many other benign (Hailey–Hailey disease, Darier's disease, Grover's disease) and malignant (actinic keratoses, basal cell carcinoma) skin conditions may exhibit this phenomenon.

Four forms of pemphigus are recognized on clinicopathological criteria, differing primarily in the site of the blister within the epidermis, and the response of the epidermis to the disease. Pemphigus vulgaris and pemphigus vegetans have suprabasalar bullae while pemphigus foliaceus and pemphigus erythematosus have more superficial subcorneal bullae (*Fig*. 17.4).

Skin eruptions clinically, histologically and immunohistochemically identical to pemphigus vulgaris and pemphigus foliaceus may be induced by the drugs D-penicillamine, rifampicin, cataprine and practolol.

4.3. Immunohistochemistry

The characteristic finding in all forms of pemphigus is the presence, on direct immunofluorescence, of immunoglobulins deposited in the intercellular spaces of the epidermis and mucosa (*Fig*. 17.5). The immunoglobulin is generally IgG but IgM and IgA may be present. In early lesional skin of patients with pemphigus, $C1_q$, C_4 and C_3 have also been reported. The immunoglobulin is found in lesional and clinically uninvolved skin and mucous membrane. By immuno-electron

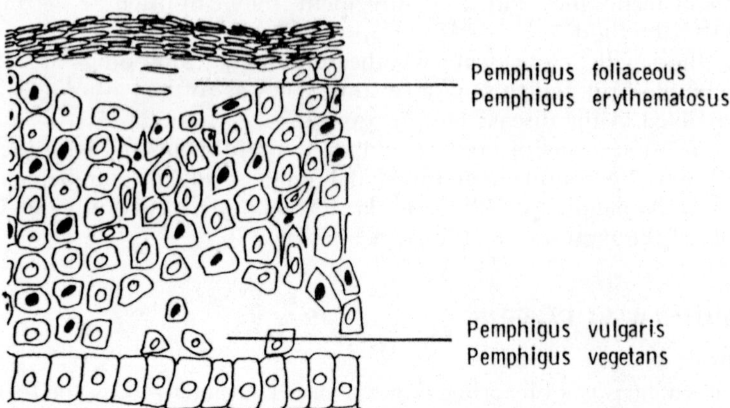

Fig. 17.4. Histopathology of the variants of pemphigus.

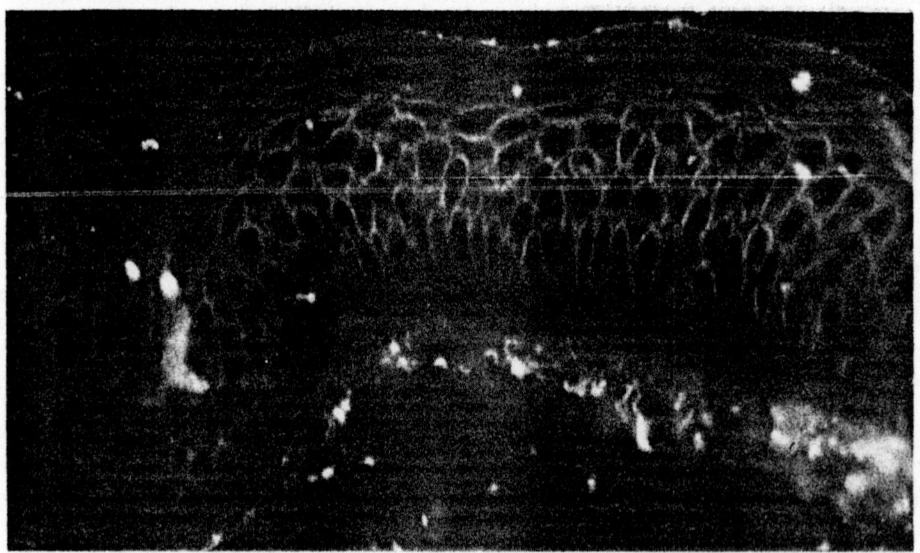

Fig. 17.5. Direct immunofluorescence of pemphigus vulgaris showing deposition of intercellular IgG.

microscopy, these antibodies have been shown to be localized to the intercellular substance of the epidermis and mucosa.[4]

Indirect immunofluorescence shows the presence of circulating IgG directed against the intercellular substance of the epidermis and mucosa. The circulating antibody may be undetectable in the earliest stages of the disease even though direct immunofluorescence is positive. The titre of the circulating antibodies frequently reflects the activity of the disease, generally falling with treatment and increasing with relapses. Measurement of the titre of the pemphigus antibody is thus widely used to monitor patients with pemphigus and disappearance of the circulating antibody has been suggested as a criterion for reducing and stopping treatment in these patients.[5] The titre of the circulating antibody, however, does not always correspond to the clinical activity of the disease. In a study of 19 patients with pemphigus at St John's Hospital, the disease activity was compared with the titre of circulating intercellular antibody over a period of 4–6 months. In 47 per cent of the patients, the correlation was good, in 32 per cent fair and in 21 per cent poor. Monitoring of the titre of circulating intercellular antibody in pemphigus may thus contribute to the management of patients with this disease, but the limitations of the correlation between disease activity and antibody titre must be recognized.

Circulating antibodies against the intercellular substance of epidermis and mucosa are characteristic of pemphigus, but similar antibodies have been described in a number of other conditions. In anti-A and anti-B blood group sera, these antibodies may be present, but are blocked by adsorption with AB substance. Circulating pemphigus-like antibodies are also present in penicillin morbilliform drug eruptions, trichophyton infections, bullous pemphigoid, cicatricial pemphigoid, thermal burns, and a variety of other conditions. The antibodies present in these conditions are usually of low titre.

In general, the immunofluorescence findings in all forms of pemphigus are the same. Bystryn et al.,[6] however, reported a case of pemphigus foliaceus with circulating antibodies reactive only with the subcorneal intercellular spaces. In pemphigus erythematosus and drug-induced pemphigus, the immunofluorescence findings may show the typical features of pemphigus but also those of lupus erythematosus with linear basement membrane zone deposition of immunoglobulin and the presence of anti-nuclear antibodies.

4.4. Pathogenesis

Pemphigus is generally considered to be an autoimmune disease characterized by serum immunoglobulin directed against the intercellular substance of the epidermis. The mechanism by which the pemphigus antibody induces acantholysis is still controversial. The pemphigus antibodies are of the IgG class and are thus capable of fixing complement. However, although $C1_q$, C_4 and C_3 have been demonstrated in early lesions of pemphigus, *in vitro* C_3 staining has failed to demonstrate that the circulating pemphigus antibody will fix complement.

Schiltz and Michael[7] have shown in organ culture systems that pemphigus antibody induced acantholysis in the absence of complement. An alternative mechanism for the pathogenesis of pemphigus was suggested by Farb et al.[8] who showed that acantholysis induced by pemphigus antibody in tissue culture systems was inhibited by soya bean trypsin inhibitor. It thus appears that the immunoglobulins in pemphigus may exert their effect on the intercellular substance by activating enzymes such as serine esterase.

5. BULLOUS PEMPHIGOID

5.1. Clinical

Bullous pemphigoid is a blistering skin disorder occurring mainly in the elderly. The bullae are characteristically tense, occurring on normal or erythematous skin predominantly in the flexures—groins, axillae and neck. The disease has a peak age incidence in the sixth and seventh decades, but may occur in any age group. Approximately 10 per cent of patients have mucosal involvement. Before the advent of corticosteroid and antibiotic therapy, up to 15 per cent of patients died as a result of metabolic disturbances or secondary infections. Treatment with corticosteroids and immunosuppressants has reduced this mortality but has generated a significant morbidity especially when used for prolonged periods.

5.2. Histopathology

Bullous pemphigoid is a subepidermal bullous disorder associated with an inflammatory infiltrate of varying intensity. The inflammatory infiltrate is predominantly perivascular consisting of mononuclear cells, eosinophils and polymorphonuclear leucocytes, and eosinophils and polymorphonuclear leucocytes may be present in the blister cavity.

Electron microscopy shows the blister to arise in the lamina lucida, between the plasma membrane and the basal lamina.

5.3. Immunohistochemistry

Direct immunofluorescence of perilesional skin of patients with bullous pemphigoid shows a linear deposition of immunoglobulin along the basement membrane zone (*Fig*. 17.6). Any immunoglobulin class may be deposited but IgG is the most frequently found. In addition, complement components, properdin, factor B and fibrin are found, suggesting that complement may be involved in the pathogenesis of the disease by either the classical or alternative pathways. In early stages of the disease, direct immunofluorescence may occasionally only demonstrate C_3 deposition at the basement membrane zone, and subsequent follow-up of the patient shows the appearance of IgG.

In indirect immunofluorescence the presence of IgG antibodies to the basement membrane zone is demonstrated in about 80 per cent of patients. Unlike pemphigus, there is no correlation between the clinical activity of the disease and the pemphigoid antibody titre.

5.4. Pathogenesis

The pemphigoid antigen is localized in the lamina lucida of the basement membrane zone of the skin and is produced by basal keratinocytes of the epidermis. A basic macromolecular protein has been characterized by Diaz et al.[9] of molecular weight 31 000 which is thought to be the putative pemphigoid antigen and will block bullous pemphigoid antibody staining of normal human skin. The same protein has also been demonstrated in saliva and urine of normal individuals.

If a single cell suspension of epidermal cells is produced by trypsinization of normal human skin, the basal keratinocytes can be identified by staining with bullous pemphigoid serum in an indirect immunofluorescence reaction. In these preparations, basal cells can be identified by the presence of a fluorescent cap at

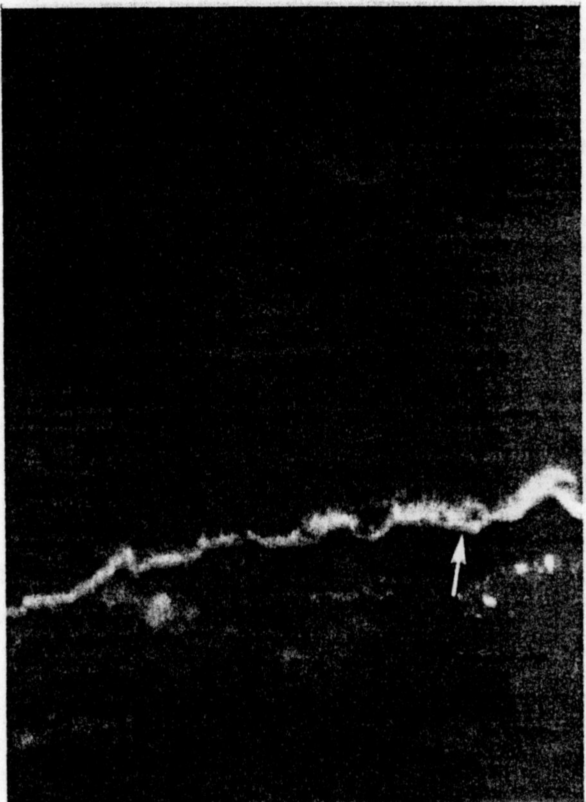

Fig. 17.6. Direct immunofluorescence of bullous pemphigoid showing linear deposition of IgG in the basement membrane zone. Note that the IgG is in the roof of the blister (→).

one pole of the cell. The pemphigoid antigen is produced by keratinocytes in tissue culture.

The bullous pemphigoid antibody is usually of the IgG class and has been shown to fix complement *in vitro*. Complement components are found in the basement membrane zone of perilesional skin of patients with bullous pemphigoid, together with properdin and factor B, suggesting that complement may be activated by both classical and alternative pathways during the pathogenesis of bullous pemphigoid.

Several authors have reported the presence of neutrophil and eosinophil chemotactic activity in bullous pemphigoid blister fluid and have attributed this to the C5a fraction of the fifth component of complement[10] and to the presence of eosinophil chemotactic factor ECF-A.[11] These studies suggest a good theoretical model for the pathogenesis of bullous pemphigoid with IgG deposition leading to complement activation resulting in damage to the basement membrane zone and the inflammatory response. Passive transfer experiments, however, have failed to demonstrate this mechanism *in vivo*. Sams and Gleich[12] infused large amounts of high-titre pemphigoid serum into monkeys. The human IgG localized to the basement membrane zone within 2 hours and remained there for several days.

None of the animals, however, developed clinical skin disease or histological evidence of blister formation.

6. BULLOUS PEMPHIGOID AND CARCINOMA

Bullous pemphigoid is regarded by many as a cutaneous marker of underlying malignancies. Many of the reports in the older literature, however, tended to be anecdotal and gave a rather confused view of this relationship. Recently, Hodge et al.[13] studied 134 patients with bullous pemphigoid and reported an incidence of malignancy of 4 per cent in patients with circulating anti-basement membrane IgG and 23 per cent in patients without circulating antibody. The former group was the largest consisting of 95 patients and carcinomas were found in 4 of these: 2 carcinoma of the bladder, 1 carcinoma of the bronchus and 1 carcinoma of the prostate. In the latter group there were 35 patients and of these 8 developed concurrent malignancies: 3 carcinoma of the bronchus, 1 carcinoma of the thyroid, 1 carcinoma of the larynx, 2 carcinoma of the cervix and 1 carcinoma of the stomach.

This study strongly suggests that patients with bullous pemphigoid who show no circulating anti-basement membrane zone IgG should be investigated for the possibility of a coexisting carcinoma.

7. HERPES GESTATIONIS

7.1. Clinical

Herpes gestationis is a blistering disease occurring in pregnancy. It is characteristically highly pruritic and individual skin lesions range from macular erythema and papules to tense bullae. The sites of predilection are the abdomen and extremities, but any area may be involved.

Herpes gestationis occurs once every 10 000–30 000 deliveries[14] and may recur with subsequent pregnancies, with the first menstrual period after the birth of the child, or with starting the contraceptive pill. It usually begins at the end of the second trimester but has been reported in the first trimester and as late as the early puerperium. The disease is self-limiting but may be associated with an increased mortality of infants born to affected mothers.

7.2. Histopathology

As with bullous pemphigoid, bullous lesions of herpes gestationis show a subepidermal cleft, but, unlike bullous pemphigoid, ultrastructurally the defects appear to be restricted to the basal keratinocytes with basal cell degeneration, disruption of the plasma membrane and accumulation of fluid between basal cells and the basal lamina. The infiltrate is perivascular around the superficial and deep dermal plexuses, and comprises histiocytes, lymphocytes, eosinophils and occasional neutrophils. On light microscopy, the characteristic feature is the presence of papillary oedema leading to subpapillary vesiculation and basal cell necrosis over the tips of the dermal papillae.

7.3. Immunohistochemistry

The most consistent finding in herpes gestationis on direct immunofluorescence is the presence of a heavy linear C_3 deposition on the basement membrane zone. Immunoglobulin deposition may or may not be detected. If present, the immunoglobulin is usually IgG, but occasionally IgA and IgM have been observed. $C1_q$, C_4 and C_5 are less often detected in herpes gestationis, but properdin and factor B are frequently found.

Direct immunofluorescence findings may be present in the skin for as long as one year after clinical lesions have resolved.

Indirect immunofluorescence demonstrates detectable IgG antibody to the basement membrane zone in 10–20 per cent of patients with herpes gestationis. The titre is usually low. However, the majority of patients with herpes gestationis can be shown to have a circulating complement-fixing antibody by the *in vitro* C_3 staining method. This antibody, the so-called herpes gestationis factor, has been shown to be an IgG which avidly fixes complement.

Yaoita et al.[15] used horseradish peroxidase in an immuno-electron microscopical technique to show that C_3 deposition in herpes gestationis occurs in the lamina lucida, and Honigsman et al.[16] demonstrated IgG deposition in herpes gestationis in a similar distribution.

7.4. Pathogenesis

The herpes gestationis factor is an IgG antibody which is directed against a component of the basement membrane zone and avidly fixes complement. In most patients with herpes gestationis, this IgG is probably present on the basement membrane zone, but is undetectable using direct immunofluorescence. It results, however, in the deposition of considerable amounts of C_3 at the basement membrane zone. Although early components of complement are not found in herpes gestationis skin, while properdin and factor B are, *in vitro* tests using C_2-deficient human serum as a source of complement in the *in vitro* C_3 staining method suggest that the herpes gestationis factor fixes complement by the classical pathway.

8. CICATRICIAL PEMPHIGOID

8.1. Clinical

Cicatricial pemphigoid is a chronic bullous dermatosis primarily involving the mucous membranes. The oral and occular mucous membranes are most frequently affected, resulting in erosions, scarring and subepidermal bullae. A more uncommon variant, the Brunsting–Perry variant, is associated with localized cutaneous vesiculobullous lesions involving predominantly the head and neck.

Severe scarring of the conjunctivae occurs in up to 60 per cent of patients leading to blindness from corneal damage in about 20 per cent of patients.

8.2. Histopathology

Cicatricial pemphigoid is characterized by subepidermal bullous formation, indistinguishable from bullous pemphigoid. Ultrastructurally, the basal lamina may show evidence of replication[17] or may be largely destroyed.

8.3. Immunohistochemistry

Direct immunofluorescence in cicatricial pemphigoid may show similar findings to those of bullous pemphigoid. Immunoglobulins, especially IgG and IgA are deposited along the basement membrane zone and $C1_q$, C_4, C_3, factor B and properdin may also be detected. Relatively few patients, 10 per cent, show circulating antibodies to the basement membrane zone by indirect immunofluorescence and some of these circulating antibodies may demonstrate restricted specificity. Some of these antibodies will show organ specificity, reacting to mucosa and not to skin, others will show species specificity reacting to human and not animal tissue, and others will show idiotype specificity reacting only with the patient's own tissue.

8.4. Pathogenesis

The clinical appearances of bullous pemphigoid and cicatricial pemphigoid are different but similarities in the histopathology and immunofluorescence findings suggest that these two diseases have similar underlying pathogenic mechanisms.

9. DERMATITIS HERPETIFORMIS

9.1. Clinical

Dermatitis herpetiformis is a chronic intensely pruritic papulovesicular dermatosis which affects the extensor surfaces of the body. The age of onset is generally in the second to fourth decades, but no age group is spared. It runs a chronic course of variable severity for an indefinite period. Most patients with dermatitis herpetiformis have an associated asymptomatic gluten-sensitive enteropathy.

Treatment of dermatitis herpetiformis with sulphones or sulphapyridine results in rapid relief from symptoms and clinical skin disease in the majority of patients.

9.2. Histopathology

Typical features of dermatitis herpetiformis are observed in early erythematous skin lesions. In these, the tips of dermal papillae show oedema and accumulation of neutrophils and a few eosinophils to form micro-abscesses. Fibrin accumulates in the papillary tips and subepidermal splits appear. The blisters are initially multilocular but with loss of cohesion of the rete ridges to the dermis, a single bullous forms. In the dermis there is an associated perivascular and subpapillary inflammatory infiltrate of neutrophils and eosinophils. Ultrastructurally, the basal lamina shows extensive damage and may be absent in more advanced lesions.

9.3. Immunohistochemistry

Perilesional and uninvolved skin of patients with dermatitis herpetiformis show a granular deposit of IgA in the papillary tips (*Fig.* 17.7). A smaller percentage of patients with dermatitis herpetiformis may show a linear basement membrane zone deposition of IgA rather than the granular deposits. Other immunoglobulin classes may occasionally be present with both granular and linear IgA but these are not a consistent feature, even in the same patient. C_3 is frequently found with the IgA.

Indirect immunofluorescence in dermatitis herpetiformis is usually negative, but in patients with linear IgA deposition, circulating IgA antibodies to the basement membrane zone may be detected.

Fig. 17.7. Dermatitis herpetiformis, IgA granular deposits in the dermal papillae on direct immunofluorescence.

Seah et al.[18] reported the presence of a circulating antibody in 16 per cent of their patients which reacted with reticulin in the connective tissue of rat and human organs. These antibodies have now been reported in 17–22 per cent of patients with dermatitis herpetiformis and have been characterized as being IgG immunoglobulins. The possible role of the antibodies to reticulin in the pathogenesis of dermatitis herpetiformis has been advocated, but the absence of IgG deposition in dermatitis herpetiformis militates against this.

9.4. Pathogenesis

Dermatitis herpetiformis is associated with a gluten-sensitive enteropathy and patients strictly adhering to a gluten-free diet may require only minimal or even no treatment for their skin disease. The IgA deposited in dermatitis herpetiformis has been shown to be of gut origin and suggestions as to why it localizes to the skin have included the possibility of gluten proteins being deposited in the skin or the possible presence of a protein in the skin which is cross reactive with the anti-gluten IgA.

Immuno-electron microscopical studies by Yaoita and Katz[19] and Stingl et al.[20] have shown that IgA in the granular form of dermatitis herpetiformis is localized in clumps in the dermal papillae in close association with dermal microfibrillar bundles which seem to be peculiar to dermatitis herpetiformis. It is thus possible that the primary defect involves the formation of these microfibrillar structures to which the IgA cross-reacts. Studies to date have failed to demonstrate gluten in the IgA deposits which further supports the suggestion that the IgA is present because of cross-reactivity with a structural protein in the skin.

IgA is capable of activating complement via the alternative pathway and complement activation could explain the inflammatory response and structural changes in dermatitis herpetiformis.

10. APPLICATION OF IMMUNOFLUORESCENCE TO THE INVESTIGATION AND DIAGNOSIS OF BULLOUS DISORDERS

In many non-immunological and genetically determined bullous skin diseases, immunoglobulin deposition is not found on direct immunofluorescence but immunofluorescent techniques may still be useful in their diagnosis. In the bullous diseases encompassed by the term epidermolysis bullosa, there are a number of clinically, genetically and prognostically different entities, united by the characteristic feature of blister formation at sites of trauma. The fundamental defects appear to be structural abnormalities in the dermo-epidermal junction area and, by electron microscopy, the different varieties can be identified by the structural defect present. Electron microscopy, however, is not a technique open to all clinicians and in situations where it is difficult or impossible to examine the tissue by electron microscopy, immunohistochemistry may provide important guides to the differentiation of the different diseases.

In epidermolysis bullosa simplex, the defect is in the basal keratinocytes and the cleft occurs through the cells (*Fig.* 17.8). This form of the disease has a good prognosis and low morbidity. In the junctional forms of epidermolysis bullosa, the defect is within the lamina lucida and clefts occur at this level. In the fetus and newborn, the junctional form of epidermolysis bullosa may be lethal, while in adults the disease has a relatively good prognosis but may be associated with severe morbidity. In the dystrophic forms of epidermolysis bullosa, the defect occurs below the basal lamina and the split is subepidermal. These forms are associated

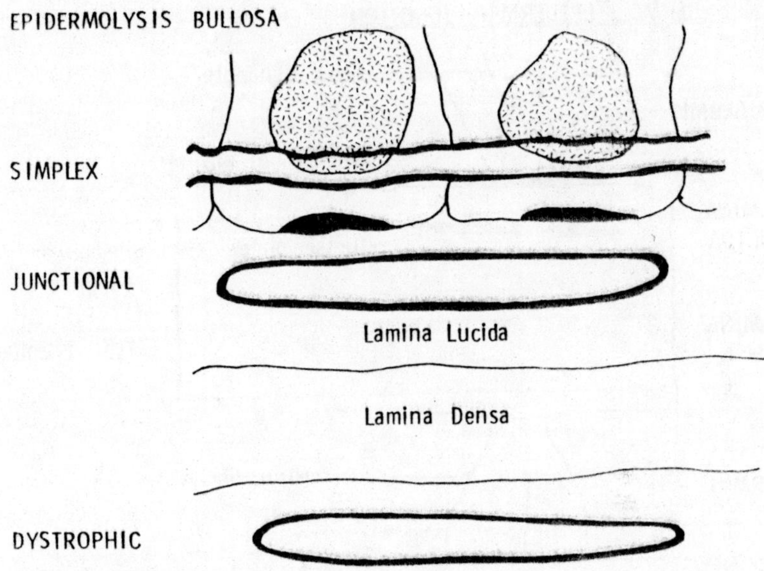

Fig. 17.8. Position of the blisters in the basement membrane zone in the various forms of epidermolysis bullosa.

with high morbidity and are often associated with gut involvement and other disease processes.

The differentiation of these different forms of epidermolysis bullosa is thus important, prognostically and also therapeutically. Using immunofluorescence techniques, the three major forms of epidermolysis bullosa can be differentiated. High titre bullous pemphigoid serum and a specific anti-type 4 collagen antiserum are required. Type 4 collagen is localized in the basal lamina and dermis and is absent from the lamina lucida and epidermis. Bullous pemphigus antigen is localized to the lamina lucida (*Fig.* 17.9). Using an indirect immunofluorescence technique, lesional skin from the patient can be examined and the bullous pemphigoid antigen and type 4 collagen localized within the blister.

In the simplex form, both bullous pemphigoid and type 4 collagen will be found in the floor of the blister. In the junctional forms, bullous pemphigoid antigen will be found in the roof of the blister and, on occasion, in the floor of the blister and type 4 collagen will be localized in the floor. In the dystrophic forms, bullous pemphigoid antigen and type 4 collagen will be present in the roof of the blister. This relatively easy technique may therefore quickly differentiate these major variants.

11. IDENTIFICATION OF T-CELLS IN CUTANEOUS TISSUE SECTIONS

11.1. Introduction

T-cells hold an important place in cutaneous pathology, and T-cells interact with the epidermis in complex and varied ways. Evidence strongly suggests that the skin is a physiological site for T-cell migration and represents a T-cell organ of the body with many similarities to the thymus.

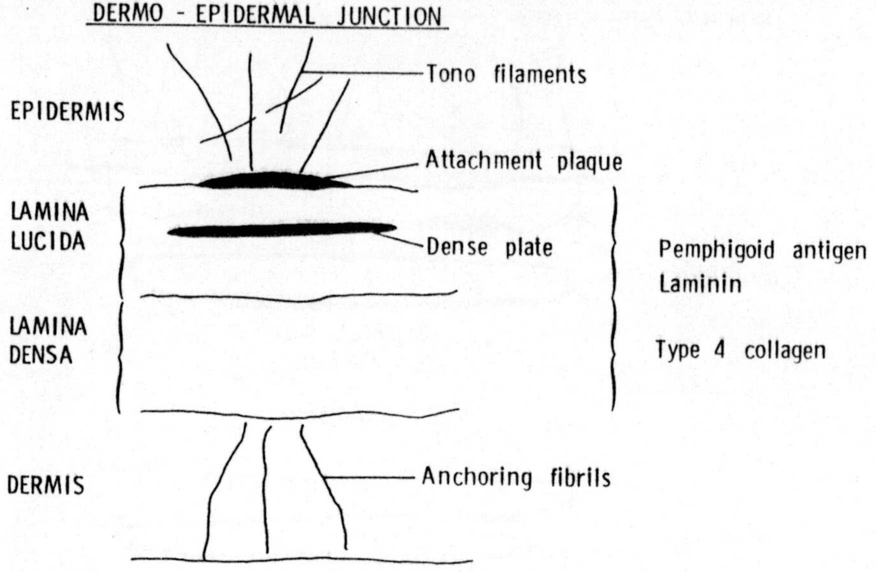

Fig. 17.9. Localization of pemphigoid antigen, laminin and type 4 collagen at the dermo-epidermal junction.

Morphologically, the skin and thymus share certain features:

a. Both possess a keratinizing epithelium—the Hassall's corpuscle in the thymus and the epidermis in the skin.

b. Both produce immunohistochemically identical keratin.

c. Both possess an immigrant population of specialized macrophages—the Langerhans cells.

d. A possible genetic link is seen in the nude mouse which lacks both thymus and an important skin appendage, hair.

Functionally, the skin has been shown to exert a variety of changes in selected T-cell subpopulations:

a. Human and murine epidermal cells have been shown to induce the enzyme terminal deoxyribonucleotidyl transferase in mature T-cells *in vitro.*[21]

b. Epidermal cells have been shown to induce the thymocyte antigen recognized by the mouse monoclonal antibody OKT 6, in malignant helper T-cells.[22]

c. Murine epidermal cells *in vitro* produce an interleukin I-like substance.[23]

d. A substance immunohistochemically identical to thymopoietin has been demonstrated in basal keratinocytes of the epidermis.[24]

In a large number of skin disorders in which lymphocytic infiltrates are present, the predominant cell type has been shown by immunohistochemistry to be the T-cell. These diseases include lichen planus, discoid lupus erythematosus, psoriasis, solar keratoses and chronic eczema. Of particular interest is the finding that the vast majority of lymphomas originating or predominantly affecting the skin are of T-cell origin.

In the ensuing section, we will describe the use of both heterologous and monoclonal antibodies in the investigation of an important group of skin malignancies, the cutaneous T-cell lymphomas.

The cutaneous T-cell lymphomas are neoplasms of T-lymphocytes which originate or predominantly involve the skin. With the advent of immunohistochemical identification of T-cells, the incidence of this group of tumours has been shown to be much greater than previously considered, and probably approximates to that of Hodgkin's disease.[25]

The disease usually starts as a non-specific skin eruption which may simulate eczema, psoriasis or one of the other superficial dermatoses. The disease evolves through a recognized sequence of stages, the skin becoming raised into plaques and then tumid. In the early stages, histology may be non-specific and multiple biopsies are often required before the diagnosis is confirmed. Similarly, in the tumour stage, the histology may be dominated by the presence of tumour cells which may be difficult to differentiate from B-cell or histiocytic tumours.

Immunohistochemistry using antisera to T-cells thus plays an important role in the diagnosis of this condition and differentiation of cutaneous T-cell lymphomas from tumours of other cell types.

11.2. Heterologous and Monoclonal Antibodies to T-Cells

The most significant recent advance in immunology has been the hybridoma method of producing monoclonal antibodies. This discovery has now made possible the large scale production of antibodies which are totally specific for a single antigen.

How do these antibodies differ from conventional heterologous antisera and what are their advantages and disadvantages?

Both heterologous and monoclonal antibodies are initially raised by immunizing animals with an antigen. In the case of antibodies to T-cells the antigen is purified T-cells or thymocytes. In heterologous antisera, the animal used is usually a rabbit or a goat, and for monoclonal antibodies, the animal is a mouse. After a repeated challenge by the T-cells, the animal becomes sensitized to the antigenic sites on the T-cells, producing antibodies to each of these sites. These antigenic sites will include species-specific antigens, HLA-related antigens and T-cell-specific antigens.

For heterologous antisera, the animal is bled and the serum is absorbed extensively with a variety of human tissues and different cell populations including monocytes and B-cells. This results in a mixture of antibodies which will react only with T-cells. These antisera will not differentiate T-cell subpopulations and their specificity will depend on the care taken in the absorption. They will, however, react with a large number of different antigenic sites on T-cells, and may thus be useful in the examination of fixed tissue in which a high percentage of the antigenic sites are masked by protein cross-linkages.

In monoclonal antibody production, once the mouse has been shown to be producing antibodies reactive with the original immunizing cell, the animal is sacrificed and the spleen removed. The majority of spleen cells in such an animal are B-cells sensitized to the antigens present on the immunizing T-cells. Each B-cell will be sensitized to a single antigen. By incubating these spleen cells with myeloma cells in the presence of polyethylene glycol, the nuclear components of B-cells and myeloma cells fuse to produce tetraploid cells. These cells acquire, from the B-cell parent, an antigen specificity and, from the myeloma cell, immortality. The hybrid cells are then screened for antibody specificity, and clones of cells are selected which produce antibodies specifically reactive with different T-cell differentiation antigens. The cells are then grown in tissue culture or in a host mouse and have the potential to produce large amounts of monospecific antibody.

Using this technique, Kung et al.[26] have been able to produce a series of monoclonal antibodies which react with different T-cell differentiation antigens. These antigens have been shown to be acquired and lost by T-cells during their ontogeny from bone marrow precursors to functionally mature T-cells. Using these monoclonal antibodies, T-cells at each stage in their intrathymic and extrathymic differentiation can be identified.

The advantages of monoclonal antibodies over conventional heterologous antisera are fairly obvious. Once the clone of cells has been selected and the antibody characterized, the antibody can be produced in large quantities and will be monospecific for a single antigen. No adsorptions with tissues or cell lines is necessary. The major disadvantages of these antibodies is that the density of the antigen on the surface of the T-cell may be so sparse that the antibody can only be used in very sensitive techniques such as radioimmunoassay. Even with antibodies recognizing an antigen which is present in larger amounts on the T-cell, fixation may significantly reduce the number of antigenic sites available for antibody binding to make fixed tissue unusable with these antibodies (*Table* 17.1).

Table 17.1 **Comparison of heterologous and monoclonal antibodies to T-cells**

Heterologous	*Monoclonal*
A mixture of antibodies reactive with multiple different antigenic sites on the T-cell surface	A single monospecific antibody reactive with a single surface antigen
Will identify T-cells but will not differentiate T-cell subpopulations	Will identify T-cells at all levels of differentiation from prothymocytes to functionally mature helper and suppressor T-cells
May be used in fixed tissue	Reactivity may be obliterated by fixation

11.3. Heterologous Antisera to T-Cells (Anti-Human T-Lymphocyte Antigen, HTLA)

These antisera are commercially available (Institute Merieux, Lyon, France) and have been used successfully in the investigation of cutaneous cellular infiltrates.[27,28] They are produced in rabbits using thymocytes as the antigen and the resulting antiserum is absorbed out on various human tissues, adherent cells and B-lymphoblastoid lines.[29] The specificity of the antisera is controlled by microlymphocytotoxicity tests and inhibition of E rosette formation.

The technique we employ, using these antisera is an indirect immunoperoxidase reaction. Peroxidase-conjugated sheep anti-rabbit globulin is used as the second antiserum, the peroxidase being developed by the Graham and Karnovsky[30] technique.

T-cells can be identified by the presence of a ring of brown reaction product on the cell surface. For details, *see* the Appendix, p. 327.

Immunoperoxidase has certain advantages over conventional immunofluorescence:

a. The preparations are permanent and can be stored indefinitely.
b. The reaction product does not fade and can be examined repeatedly.
c. An ordinary light microscope is used, and the cells are seen in the overall context of the architecture of the skin and the infiltrate.

11.3.1. *Immuno-Electron Microscopy*

Using heterologous antisera, the indirect immunoperoxidase technique can be extended to examine the tissue ultrastructurally. Fresh frozen tissue cannot, however, be used, as freezing causes disruption of the cytoplasmic membrane and such poor preservation that the material is unsuitable for examination at ultrastructural level. For immuno-electron microscopy, therefore, the tissue must be fixed before freezing, which helps to preserve cellular integrity. For details, *see* the Appendix, p. 327.

T-cells are identified by the presence of a granular black reaction product which is restricted to the cytoplasmic membrane of the cells. In counterstained sections, the micromorphology of the cell can be examined including the nuclear characteristics (*Fig.* 17.10).

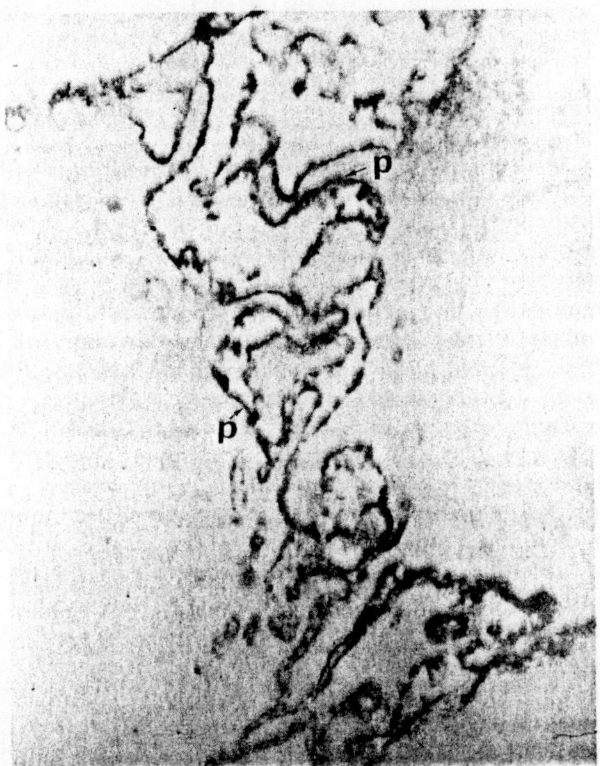

Fig. 17.10. Cutaneous T-cell lymphoma—immuno-electron microscopy using anti-HLA showing two lymphoid cells in opposition (P, peroxidase reaction product).

11.4. Monoclonal Antibodies to T-Cells

The monoclonal antibodies produced by Kung et al.[26] are now commercially available as the orthoclone antibodies (Ortho Pharmaceuticals, Rariton, New Jersey). These are designated OKT (Ortho, Kung, T-cell) and the following have been used in our laboratory: OKT 1 and OKT 3 react with all circulating T-cells and the immunocompetent medullary thymocytes. OKT 4 is present on all thymocytes but selectively reacts with the helper subpopulation of mature circulating T-cells. OKT 8 is present on all thymocytes, but selectively reacts with the suppressor T-cell subpopulation in the blood. OKT 6 is present only on intrathymic cortical thymocytes, but has been found to identify Langerhans cells in the epidermis (*see* Section 12). OKT 10 is not commercially available but is of interest. This antibody reacts with all thymocytes but is lost when the T-cells are peripheralized and less than 5 per cent of circulating mononuclear cells are reactive with OKT 10 (*Table* 17.2).

Table 17.2. ***T-cell ontogeny***

Stage	*Antibody reactivity*
Prothymocyte	OKT 10
Cortical thymocyte	OKT 4, OKT 6, OKT 8, OKT 10
Medullary thymocyte	OKT 1, OKT 3, OKT 4, OKT 8, OKT 10
Helper T-cell	OKT 1, OKT 3, OKT 4
Suppressor T-cell	OKT 1, OKT 3, OKT 8

OKT 1, OKT 3, OKT 6, OKT 8 and OKT 10 react with antigens which are present in high density on the T-cell cytoplasmic membranes. They are thus easily used in immunohistochemistry on tissue section using either direct or indirect immunoperoxidase reactions. OKT 4, however, is reactive to an antigen present at a much lower density and is only usable in indirect or triple-layer techniques. For use in direct immunoperoxidase reaction, the antibodies are first conjugated to horseradish peroxidase, using a modification of the technique of Nakane and Kawoai.[31] The direct technique has the advantage of high specificity and low background, although it is not as sensitive as indirect techniques.

T-cells are recognized by the presence of a ring of dark brown reaction product on the cell surface (*Fig.* 17.11). The intensity of staining varies from cell to cell, reflecting the surface density of the antigen on the cell's surface.

In a recent study of 91 patients with cutaneous T-cell lymphoma using the monoclonal antibodies, three main patterns were observed:

a. 64 per cent of patients showed a fairly homogeneous distribution of the different T-cell subpopulations, of which 60 per cent were OKT 1 +ve, 54 per cent OKT 4 +ve (helper) and 8 per cent were OKT 8 +ve (suppressor). Many patients showed an increase of OKT 8 +ve cells up to 60 per cent at the margins of the infiltrate.

b. 21 per cent of patients showed selective loss of OKT 1 antigen.

c. 15 per cent of patients showed large numbers of OKT 8 +ve cells (50–90 per cent) but the percentages of OKT 1 +ve and OKT 4 +ve cells were within the ranges seen in group 1, suggesting the presence of an immature population of T-cells which were OKT 4 +ve and OKT 8 +ve.

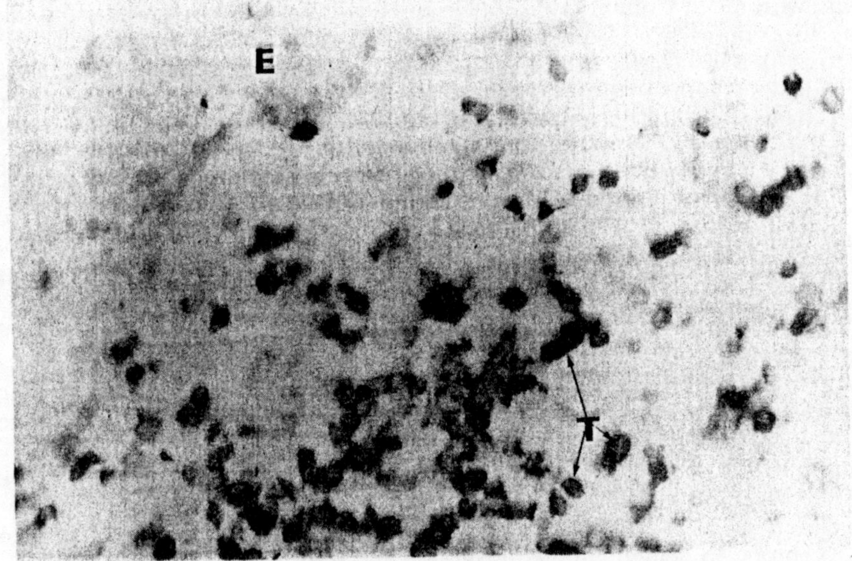

Fig. 17.11. Cutaneous T-cell lymphoma skin—direct immunoperoxidase reaction with OKT 1 (E, epidermis; T, OKT 1-reactive T-cells).

In the epidermis the majority of lymphoid cells were OKT 1 +ve and OKT 4 +ve, but in 26 per cent of patients OKT 8 +ve cells were also identified.

The results of this study suggest that in the typical plaque stage of cutaneous T-cell lymphoma, this technique is only of limited diagnostic value, as similar results may be found in benign dermatoses. Helpful features in differentiating cutaneous T-cell lymphoma from benign disorders using these techniques are:

a. The homogeneous distribution of T-cell subpopulations in the infiltrate. In benign lymphocytic infiltrates, the lymphoid cells are often arranged in a nodular fashion.

b. Selective loss of the OKT 1 antigen.

c. Presence of immature cells which are OKT 4 +ve and OKT 8 +ve.

In tumid lesions, where a monomorphous infiltrate of tumour cells is present, these techniques are of great value in identifying the cells as T-cells.

Recently, Berger et al.[32] have produced two tumour-specific monoclonal antibodies, BE 1 and BE 2. These antibodies recognize cutaneous T-cell lymphoma cells but not normal thymocytes or peripheral blood T-cells. The use of these antibodies will hopefully provide the criteria for the early diagnosis of cutaneous T-cell lymphoma.

12. OKT 6, A LANGERHANS CELL MARKER

OKT 6, as has previously been described, is a mouse monoclonal antibody raised against human thymocytes and reactive with the cortical immunocompetent thymocytes. A chance finding, however, has made this antibody one of the most important reagents in the investigation of cutaneous disease.

Fithian, working at Columbia Presbyterian Hospital, New York, was using this antibody to examine skin from patients with cutaneous T-cell lymphoma, and noticed that dendritic cells in the epidermis were reactive with this antibody. Double labelling with an anti-HLA-DR antibody showed that these OKT 6 reactive cells were also HLA-DR positive and were thus Langerhans cells.[33]

Langerhans cells are an immigrant population of epidermal cells which are considered to be of bone marrow origin.[34] They are immunocompetent cells which are able to replace monocytes in presenting antigen to T-cells and in the mixed lymphocyte culture.[35] The Langerhans cells are central in the pathogenesis of contact allergic sensitization[36] and may be the target cells in allogenic graft rejection.[37] The cells have also been implicated in the pathogenesis of cutaneous T-cell lymphoma and graft versus host disease.

Langerhans cells possess a number of membrane and cytoplasmic characteristics which help to distinguish them from other epidermal cells. The cells are aureophilic and will selectively take up gold, and express readily identifiable surface ATPase activity. They possess receptors for the Fc fragment of IgG, the split product of the third component of complement, and have membrane-associated Ia antigen. These markers, however, do not allow consistent and reliable differentiation from other mononuclear cells, as B-cells and monocytes also possess Fc receptors, Ia antigen and C_3 receptors. In benign cutaneous disease and in normal healthy individuals, OKT 6 is not found in extrathymic sites. A recent study by Griffiths-Chu et al.[38] has shown the presence of OKT 6-reactive T-cells in neonatal blood, but this appears to be related to a physiologically increased thymocyte division rate associated with normal fetal development.

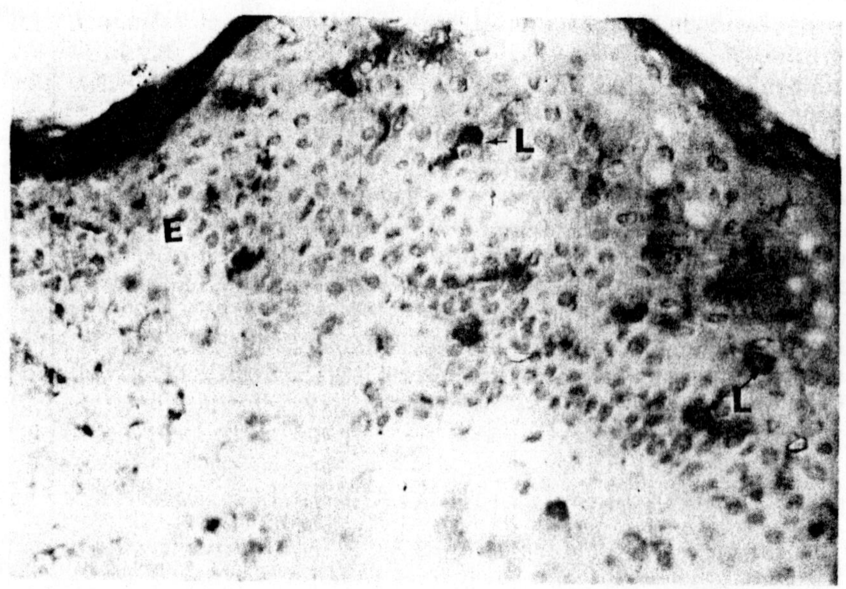

Fig. 17.12. Normal human skin—direct immunoperoxidase reaction with OKT 6 (E, epidermis; L, Langerhans cells).

OKT 6 is thus a specific and reliable marker for Langerhans cells, and is an important reagent not only for the investigation of cutaneous disease but also in the purification and examination of Langerhans cells.

OKT 6 may be used in direct immunoperoxidase reactions. In cutaneous T-cell lymphoma, 1–5 per cent of the cells in the dermal infiltrate are OKT 6 +ve which correlates well with ultrastructural studies which have shown Langerhans cells to be present in the dermal infiltrates of cutaneous T-cell lymphoma (*Fig.* 17.12).

13. IMMUNO-ELECTRON MICROSCOPY USING MONOCLONAL ANTIBODIES

Fixation of tissue with the majority of fixatives: formalin, paraformaldehyde, glutaraldehyde, obliterates reactivity with all the OKT series of antibodies. The immuno-electron microscopical techniques using heterologous antisera, are thus not suitable for use with monoclonal antibodies. Until an appropriate fixative is found, fresh tissue is necessary when monoclonal antibodies are used in immuno-electron microscopy.

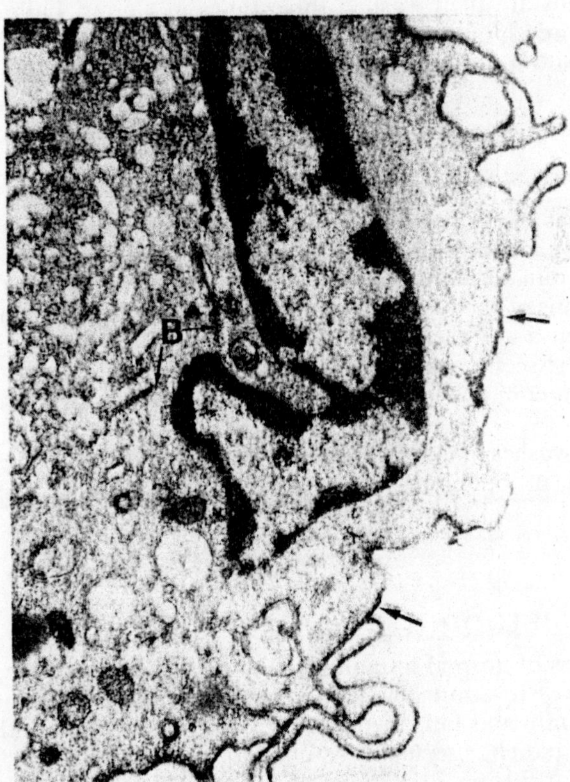

Fig. 17.13. Immuno-electron microscopy on epidermal cell suspension showing a Langerhans cell labelled with OKT 6 (→ peroxidase reaction product; B, Birbeck granules).

13.1. Immuno-Electron Microscopy of Langerhans Cells with OKT 6[39]

In this technique, Langerhans cells are examined in a cell suspension, and sensitivity to fixation is circumvented by using fresh cells. Epidermal cell suspensions are produced using cold trypsinization.[40] For method, *see* the Appendix, p. 330.

Langerhans cells can be identified by the black reaction product on the cytoplasmic membrane, and in counterstained sections Birbeck granules can be seen (*Fig.* 17.13).

Appendix

I. DIRECT IMMUNOFLUORESCENCE METHOD

For direct immunofluorescence, five cover slips are needed for each skin biopsy. 5-μm sections are fan dried for 10 min and washed in Coons' buffer (Formula III) for 10 min. The sections are then dried for 10 min and covered with an appropriate dilution of one of the following fluorescein-conjugated antisera: anti-IgM, anti-IgG, anti-IgA, anti-C_3 and anti-fibrin, and incubated at 37 °C for 30 min in a moist chamber. The sections are then washed three times in Coons' buffer for 10 min each time, fan dried and mounted in buffered glycerol (Formula II). The sections are then examined with a fluorescence microscope.

II. INDIRECT IMMUNOFLUORESCENCE METHOD

5-μm cryostat sections of the tissue substrate are cut onto cover slips. Two cover slips are required for each serum. The sections are fan dried for 10 min, washed in Coons' buffer for 10 min and fan dried for 10 min. Each cover slip is then covered by one of two dilutions of the patient's serum—1 in 10 and 1 in 80, and incubated for 30 min at 37 °C in a moist chamber. After three washes in Coons' buffer for 10 min each wash, the sections are fan dried and incubated with an appropriate dilution of fluorescein-conjugated anti-human IgG at 37 °C in a moist chamber for 30 min.

After a final three washes in Coons' buffer for 10 min each wash, the sections are fan dried, mounted in buffered glycerol and examined with a fluorescence microscope.

III. INDIRECT C_3 METHOD

5-μm cryostat sections of normal human skin are cut onto cover slips. Two cover slips are required for each serum. The sections are fan dried for 10 min, washed in Coons' buffer for 10 min and fan dried for 10 min. The sections are then covered either with undiluted patient's serum or serum diluted 1 in 4 in Coons' buffer, and incubated at 37 °C for 30 min in a moist chamber. The sections are washed in Coons' buffer for 30 min (three changes), fan dried and incubated with complement. The source of complement we generally use is normal human serum diluted 1 in 5 in complement-diluting buffer (Formula IV). The sections are incubated at 37 °C for 30 min in a moist chamber and then washed three times for 10 min each wash in Coons' buffer. The sections are fan dried for 10 min and then covered with an appropriate dilution of a fluorescein-conjugated anti-C_3 and incubated at 37 °C for 30 min in a moist chamber. After a final three washes at 10 min a wash in Coons' buffer, the sections are fan dried, mounted in buffered glycerol and examined with a fluorescence microscope.

IV. INDIRECT IMMUNOPEROXIDASE REACTION (HETEROLOGOUS)

4-μm cryostat sections are cut onto cover slips and fixed for 10 min in 3 per cent paraformaldehyde in phosphate-buffered saline. After a 30-min wash, the sections are incubated with the antibody to human T-lymphocytes at an appropriate dilution for 30 min at room temperature in a moist chamber. After a 30-min wash the sections are incubated for a further 30 min with the peroxidase-conjugated sheep anti-rabbit antiserum at an appropriate dilution at room temperature in a moist chamber. After a further 30-min wash the sections are incubated with the Graham-Karnovsky medium[14] (Formula V) for 10 min in the dark, washed for 30 min, dehydrated through graduated alcohols, cleared in xylene and mounted in DPX.

The intensity of staining can be increased by counter-staining with one of the blue dyes—methylene blue, toluidine blue, or treating the sections with 1 per cent osmium tetroxide (Formula VI) for 10 min after the final wash.

V. IMMUNO-ELECTRON MICROSCOPY

After biopsy the skin is cut into 2-mm cubes and fixed for 1 h in 3 per cent paraformaldehyde in 0·1 M phosphate-buffered saline (PBS) pH 7·5 containing 8·5 per cent sucrose at 4 °C. The specimens are then washed in 0·1 M PBS (pH 7·5) containing 8·5 per cent sucrose for 16 h at 4 °C, blotted dry and frozen as previously described.

VI. INDIRECT IMMUNOPEROXIDASE FOR ELECTRON MICROSCOPY

20-μm cryostat sections are cut onto coverslips and air dried for 30 min. The sections are washed for 30 min and incubated with the human T-lymphocyte antibody for 30 min at room temperature in a moist chamber. After a wash for 30 min, the sections are incubated with the peroxidase-conjugated anti-rabbit IgG at an appropriate dilution for 30 min at room temperature in a moist chamber.

The sections are washed for a further 30 min and incubated with the Graham–Karnovsky medium for 30 min at room temperature in the dark. After a final wash for 30 min the sections are post-fixed in 1 per cent osmium tetroxide (Formula VI) for 10 min, washed in distilled water from a wash bottle and dehydrated in alcohol: 5 min each in 25, 50, 70 and 95 per cent alcohol and two 10-min baths in 100 per cent alcohol. The sections are then washed in two 5-min baths of propylene oxide and a 50 per cent solution of propylene oxide and resin (TAAB Laboratories) is layered over the sections for 20 min. The sections are drained by putting them on their sides for 5 min and are then layered with resin alone overnight at 4 °C. The resin is drained off and fresh resin is replaced for 4 h at room temperature. Gelatin capsules are then filled with the resin and inverted over the sections and the resin is polymerized at 60 °C for 16 h (*Fig.* 17.14). The capsules are then separated from the cover slips by plunging them into liquid nitrogen, which results in a clean separation of the cover slip from the resin block which contains the section.

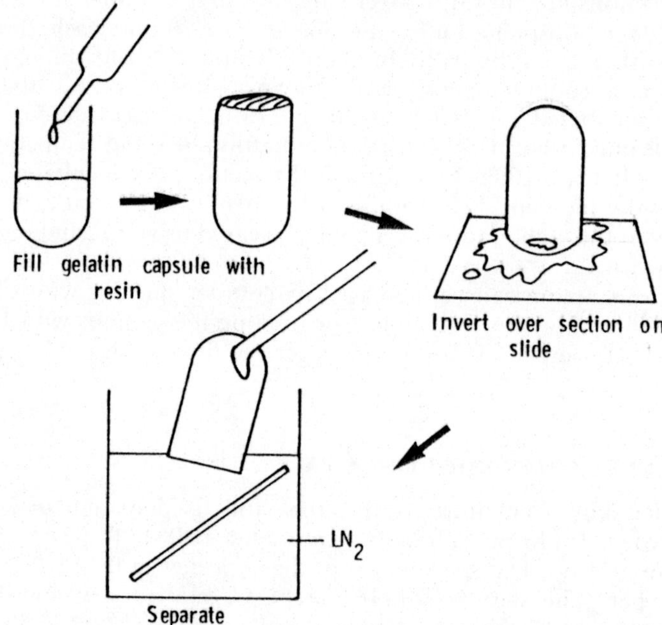

Fig. 17.14. Resin embedding for immuno-electron microscopy.

The block is then trimmed and cut on an ultramicrotome and thin sections are examined unstained at 50 kV, or counterstained with uranyl acetate and lead citrate at 75 kV in the electron microscope.

VII. PEROXIDASE CONJUGATION OF ANTIBODIES

The horseradish peroxidase must have a high degree of purity [Reinheitszahl (RZ) of 3 or more].

The quantities given can be increased or decreased proportionally depending on the original weight of the horseradish peroxidase.

Dissolve 16·7 mg horseradish peroxidase in 3 ml distilled water. Add 0·33 ml 1 M bicarbonate buffer pH 9·6. Add 0·15 ml 1 per cent phenylisothiocyanate in absolute alcohol and stir using a magnetic stir bar at room temperature for 1 h.

Add 0·33 ml 1 M acetate buffer pH 4 to readjust the pH to about 6. Add 3·3 ml 0·06 M sodium metaperiodate, freshly made, dropwise. Stir for 20 min at room temperature at which time the solution will have turned from brown to green.

Add 0·033 ml ethylene glycol and stir for 15–30 min at room temperature. The solution turns brown and formaldehyde is released.

Dialyse against 1 litre 1 mM acetate buffer pH 4·5 at 4 °C with three changes for 12 h.

The resulting solution will be 8·1 ml. To this add 0·405 ml 1 M bicarbonate buffer pH 9·5 and add the purified immunoglobulin in the ratio 5 mg enzyme to 7·5 mg antibody. Stir for 1 h at room temperature. To this add a freshly made solution of sodium borohydride (5 mg/ml) at the ratio 1 mg sodium borohydride to 10 mg peroxidase and stir for 20 min at room temperature. During this period, hydrogen will be released and will form a sheet of bubbles on the surface. Haem groups present will be reduced and will give the solution a reddish tinge.

Add an equal volume of saturated ammonium sulphate which will precipitate the immunoglobulin conjugate but will leave free peroxidase in solution. Spin down the pellet at 2000 rev./min, and resuspend in phosphate-buffered saline. Dialyse for 12 h at 4 °C against 1 litre of phosphate-buffered saline with three changes.

To check the success of the conjugation, measure the absorbance with a spectrophotometer at 405 and 280 nm. The ratio 405/280 should approximate 0·5–0·7. If it is over 0·7, there will be problems with background.

VIII. DIRECT IMMUNOPEROXIDASE REACTION (MONOCLONAL)

5-µm cryostat sections are cut onto cover slips, and fixed in acetone at room temperature for 5 min. The sections are washed for 30 min, fan dried and incubated with an appropriate dilution of the peroxidase-conjugated monoclonal anti-T-cell antibody for 30 min at room temperature in a moist chamber. The sections are washed in two 15-min changes of buffer, and then incubated with the Graham–Karnovsky medium for 10 min in the dark. After a final 30-min wash, the sections are dehydrated through graduated alcohols, cleared in xylene and mounted in DPX.

IX. INDIRECT IMMUNOPEROXIDASE REACTION (MONOCLONAL)

5-µm cryostat sections are cut onto cover slips and fixed in acetone for 5 min at room temperature. The sections are washed for 30 min and then incubated with an appropriate dilution (1 in 40 of commercially available reagents) of the monoclonal antibodies to T-cells for 30 min at room temperature in a moist chamber. After two 15-min washes the sections are fan dried and incubated for 30 min at room temperature in a moist chamber. After two 15-min washes, the sections are incubated with the Graham–Karnovsky medium for 10 min in the dark, washed for 30 min, dehydrated through graduated alcohols, cleared in xylene and mounted in DPX.

X. IMMUNO-ELECTRON MICROSCOPY OF LANGERHANS CELLS WITH OKT 6

Fresh skin is cut into small strips and washed in 1 : 5000 EDTA. The strips are then incubated for 16 h at 4 °C in crude trypsin 0·25 per cent. The epidermis is then separated from the dermis using fine forceps, and the epidermal sheets are agitated in trypsin–EDTA. The resulting cell suspension is filtered through gauze onto 5 ml of fetal calf serum. The cells are washed and spun at 1200 rev./min for 6 min to pellet the cells. The cells are then incubated with the peroxidase-conjugated OKT 6 for 30 min at 4 °C. After three washes, the cells are fixed in 2 per cent glutaraldehyde in cacodylate buffer for 10 min, and then washed three times. The cells are then incubated with the Graham–Karnovsky medium for 10 min at room temperature, washed three times and post-fixed for 10 min in 1 per cent osmium tetroxide. The cells are then dehydrated through graduated alcohols and washed in propylene oxide. The cells are impregnated with resin by incubation in 50 per cent resin in alcohol for 30 min, then resin alone for 16 h, and polymerized in fresh resin at 60 °C for 16 h. Dehydration and resin embedding is best conducted in a BEEM capsule as cell loss is reduced.

The blocks can then be cut on an ultramicrotome and thin sections examined unstained at 50 kV or counterstained with lead citrate and uranyl acetate at 75 kV.

Formula I. Liquid fixative for direct immunofluorescence

Stock solution may be stored indefinitely but buffer and fixative should be made fresh each week.

Buffer

1 M potassium citrate buffer (pH 7·0)	2·5 ml
0·1 M magnesium sulphate	5·0 ml
0·1 M methylmaleimide	5·0 ml
Distilled water	87·5 ml

Adjust pH to 7·0 with 1 M potassium hydroxide

Fixative

Buffer	100 ml
Ammonium sulphate	55 g

Formula II. Buffered glycerol pH 8·5

$NaHCO_3$	0·0715 g
Na_2CO_3	0·016 g
Distilled water	10 ml
Glycerol	90 ml

Formula III. Coon's phosphate-buffered saline pH 7·4

NaCl	20·25 g
Na_2HPO_4	3·20 g
$NaH_2PO_4 \cdot 2H_2O$	0·39 g
Distilled water	2·5 litre

Formula IV. Complement-diluting buffer

Barbitone	0·575 g
NaCl	8·5 g
$MgCl_2$	0·168 g
$CaCl_2$	0·028 g
Barbitone soluble	0·185g
Distilled water	1 litre
pH 7·2	

This buffer is commercially available as Oxoid complement fixation test diluent tablets.

Formula V. Graham–Karnovsky medium

10 mg 3′,3′-diaminobenzidine tetrahydrochloride
10 ml phosphate-buffered saline pH 7·4
Stir at 80 °C for 15 min (in water bath)
Filter through a 0·45-μm millipore filter
Before use, add 1 drop 3 per cent hydrogen peroxide

Formula VI. 1 per cent osmium tetroxide

2 per cent osmium tetroxide in distilled water	1 ml
Sodium cacodylate, 0·4 M	0·5 ml
Glucose, 10·8 per cent	0·5 ml

REFERENCES

1. Chayen J., Bitensky L., Butcher R. and Poulter L. *A Guide to Practical Histochemistry*. Edinburgh, Oliver and Boyd, 1969.
2. Michel B., Milner Y. and David K. Preservation of tissue-fixed immunoglobulins in skin biopsies of patients with lupus erythematosus and bullous diseases. Preliminary report. *J. Invest. Dermatol.* 1973, **59**, 449–452.
3. Skeete M. V. H. and Black M. M. The evaluation of a special liquid fixative for direct immunofluorescence. *Clin. Exp. Dermatol.* 1977, **2**, 49–56.
4. Wolff K. and Schreiner E. Ultrastructural localisation of pemphigus auto-antibodies within the epidermis. *Nature* 1971, **229**, 59–61.
5. O'Loughlin S., Goldman G. C. and Provost T. T. Fate of pemphigus antibody following successful therapy. *Arch. Dermatol.* 1978, **114**, 1769–1772.
6. Bystryn J. C., Abel E. and Defeo C. Pemphigus foliaceus—subcorneal intercellular antibodies of unique specificity. *Arch. Dermatol.* 1974, **110**, 857–861.
7. Schiltz J. R. and Michel B. Production of epidermal acanthosis in normal human skin *in vitro* by the IgG fraction from pemphigus serum. *J. Invest. Dermatol.* 1976, **67**, 254–260.
8. Farb R. M., Dykes R. and Lazarus G. S. Anti-epidermal cell surface pemphigus antibody detaches viable epidermal cells from culture plates by activation of proteinase. *Proc. Natl Acad. Sci. USA* 1978, **75**, 459–463.
9. Diaz C. A., Calvanico N. J., Tomasi T. B. Jr and Jordan R. E. Bullous pemphigoid antigen: isolation from normal human skin. *J. Immunol.* 1977, **118**, 455–460.
10. Diaz-Perez J. L. and Jordan R. E. The complement system in bullous pemphigoid. IV. Chemotactic activity in blister fluid. *Clin. Immunol. Immunopathol.* 1976, **5**, 360–370.

11. Wintrobe B. U., Soter N. A., Mihm M. C. Jr, Goetzel E. J. and Austin K. F. Functional and morphologic evidence of mast cell involvement in bullous pemphigoid. *Clin. Res.* 1977, **25**, 288 A.
12. Sams W. M. Jr and Gleich G. J. Failure to transfer bullous pemphigoid with serum from patients. *Proc. Soc. Exp. Biol. Med.* 1971, **136**, 1027–1031.
13. Hodge L., Marsden R. A., Black M. M., Bhogal B. and Corbett M. F. Bullous pemphigoid: the frequency of mucosal involvement and concurrent malignancy related to indirect immunofluorescent findings. *Br. J. Dermatol.* 1981, **105**, 65–69.
14. Koldony R. C. Herpes gestationis: a new assessment of incidence, diagnosis and fetal prognosis. *Am. J. Obstet. Gynecol.* 1969, **104**, 39–45.
15. Yaoita H., Gullino M. and Katz S. I. Herpes gestationis. Ultrastructure and ultrastructural localisation of *in vivo*-bound complement. *J. Invest. Dermatol.* 1976, **66**, 383–388.
16. Hönigsmann H., Stingl G., Holubar K. and Wolff K. Herpes gestationis: fine structural pattern of immunoglobulin deposits in the skin *in vivo*. *J. Invest. Dermatol.* 1976, **66**, 389–392.
17. Susi F. R. and Shklar G. Histochemistry and fine structure of oral lesions of mucous membrane pemphigoid. *Arch. Dermatol.* 1971, **104**, 244–253.
18. Seah P. P., Fry L., Holborow E. J., Rossiter M. A., Doe W. E., Magalhaes A. F. and Hoffbrand A. V. Antireticulin antibody: incidence and diagnostic significance. *Gut* 1973, **14**, 311–315.
19. Yaoita H. and Katz S. I. Immunoelectron microscopic localisation of IgA in skin of patients with dermatitis herpetiformis. *J. Invest. Dermatol.* 1976, **67**, 502–506.
20. Stingl G., Hönigsmann H., Holubar K. and Wolff K. Ultrastructural localisation of immunoglobulins in dermatitis herpetiformis. *J. Invest. Dermatol.* 1976, **67**, 507–512.
21. Rubenfeld M., Silverstone A., Knowles D., Halper J., De Sostoa A., Fenoglio C. and Edelson R. Induction of lymphocyte differentiation by epidermal cultures. *J. Invest. Dermatol.* 1981, **71**, 221–225.
22. Chu A. C., Berger C. L. and Edelson R. L. Induction of a thymocyte phenotype in malignant helper T-cells by epidermal cell cultures. *Br. J. Dermatol.* 1982, **106**, 736.
23. Luger T. A., Stadler B. M., Katz S. I. and Oppenheim J. J. Epidermal cell (keratinocyte)—derived thymocyte-activating factor (ETAF). *J. Immunol.* 1981, **127**, 1493–1498.
24. Chu A. C., Patterson J., Goldstein G., Berger C., Takezaki S. and Edelson R. Thymopoietin-like substance in human skin. 1982, submitted for publication.
25. Edelson R. L. Cutaneous T cell lymphoma: mycosis fungoides, Sezary syndrome and other variants. *J. Am. Acad. Dermatol.* 1980, **2**, 89–106.
26. Kung P. C., Goldstein G., Reinherz E. L. and Schlossman S. F. Monoclonal antibodies defining distinctive human T cell surface antigens. *Science* 1979, **206**, 347–349.
27. Chu A. C. and Macdonald D. M. Identification in situ of T lymphocytes in the dermal and epidermal infiltrates of mycosis fungoides. *Br. J. Dermatol.* 1979, **100**, 177–182.
28. Alario A., Ortonne J. P., Schmitt D. and Thivolet J. Lichen planus: study with anti-human T lymphocyte antigen (anti-HTLA) serum on frozen tissue sections. *Br. J. Dermatol.* 1978, **98**, 601–604.
29. Touraine J. L., Touraine F., Kiszkiss D. F., Choi Y. S. and Good R. A. Heterologous specific antiserum for identification of human T cells. *Clin. Exp. Immunol.* 1974, **16**, 503–520.
30. Graham R. C. and Karnovsky M. J. The early stages of absorption of injected horseradish peroxidase in the proximal tubule of the mouse kidney: ultrastructural evidence by a new technique. *J. Histochem. Cytochem.* 1966, **14**, 291–302.
31. Nakane P. K. and Kawaoi A. Peroxidase-labelled antibody. A new method of conjugation. *J. Histochem. Cytochem.* 1974, **22**, 1084–1092.
32. Berger C. L., Takezaki S., De Pietro W., Chu A., Morrison S. and Edelson R. Production of monoclonal antibodies reactive with the neoplastic lymphocytes of cutaneous T cell lymphoma. *Clin. Res.* 1981, **29**, 588A.
33. Fithian E., Kung P., Goldstein G., Rubenfeld M., Fenoglio C. and Edelson R. Reactivity of Langerhans cells with hybridoma antibody. *Proc. Natl Acad. Sci. USA* 1981, **78**, 2541–2544.
34. Tamaki K. and Katz S. Ontogeny of Langerhans cells. *J. Invest. Dermatol.* 1980, **75**, 12–13.
35. Stingl G., Katz S. I., Clement L., Green I. and Shevach E. M. Immunological functions of Ia-bearing epidermal Langerhans cells. *J. Immunol.* 1978, **121**, 2005–2012.
36. Silberberg I., Baer R. L. and Rosenthal S. A. The role of Langerhans' cells in contact allergy. I. An ultrastructural study in actively induced contact dermatitis in guinea pigs. *Acta Derm. Venereol.* 1974, **54**, 321–331.
37. Rowden G. R. Expression of Ia antigens on Langerhans cells in mice, guinea pigs and man. *J. Invest. Dermatol.* 1980, **75**, 22–31.

38. Griffiths-Chu S. M., Patterson J., Chu A. C., Berger C. L., Goldstein G. and Edelson R. L. A population of immature T cells in normal neonatal blood. 1982, submitted for publication.
39. Chu A. C., Eisinger M., Lee J. S., Takezaki S., Kung P. C. and Edelson R. L. Immuno-electron microscopic identification of Langerhans cells using a new antigenic marker. *J. Invest. Dermatol.* 1982, **78**, 177–180.
40. Eisinger M., Lee J. S., Hefton J. M., Darzynkiewicz Z., Chiao J. W. and De Harven E. Human epidermal cell cultures. Growth and differentiation in the absence of dermal components or medium supplements. *Proc. Natl Acad. Sci. USA* 1979, **76**, 5430–5344.

18

Immunocytochemistry in Histopathology with particular reference to Renal Disease

D. J. Evans

Although the achievements of immunocytochemistry over the past decades have been impressive, the advances have made a major impact on only a few areas of diagnostic histopathology. I thought it might be of interest to examine the reasons for this, and to speculate on the possible extensions to its use. Naturally, I shall need to cover some of the topics previously reviewed, but I will be doing this with the cynical eye of the practising pathologist rather than with the optimism of the researcher.

1. TUMOUR DIAGNOSIS

One may get the impression from the enthusiast, that the diagnostic histopathologist spends much of his time being bewildered by tumours of obscure origin or histogenesis. This is a serious over-statement of the case; probably more than 90 per cent of tumours present with an obvious primary site. Where presentation is with metastases, the question of the nature of the primary tumour is only of relevance to the patient and to the clinician where the tumour is one of those which responds to therapy with hormones, cytotoxics or radiotherapy, and this is unfortunately all too rare.

One such tumour is the seminoma, which may originate in the testis or in the mediastinum. This is exquisitely radiosensitive and it may be successfully treated by radiotherapy or cytotoxic agents even in the presence of metastases. It is relatively readily recognized morphologically, which is just as well, since no reliable marker exists for it. Another example of a tumour which responds to cytotoxic agents is shown in *Fig.* 18.1. A young man who presented with haemoptysis was found to have extensive lung shadowing on chest X-ray and was referred with a suggested diagnosis of Goodpasture's syndrome. The lung lesions proved to be due to haemorrhagic tumour, histologically a choriocarcinoma. The presence of chorionic gonadotrophin was confirmed in the tumour by immunostaining (*Fig.* 18.2) and elevated levels were found in the urine and blood. The tumour proved to have originated from a mediastinal primary and the patient had Klinefelter's syndrome.[1,2] Despite the moribund state of the patient and the presence of a cerebral metastasis, after 3 years the patient is alive and tumour free following chemotherapy. In this case morphology alone sufficed to make the diagnosis and since this could be checked by serological methods it would be hard to argue that immunocytochemistry had made an important contribution. A similar argument could be applied to endodermal sinus tumours (where alpha-fetoprotein is the marker), to prostatic carcinoma (acid phosphatase) and to medullary carcinoma of the thyroid (calcitonin).

With follicular and papillary carcinomas of the thyroid, the case for immunocytochemistry is substantially stronger, since alterations in the serum are likely to be absent. Cases have been reported, particularly with papillary carcinomas of the thyroid, where positive recognition could only be achieved by the identification of thyroglobulin production.[3]

In my view, with the exception of calcitonin, identification of peptide hormones in tumours has so far proved of little practical diagnostic use although other pathologists may take a different view (*see* Chapter 20). It is often extremely difficult to demonstrate a peptide in a tumour which is actively secreting a hormone; for instance, in a case of appendiceal carcinoid which was associated with Cushing's syndrome and produced ectopic adrenocorticotropic hormone (ACTH) as demonstrated by serum assays, only an occasional cell showed positive staining. This is, of course, not always the case: bronchial carcinoids producing Cushing's syndrome may show widespread staining for ACTH. The classification of the tumour here is based on morphological features and does not depend upon the presence of the peptide hormone. You will see at this stage why I am prejudiced against the use of the term APUDoma, which is often less informative clinically and prognostically than the conventional histopathological name: for instance, both oat cell carcinoma of the lung and carcinoid of the appendix shelter under this umbrella term, but it is hardly necessary for me to point out the differences in the prognosis in the two cases.

In carcinomas, the list of antigens identified grows steadily longer. Carcinoembryonic antigen, casein, and epithelial membrane antigen are all acquiring an extensive literature (*see* Chapter 15) but other, less popular, antigens have been described.[4,5] The major disadvantage to most of them is their lack of specificity, though they will often narrow the range of diagnostic possibilities as to the primary origin of a metastatic carcinoma, and they may aid the recognition of tumour cells, for instance in bone marrow biopsies, when for various reasons this is morphologically difficult.

Two problems which do occasionally confront the histopathologist are the distinction of a poorly differentiated carcinoma from a lymphoma and the recognition of a sarcoma from a pseudosarcomatous carcinoma. It seems that the majority of lymphomas express HLA-DR antigens and that they contain vimentin as the intermediate filament in their cytoplasm. Most sarcomas also contain vimentin, while in carcinomas prekeratin is the intermediate filament normally

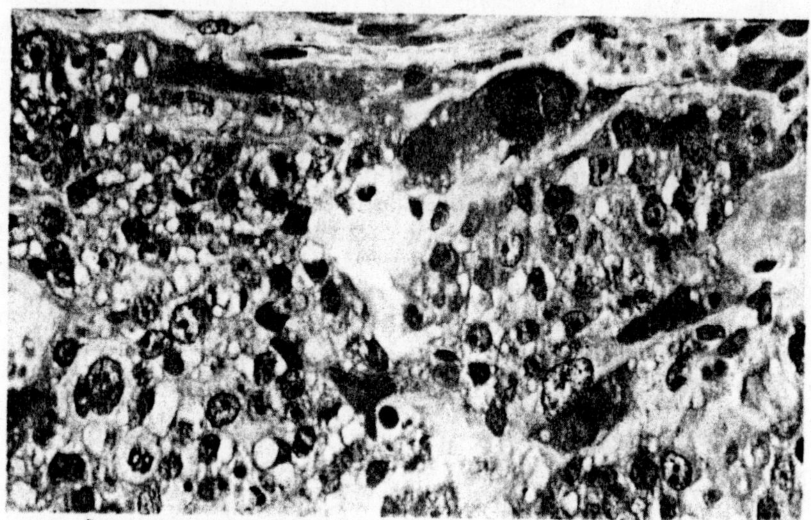

Fig. 18.1. Metastatic choriocarcinoma in lung. Haematoxylin and eosin.

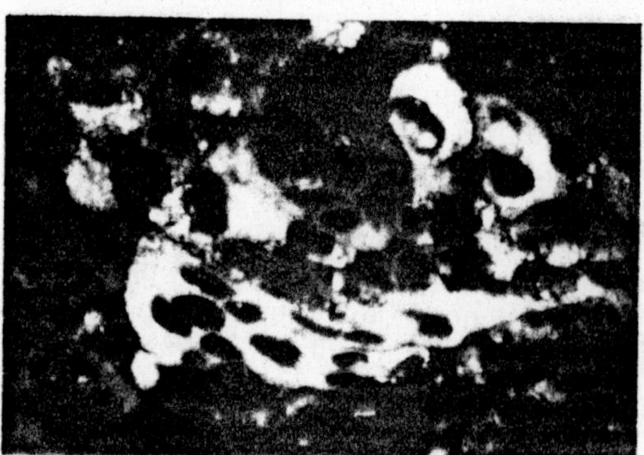

Fig. 18.2. Tissue from metastatic choriocarcinoma stained for β-HCG.

present. These sera are, however, not as yet widely available, and it remains to be seen whether they will acquire a widespread use (*see* Chapter 16).

There is no doubt that cell surface markers may contribute to the classification of lymphomas and help to distinguish lymphomas from pseudolymphomas (*see* Chapter 14). There is however an organizational difficulty in that fresh frozen material is required since satisfactory results can rarely be obtained from processed tissue. This means that the diagnostic problem must have been foreseen prior to the time of the biopsy. Not all our colleagues, alas, seem capable of this foresight.

2. CELL IDENTIFICATION

The development of monoclonal antibodies has rendered it possible to recognize subsets of lymphocytes in suitably frozen tissue specimens. This might potentially have applications in the areas of skin, renal and gastrointestinal biopsy, though I am not as yet convinced, since it seems usually possible to identify lichen planus without characterizing the lymphocytes as T-cells, and to recognize transplant rejection without subcategorizing the infiltrate.

The histiocyte markers (R490, trypsin, α-1-antitrypsin, lysozyme) seem to have been chiefly used in trying to pin down tumours of histiocytic origin and it is indeed difficult to find other practical problems that require their use.

If cell differentiation is controlled through the nucleus it might in theory be possible to find nuclear constituents which are characteristic of individual cell types. Spurred on by the existence of spontaneous antinuclear factors specific for polymorphs, we succeeded in producing an antibody experimentally.[6] However, this antigen proved extremely fragile to fixatives, and we have not so far succeeded in producing comparable antisera to epithelial cell nuclei.

3. SECRETORY PRODUCTS

It often seems to happen that a marker is first thought to be specific for a single group of cells, and is subsequently found to be present in a wider range of locations. An example of this is α-1-antitrypsin, which was first found to be present in the serum, then identified in the globular inclusions in liver cells of patients with a biochemical deficiency of the enzyme. Antisera are now used much more frequently in the study of histiocytes or Paneth cells (which also contain the enzyme) than for the rather uncommon purpose of identifying the liver inclusions. Recently the existence of angiotensin-converting enzyme has been described in the cells of sarcoid granulomas.[7] It is said to be absent in the cells of other types of granulomas and, if confirmed, would provide a valuable technique for positively identifying sarcoid, the diagnosis of which at the moment is essentially one of exclusion. The elimination of tuberculosis as a differential diagnosis can be a matter of great difficulty, especially in patients of Asian origin.

4. INFECTIVE AGENTS

With infective agents, immunocytochemical methods are largely restricted to a confirmatory role. There is a long list of agents potentially recognizable, ranging from viruses to fungi.

Through immunochemical staining, *Legionella* is readily recognizable morphologically in what may otherwise appear a non-specific lung lesion, and the antigen is readily identified in routinely processed material[8] (*Plate* 23). On the other hand, in order to demonstrate the spirochaetes of syphilitic lesions the tissue must be optimally fixed[9] (*Plate* 24). Whipple's disease is a putative infection, but the organism involved has so far resisted identification. In the intestine and mesenteric nodes, the disease shows a characteristic histological appearance with lipid-containing spaces in the tissue and many macrophages which have an eosinophilic cytoplasm showing strong granular positivity on staining with the periodic acid–Schiff (PAS) method. The diagnosis is considerably more difficult in the rectal biopsy, or in the peripheral lymph nodes, where there may be only scattered aggregates of macrophages. It appears that the PAS-positive macrophages contain a polysaccharide in which rhamnose is an important component. This may be recognized by antisera to type B and type G streptococci. This allows the macrophages to be confidently identified in a wide variety of tissues (*Plate* 25). This antigen too is stable to fixation and paraffin embedding.[10,11]

Amoebae in rectal biopsy material are sometimes difficult to distinguish from degenerating cells, though their positivity with the PAS stain is a helpful discriminatory point. Positive identification can be made on paraffin sections using antiserum obtained from an infected patient.

Specific identification of fungi is not usually the domain of the histopathologist and indeed morphological methods are much inferior to culture. However, immunofluorescence has for many years been used in this field and may narrow the field of possibilities even when it does not enable a species identification to be made.[12,13]

The diagnosis of toxoplasmosis is fairly frequently and accurately made by surgical pathologists although they may ask for a confirmatory dye test to be performed on the serum. The intact organism is rarely demonstrable in tissue sections but there is one report claiming that *Toxoplasma* antigen may be identified in the nodes showing the characteristic histology.[14]

The diagnosis of virus infections is only rarely in the province of the biopsy pathologist with the occasional exceptions of cytomegalovirus, herpes simplex and the virus of subacute sclerosing panencephalitis. It is usually the virologist who will employ the immunochemical methods here although in herpes encephalitis it may be easier to stain viral antigens in infected cells than to find the intranuclear inclusions. It has recently been claimed that antigens of cytomegalovirus may be recognizable in tissue sections of infected but morphologically normal lymphoid cells[15] though this observation has not as yet been confirmed.

5. EXTRACELLULAR COMPONENTS

Although it is possible to recognize immunochemically a considerable number of extracellular components (for example various types of collagen, fibronectin, laminin), few practical applications have as yet appeared. Amyloid is perhaps the most striking extracellular material encountered in the course of pathological processes. Although it inevitably appears to have P component this is not reliably specific, as it is associated with elastic tissue and is also found in the glomerular basement membrane. Antisera to amyloid protein A (AA protein), can be used to confirm the secondary nature of an amyloid deposit, and this is more intellectually satisfying and less empirical than the sensitivity of Congo Red staining to pre-oxidation with acid permanganate. Because immunoglobulin-derived amyloid is related to the variable part of the light chain it is not possible to obtain an antiserum which will consistently stain it.

Other extracellular materials may deposit from circulating elements of the blood, such as immunoglobulins, complement components and fibrin, and they have been mainly studied by pathologists interested in the skin and kidney though there is increasing interest in the role of humoral immunity in liver disease.[16]

Apart from pemphigus, where the acantholytic bullus makes the diagnosis simple on morphological grounds, other bullous diseases are difficult to categorize accurately on histology alone and accuracy is much improved with immunological methods. Immunostaining is also a boon in the skin rashes of systemic lupus erythematosus where on conventional sections changes are often surprisingly slight and deposition of immunoglobulin at the dermo-epidermal junction clearcut (*see* Chapter 17).

5.1. Renal Disease

It is in the area of renal disease where immunostaining has become accepted as a necessary prerequisite for diagnosis, though it has raised problems as well as solving them. Several classifications of glomerular disease exist and the one that I have used has been modified from that of Heptinstall.[17]

Each of these diagnostic categories may have a characteristic pattern of staining with immunological reagents (*see Table* 18.1). Thus, in membranous nephropathy there is a rather uniform granular staining for IgG on the outside of the basement membrane (*Fig.* 18.3) though this pattern may be lost late in the course of the disease. The chief differential diagnosis of membranous nephropathy is from lipoid

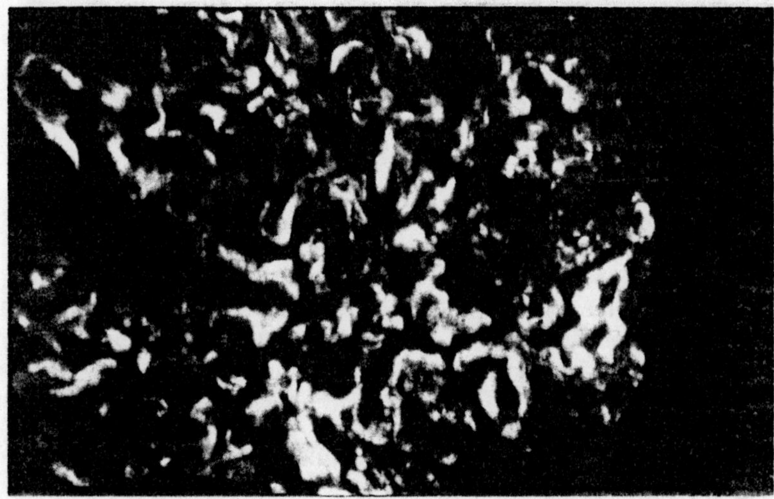

Fig. 18.3. Membranous nephropathy showing extensive granular IgG.

nephrosis and this can be a matter of some difficulty where the basement membrane is not thickened and has not developed spikes visible on staining with periodic acid–silver methenamine. The presence of C_3 in the deposits and of IgA is less reliable and so far no significance has been attached to their presence or absence. It is possible to detect subepithelial protein deposits electron microscopically or by semithin plastic sections, but for most laboratories immunostaining is more convenient and it is probably the most sensitive technique available.

Mesangial proliferation is rarely seen to any significant degree in membranous nephropathy and it is therefore unusual to encounter difficulty in distinguishing it from mesangiocapillary glomerulonephritis. Where difficulty does arise the distribution of the immunoglobulin and complement in the glomeruli will resolve the problem since in mesangiocapillary nephritis a uniform granular staining pattern is never seen. The most common form of the disease (Type 1) shows large irregular subepithelial deposits in which C_3 can be demonstrated but in which immunoglobulins are less reliably present[18] (*Fig.* 18.4).

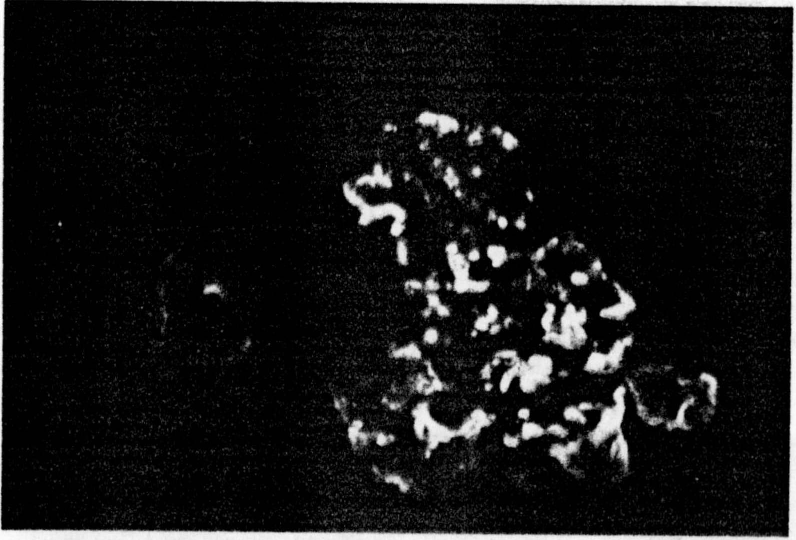

Fig. 18.4. Mesangiocapillary glomerulonephritis showing large, irregular C_3 deposits with a lobular outline.

Another area of practical difficulty is the distinction of mesangiocapillary from acute nephritis. Heavy polymorph infiltration in mesangiocapillary nephritis is not unusual though it is often remarkably patchy. Although in acute glomerulonephritis there is no mesangial interposition, it may not be easy to demonstrate this in paraffin sections. This problem too may be resolved by the use of semithin sections or immunostaining: the classical picture of lumpy deposits in acute glomerulonephritis which often contain IgG as well as C_3, is easily distinguishable from the pattern of mesangiocapillary nephritis described above.

In the case of crescentic nephritis it is now quite clear that we are not dealing with a single entity. The rupture of glomerular capillaries and the formation of a fibrin clot in Bowman's space will initiate the crescent: the underlying glomerular lesion may be one associated with a systemic disease (such as Henoch–Schönlein purpura, systemic lupus erythematosus, polyarteritis nodosa, Goodpasture's syndrome) or may occur in isolation. In the latter case the glomeruli may be damaged by immune complex type of disease (acute, mesangiocapillary or even, rarely, membranous nephropathy) or may be damaged by antibodies to glomerular basement membrane (*Fig.* 18.5). Distinguishing between these possibilities on light microscopy of paraffin sections is an impossible task. The recognition of the patterns of immunoglobulin deposition however, usually makes classification of the disease process possible (*see Table* 18.1).

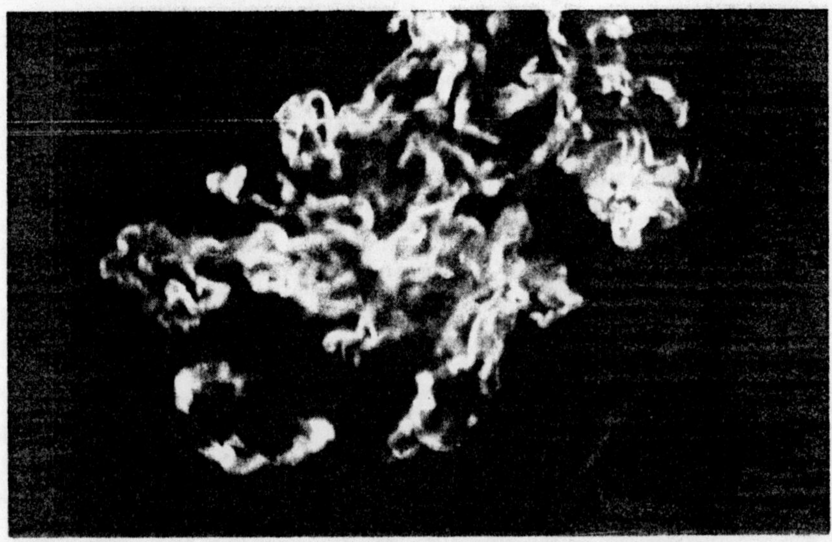

Fig. 18.5. Rapidly progressive glomerulonephritis due to antibodies to glomerular basement membrane. There is uniform staining for IgG in the capillary loop and in Bowman's capsule.

Contrary to the impression one would derive from the textbooks, in quite a substantial proportion of cases the renal disease is not easy to classify except under the generic name of 'focal nephritis'. It was in this group that Berger[19] succeeded in defining a subpopulation in which all the glomeruli showed deposits of IgA restricted to the mesangium (*Fig.* 18.6), usually in association with C_3 and often with IgG. This finding is particularly frequent in males with asymptomatic microscopic haematuria but may be seen in patients with a variety of presentations ranging from hypertension to the nephrotic syndrome. Although the light microscopical appearance seen in such cases is usually described as focal nephritis, this is not always the case. Some cases show no abnormality to light microscopy and others may have a chronic nephritic picture. This raises the question as to whether cases should not be classified primarily on the immunopathology. If the mesangial deposition of IgA is the primary pathogenetic event, perhaps it would be logical to call all such cases 'mesangial IgA disease' and to use the histopathological classification as a guide to the prognosis of the individual patient. There is as yet no formal proof that the aetiology is identical in all such cases. Different views have been expressed in the literature as to whether the

Table 18.1. Immunohistology of glomerulonephritis

Disease	*Immunohistology*
Acute proliferative nephritis	Granular C_3 as scattered deposits mainly subendothelial and subepithelial Usually granular IgG in similar distribution Variable IgA and IgM; little fibrin
Membranous nephropathy	Uniform granular subepithelial IgG C_3 usually similar to IgG IgA sometimes similar to IgG IgM, fibrin usually absent except for small patches
Mesangiocapillary (Type 1)	C_3 present in large subendothelial clumps, rather irrregular Immunoglobulins may be present in similar distribution; IgA is less common than IgG and IgM Fibrin usually absent
Rapidly progressive (crescentic) nephritis	Fibrin always present in crescent—other findings vary with pathogenesis
Mesangial IgA disease	Mesangial deposits of IgA usually with IgG and C_3; fibrin occasionally seen; IgM only in minute amounts
Focal sclerosis	Segmental granular IgM and C_3; less commonly IgG; IgA and fibrin usually absent

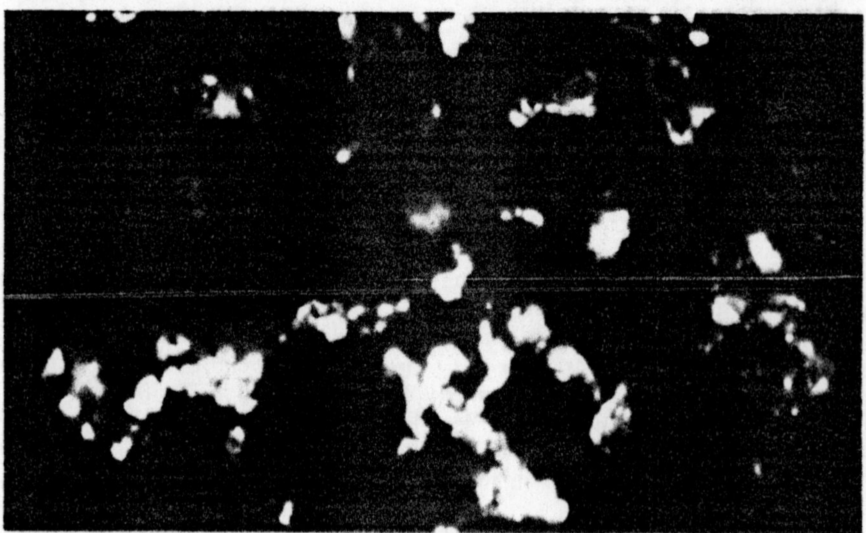

Fig. 18.6. Berger's disease showing IgA deposits confined to the mesangium.

absence of IgM in the mesangium is an important diagnostic criterion, but this may in part be due to differences in the technique used as the quantities of IgM present are often small and more easily detected by indirect immunofluorescence than by a single layer technique. There are no good reasons for believing that 'mesangial IgA disease' is any more heterogeneous than any of the other diagnostic categories which have been identified. Similar arguments about classification could be applied to glomerulonephritis due to antibodies to glomerular basement membrane. The glomerular disease may be focal or diffuse and there may be associated lung haemorrhage (Goodpasture's syndrome). Although the prognosis is influenced by the extent of renal disease the antibody has a restricted range of specificities, and the tendency to develop the disease seems to be influenced by genetic factors (in particular the presence of HLA-DR2).

In some cases of nephritis immunopathological findings have not been helpful. Findings in many cases of focal nephritis are patchy and apparently secondary to damage, and in lipoid nephrosis a wide variety of minor changes has been noted. There has recently been an attempt to define a group of patients with mesangial IgM deposition[20] though doubt remains as to whether they differ significantly in clinical or prognostic features from lipoid nephrosis.[21]

It should not be forgotten that systemic disease accompanying glomerular nephritis is not always overt. Subacute bacterial endocarditis and systemic lupus erythematosus, in particular, may present primarily as renal disease and a diagnosis may be possible on the combined morphological and immunological findings even in the absence of clinical indications.[22]

We have recently investigated the use of fluorescent antisera to amyloid P component in glomerular disease.[23] In the normal individual this forms a bright continuous line (*Fig.* 18.7) (related to the lamina rara interna as shown ultrastructurally by immunoperoxidase). This normal pattern is lost in many forms of glomerular disease including some cases of lipoid nephrosis; a number of changes may be seen varying with the type of disease and including bright granular staining in membranous nephropathy (*Fig.* 18.8). We are still trying to assess the

Fig. 18.7. Normal human glomeruli stained with antiserum against amyloid P component. There is a smooth outlining of glomeruli.

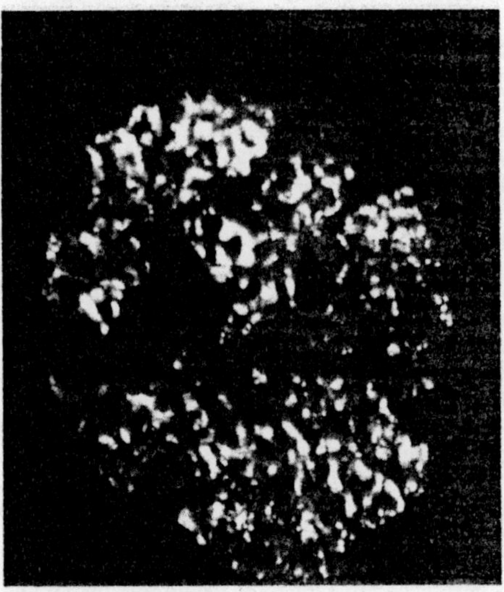

Fig. 18.8. Membranous nephropathy stained with antiserum to amyloid P component. Note the granular appearance.

significance of the findings, but our original hope that the presence of P component was directly related to the filtration barrier can no longer be maintained, since it is absent from the neonate and infant.[24]

6. CONCLUSION

Immunological methods have already made some impact on histopathological practice. They are most likely to win widespread acceptance when they are applicable to conventionally fixed tissues, or when their value can be predicted prior to taking the material. One has, however, to bear in mind that histopathology is already a very cost-effective and accurate technique, and the role of immunocytochemistry is likely to remain marginal in most non-specialist units.

REFERENCES

1. Sogge M. R., McDonald S. D. and Cofold P. B. The malignant potential of the dysgenetic germ cell in Klinefelter's syndrome. *Am. J. Med.* 1979, **66**, 515–518.
2. Weetman A. P. and Borysiewicz L. Androgen production in a patient with Klinefelter's syndrome and choriocarcinoma. *Br. Med. J.* 1980, **281**, 585–586.
3. Burt A. and Goudie R. B. Diagnosis of primary thyroid carcinoma by immunohistochemical demonstration of thyroglobulin. *Histopathology* 1979, **3**, 279–286.
4. Hobbs J. R., Branfoot A. C. and Knapp M. L. Pancreatic oncofoetal antigen in pancreatic cancer. *Practitioner* 1980, **224**, 1023–1030.
5. Leathem A. Breast-specific antigen recognition by non-mammalian antibodies. *J. Pathol.* 1980, **131**, 250.
6. Lampert I. A., Evans D. J. and Chesterton J. Leucocyte-specific antinuclear antibodies. *Virchows Arch. (Cell Pathol.)* 1975, **18**, 331–335.
7. Silverstein E., Pertschuk L. P. and Friedland J. Immunofluorescent localisation of angiotensin converting enzyme in epithelioid and giant cells of sarcoid granulomas. *Proc. Natl Acad. Sci. USA* 1979, **76**, 6646–6648.
8. Broome C. V., Cherry W. B., Winn W. C., Jr and MacPherson B. R. Rapid diagnosis of Legionnaire's disease by direct immunofluorescent staining. *Ann. Intern. Med.* 1979, **90**, 1–4.
9. AL-Samarrai H. T. and Henderson W. G. Immunofluorescent staining of *Treponema pallidum* and *Treponema pertenue* fixed in formalin and embedded in paraffin wax. *Br. J. Vener. Dis.* 1977, **53**, 1–11.
10. Keren D. F., Weisburger W. R., Yardley J. H., Salyer W. R., Arthur R. R. and Charache P. Whipple's disease: demonstration by immunofluorescence of similar bacterial antigens in macrophages from 3 cases. *Johns Hopkins Med. J.* 1976, **139**, 51–59.
11. Ali M. and Evans D. J. Diagnosis of Whipple's disease. In preparation.
12. Kaplan W. and Kraft D. E. Demonstration of pathogenic fungi in formalin fixed tissues by immunofluorescence. *Am. J. Clin. Pathol.* 1969, **52**, 420–423.
13. El-Nageeb S. and Hay R. J. Immunoperoxidase staining in the recognition of Aspergillus infections. *Histopathology* 1981, **5**, 437–444.
14. Matossian R. M., Nassar V. H. and Basmadji A. Direct immunofluorescence in the diagnosis of toxoplasmic lymphadenitis. *J. Clin. Pathol.* 1977, **30**, 847–850.
15. Abramovitz A., Livni N., Morag A. and David Z. Immunoperoxidase study of cytomegalovirus mononucleosis. *Arch. Pathol. Lab. Med.* 1982, **106**, 115–118.
16. Hopf U., Meyer zum Büschenefelde K.-H. and Arnold W. Detection of a liver membrane autoantibody in HBsAg-negative chronic active hepatitis. *N. Engl. J. Med.* 1976, **294**, 578–582.
17. Heptinstall R. H., *Pathology of the Kidney*, 2nd ed. Boston, Little Brown, 1974: 319–330.
18. Morel Maroger L., Leathem A. and Richet G. Glomerular abnormalities in non-systemic disease. Relationship between findings by light microscopy and immunofluorescence in 443 renal biopsy specimens. *Am. J. Med.* 1972, **53**, 170–184.
19. Berger J. and Hinglais N. Les dépôts intercapillaires d'IgA-IgG. *J. Urol. Néphrol.* 1968, **74**, 694–695.

20. Cohen A., Border W. and Glassock R. Nephrotic syndrome with glomerular mesangial IgM deposits. *Lab. Invest.* 1978, **38**, 610–619.
21. Vilches A. R., Turner D. R., Cameron J. S., Ogg C. S., Chantler C. and Williams D. G. Significance of mesangial IgM deposition in 'minimal change' nephrotic syndrome. *Lab. Invest.* 1982, **46**, 10–16.
22. Boulton Jones J. M., Sissons J. G. P., Evans D. J. and Peters D. K. Renal lesions of subacute infective endocarditis. *Br. Med. J.* 1974, **2**, 11–14.
23. Dyck R. F., Evans D. J., Lockwood C. M., Rees A. J., Turner D. and Pepys M. Amyloid P component in human glomerular basement membrane. *Lancet* 1980, **2**, 606–609.
24. Evans D. J. and Pepys M. Unpublished results, 1981.

19

Indirect Immunofluorescence in the Study and Diagnosis of Organ-specific Autoimmune Diseases

W. A. Scherbaum, R. Mirakian, R. Pujol-Borrell, B. M. Dean and G. F. Bottazzo

1. INTRODUCTION

The detection of autoantibodies is an essential diagnostic criterion to establish the autoimmune origin of diseases previously considered of unknown aetiology. For this purpose the indirect immunofluorescence technique is still the test of choice. It is a simple and reliable method in the hands of the expert for detecting serum antibodies directed against a variety of tissue structures or cell components. This test is carried out on cryostat sections of unfixed tissue with the clear advantage of detecting autoantibodies to unaltered antigens. The antibody test requires no more than a small amount of serum from the patient and a positive result provides a valuable diagnostic tool and the basis for the study of the complex role of cellular and humoral reactions which might lead to overt disease.

Immunofluorescence also helps to identify new autoantibodies and these may be so specific that they react only with one cell type within an organ. This is the case for the pancreatic islets, the pituitary gland and the gut mucosa. These organs are composed of different cells which cannot be distinguished by light microscopy. For the exact characterization of single cell antibodies within these organs we have developed a four-layer double-fluorochrome immunofluorescence sandwich test (*Fig.* 19.1) performed on the same tissue section which enables us to determine precisely the endocrine nature of positive cells. Applying this test we were able to demonstrate specific antibodies reacting with pituitary prolactin[1] and growth hormone cells,[2] antibodies to glucagon and somatostatin cells of the pancreatic islets,[3] and antibodies to gastrin-,[4] secretin- and gastric inhibitory polypeptide (GIP)-,[5] somatostatin- and glucagon-producing cells of the gut.[6]

The spectrum of the known organ-specific autoimmune diseases is shown in

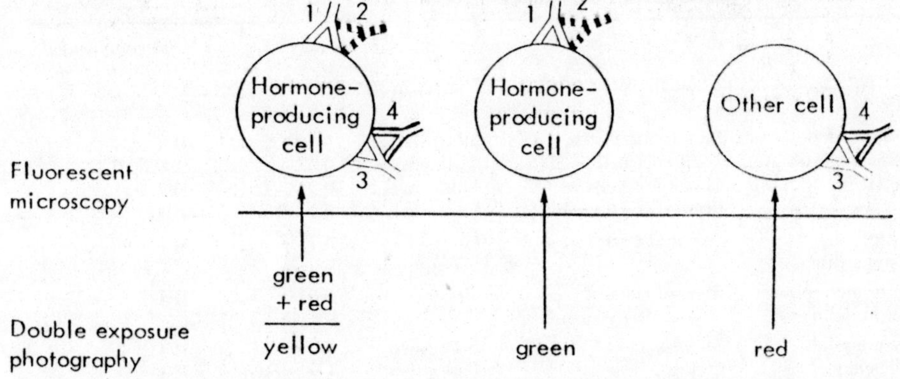

Fig. 19.1. Diagram of double immunofluorescence technique used for characterization of autoantibodies to single cells in complex endocrine organs. Each section is sequentially stained with: 1. Anti-hormone serum (e.g. raised in rabbits), 2–6 h, depending on the affinity and avidity of the antiserum. 2. Fluoresceinated anti-rabbit-Ig (e.g. raised in goats), 30–40 min. 3. Patient's serum, 2–8 h, depending on the organ. 4. Rhodaminated anti-human Ig (raised in goats), 30 min. Between every step the sections are cautiously washed in PBS (not shaken) and the reaction is finally read using a fluorescence microscope fitted with epi-illumination and an automatic camera.
Left column: Final reaction when the human autoantibody and the animal anti-hormone serum are directed against the same cells. The final yellow colour is obtained by double exposure photography of the same field.
Middle and right columns: Final results when the rabbit anti-hormone serum and the patient's serum react with different cells. Double exposure photographs show separate green and red cells. This technique may be relied on because the human antibodies are directed to intracellular membranes and not against the hormones themselves.
The filters used are: for fluorescein, Zeiss set 09 blue excitation, 450–490 with interference red barrier filter KP 560 18 × 2, and for rhodamine, set 15, green excitation H546.

Table 19.1 and it is intriguing that they affect mainly the endocrine system. Their corresponding autoantibodies usually react with cytoplasmic antigens which segregate with the microsomal fraction of the target cells and appear to be integral membrane-bound receptor proteins probably involved in the intracellular transport of hormones.

Until recently it was thought that microsomal antibodies reacted only with *intracytoplasmic* membranes, but immunofluorescence studies on monolayer cultures of viable thyroid,[7] adrenocortical[8] and gastric parietal cells[9] have now shown that the internal autoantigens are also represented on the plasma membrane. This makes the microsomal antigens accessible to sensitized immunocompetent cells or antibody-dependent cytotoxic mechanisms so that a pathogenic role of organ-specific autoantibodies is becoming increasingly acceptable. It is of interest that similar close correlations between surface immunofluorescence and cytoplastic immunofluorescence have not been found in the study of islet cell antibodies.[10]

2. CHOICE OF TISSUE

Human tissues are always preferable but, if they are unobtainable, monkey tissues may be employed. Rodent organs often show considerable non-specific flu-

Table 19.1. Spectrum of organ-specific autoimmune diseases

Disease	*Antibodies to*	*Frequency of antibodies (%) Patients*	*Controls*	*Recommended immunological test*
Insulin-dependent diabetes mellitus (Type I)	Pancreatic islets all islet cells	80 at diagnosis	0·5	IFL and CF–IFL
	Glucagon cells	Unknown	0·2	IFL
	Somatostatin cells	Unknown	0·2	IFL
Idiopathic Addison's disease	Adrenal cortex	60 at diagnosis	0·1	IFL and CF–IFL
Premature menopause with adrenalitis	Steroid cells of gonads + placenta	80	0·01	IFL
Primary gonadal insufficiency	Sperms	Unknown	7	IFL
	Ovum	Unknown	Unknown	IFL
Hashimoto's thyroiditis	Thyroglobulin	*see* Refs 40, 74	Dependent on age	Passive HA
Primary myxoedema	Thyroid microsomes	41, 75	Sex (F > M)	Passive HA
Graves' thyrotox.	TSH-receptors	44, 45, 76	Test	44, 45, 76
Endocrine exophthalmos	Extra-ocular muscle membranes	77	unknown	77
Pernicious anaemia	Intrinsic factor	i.s. 56	Unknown	RIA
	Parietal cells	90	F > M 0–16	IFL
Myasthenia gravis (acquired form)	Acetylcholine receptors	45* 90†	0	alpha-Bgt assay‡
	Muscle fibrils	20–40	0 2	
Fundal gastritis (Type A)	Parietal cells	60	*see above*	IFL
Antral gastritis (Type B)	Gastrin cells	10	0·1	IFL
Primary hypoparathyroidism	Chief cells of parathyroid gland	Very rare	Unknown	IFL
Type I diabetes in early phase	Pituitary PRL/hGH and other cells	17	0	IFL
Idiopathic diabetes insipidus	AVP cells of the hypothalamus	Unknown	Unknown	IFL

* Ocular form of myasthenia gravis (Type I).
† Generalized form of myasthenia gravis (Types II–IV).
‡ Alpha-bungarotoxin immunoprecipitation assay using human peripheral muscle as antigen.
AVP, arginine vasopressin.
CF–IFL, complement fixation using the IFL technique.
HA, haemagglutination.
hGH, human growth hormone.
IFL, indirect immunofluorescence.
i.s., in the serum.
PRL, prolactin.
RIA, radioimmunoassay.

orescence and when human sera are applied heterophile antibodies may appear and give non-specific results. Also, the cross-reactivity of human autoantibodies on animal tissue varies widely. The thyroid antigen–antibody systems show more restriction with regard to animal species outside the primates, whereas the parietal cell antibody cross-reacts strongly even with the 'all-purpose' gastric mucosal cells of amphibians.

Some tissue antigens are quickly autolysed after death and therefore it is advisable to use tissues obtained at operation or from kidney transplant donors. For the detection of microsomal antibodies, unfixed tissues should always be used in the first instance. Some fixatives may destroy the putative autoantigen or reveal new antigenic sites which can give false positive reactions as is now well illustrated in the use of Bouin-fixed pancreas for the detection of islet-cell antibodies.[11]

Antibodies like those directed to pancreatic islets, pituitary, hypothalamic vasopressin cells and single endocrine cells of the gut are usually of a low titre and they sometimes give only a very dim fluorescence compared with the strong thyroid or gastric antibody reaction. For this reason it is important to test the sera undiluted in the first place, to have an ultraviolet microscope equipped with epi-illumination and to look at the sections at high magnification.

3. AUTOANTIBODIES TO PANCREATIC ISLET CELLS

3.1. Cytoplasmic Islet Cell Antibodies

3.1.1. *Islet Cell Antibodies Detected with Anti-IgG Conjugate (Islet cell antibody–IgG)*

The increased coexistence of Type I (insulin-dependent) diabetes with Addison's disease,[12] thyroid autoimmune diseases,[13] pernicious anaemia and primary hypoparathyroidism[14] suggested that this type of diabetes might be of autoimmune origin. Specific autoantibodies to pancreatic islet cells were first described in polyendocrine patients.[15] Lendrum et al.[16] were able to show that over 80 per cent of children with sudden-onset diabetes had islet cell antibodies at the time of diagnosis and the incidence decreased to 50 per cent within two months and to 15–20 per cent after the first five years. Patients in whom these antibodies persist usually suffer from polyendocrine autoimmune diseases.[17] Islet cell antibodies are not a feature of classical Type II diabetes, but when they are present in these cases we also find other organ-specific autoantibodies.

For the islet cell antibody–IgG test it is crucial to use a fresh gland obtained from kidney transplant donors and it is essential to use blood group O tissue. Blood group antigens are secreted by the exocrine pancreas and high titres of ABH isoagglutinins in the test sera may produce disturbing immunofluorescence patterns in the exocrine acini which interfere with the reading on the islets.

Several pancreases should be collected because different specimens may possess variable amounts of autoantigens and vary in their contents of the different islet cell types.[18] There are also important variations in the distribution of the different endocrine cells within the gland[19] so that it is advisable to take blocks from the head and tail of the pancreas.

The antigen of islet cell antibodies is common to all types of islet cells[20] and staining of the entire islet is obtained (*Fig.* 19.2). Yet it is only the beta-cells which are selectively destroyed in Type I diabetes, so islet cell antibodies appear not to be responsible for the islet-cell damage, despite their great value as a serological marker.

Absorption of islet cell antibody-positive sera with excess insulin, glucagon and somatostatin does not affect the islet immunofluorescence[21] which demonstrates that these antibodies do not react with the pancreatic hormones but with the intracellular membranes in the islet cells.

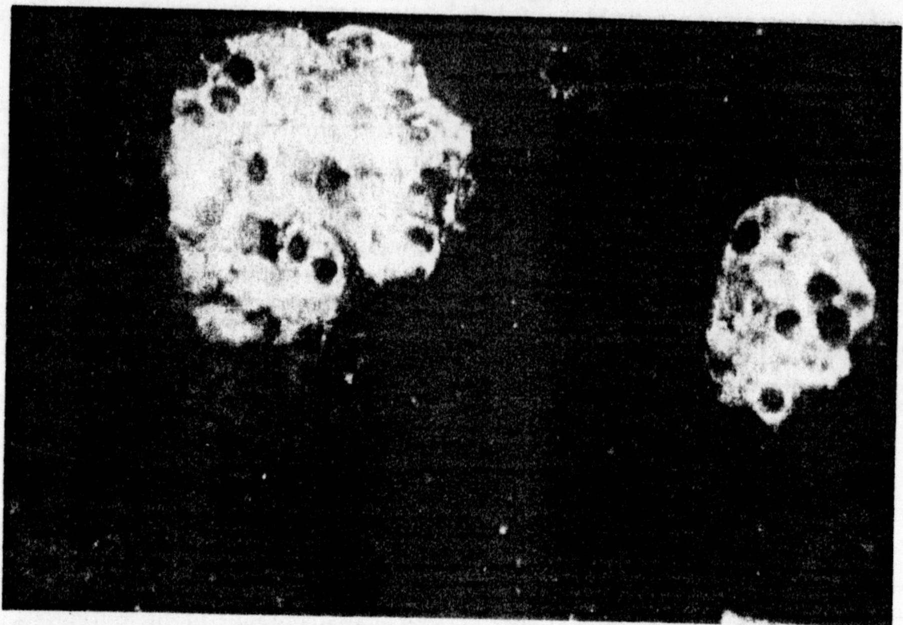

Fig. 19.2. Cryostat section of blood group O human pancreas treated with the serum of a newly diagnosed insulin-dependent diabetic patient and stained with FITC anti-human IgG conjugate. The serum contains the conventional islet-cell antibodies which stain the entire islet of Langerhans.

3.1.2. *Complement-fixing Islet Cell Antibodies*

In our laboratory we have developed a simple complement-fixation test which is performed by an immunofluorescence technique. In the complement-fixation–immunofluorescence test the patient's undiluted serum is applied to the tissue section for 30 min at room temperature. After washing, a drop of fresh normal human serum is added as a source of complement, and the reaction is revealed by adding anti-human C_3 conjugate as the third layer. When this test is done on human pancreas concurrently with the standard IgG test, 50–55 per cent of islet cell antibody-positive sera fix complement.[21]

Complement-fixing islet cell antibodies are independent of the subclass composition of the islet cell antibodies and sometimes produce selective staining of portions of the islets. Some non-diabetics with polyendocrine autoimmune diseases and some healthy relatives of Type I diabetics have islet cell antibodies in their serum and in these cases it is important to test also for complement-fixing islet cell antibodies because this subspecies seems to be more closely related to the expression of disease.[22,23]

3.2. Pancreatic Islet Cell Surface Antibodies

The immunofluorescence test can be applied to detect autoantibodies reacting with the surface of islet cells in serum from insulin-dependent diabetic patients. These antibodies were first demonstrated with cultured insulinoma cells[24] where nearly 90 per cent of recently diagnosed young onset diabetics were positive. With ob/ob mouse islet and with dispersed rat islet cells,[25] 32 per cent of such sera were islet cell surface antibody-positive compared with 4 per cent of sera from healthy controls. Recently surface islet cell antibodies were demonstrated using human fetal cell cultures[10] and it is now clear that these antibodies are mostly restricted to Type I diabetics and some of their islet cell antibody-positive first degree relatives. Islet cell surface antibodies are mostly of IgG and sometimes of IgM class and their presence does not correlate with cytoplasmic islet cell antibodies.

3.3. Pancreatic Glucagon Cell and Somatostatin Cell Antibodies

The pancreatic islets contain at least four types of endocrine cells which produce insulin, glucagon, somatostatin and human pancreatic polypeptide, respectively and can be recognized on the basis of specific antisera raised in animals[26,27] to purified hormones extracted from the pancreas or synthesized chemically.[28] Specific human antibodies to glucagon cells and somatostatin cells were identified by the double-fluorochrome immunofluorescence technique.[3] The antibodies are found in 1–4 per cent of sera from a variety of diseases and it is of interest that about half of these antibodies cross-react with the equivalent cells in the gut.

Glucagon cell and somatostatin cell antibodies are more frequent in diabetic patients, especially when there is evidence of polyendocrine autoimmunity. Yet it is difficult to establish their prevalence in the presence of islet cell antibodies unless they are of a higher titre or of different Ig class.

4. ANTIBODIES TO ADRENAL CORTEX

Since tuberculosis is controlled, autoimmune destruction is the most common cause of adrenocortical insufficiency.[29] Immunofluorescence provides the simplest and most sensitive test available for the detection of adrenal antibodies. These antibodies are positive in 60 per cent of patients with newly diagnosed idiopathic adrenocortical failure.[30] Cases with other associated autoimmune diseases such as Schmidt's or *Candida* endocrinopathy syndromes are nearly always positive.[31] The antibodies are usually negative in patients with symptomatic Addison's disease, and in healthy controls they occur in less than 1 in 1000.[32]

The majority of positive sera react with all three layers of the adrenal cortex. Like islet cell antibodies, the titres of adrenal antibodies are usually low (1 : 1 to 1 : 64), and it has been observed that they tend to decline over the years[33] and some become negative.[34]

Adrenal antibodies are positive in 1–3 per cent of patients without overt organ-specific autoimmune diseases and in about 7 per cent of polyendocrine cases without clinical signs of adrenocortical failure. We have recently followed up 30 of such patients over a 3-year period. During this period, 2 patients with Graves' thyrotoxicosis developed biochemical signs of adrenocortical failure and both of them had high titres of adrenal antibodies which were also complement-fixing at the time of entry to the study,[35] so it is advisable to perform the complement-fixation–immunofluorescence test when the outcome of an adrenal antibody-positive non-Addisonian patient is to be evaluated.

5. ANTIBODIES TO STEROID-PRODUCING CELLS OF THE GONADS AND THE PLACENTA

Steroid-cell antibodies are a group of antibodies cross-reacting with steroid-secreting cells in the ovary, testis and placenta.[36] These antibodies can only be recognized by immunofluorescence and they have only been detected so far in sera that also react with human adrenal cortex. Steroid cell antibodies have various patterns on corpus luteum and other ovarian structures[18] and nearly all of them stain the Leydig cells in the testis.[37]

In cases of isolated Addison's disease, steroid cell antibodies are present in 22 per cent but they are positive in over half of Addisonian patients with other endocrine disorders.[32,34] Patients with these antibodies show an increased tendency to premature menopause and ovarian failure or infertility. These antibodies may also appear in the serum several years before the onset of clinical symptoms and these patients are often polyendocrine cases. About 20 per cent of them have islet cell antibodies and are prone to sudden onset of ketotic diabetes, many have an autoimmune thyroid disease and some have overt or latent pernicious anaemia so that an extended clinical check-up is justified when the appropriate antibodies are detected.

6. THYROID ANTIBODIES

The detection of thyroid autoantibodies in Hashimoto's thyroiditis[38] gave the key impulse to investigate autoimmunity applied to human disease and thyroid autoimmunity is still the most extensively investigated of all organ-specific autoimmune diseases.

Thyroglobulin antibodies and antibodies to the 'second colloid antigen' can be detected by immunofluorescence on thyroid cryostat sections prefixed in ethanol for 30 min at 56 °C.[39] Thyroglobulin antibodies give a floccular immunofluorescence pattern and 'second colloid antigen' antibodies show a uniform fluorescence. 'Second colloid antigen' antibodies have no diagnostic significance and the immunofluorescence test for the detection of thyroglobulin antibodies is now replaced by the haemagglutination test for routine investigations.[40] Haemagglutination titres of 1 : 20 and above can be considered positive.

Thyroid microsomal antibodies can be detected by immunofluorescence on unfixed cryostat sections and the fluorescence is brightest on the microvilli near the colloid edge of the cells. The results obtained by immunofluorescence and the haemagglutination test correlate well (*Fig.* 19.3). A positive immunofluorescence reaction at 1 : 1 corresponds to 10^{-2} (i.e. 1 : 100) in the haemagglutination test and this relationship still holds up to 1 : 1280 by immunofluorescence and 1 : 128 000 by haemagglutination. The haemagglutination test is now the method of choice for detecting thyroid microsomal antibodies[40] and significant titres start at 1 : 100.[41] Occasionally thyrotoxic patients may have false negative thyroid microsomal and thyroglobulin antibody results in the haemagglutination test due to the presence of a non-specific haemagglutination inhibitor related to the blood lipoproteins.[42] When a double-antibody precipitation technique is used, thyroid antibodies are still positive in these cases suggesting that antibody binding and immunofluorescence are not affected by these inhibitors.

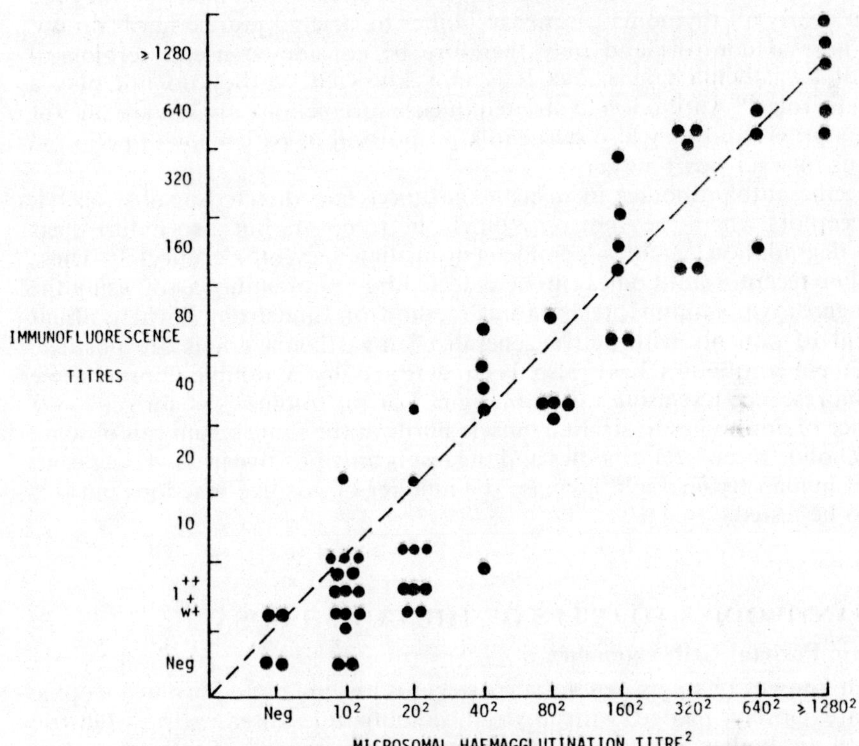

Fig. 19.3. The close correlation between the results obtained in the immunofluorescence and the haemagglutination test suggests that they detect very similar 'microsomal' antibodies. The haemagglutination test is fast and simple, it gives qualitative and semiquantitative results and can easily be combined with the determination of thyroglobulin antibodies. It is therefore the method of choice for the routine diagnosis of autoimmune thyroid disease.

Antibodies reacting with the thyroid surface were first detected by Fagraeus and Jonsson[43] on viable suspensions of human thyroid cells where they give an interrupted line of staining around each cell. Recently Khoury et al.[7] demonstrated by immunofluorescence studies on cultured human thyroid cell monolayers that the microsomal antigen is also represented on the surface of thyroid cells and this finding is of wide ranging biological significance.

Thyroid-stimulating immunoglobulins cannot be seen by immunofluorescence owing to the small number of thyroid-stimulating hormone receptors per cell. These antibodies[44,45] as well as thyroid growth-stimulating[46] and blocking antibodies[47] are detected by other methods.

7. IMMUNOFLUORESCENCE IN MYASTHENIA GRAVIS

A screening immunofluorescence test on several tissues should be performed in every patient with acquired myasthenia gravis because a number of these patients have other associated autoimmune diseases.[48] Antibodies to striated muscle fibrils[49] can be detected by immunofluorescence in about 40 per cent of the patients, their prevalence is especially high in older myasthenic men and they are positive in nearly all thymoma cases. Antibodies to striated muscle fibrils do not occur in normal controls and may therefore be considered useful serological markers of myasthenia gravis. Yet it is now known that they do not play a pathogenetic role.[50] Antibodies to striated muscle sarcolemma are less specific for myasthenia gravis and they also occur in a proportion of patients with peri- and myocarditis or after heart surgery.

The specific autoantibodies in myasthenia gravis are directed against acetylcholine receptors where they not only block the receptors but also induce their increased degradation[51] and a complement-mediated cytotoxic effect.[52] These acetylcholine receptor antibodies can be detected by radioimmunoassay using the alpha-bungarotoxin–immunoprecipitation method of Lindstrøm[53] where about 90 per cent of patients with active generalized myasthenia gravis are positive. Recently these antibodies have also been detected by a double-fluorochrome immunofluorescence test using rat diaphragm[55] or rat tongue[56] as antigens. Yet the presence of antibodies to striated muscle fibrils in the same serum can obscure the acetylcholine receptor antibodies and the test is only positive in 21–42 per cent. The use of human tissue might increase the number of positive reactions but this has still to be tested.

8. AUTOANTIBODIES TO CELLS OF THE GUT MUCOSA

8.1. Gastric Parietal Cell Antibodies

It has been known for many years that pernicious anaemia and chronic atrophic gastritis are closely related with thyroid autoimmune diseases.[56,57] Gastric parietal cell antibodies are detected by immunofluorescence but intrinsic factor antibodies are demonstrated by radioimmunoassay. Patients with parietal cell antibodies were shown to have various degrees of atrophy of the fundal mucosa (Type A gastritis), frequent achlorhydria tending to evolve towards latent or overt pernicious anaemia and a high association with endocrine autoimmune diseases.[58] Gastrin levels and gastrin cell counts were found to be increased in pernicious anaemia and subclinical fundal gastritis.[59]

8.2. Gastrin Cell Antibodies

In antral (Type B) gastritis parietal cell antibodies are usually absent, there is no association with endocrine autoimmune diseases in contrast to Type A gastritis, and the gastrin levels are greatly diminished.[60] About 8 per cent of these cases have antibodies to gastrin cells as characterized by the double immunofluorescence technique.[4]

For the detection of gastrin cell antibodies it is crucial to select a piece of human antrum quite devoid of parietal cells yet containing a sufficient number of gastrin cells. The patients' sera are first tested by conventional immunofluorescence on unfixed tissue and to identify the reacting cells the positive sera are retested by the four-layer double-fluorochrome immunofluorescence test (*Plates* 26,27) applying a specific antiserum to gastrin. Gastrin cell antibodies are of IgG class, they are of low titre, they fix complement and they tend to persist over the years. Gastrin cell antibodies are rarely found in patients with pernicious anaemia or simple fundal gastritis and are not found in polyendocrine autoimmune diseases. Gastrin cell antibodies were recently also demonstrated using the avidin–biotin-complex antibody method.[61]

Patients with gastrin cell antibodies showed a diminished response of the serum gastrin to standard protein meals, suggesting that the antibodies reflect a loss of gastrin cells in the autoimmune process. The gastrin cell antibodies may therefore be considered as markers for selective destruction of gastrin cells in cases of primary autoimmune antral gastritis.

8.3. Autoantibodies in Coeliac Disease

Our first suggestion that autoantibodies to gastric inhibitory polypeptide- and secretin-producing cells of the gut might exist is based on the finding of Besterman et al.[62] who noticed a marked reduction in the release of secretin after duodenal acidification and a diminished output of gastric inhibitory polypeptide after a standard protein meal in a proportion of patients with coeliac disease. Duodenal mucosa was chosen to test sera of patients with proven coeliac disease because this tissue is particularly rich in endocrine cells.[63] Of 173 sera tested 16 (18 per cent) gave positive cytoplasmic immunofluorescence on single cells whereas 73 polyendocrine sera and the sera of 30 healthy blood donors were negative. Four of the positive sera reacted with gastric inhibitory polypeptide cells, 1 with secretin cells and 21 with both gastric inhibitory polypeptide and secretin cells as tested by the four-layer double-fluorochrome test.[5] Extensive metabolic studies and a better understanding of the exact physiology of the gastrointestinal hormones are required to establish the full significance of these antibodies.

8.4. Antibodies to Single Endocrine Cells of the Gut Mucosa in Type II (Non-Insulin-Dependent) Diabetes

Some patients with Type II diabetes have a poor gastric inhibitory polypeptide response to protein meals and when the sera of such patients were investigated for antibodies to gut mucosa, 31 out of 154 sera tested had single cell antibodies. With the double-fluorochrome immunofluorescence test, 24 were shown to be directed against gastric inhibitory polypeptide cells, and only 3 sera had antibodies to other endocrine cells of the gastrointestinal tract.[6]

Gastric inhibitory polypeptide is known to affect the insulin secretion that follows the ingestion of glucose[64] and a defective gastric inhibitory polypeptide secretion might influence the mechanism of insulin release contributing to the pathogenesis of Type II diabetes in cases where the pancreatic beta-cells still contain sufficient amounts of insulin.

8.5. Autoantibodies to Gut Epithelium in Protracted Diarrhoea

Recently new antibodies to gut epithelium have been detected independently by Walker-Smith et al.[65] and our group[66] in three infants with severe diarrhoea. The two children observed by ourselves had specific complement-fixing autoantibodies reacting by immunofluorescence with the cytoplasm of human duodenum, jejunum and colon mucosal epithelium. One of our cases had severe unresponsive enteropathy with persistent gut epithelial antibodies present until death. The other had transient mucosal damage, and clinical and histological improvements were associated with antibody disappearance. In a few cases of unexplained protracted diarrhoea mucosal damage is probably the direct consequence of an autoimmune process which can only be revealed by the antibody test.

9. AUTOANTIBODIES TO PITUITARY CELLS

9.1. Prolactin Cell Antibodies

These antibodies were first detected in polyendocrine autoimmune disease and they were finally characterized applying the double-fluorochrome test using specific anti-pituitary hormone antisera.[1] Prolactin cell antibodies have never been found in severe panhypopituitarism, but they are now known to occur in a few patients with partial pituitary defects, some of whom presented with amenorrhoea or infertility. The prolactin responses to thyrotropin-releasing hormone in patients who possess these antibodies are either diminished or raised confirming that an autoimmune process does not always lead to destructive mechanisms.

9.2. Growth-Hormone Cell Antibodies

These antibodies were first identified in a young girl with arrested growth whose mother had autoimmune adrenalitis and thyroiditis.[2] In this patient it proved possible to demonstrate a partial defect of human growth hormone secretion in response to stimulation tests. When the sera of a selected group of 220 children with growth defects were tested, 10 per cent reacted with isolated cells in the anterior pituitary.[18] The majority were directed against prolactin cells, three sera reacted with human growth hormone cells and a few remain to be identified.

9.3. ACTH Cell Antibodies

There are special difficulties in detecting ACTH cell antibodies because these cells appear to have Fc receptors on their surface and in their cytoplasm so that all normal immunoglobulins react with these cells by attachment of their Fc part which gives a uniform cytoplasmic fluorescence of the oval-shaped ACTH cells.[67]

9.4. Multiple Cell Immunofluorescence Pattern

There have been a number of reports of children with unusually tall stature presenting with Type I diabetes. Boys of prepubertal age are tall at onset of diabetes and in both sexes skeletal maturity is found to be significantly advanced, suggesting a prediabetic metabolic or hormonal abnormality.[68] Thus pituitary and other endocrine effects may play a fundamental role in the early pathogenesis of

Type I diabetes and we have looked for pituitary antibodies in recently diagnosed diabetics and in their genetically predisposed relatives during the 'latency period'. Of the islet cell antibody-positive diabetics, 17 per cent turned out to be positive for antibodies mostly directed to more than one cell type of the pituitary. Surprisingly, these antibodies were also positive in 12 (36·4 per cent) of 33 selected islet cell antibody-positive sera from their healthy genetically predisposed relatives of whom 7 became diabetic during a 3-year follow-up period.[69] The serum of only 1 out of 48 patients with long-standing insulin-dependent diabetes was positive and pituitary antibodies were not found in 48 normal controls.

These results obtained by immunofluorescence studies suggest that in some cases temporary autoimmune processes can involve the anterior pituitary and the endocrine pancreas simultaneously. These antibodies might either represent an expression of a destructive process leading to hypophysitis or more likely they are markers for a stimulatory process at the pituitary level.

10. AUTOANTIBODIES TO VASOPRESSIN-PRODUCING HYPOTHALAMIC CELLS IN IDIOPATHIC DIABETES INSIPIDUS

The blood–brain barrier is permeable for proteins at the level of the basal hypothalamus[70] and antigens and antibodies are thus able to reach these parts of the brain. In analogy with other 'idiopathic' hormone deficiencies we considered the possibility that there might exist an autoimmune diabetes insipidus. The sera of diabetes insipidus patients were therefore tested by immunofluorescence on unfixed rat, fresh human post-mortem, and human fetal hypothalamus. We were able to demonstrate autoantibodies in the serum of patients with idiopathic diabetes insipidus reacting with the cytoplasm of large secretory cells in the supraoptic and paraventricular nuclei.[71] Applying the four-layer double-fluorochrome immunofluorescence test the same cells were shown to react with specific anti-vasopressin hormone sera raised in rabbits. The antibodies were not found in controls, polyendocrine sera and all the cases of symptomatic diabetes insipidus tested so far.

By the determination of hypothalamic vasopressin cell antibodies we can now separate an autoimmune form of diabetes insipidus from the 'idiopathic' cases which sometimes later turn out to be symptomatic as, for example, in unrecognized carcinomatosis.

11. FUTURE PROSPECTS

Autoantibodies have been found directed against the cytoplasm of nearly all endocrine glands, scattered endocrine cells and a number of specialized organs. Through the introduction of the double-fluorochrome immunofluorescence and the development of specific hormone antisera it has become possible to characterize the nature of single cells reacting with the antibodies even when the form and shape of these cells is no different from the surrounding ones. Extensive metabolic and histopathological studies will be necessary to establish the clinical significance of the newly detected autoantibodies to pituitary, hypothalamic and gut cells.

Every endocrine cell is connected with the immune system through an enormous number of surface receptors and the presence of microsomal antigens on the cell surface can now explain the pathogenetic significance of cytoplasmic antibodies. To study this directly, it will be necessary to investigate the interactions of immunoglobulins with sparse components of the cell surface which do not represent enough binding sites to show up in immunofluorescence. Far more sensitive methods are required for the detection of receptor antibodies. When the purified or synthesized antigen is available, an immunoenzymatic method (ELISA) can be employed. However most antibody or membrane fractions of cells are integrated in the lipid bilayer and have not been purified. To study these structures the receptors must be solubilized in detergents and then studied as soluble proteins.

The receptor protein can be labelled biosynthetically by growing the cells in the presence of amino acids or sugars that contain radioactive or heavy isotopes. The hormone can be tagged with fluorescent, radioactive, or electron-dense moieties like iron or gold, in conjunction with ultrasensitive light detectors or radio-autography and electron microscopy. These hormone preparations can be used to track the hormone or the antibody after it makes contact with the receptor on the cell.[72] For the detailed study of receptors it is also essential to apply monoclonal antibodies directed against components of the receptors.[73] The exploitation of these new methods will lead to much deeper biological insights than have been envisaged so far.

Acknowledgements

We wish to thank Professor Deborah Doniach for helpful suggestions and advice. Miss Queenie Jayawardena kindly typed the manuscript.

REFERENCES

1. Bottazzo G. F., Pouplard A., Florin-Christensen A. and Doniach D. Autoantibodies to prolactin-secreting cells of human pituitary. *Lancet* 1975, **ii**, 97–101.
2. Bottazzo G. F., McIntosh C. W., Stanford W. and Preece M. Growth-hormone cell antibodies and partial growth-hormone deficiency in a girl with Turner's syndrome. *Clin. Endocrinol.* 1980, **12**, 1–9.
3. Bottazzo G. F. and Lendrum R. Separate autoantibodies to human pancreatic glucagon and somatostatin cells. *Lancet* 1976, **ii**, 873–876.
4. Vandelli C., Bottazzo G. F., Doniach D. and Francheschi F. Autoantibodies to gastrin producing cells in antral (type B) chronic gastritis. *N. Engl. J. Med.* 1979, **300**, 1406–1410.
5. Mirakian R., Bottazzo G. F. and Doniach D. Autoantibodies to duodenal gastric-inhibitory peptide (GIP) cells and to secretin (S) cells in patients with coeliac disease, tropical sprue and maturity onset diabetes. *Clin. Exp. Immunol.* 1980, **41**, 33–42.
6. Mirakian R., Richardson C. A., Bottazzo G. F. and Doniach D. Humoral autoimmunity to gut-related endocrine cells. *Clin. Immunol. Newsletter* 1981, **2**, 161–167.
7. Khoury E. L., Hammond L., Bottazzo G. F. and Doniach D. Presence of the organ-specific 'microsomal' antigen on the surface of human thyroid cells in culture: its involvement in complement-mediated cytotoxicity. *Clin. Exp. Immunol.* 1981, **45**, 316–328.
8. Khoury E. L., Hammond L., Bottazzo G. F. and Doniach D. Surface-reactive antibodies to human adrenal cells in Addison's disease. *Clin. Exp. Immunol.* 1981, **45**, 48–58.
9. Masala C., Smurra G., Di Prima M. A., Amendolea M. A., Celestino D. and Salsano F. Gastric parietal cell antibodies: demonstration by immunofluorescence of their reactivity with the surface of the gastric parietal cells. *Clin. Exp. Immunol.* 1980, **41**, 271–280.
10. Pujol-Borrell R., Khoury E. L. and Bottazzo G. F. Islet cell surface antibodies in Type I (insulin-dependent) diabetes mellitus: use of human fetal pancreas cultures as substrate. *Diabetologia* 1982, **22**, 89–95.

11. Bottazzo G. F., Mirakian R., Dean B. M., McNally J. M. and Doniach D. How immunology helps to define heterogeneity in diabetes mellitus. In: Tattersall R. B. and Koberling J. K. (ed.) *Genetics of Diabetes Mellitus*. London, Academic Press, 1982, in press.

12. Nerup J. Addison's disease—clinical studies. A report of 108 cases. *Acta Endocrinol.* 1974, **76**, 127–141.

13. Solomon N., Carpenter C. C. J., Bennett J. L. and McGehee Harvey A. Schmidt's syndrome (thyroid and adrenal insufficiency) and coexistent diabetes mellitus. *Diabetes* 1965, **14**, 300–304.

14. Blizzard R. M., Chee D. and Davies W. The incidence of parathyroid and other antibodies in the sera of patients with idiopathic hypoparathyroidism. *Clin. Exp. Immunol.* 1966, **1**, 119–128.

15. Bottazzo G. F., Florin-Christensen A. and Doniach D. Islet-cell antibodies in diabetes mellitus with autoimmune polyendocrine deficiencies. *Lancet* 1974, **ii**, 1279–1283.

16. Lendrum R., Walker J. G. and Gamble D. R. Islet-cell antibodies in juvenile diabetes mellitus of recent onset. *Lancet* 1975, **i**, 880–882.

17. Doniach D., and Bottazzo G. F. Polyendocrine autoimmunity. In: Franklin E. C. (ed.) *Clinical Immunology Update*. New York, Elsevier, 1981: 95–121.

18. Doniach D., Bottazzo G. F. and Drexhage H. A. The autoimmune endocrinopathies. In: Lachmann P. A. and Peters D. K. (ed.) *Clinical Aspects of Immunology*, Vol. 2. Oxford, Blackwell Scientific Publications, 1982: 903–937.

19. Orci L., Malaisse-Lagae F., Baetens D. and Perrelet A. Pancreatic polypeptide-rich regions in human pancreas. *Lancet* 1978, **ii**, 1200–1201.

20. Bottazzo G. F. and Doniach D. Islet-cell antibodies (ICA) in diabetes mellitus: Evidence of an autoantigen common to all cells in the islet of Langerhans. *Ric. Clin. Lab.* 1978, **8**, 29–38.

21. Bottazzo G. F., Pujol-Borrell R. and Doniach D. Humoral and cellular immunity in diabetes mellitus. *Clin. Immunol. Allergy* 1981, **1**, 139–159.

22. Bottazzo G. F., Dean B. M., Gorsuch A. N., Cudworth A. G. and Doniach D. Complement-fixing islet-cell antibodies in Type I diabetes: possible monitors of active beta cell damage. *Lancet* 1980, **i**, 668–672.

23. Gorsuch A. N., Spencer K. M., Lister J., McNally J. M., Dean B. M., Bottazzo G. F. and Cudworth A. G. The natural history of Type I (insulin-dependent) diabetes mellitus: evidence for a long pre-diabetic period. *Lancet* 1981, **ii**, 1363–1365.

24. MacLaren N. K., Huang S. W., Fogh J. Antibody to cultured insulinoma cells in insulin-dependent diabetes. *Lancet* 1975, **i**, 997–1000.

25. Lernmark A., Freedman Z. R., Hofmann C., Rubenstein A. H., Steiner D. F., Jackson R. L., Winter R. J. and Traisman H. S. Islet-cell surface antibodies in juvenile diabetes mellitus. *N. Engl. J. Med* 1978, **299**, 375–380.

26. Larsson L.-I. Gastrointestinal cells producing endocrine, neurocrine and paracrine messengers. In: Crentzfeld W. (ed.) *Clinics in Gastroenterology*. London, Saunders, 1980: 485–516.

27. Ravazzola M. and Orci L. Glucagon and glicentin immunoreactivity are topologically segregated in the alpha granule of the human pancreatic A cell. *Nature* 1980, **284**, 66–67.

28. Polak J. M., Adrian T. E., Bryant M. G., Bloom S. R., Heitz P. U. and Pearse A. G. E. Pancreatic polypeptide in insulinomas, gastrinomas, vipomas and glucagonomas. *Lancet* 1976, **i**, 328–330.

29. Stuart-Mason A., Maede T. W., Lee J. A.-H. and Morris J. N. Epidemiological and clinical picture of Addison's disease. *Lancet* 1968, **ii**, 744–747.

30. Scherbaum W. A. and Berg P. A. Bedeutung von Antikörpern in der Diagnostik endokrinologischer Erkrankungen. *Dtsch. Med. Wochenschr.* 1981, **106**, 308–313.

31. Doniach D., Cudworth A. G., Khoury E. L. and Bottazzo G. F. Autoimmunity and the HLA-system in endocrine diseases. In: *Recent Advances in Endocrinology and Metabolism*, 1982, **2**, 99–132.

32. Irvine W. J. Autoimmunity against steroid producing organs. In: Miescher P. A. et al. (ed.) *Menarini Series on Immunopathology*. Basel, Schwabe, 1978: 35–49.

33. Wuepper K. D., Wegienka L. C. and Fudenberg H. H. Immunologic aspects of adrenocortical insufficiency. *Am. J. Med.* 1969, **46**, 206–216.

34. Sotsiou F., Bottazzo G. F. and Doniach D. Immunofluorescence studies on autoantibodies to steroid-producing cells, and to germline cells in endocrine disease and infertility. *Clin. Exp. Immunol.* 1980, **39**, 97–111.

35. Scherbaum W. A. and Berg P. A. Development of adrenocortical failure in non-Addisionian patients with antibodies to adrenal cortex. A clinical follow-up study. *Clin. Endocrinol.* 1982, **16**, 345–352.

36. Anderson J. R., Goudie R. B., Gray K. G. and Stuart-Smith D. A. Immunological features of idiopathic Addison's disease: an antibody to cells producing steroid hormones. *Clin. Exp. Immunol.* 1968, **3**, 107–117.

37. Kamp P., Platz P. and Nerup J. Steroid-cell antibody in endocrine diseases. *Acta Endocrinol.* 1974, **75**, 729–740.
38. Roitt I. M., Doniach D., Campbell P. N. and Hudson R. V. Autoantibodies in Hashimoto's disease (lymphadenoid goitre). *Lancet* 1956, **ii**, 820–822.
39. Balfour B. M., Doniach D., Roitt I. M. and Couchman K. G. Fluorescent antibody studies in human thyroiditis: autoantibodies to an antigen of the thyroid colloid distinct from thyroglobulin. *Br. J. Exp. Pathol.* 1961, **42**, 307–316.
40. Doniach D. and Bottazzo G. F. Thyroid autoimmunity In: Miescher P. A. et al. (ed.) *Menarini Series on Immunopathology*. Basel, Schwabe, 1978: 22–33.
41. Scherbaum W. A., Stöckle G., Wichmann J. and Berg P. A. Immunological and clinical characterization of patients with untreated euthyroid and hypothyroid autoimmune thyroiditis. *Acta Endocrinol.* 1982, **100**, 373–381.
42. Wilkin T. J., Swanson Beck J., Hayes P. C., Potts R. C. and Young R. J. A passive haemagglutination (TRC) inhibitor in thyrotoxic serum. *Clin. Endocrinol.* 1979, **10**, 507–514.
43. Fagraeus A. and Jonsson J. Distribution of organ antigens over the surface of thyroid cells as examined by the immunofluorescence test. *Immunology* 1970, **18**, 413–416.
44. McKenzie J. M., Zakarija M. and Sato A. Humoral immunity in Graves' disease. *Clin. Endocrinol. Metabol.* 1978, **7**, 31–45.
45. Hall R. and Rees Smith B. Role of thyroid autoantibodies in thyrotoxicosis. In: Franklin C. P. (ed.) *Clinical Immunology Update*. New York, Elsevier, 1979: 291–306.
46. Drexhage H. A., Bottazzo G. F., Bitensky L., Chayen J. and Doniach D. Evidence for thyroid-growth-stimulating immunoglobulins (TGI) in some goitrous thyroid diseases. *Lancet* 1980, **ii**, 287–292.
47. Drexhage H. A., Bottazzo G. F., Bitensky L., Chayen J. and Doniach D. Thyroid growth-blocking antibodies in primary myxoedema. *Nature* 1981, **289**, 594–596.
48. Oosterhuis H. J. G. H. Studies in myasthenia gravis. Part I. A clinical study of 180 patients. *J. Neurol. Sci.* 1964, **1**, 512–546.
49. Strauss A. J. L., Seegal B. C., Hsu K. C., Burkholder P. M., Nastuk W. L. and Osserman K. E. Immunofluorescence demonstration of a muscle-binding complement-fixing serum globulin fraction in myasthenia gravis. *Proc. Soc. Exp. Biol. Med.* 1960, **105**, 184–191.
50. Vincent A. Immunology of myasthenia gravis: recent developments. *Clin. Immunol. Allergy* 1981, **1**, 161–179.
51. Drachman D. B. Myasthenia gravis. *N. Engl. J. Med.* 1978, **298**, 136–142.
52. Sahashi K., Engel A. G., Lambert E. H. and Howard F. M. Ultrastructural localization of the terminal and lytic 9th complement component (C_9) at the motor endplate in myasthenia gravis. *J. Neurol. Exp. Neuropathol.* 1980, **39**, 160–172.
53. Lindstrøm J. An assay for antibodies to human acetylcholine receptor in serum from patients with myasthenia gravis. *Clin. Immunol. Immunopathol.* 1977, **7**, 36–43.
54. Sondag-Tschroots J. R. J. M., Schulz-Raateland R. C. M., Van Walbeek H. K. and Feltkamp T. E. W. Antibodies to motor endplates demonstrated with the immunofluorescence technique. *Clin. Exp. Immunol.* 1979, **37**, 323–327.
55. Storch W. and Trautmann B. Doppelimmunofluoreszenzoptischer Nachweis von Antikörpern gegen motorische Endplatten des Skeletmuskels bei Myasthenia gravis. *Acta Histochem.* 1981, **69**, 18–22.
56. Doniach D. and Roitt I. M. An evaluation of gastric and thyroid autoimmunity in relation to hematologic disorders. *Semin. Hematol.* 1964, **1**, 313–343.
57. Doniach D., Roitt I. M. and Taylor K. B. Autoimmunity in pernicious anemia and thyroiditis: a family study. *Ann. N.Y. Acad. Sci.* 1965, **124**, 605–625.
58. Chanarin I. Gastric autoimmunity. In: Miescher P. A. et al. (ed.) *Menarini Series on Immunopathology*. Basel, Schwabe, 1978: 79–83.
59. Stockbrügger R., Larsson L.-I., Lundqvist G. et al. Antral gastrin cells and serum gastrin in achlorhydria. *Scand. J. Gastroenterol.* 1977, **12**, 209–213.
60. Bordi C., Gabrielli M. and Missale G. Pathological changes of endocrine cells in chronic atrophic gastritis: an ultrastructural study on peroral gastric biopsy specimens. *Arch. Pathol. Lab. Med.* 1978, **102**, 129–135.
61. Uibo R. M. and Krohn K. J. E. Demonstration of gastrin cell autoantibodies in antral gastritis with avidin–biotin-complex antibody technique. *Gut* 1982, In press.
62. Besterman H. S., Bloom S. R., Sarson D. L., Blackburn A. M., Johnston D. J., Patel H. R., Stewart J. S., Modigliani R., Guerin S. and Mallinson C. N. Gut hormone profile in coeliac disease. *Lancet* 1978, **i**, 785–788.

63. Bryant M. G. and Bloom S. R. Distribution of the gut hormones in the primate intestinal tract. *Gut* 1979, **20**, 653–659.
64. Brown J. C. and Otte S. C. GIP and the entero-insular axis. *Clin. Endocrinol. Metab.* 1979, **8**, 365–377.
65. Walker-Smith J. A., Unsworth D. J., Hutchins P., Phillips A. D. and Holborow E. J. Autoantibodies against gut epithelium in child with small-intestine enteropathy. *Lancet* 1982, **i**, 566–567.
66. Savage M. O., Mirakian R., Harries J. T. and Bottazzo G. F. Could protracted diarrhoea of infancy have an autoimmune pathogenesis? *Lancet* 1982, Letter. **ii**, 966–967.
67. Pouplard A., Bottazzo G. F., Doniach D. and Roitt I. M. Binding of human immunoglobulins to pituitary ACTH cells. *Nature* 1976, **261**, 142–144.
68. Edelsten A. D., Hughes I. A., Oakes S., Gordon I. R. S. and Savage D. C. L. Height and skeletal maturity in children with newly-diagnosed juvenile-onset diabetes. *Arch. Dis. Child.* 1981, **56**, 40–44.
69. Mirakian R., Cudworth A. G., Bottazzo G. F., Richardson C. A. and Doniach D. Autoimmunity to anterior pituitary cells and the pathogenesis of Type I (insulin-dependent) diabetes mellitus. *Lancet* 1982, **i**, 755–759.
70. Broadwell R. D. and Brightman M. W. Entry of peroxidase into neurons of the central and peripheral nervous system from extracerebral and cerebral blood. *J. Comp. Neurol.* 1976, **166**, 257–284.
71. Scherbaum W. A., Bottazzo G. F. and Slater J. D. H. Autoantibodies to Vasopressin-producing cells of human hypothalamus in idiopathic diabetes insipidus. American Endocrine Society 1982, Abstract. No. 18.
72. Roth J. and Grunfeld C. Endocrine systems: mechanisms of disease, target cells, and receptors. In: Williams R. H. (ed.), *Textbook of Endocrinology*, 6th ed. London. Saunders, 1981: 15–72.
73. Kohn L. D., Aloj S. M., Beguinot F., Vitty P., Yavin E., Yavin Z., Laccetti P., Grollman E. F. and Valente W. A. Molecular interactions at the cell surface. Role of glycoconjugates and membrane lipids in receptor recognition processes. *Dight Foundation Symposium. Human Genetics and Membrane Biology*, 1982. In press.
74. Doniach D., Bottazzo G. F. and Rusell R. C. G. Goitrous autoimmune thyroiditis (Hashimoto's disease). *Clin. Endocrinol.* 1979, **8**, 63–80.
75. Scherbaum W. A., Rosenau K. O. and Seif F. J. Antikörper gegen Schilddrüsenmikrosomen und Thyreoglobulin beim Morbus Basedow und anderen Erkrankungen. *Med. Welt (Stuttg.)* 1979, **30** *N.F.*, 1401–1406.
76. Doniach D. and Marshall N. J. Autoantibodies to the thyrotrophin (TSH) receptor on thyroid epithelium and other tissues. In: Talal N. (ed.) *Autoimmunity*. London, Academic Press, 1977: 621–642.
77. Winand R. J., De Wolf M., Consiglio E. and Kohn L. D. Autoimmunity, the thyrotrophin receptor and exophthalmos. In: Klein E., Horsker F. A. and Beysel D. (ed.) *Autoimmunity in Thyroid Diseases*. Stuttgart, Schattauer, 1979, 125–146.

20 Immunocytochemistry in Endocrine Pathology

P. U. Heitz

1. INTRODUCTION

The number of applications of immunocytochemistry in histology, histopathology, bacteriology, virology and other fields is growing at an almost exponential rate. Immunocytochemistry has contributed decisively to our present knowledge of the pathogenesis and diagnostic possibilities of many diseases, e.g. diseases of the skin and kidney, and in the diagnosis of hepatitis[1] and malignant lymphomas (*see* Chapters 14, 17 and 18).

At present, one of the most rapidly expanding fields of application of immunocytochemical techniques is *neuroendocrinology*. Not only have many peptides been found to occur in the central nervous system and (often in a slightly modified form) in the gastrointestinal tract, but it is now obvious that the central nervous system contains a large number of peptides such as gastrin, vasoactive intestinal peptide, somatostatin, thyrotropin-releasing hormone, motilin and others,[2–6] possibly acting as neurotransmitters (*see* Chapter 11).

Immunocytochemistry is successfully and increasingly used in *endocrine pathology*. The production of antibodies to hormones has led to the development of immunocytochemical and immunoassay techniques sensitive enough to detect hormones at the cellular or subcellular level and to determine hormone concentrations in blood and tissue extracts. The classification of diseases of neuroendocrine organs has been and is still being much revised enabling the endocrine pathologist to use the same nomenclature as the biochemist and the clinician; thus he may refer to an insulinoma rather than to a pancreatic endocrine cell tumour, possibly a B-cell tumour.

In conventional histology, i.e. in frozen sections and in paraffin sections, stained with tinctorial stains, the pathologist is able to diagnose an endocrine tumour, but is not able to analyse it at the cellular level due to the lack of sensitivity and specificity of tinctorial stains. Electron microscopy is time consuming and prone to sampling errors and leads to a delay of diagnosis of several days. In addition, because of the marked polymorphism of many cellular organelles in tumours, particularly cytoplasmic secretory granules, this technique does not often allow a precise diagnosis to be made.

Immunocytochemical techniques have now been refined to such a degree that many peptide antigens can be detected in conventionally fixed (i.e. liquid formaldehyde, Bouin's fluid) paraffin-embedded tissue. Some antigens (e.g. prolactin, corticotropin) can even be visualized after fixation in glutaraldehyde and osmium tetroxide (*Plate* 28).[7–9] If polyclonal highly specific antisera are used at optimum dilution, and if appropriate technical controls are systematically carried out, the reactions obtained are reliable.[10] In tumours, the specificity of immunocytochemical reactions can be checked by radioimmunoassay of hormones in tissue extracts.[11,12] In addition, there are now a large number of antisera of good quality available.

2. ANALYSIS OF NORMAL ENDOCRINE TISSUE

The advances in histopathological diagnosis of endocrine tumours and hyperplasia brought about by systematic use of immunocytochemistry can be summarized as follows.

The analysis of normal endocrine tissue has led to the discovery of previously unsuspected *new endocrine cell systems*, i.e. in the gastrointestinal tract, and to the realization of the very complex neuroendocrine design of endocrine glands and cell systems (for review *see* Grossman et al.[13] and Bloom and Polak[14] and Chapter 11). In addition, endocrine glands have been discovered to be extremely complicated cellular systems, e,g. the pituitary and the endocrine pancreas. It has been found that there are constant distribution patterns of endocrine cells in these organs[15,16] and there appear to be particular relationships between the various cell types, especially in the endocrine pancreas.[17] The phylogeny and ontogeny of neuroendocrine cells are being extensively investigated using immunocytochemical and immunoassay methods.[18,19]

2.1. Neuroendocrine Design and Extent of Endocrine Tissue

Knowledge of the neuroendocrine design and the extent of endocrine tissue under physiological conditions enables the pathologist to define hypo- and hyperplasia of endocrine cells. It has been found that many pituitary micronodules detected at autopsy are probably multicellular hyperplastic nodules, not tumours.[20] In persistent hyperinsulinaemic hypoglycaemia of the infant a precise assessment of the extent to which the cells of the endocrine pancreas have proliferated is possible only by using immunocytochemistry.[21] It has been found that the disease is always a multicellular one, not an isolated B-cell proliferation. Quantification by morphometric techniques, in combination with radioimmunoassay, has revealed a decrease in somatostatin content and in the ratio of somatostatin to insulin cells in

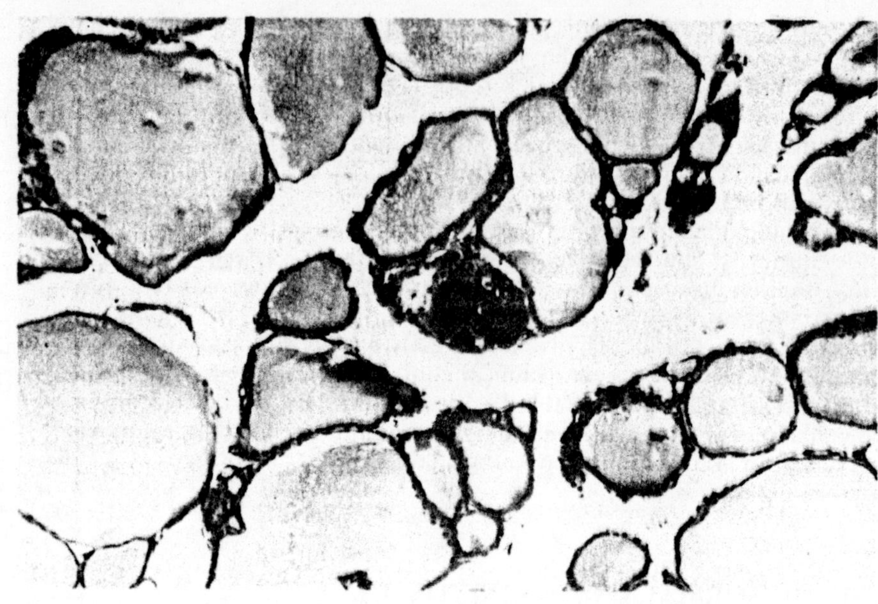

a

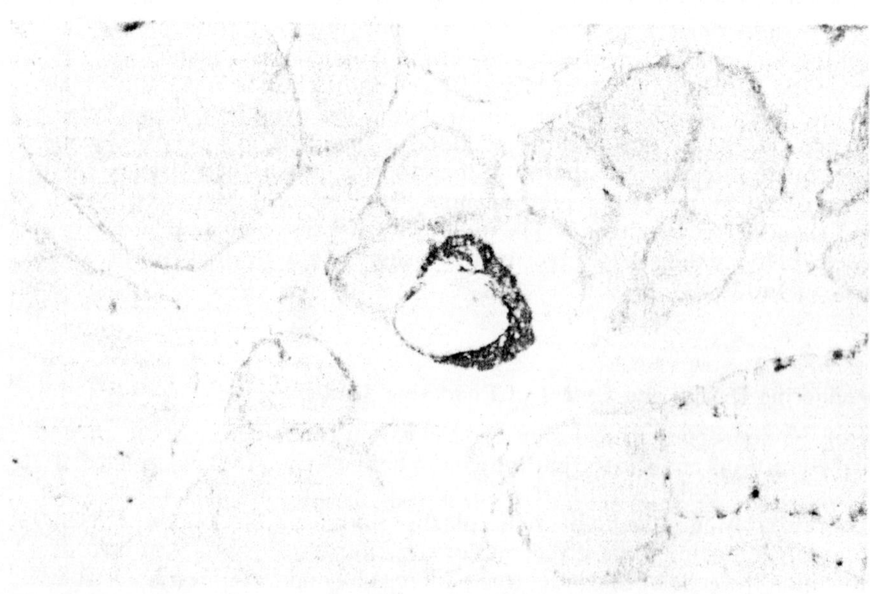

b

Fig. 20.1. C-cell hyperplasia in familial medullary carcinoma (multiple endocrine neoplasia 2a).
a. Small group of cells between thyroid follicles (centre). Haematoxylin and eosin.
b. Same area: calcitonin producing C-cells. Fixation in liquid formaldehyde, embedding in paraffin. Unlabelled antibody enzyme method for calcitonin.

some cases.[22,23] In rare cases, only a focal proliferation of endocrine cells was present in the pancreas[24] and, in still other cases, no proliferation of endocrine cells at all was found.[25,26] Thus it now appears that nesidioblastosis or nesidiodysplasia[26] is a heterogeneous disease, possibly arising from different pathogenetic pathways.[27]

Similar analyses of calcitonin-producing C-cells of the thyroid in kindred with familial medullary thyroid carcinoma disclosed a wide spectrum of C-cell pathology, previously unknown, ranging from mild diffuse C-cell hyperplasia (*Fig.* 20.1) to small and large invasive carcinoma.[28]

3. ENDOCRINE TUMOURS

Systematic immunocytochemical investigation into the pathology of endocrine tumours has revealed a world of new information. Precise diagnosis must be based on immunocytochemical analysis of these tumours. If possible, the results should be corroborated by radioimmunoassay on tumour extracts. The simple histological classification of pituitary tumours has been revolutionized by systematic use of immunocytochemistry[29] and it is now current knowledge that the previously unsuspected prolactinoma is by far the most common pituitary tumour (*Fig.* 20.2.). *Carcinoids* of the gastrointestinal tract and of the bronchial tree, formerly thought to be tumours of the enterochromaffin cells, were shown to be endocrine tumours, able to produce a range of peptides, i.e. gastrin, somatostatin, pancreatic polypeptide and substance P, in addition to biogenic amines.[30] The classification of *pancreatic endocrine tumours* is much more accurate than it was ten years ago when only B- and non-B-cell tumours could be distinguished. In addition, many endocrine tumours have been found to be multicellular and multihormonal and it is obvious that they are complex biological systems, not simply proliferations of a given cell type.[12]

Using immunocytochemistry, the *grade of differentiation*, i.e. the production of hormones, can be assessed at tissue level. It has been shown that differentiated thyroid carcinoma, that is follicular and papillary cancer, usually retains the ability to produce thyroglobulin, in contrast to less differentiated tumours such as anaplastic carcinoma. In addition, autonomous thyroglobulin-producing areas of the thyroid, i.e. autonomous adenomas or hyperplastic nodules, can be defined.[31]

It is notoriously difficult or impossible to assess the *biological behaviour of pancreatic endocrine tumours* by conventional histology in the absence of metastases. Some years ago, malignánt functioning insulinomas were shown to secrete the α-chain common to glycoprotein hormones.[32] By immunocytochemistry this finding could be confirmed and extended to the other types of pancreatic endocrine tumours (*Fig.* 20.3).[33] Tumours producing the α-chain are virtually always malignant.

Ectopic hormone secretion by tumours is sometimes difficult to prove by clinical diagnostic means. Ectopic hormones can often be localized in tumours by immunocytochemistry, provided that the molecular form of the substance produced is immunologically reactive with the available antisera.

A major advance has been the introduction of antisera specifically directed against *neurone-specific enolase* (*see* Chapter 11). The application of this immunoreaction allows the immediate recognition of neuroendocrine cells and

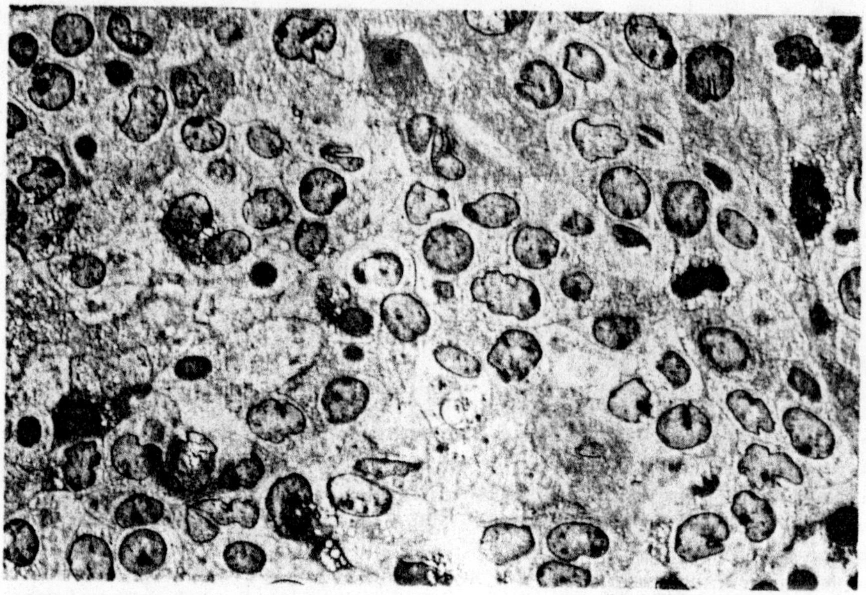

a

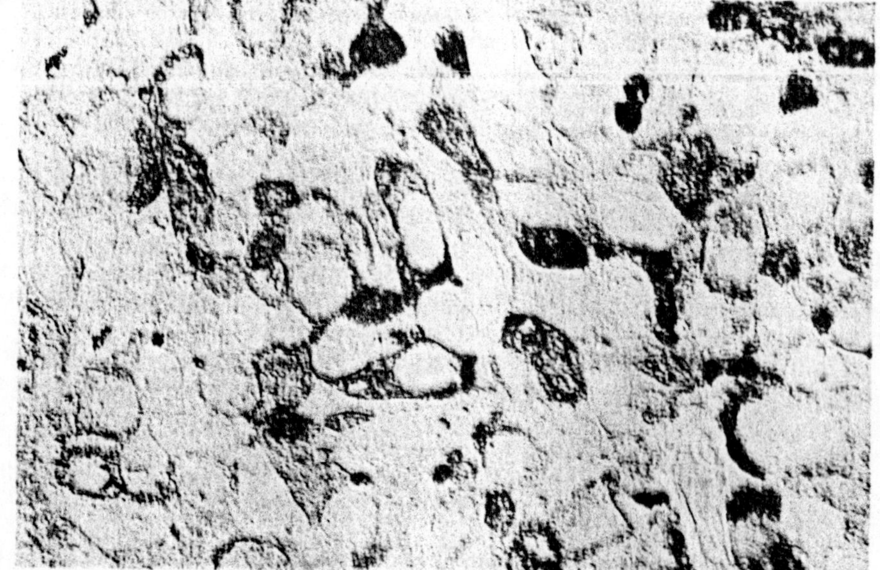

b

Fig. 20.2. Prolactinoma of human pituitary.
a. Irregularly shaped tumour cells with secretory granules present in the cytoplasm. Semithin section (1 μm), toluidine blue staining.
b. Immunoreaction for prolactin in the cytoplasm of many tumour cells. Fixation in glutaraldehyde and osmium tetroxide, embedding in Epon, post-embedding staining of semithin sections (1 μm), differential interference contrast optics.

Fig. 20.3 Malignant VIPoma of the pancreas. Production of the alpha-chain of glycoprotein hormones in some tumour cells. Fixation in liquid formaldehyde, embedding in Epon. Post-embedding staining on semithin sections (1 μm), differential interference contrast optics.

neuroendocrine tumours independently of the hormones or neurotransmitters produced,[11,35] including neuroendocrine carcinomas of the skin (*Fig.* 20.4).[36,38]

4. CONCLUSION

Immunocytochemistry has considerably enlarged our knowledge of the design of the important neuroendocrine control system of homeostasis in many species, including man. The various immunocytochemical techniques are already most valuable tools for the precise functional classification of endocrine tumours, the assessment of their grade of differentiation and, in pancreatic endocrine tumours, of their biological behaviour. The techniques can often be used in combination with routine tissue preparation, with electron microscopy and, in addition, with cell cultures (*Fig.* 20.5). Moreover, quantification of the extent of endocrine cell systems under normal and pathological conditions, and of protein synthesis[39,40] has become possible using immunocytochemical techniques combined with morphometric methods. Further progress in research and diagnostic procedures can be expected from the systematic use of monoclonal antibodies.

The refinement of the techniques also enables the pathologist to conduct thorough retrospective analyses of tissues routinely fixed and embedded in paraffin and stored for many years.

The number of applications of immunocytochemical techniques is virtually unlimited, as can be seen from the various chapters of this book. It is obvious that immunocytochemistry has made, and continues to make, major contributions to diagnostic pathology as well as to biological research.

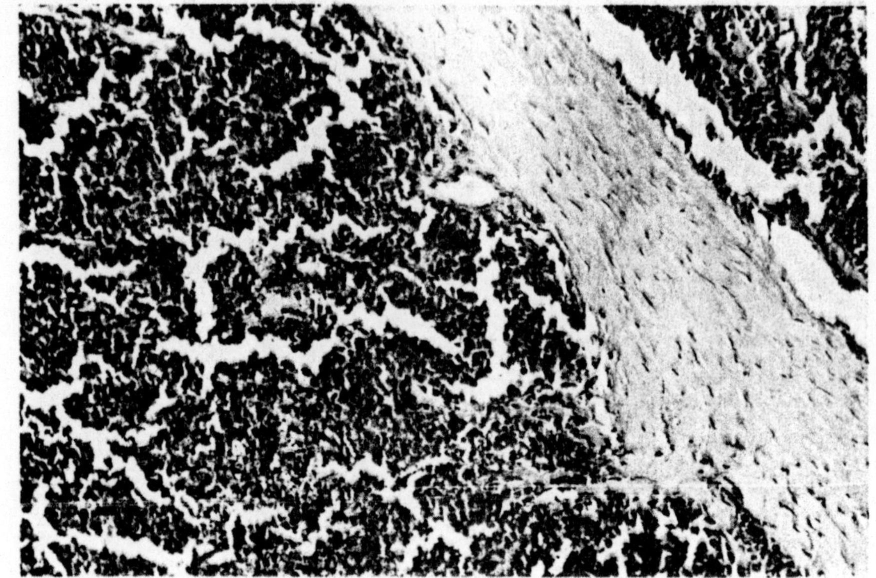

a

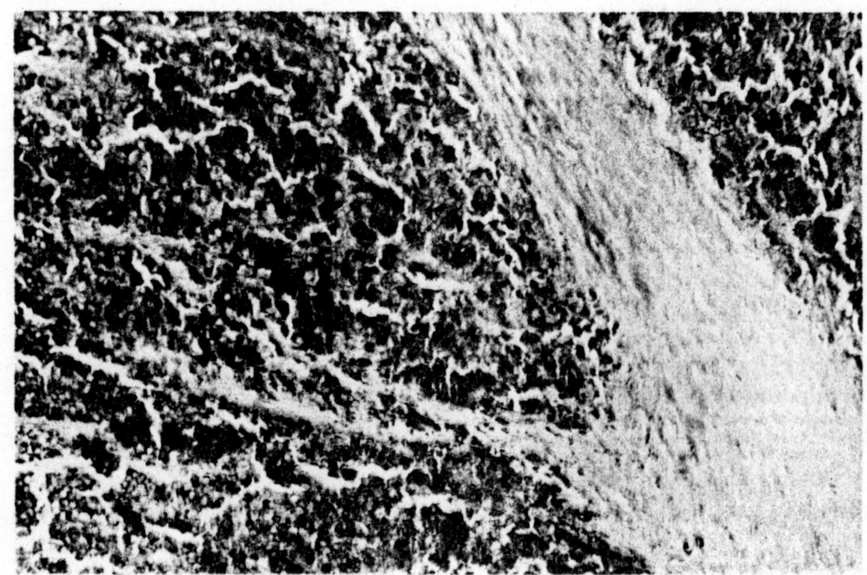

b

Fig. 20.4. Neuroendocrine carcinoma of the skin.
a. Tightly packed cells of the tumour invading subcutaneous tissue. Haematoxylin and eosin.
b. Same area. Many tumour cells contain neurone-specific enolase. Fixation in liquid formaldehyde, embedding in paraffin. Unlabelled antibody enzyme method for neurone-specific enolase.

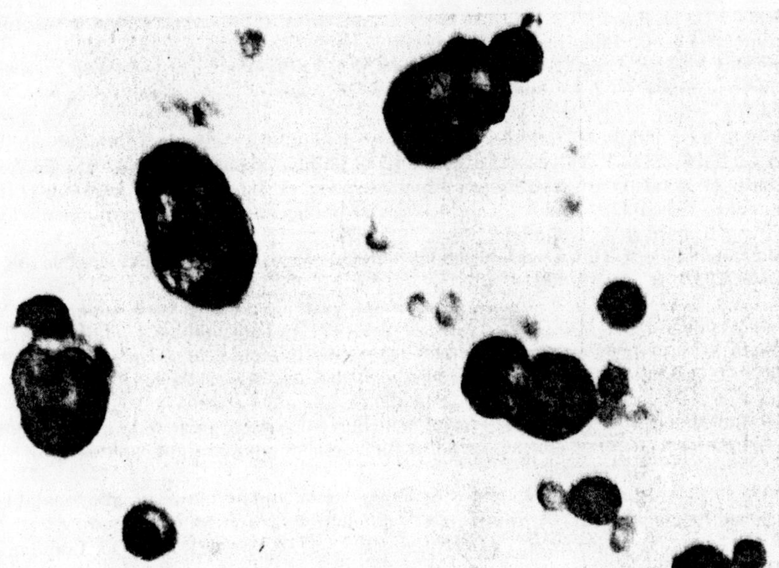

Fig. 20.5. Cell culture (14 days) of a malignant insulinoma. The cells have retained their neuroendocrine characteristics, containing neurone-specific enolase. Fixation in a mixture of liquid formaldehyde and glutaraldehyde, pre-embedding staining, embedding in Epon. Thick section, differential interference contrast optics. The cell cultures were established by Dr U. Otten, Biocenter, University of Basel.

REFERENCES

1. Bianchi L. and Gudat F. Immunopathology of hepatitis B. In: Popper H. and Schaffner F. (ed.) *Progress in Liver Diseases*, Vol. 6. New York, Grune and Stratton, 1979: 371–392.
2. Dimaline R. and Dockray G. J. Multiple immunoreactive forms of VIP in human colonic mucosa. *Gastroenterology* 1978, **75**, 387–392.
3. Polak J. M. and Buchan A. M. J. Motilin. Immunocytochemical localization indicates a possible molecular heterogeneity or the existence of a motilin family. *Gastroenterology* 1979, **76**, 1065–1066.
4. O' Donohue T. L., Beinfeld M. C., Chey W. Y., Chang T. M., Nilaver G., Zimmerman E. A., Yajima H., Adachi H., Poth M., McDevitt R. P. and Jacobowitz D. M. Identification, characterization and distribution of motilin immunoreactivity in the rat central nervous system. *Peptides* 1981, **2**, 467–477.
5. Jacobowitz D. M., O'Donohue T. L., Chey W. Y. and Chang T. M. Mapping of motilin-immunoreactive neurons of the rat brain. *Peptides* 1981, **2**, 479–487.
6. Pradayrol L., Jörnvall H., Mutt V. and Ribet A. N-Terminally extended somatostatin: the primary structure of somatostatin-28. *FEBS. Lett.* 1980, **109**, 55–58.
7. Baskin D. G., Erlandsen S. L. and Parson J. A. Immunocytochemistry with osmium-fixed tissue. I. Light microscopic localization of growth hormone and prolactin with the unlabeled antibody enzyme method. *J. Histochem. Cytochem.* 1979, **27**, 867–872.
8. Erlandsen P. L., Parsons J. A. and Rodning C. B. Technical parameters of immunostaining of osmicated tissue in epoxy sections. *J. Histochem. Cytochem.* 1979, **27**, 1286–1289.
9. Heitz P. U. and Wegmann W. Identification of neoplastic Paneth cells in an adenocarcinoma of the stomach using lysozyme as a marker, and electron microscopy. *Virchows Arch. Abt. A. Pathol. Anat.* 1980, **386**, 107–116.
10. Heitz P. U. Immunocytochemistry-theory and application. *Acta Histochemica* 1982, Suppl. 25, 17–35.
11. Tapia F. J., Polak J. M., Barbosa A. J. A., Bloom S. R., Marangos P. J., Dermody C. and Pearse A. G. E. Neuron-specific enolase is produced by neuroendocrine tumours. *Lancet* 1981, **i**, 808–811.

12. Heitz P. U., Kasper M., Polak J. M. and Klöppel G. Pancreatic endocrine tumors: immunocytochemical analysis of 125 tumors. *Hum. Pathol.* 1982, **13**, 263–271.
13. Grossman M. I., Brazier M. A. B. and Lechago J. (ed.) *Cellular basis of chemical messengers in the digestive system. UCLA Forum in Medical Sciences Number 23.* Academic Press, New York, 1981. 1981.
14. Bloom S. R., Polak J. M. (ed.) *Gut Hormones.* Churchill Livingstone, Edinburgh, 1981.
15. Orci L., Baetens D., Ravazzola M., Stefan Y. and Malaisse-Lagae F. Pancreatic polypeptide and glucagon: non-random distribution in pancreatic islets. *Life Sci.* 1976, **19**, 1811–1816.
16. Bommer G., Friedl U., Heitz P. U. and Klöppel G. Pancreatic PP cell distribution and hyperplasia. *Virchows Arch. Abt. A. Pathol. Anat.* 1980, **387**, 319–331.
17. Orci L. and Unger R. H. Functional subdivision of islets of Langerhans and possible role of D cells. *Lancet* 1975, **ii**, 1243–1244.
18. Falkmer S. and Östberg Y. Comparative morphology of pancreatic islets in animals. In: Volk B. W. and Wellmann K. F. (ed.) *The Diabetic Pancreas.* London, Baillière Tindall, 1977: 15–59.
19. Pearse A. G. E. The APUD concept and hormone production. *Clin. Endocrinol.* 1980, **9**, 211–222.
20. Heitz P. U. Multihormonal pituitary adenomas. *Hormone Res.* 1979, **10**, 1–13.
21. Heitz P. U., Klöppel G., Haecki W. H., Polak J. M. and Pearse A. G. E. Nesidioblastosis: the pathologic basis of persistent hyperinsulinemic hypoglycemia in infants. Morphologic and quantitative analysis of seven cases based on specific immunostaining and electron microscopy. *Diabetes* 1977, **26**, 632–642.
22. Polak J. M. and Bloom S. R. Decrease of somatostatin content in persistent neonatal hyperinsulinaemic hypoglycaemia. In: Andreani D., Lefebvre P. J. and Marks V. (ed.) *Current Views on Hypoglycemia and Glucagon. Proceedings of the Serono Symposia,* Vol. 30. London, Academic Press, 1980: 367–378.
23. Falkmer S., Søvik O. and Vidnes J. Immunohistochemical, morphometric, and clinical studies of the pancreatic islets in infants with persistent neonatal hypoglycemia of familial type with hyperinsulinism and nesidioblastosis. *Acta Biol. Med. Germ.* 1981, **40**, 39–54.
24. Klöppel G., Altenähr E. and Menke B. The ultrastructure of focal islet cell adenomatosis in the newborn with hypoglycemia and hyperinsulinism. Contribution to the classification of neonatal insulinomas. *Virchows Arch. Abt. A. Pathol. Anat.* 1975, **366**, 223–236.
25. Jaffe R., Hashida Y. and Yunis E. J. Pancreatic pathology in hyperinsulinemic hypoglycemia of infancy. *Lab. Invest.* 1980, **42**, 356–365.
26. Gould V. E., Memoli V. A., Dardi L. E. and Gould N. S. Nesidiodysplasia and nesidioblastosis of infancy. *Scand. J. Gastroenterol.* 1981, **16**, Suppl 70, 129–142.
27. Dahms B. B., Landing B. H., Blaskovics M. and Roe T. F. Nesidioblastosis and other islet cell abnormalities in hyperinsulinemic hypoglycemia of childhood. *Hum. Pathol.* 1980, **11**, 641–649.
28. DeLellis R. A. and Wolfe H. J. Calcitonin immunohistochemistry. In: DeLellis R. A. (ed.) *Diagnostic Immunohistochemistry.* New York, Masson Publishing USA, Inc., 1981: 61–74.
29. Kovacs K., Horvath E. and Ryan N. Immunocytology of the human pituitary. In: DeLellis R. A. (ed.) *Diagnostic Immunohistochemistry.* New York, Masson Publishing USA, Inc., 1981: 17–35.
30. Wilander E., Grimelius L., Lundqvist G. and Skoog V. Polypeptide hormones in argentaffin and argyrophil gastroduodenal endocrine tumours. *Am. J. Pathol.* 1979, **96**, 519–530.
31. Böcker W., Dralle H. and Dorn G. Thyroglobulin: an immunohistochemical marker in thyroid disease. In: DeLellis R. A. (ed.) *Diagnostic Immunohistochemistry.* New York, Masson Publishing USA, Inc., 1981; 37–59.
32. Kahn R., Rosen S. W., Weintraub B. D., Fajans S. S. and Gorden P. Ectopic production of chorionic gonadotropin and its subunits by islet cell tumours. *N. Engl. J. Med.* 1977. **297**, 565–569.
33. Heitz P. U., Kasper M., Klöppel G., Polak J. M. and Vaitukaitis J. L. Glycoprotein–hormone alpha-chain production by pancreatic endocrine tumors—a specific marker for malignancy Immunocytochemical analysis of tumors of 155 patients. *Cancer* 1982, in press.
34. Heitz P. U., Klöppel G., Polak J. M. and Staub J. J. Ectopic hormone production by endocrine tumors. Localization of hormones at the cellular and subcellular level by immunocytochemistry. *Cancer* 1981, **48**, 2029–2037.
35. Schmechel D. E., Marangos P. J. and Brightman M. W. Neurone-specific enolase as a marker for central and peripheral neuroendocrine cells. *Nature* 1978, **276**, 834–836.
36. Gould V. E., Dardi L. E., Memoli V. A. and Johannessen J. V. Neuroendocrine carcinomas of the skin: light microscopic, ultrastructural, and immunohistochemical analysis. *Ultrastruct. Pathol.* 1980, **1**, 499–509.

37. Johannessen J. V. and Gould V. E. Neuroendocrine skin carcinoma associated with calcitonin production: a Merkel cell carcinoma? *Hum. Pathol.* 1980, **11**, 586–589.
38. Gloor F., Heitz P. U., Hofmann E., Höfler H. and Maurer R. Das neuroendokrine Merkelzellkarzinom der Haut. *Schweiz. Med. Wochenschr.* 1982, **112**, 141–148.
39. Bendayan M., Roth J., Perrelet A. and Orci L. Quantitative immunocytochemical localization of pancreatic secretory proteins in subcellular compartments of the rat acinar cell. *J. Histochem. Cytochem.* 1980, **28**, 149–160.
40. Bendayan M. Double immunocytochemical labeling applying the protein A–gold technique. *J. Histochem. Cytochem.* 1982, **30**, 81–85.
41. Cuello A. C., Wells C., Chaplin A. J. and Milstein C. Serotonin immunoreactivity in carcinoid tumours demonstrated by a monoclonal antibody. *Lancet* 1982, **i**, 771–773.

Acknowledgements

Commercial Suppliers

For information on commercial sources of antibodies, lectins, peptides etc., readers are referred to the new and necessary publications, the *Directory of Biologicals*, 1982, published by *Nature*, and *Linscott's Directory of Immunological and Biological Reagents*, 2nd edition, 1982/1983, which cover every aspect of the field.

Acknowledgements

We acknowledge with gratitude the generous contributions to the cost of colour reproductions for this book made by the firms listed below.

BDH Chemicals Ltd,
Broom Road,
Poole BH12 4NN, UK.

Beckman Instruments Inc (Bioproducts),
1050 Page Mill Road,
Palo Alto,
California 94304, USA.

Beckman-RIIC Ltd,
Sands Industrial Estate,
Lane End Road,
High Wycombe,
Bucks HP12 4JL, UK.

Cambridge Research Biochemicals Ltd,
Button End Industrial Estate,
Harston,
Cambridgeshire CB2 5NX, UK.

Flow Laboratories Ltd,
PO Box 17,
Second Avenue Industrial Estate,
Irvine KA12 8NB, UK.

Guildhay Antisera,
University of Surrey,
Guildford,
Surrey GU2 5XH, UK.

Hoechst UK Ltd,
Pharmaceutical Division,
Salisbury Road,
Hounslow,
Middlesex TW4 6JH, UK

Immulok Inc,
1019 Mark Avenue,
Carpinteria,
California 93013, USA.

Mercia Brocades Ltd,
Brocades House,
Pyrford Road,
West Byfleet,
Weybridge,
Surrey KT14 6RA, UK.

Milab,
Malmö Immunolaboratorium AB,
Box 20047
S-200 74 Malmö 20, Sweden.

Miles Laboratories Ltd,
Research Products Division,
PO Box 37,
Stoke Court,
Stoke Poges,
Slough SL2 4LY, UK.

Orthodiagnostic Systems Ltd.
Denmark House,
Denmark Street,
High Wycombe,
Bucks HP11 2ER, UK.

Polysciences Inc,
Paul Valley Industrial Park,
Warrington,
Pennsylvania 18976, USA.

Seward Laboratory,
UAC House,
Blackfriars Road,
London SE1 9UG, UK.

UCB—Bioproducts S.A.
rue Berkendael 68,
B-1060 Bruxelles, Belgium.

Index